SCIENTIFIC BASIS FOR SOIL PROTECTION IN THE EUROPEAN COMMUNITY

Proceedings of a symposium organised by

—the Commission of the European Communities, Directorate-General
  Science, Research and Development
—the Senate of Berlin

under the patronage of

—Dr Karl-Heinz Narjes, Vice-President of the Commission of the
  European Communities

and

—Dr Walter Wallmann, Bundesminister für Umwelt, Naturschutz und
  Reaktorsicherheit, Bundesrepublik Deutschland

and held in Berlin 6–8 October 1986

SCIENTIFIC BASIS FOR SOIL PROTECTION IN THE EUROPEAN COMMUNITY

Edited by

H. BARTH and P. L'HERMITE

Commission of the European Communities,
Brussels, Belgium

ELSEVIER APPLIED SCIENCE
LONDON and NEW YORK

ELSEVIER APPLIED SCIENCE PUBLISHERS LTD
Crown House, Linton Road, Barking, Essex IG11 8JU, England

Sole distributor in the USA and Canada
ELSEVIER SCIENCE PUBLISHING CO., INC.
52 Vanderbilt Avenue, New York, NY 10017, USA

WITH 104 TABLES AND 97 ILLUSTRATIONS

© ECSC, EEC, EAEC, BRUSSELS AND LUXEMBOURG, 1987

British Library Cataloguing in Publication Data
Scientific basis for soil protection
in the European Community.
1. Soil conservation—European Economic
Community countries
I. Barth, H. II. L'Hermite, P.
631.4'94   S625.E87

Library of Congress Cataloging in Publication Data
Scientific basis for soil protection in the European
Community.
Includes index.
1. Soil protection—European Economic Community
countries—Congresses. 2. Soil protection—Congresses.
I. Barth, H. II. L'Hermite, P. (Pierre), 1936–
S625.E88S35   1987   363.7'396   87-9227

ISBN 1-85166-109-3

Publication arrangements by Commission of the European Communities, Directorate-General Telecommunications, Information Industries and Innovation, Luxembourg

EUR 10870

LEGAL NOTICE

Neither the Commission of the European Communities nor any person acting on behalf of the Commission is responsible for the use which might be made of the following information.

Printed in Great Britain by Galliard (Printers) Ltd, Great Yarmouth

CONTENTS

Introduction

PH. BOURDEAU

*Directorate-General Science, Research and Development,
Commission of the European Communities, Brussels, Belgium*

We are living on a unique planet, the only one in the solar system where life exists. The very existence of life has modified the physical and chemical environment of the earth, its atmosphere and oceans, in a way that makes life sustainable. This system with its complex cybernetic mechanisms has been named GAIA by Lovelock. Man has always interfered with it on a more or less limited scale. This interference is now reaching global proportions such as climate modifications resulting from CO_2 and trace gas accumulation in the atmosphere or the destruction of stratospheric ozone, not to speak of global radioactive contamination. GAIA will probably prevail as a living system but it probably does not give much importance to man's survival as such, and it is man that has to take care of his own survival.

In the ecosystem of Planet Earth, soils are the thin interface between lithosphere and atmosphere which constitutes the essential substrate for the terrestrial biosphere, the productivity of which far exceeds that of the oceans, even though the latter cover a much larger area than the continents.

Soils themselves are complex systems. They develop through weathering of minerals, are colonised by living organisms which in turn modify their substrate making it suitable for other organisms. This induces a primary ecological succession which eventually reaches a climax, in equilibrium between climate, soil and the biological communities.

Soils play also an essential role in the biogeochemical cycles, acting as sinks and sources, and filtering the flow of aqueous solutions to the water table.

Man has disturbed ecological equilibria in soils by burning the

1

Ph. Bourdeau

vegetation, ploughing for crop cultivation, grazing animals, exploiting forests and using increasingly larger areas for urban and industrial development, for road building, for mining operations, etc. This has led to disasters, some of them going way back in the past.

More recently human activities have become more threatening: modern agriculture has changed the grouping of small fields into large units for single crop farming which are more prone to erosion. The intensive use of fertilisers and pesticides has disturbed the chemical and biological equilibria of the various soil horizons. Heavy machinery has compacted the upper layers modifying structural properties. Intensive livestock production has produced large amounts of organic wastes in some places whereas in other areas the humus content of soils has decreased as manure is no longer available.

Other human activities, not directly linked to agricultural production, have also generated problems for soils over increasing areas: the disposal of solid or liquid waste and particularly toxic waste in controlled, or quite often uncontrolled, tips, the widespread atmospheric pollution from burning fossil fuels in power plants and in motor vehicles having effects over large distances. Even pollution abatement measures have their deleterious effects, e.g. refuse incineration spreading metals and organic compounds of high toxicity, disposal of sewage sludge which has both beneficial and detrimental effects. If, in the future, as is likely, there will be more desulphurisation, we will have to dispose of very large amounts of slurries or drier wastes. The use of coal particularly brings with it the problem of fly ash disposal.

Then, there are industrial or nuclear accidents; remember Seveso and the wide pollution of soils and the tremendous costs of clean-up; Palomares where great efforts had to be expended to remove plutonium from soil contaminated by the inadvertent dropping of a bomb. Chernobyl, of course, is in everybody's memory. Rehabilitating a contaminated soil is really a very large and expensive operation.

Even recreational activities can have their negative impacts on soils, such as cross-country motorcycling which may ruin in a few moments what centuries have done in accumulating humus and developing a favourable structure.

The public and governments are recognising the growing importance of the issue of soil protection. Several member states in the EEC, particularly the Federal Republic of Germany, and other European countries such as Switzerland, have adopted national policies which will be presented here.

The objective of this symposium, which has been organised jointly by the

Senate of Berlin and the Commission of the European Communities, is to contribute to the scientific basis for soil protection, and this in three ways:

—By assessing the state of soils today throughout the range of conditions prevailing in the European Community, to see whether soils are being degraded, in which way, to what extent and with what consequences.
—By recommending measures to be taken in order to protect and, as the case may be, to rehabilitate them, and finding how to implement and monitor these measures.
—By identifying gaps in knowledge and determining research priorities.

The timing of this symposium is particularly opportune. The revising of the EEC Treaty which took place last December formally recognised protection of the environment as one of the goals of the EEC, as well as the promotion of science and technology.

With regard to environmental protection, the Commission has adopted recently its Fourth Environmental Action Programme, now presented to the Council of Ministers, which includes the development of an overall Community policy to protect soils. Specific actions are proposed to tackle the three main causes of soil degradation, i.e. the contamination by harmful substances of various origins (urban, industrial and agricultural waste, agro-chemical products, widespread atmospheric pollution), the degradation of the physical structure and of the chemical and biological status of soils, including erosion, natural hazards, compaction, etc., and the fact that soil resources are often misused and wasted by being devoted to activities that consume space and perhaps should be located elsewhere.

The actions envisaged in this Action Programme include:

—The reinforcement of the links between various Community Policies, in particular the agricultural, regional and environmental policies; soil protection must be considered in all the policies.
—The encouragement of less intensive livestock production systems and use of chemicals, of proper management of agricultural waste, of prevention of soil erosion and excessive run-off, the identification and clean-up of waste disposal sites and the reduction of the hazards to soil from current disposal practices, the encouragement of the recovery and reuse of contaminated or derelict land.
—The development of innovative soil protection techniques and the transfer of the available know-how.

It is timely to mention here that the EEC is going to launch a five-year Framework Programme for Research and Technological Development.

This is an overall plan to be implemented by specific programmes into which the Environment Research Programme, among others, will be inserted. It is therefore of interest to note that the Commission decided to put as the first line of action in the Framework Programme the subject of quality of life, including environmental protection and health, in recognition of the importance of the European life-model in which social and economic considerations have to be closely intertwined.

The EEC environmental research programme is open to other European countries in the COST-framework. The new programme (1986–1990), for which a call for proposals has been issued, is implemented in great part by cost-shared research contracts as well as by concerted actions. The call for proposals has two deadlines: the second one, 15 December 1986, includes the topic of soil protection.

During this symposium we expect substantial information to be brought together from the invited papers and from the discussions. Rapporteurs will assess gaps in knowledge and identify research needs and priorities.

The symposium continues a tradition established several years ago in order to develop jointly the scientific basis for key environmental problems. There have been conferences on acid deposition, agriculture and the environment, the problems of waste, health and environment.

The Commission of the European Communities thanks the Senate of Berlin, the Bundesministerium für Umwelt and the Umweltbundesamt for making this gathering possible.

Soil Protection Strategy in the Community

G. SCHNEIDER

Directorate-General for the Environment,
Consumer Protection and Nuclear Safety,
Commission of the European Communities, Brussels, Belgium

1. INTRODUCTION

As indicated in the introduction to the programme for the symposium, not only the Directorate-General for Agriculture, but also the Directorate-General for the Environment, Consumer Protection and Nuclear Safety of the Commission of the European Communities is interested in the results of this international symposium.

To our knowledge, the Berlin symposium is the third important international conference on soil protection to be held in Germany this year, the other two being the 'ENVITEC—Düsseldorf' on 17 February during the environmental technology fair and the working conference organised by the Federal Ministry of Agriculture in Bonn on 10 and 11 June. This is a clear sign that Germany, too, regards soil protection as one of the key issues that no country can hope to solve on its own in a world of economic and environmental interdependence and foreign policy constraints. Both of these conferences, in which the Environment Directorate took an active part, made a valuable contribution to a problem of acute concern.

The results of this symposium will undoubtedly be of an equally high calibre and play an important part in securing the practical application of the measures already taken or planned at Community level under the Fourth Community Environmental Action Programme (1987–1992).

The fact that this symposium is attempting a scientific stocktaking of soil protection in the Community on the basis of experience in some Member States which are undoubtedly pioneers in this field is of immense value to the Commission.

It is an added bonus—doubtless not only for the Commission—that non-Community countries including representatives from the United States and Israel have been invited to participate.

The fact that soil protection is to be on the agenda for the next meeting of the European Conference of Regional Planning Ministers (Council of Europe) in Switzerland and other important international conferences is a measure of the political importance of increased international cooperation in this field.

My chapter will concentrate on a few aspects of the problem and attempt in particular to clarify the Community's scope for action on soil protection. This will not only make it easier to draw up a list of priority measures and identify the research requirements, but also to plot the framework into which a European soil protection strategy should be inserted.

2. SOIL PROTECTION AND COMMUNITY STRATEGY

Really, it is astonishing that soil protection was not an issue of environment policy in our countries until relatively late in the day. This is despite the fact that the finite nature of the resource 'land', which limits the supply of food to the population, formed the basis for the theory of the 'iron law of wages' advanced by the British economist Malthus at the beginning of the last century.

The Greek philosophers for their part attributed a fundamental importance to land, far greater than the pure dependence of mankind and living nature on land as a production factor. There are many examples all around us which illustrate the profound truth of this observation.

Contamination of the soil is on the brink of becoming one of the most serious problems of our time.

In contrast to the spectacular crimes against the environment that have come to light, such as the Seveso affair or some of the more recent dramatic accidents, adverse changes in the soil often pass unnoticed. When the consequences suddenly become apparent, costly and time-consuming measures have to be taken to remedy the situation, or it may even be impossible to repair the damage because of the irreversibility of some processes.

However, purely remedial action by the planning departments concerned is totally inadequate if the scarce, finite resource 'land' is to be protected in the long term; thus a soil protection strategy which merely treats the symptoms must be replaced by a preventive policy. It is true that the

Commission, where it has direct or indirect competence in this field, has by no means always set a shining example in the past.

Nevertheless, as far as European environment policy is concerned, it can be said that the Community already included soil in the overall concept of protection and rational use of natural resources in the first action programme.

However, the specific measures to protect and conserve the soil that resulted from this commitment, which by the way was largely due to two interesting Italian and German memoranda (1974/1975), were mainly of a sectoral nature and, in addition, primarily concentrated on agriculture.

This largely sectoral or *ad hoc* Community environment policy of the 1970s, applied more by accident than design, did not really change until the launch of the third action programme (1982–1986) particularly as far as the strategic thrust of our policy is concerned. This change mainly came about because the spatial-planning dimension of environment policy became increasingly important and the Commission, committed to developing a comprehensive environmental strategy for the Community, had to broaden its planning base to consider ways of getting to grips more effectively than in the past with the multi-sectoral and multi-media problems of soil protection.

I would say that soil is a perfect example of the direction of the new Community action programme.

For your information, the European Parliament has declared its explicit support for our efforts in this field, notably on the occasion of the debate on the Gendebien Report. It is therefore not surprising that soil protection has been given great prominence as an indispensable component of a development policy to conserve resources and hence as a central concern of the Community in the Commission proposal currently under discussion for the fourth action programme on the environment.

I propose to illustrate this by discussing three main approaches to action.

1. The global strategy of the fourth action programme is initially based on the premise that the alleged conflict between economy and ecology can be resolved by combining the ecologically appropriate with the economically necessary.

In terms of resource conservation and, by analogy, soil protection, this means that the possible and necessary package of Community measures with a soil-protection component must be specially designed to exclude as far as possible the potential threat to the environment and resources of an economically motivated development policy.

It is propitious that recent events on the European and international stage—notably the European Council (of Heads of State and Government), the World Economic Summit and the OECD Conference of Ministers for the Environment—have induced political decision-makers to proclaim with remarkable unity a problem-solving strategy geared towards a line of argument that is increasingly also being accepted by economic circles. The reasoning is as follows.

A prosperous economy requires effective environmental protection and vice versa. In the long run it is impossible to secure growth and jobs without a policy of conservation, as natural resources constitute both the basis for, and the limits to, economic development. Environment policy is therefore an important part of any structural policy; it must be pursued consistently, independent of the business cycle. Wasting of finite production factors, natural resources and land, can be prevented only if structural adjustments take place at the right time. Ecology, then, is no more and no less than sensible long-term economic planning.

This is particularly clear in the case of land. There is a very good book by Erik P. Eckholm called *Losing Ground* (1976) which documents cases of soil destruction past and present. You could call the descriptions in this book fascinating, if they did not concern the destruction of the livelihood of millions of people and whole areas of the world. Two generations ago in the USA, the soil structure of the Great Plains was devastated to such an extent that, apart from the damage to the land, large areas of countryside became a dust bowl; 350 million tonnes of particles of valuable soil darkened the sun. The repercussions were widespread, including damage to health, and enormous costs were incurred in attempting to repair the damage; this period saw the creation of the US 'soil bank'.

In our own time we hear of restrictions on agriculture in the vicinity of industrial plants that emit heavy metals. Whereas the prosperity of Ancient Egypt depended on the flooding of the Nile to bring fertility to the land, the effect on our soils in western Europe of a river bursting its banks are often catastrophic. The land is so intensively cultivated and devoid of protection that the soil is washed away. In addition, the mud left behind by the flood waters is often heavily contaminated with pollutant substances.

These painful experiences lend weight to a principle which is the cornerstone of the Commission of the European Communities' environment policy, namely the principle that prevention is better than cure. In extreme cases, there is no way of restoring the fertility of soil once it has suffered erosion. Soil is built up over decades, sometimes even centuries.

2. The principle of preventive action consequently implies, to follow the logic of my previous remarks, that effective soil protection is only possible in the context of an integrated multi-sectoral environment protection policy. At a later juncture, I shall explain with the aid of specific examples the criteria which must be met in the Commission's view if systematic, more effective integration of the environment dimension into the Community's structural policies (agricultural, regional, transport, industrial and social policy) is to result in future development policies giving greater emphasis to the rational use of land.

3. Furthermore all specialist departments must be directed to make adequate provision for effective, long-term soil protection, both as part of the process of integrating the environment dimension into structural policy and by synchronising multimedia lines of action within a coherent package of specific measures.

Many phenomena such as acid rain, nitrate pollution, toxic wastes, sewage sludge, residues from intensive cultivation and factory farming, etc., can be assumed or have been proved beyond doubt to be part of an unfortunate chain of cause and effect resulting in soil damage.

For the soil is a part or compartment of the various ecosystems. Its numerous links with other compartments such as the atmosphere, hydrosphere, biosphere or lithosphere demonstrate that it is impossible to protect the soil in isolation. There are a multitude of interrelationships such as pollution of the soil from the atmosphere, or leaching of pollutants from the soil by seepage, which mean that effective soil protection requires a multimedia approach.

It is therefore necessary to control all potential sources of soil pollution as well as the traditionally acknowledged risks from agriculture and forestry. A wide range of measures is necessary in addition to changes in cultivation methods, in order to minimise the sum of negative influences on the soil.

There is a great deal of discussion on the subject of interdependence and ecological networks. There is no question that, for the purposes of soil protection and other conservation tasks, we need a whole network of measures. Two features of this network are important:

—the network of actions must be closely knotted. If one of the knots is missing, success will slip through the mesh;
—the cords of this net, i.e. the links between the actions, must not be too weak, otherwise the net will tear and efforts will have been in vain.

3. THE ROLE OF THE EUROPEAN COMMUNITY

3.1. What is the role of the Community with regard to soil protection?

It follows from the golden Community rule that all environmental measures must always be implemented at the most appropriate geographical level that the European action programme should not be regarded as yet another programme alongside the individual national environment policies. A policy or measure does not suddenly become appropriate simply by being transferred from national or regional competence to the Community level. As a result only those measures that can be implemented most effectively at the broadest geographical level are incorporated in Community programmes. Let me give you some examples.

It is worth remembering that the resource 'soil' is very unevenly distributed in the Community. In Greece, for example, only 10–15% of the surface area has soil of good agricultural quality. This imbalance has not changed much with the accession to the Community of Spain and Portugal. In addition, fate—or should we rather say incorrect regional or structural policy decisions in the past as to the location of industrial or transport infrastructure—have resulted in the relatively good soils throughout the Community being buried under concrete.

Extraction of raw materials also removes large areas of sometimes highly fertile land from agricultural and forestry use for many years at least. Above all, however, the available land area is constantly being reduced by human settlement. Good soil has always been linked with high population density.

In large parts of Europe and the world today, urbanisation processes are still particularly intense on relatively fertile land which, by the way, is usually also technically better-suited for building. In other words this trend, hostile to resource conservation, has still not been generally halted.

When considering the problem of rational use of land, it is important not to forget the natural dangers which, however, are frequently aggravated by man. Soil erosion, for example, an age-old problem and constant challenge in the Mediterranean area, is increasingly becoming a self-inflicted problem in many other parts of Europe.

In many places the land has been denuded of its protective vegetation and mercilessly exposed to wind and weather. In other cases, long experience or elementary physical laws have been ignored and cultivation methods changed or new crops introduced.

In order to push up yields in the short term, the soil is mechanically worked to such extremes that stability suffers.

Plant species are cultivated without respect for physical soil properties. Wholesale frequently wasteful use of chemicals is giving rise to increasing misgivings, including in farming circles.

The particular problem of intensive stock-rearing, regionally concentrated on the lowland soils of the Federal Republic of Germany, the Netherlands, Brittany, East England and in the Po Valley, has set a chemical chain reaction in train that has literally exhausted the buffer capacity of soil and groundwater.

Of course there is no denying the effect of the Common Agricultural Policy on the resource 'soil'. However, during the last few years, a costly learning process has set in, not least as a result of the growing pressure of environment policy. This has brought initial changes, particularly in the field of structural policy, towards a more rational use of land. They affect in particular rural areas of the Community designated as disadvantaged or environmentally sensitive.

Unfortunately, the soil is not only at risk from the natural and agricultural hazards I have briefly mentioned. Emissions and the dumping of all sorts of products, residues and wastes are a further source of damage. In large part this is a legacy of the past.

It is true that it is usually no longer possible to attribute blame for the sins of the past. On the other hand, eliminating the dangers such inherited waste problems pose for the soil and the environment as a whole is an urgent political imperative. What is certain is that tackling this iceberg, of which we have so far glimpsed only the tip, is a thankless and uphill task.

However, in order to preclude further damage, we in the Community must begin to face up to this problem and press for a properly organised, systematic search for forgotten dumps of hazardous wastes.

It is highly probable that there will be a problem of equal dimensions in the case of abandoned waste disposal sites from which pollutants have initially escaped into the atmosphere or the water and from there have got into the soil. And the only way to neutralise the potential time bombs in the soil is systematic search work on the spot followed up by appropriate action. Starting from the hazardous production plant, which of course would include previous installations, the substance cycles would have to be re-traced through the environment media and a careful search for residues made at every point where an emission to the soil may have occurred.

The problem of inherited wastes emphasises the need for a preventive policy to avoid direct or indirect soil contamination from the outset. The phenomenon of forest damage shows that soil protection also means prevention of emissions. This example, like others, demonstrates that a wide range of materials and sources is involved: dangerous substances such as heavy metals which are increasingly being found in the vicinity of production installations, flue gases from combustion plant which sometimes have repercussions over enormous distances, and pollutants from mobile sources, i.e. transport.

3.2. Community fields of action

I should now like to follow on from this picture, by no means complete, of soil conservation problems with a discussion of several important fields of Community action. These include research and information, training and advice and Community legal instruments.

As far as the first area is concerned, knowledge of the type and extent of existing or potential soil pollution is a fundamental prerequisite for an effective soil protection policy. Unfortunately there are still considerable gaps in knowledge of the functioning of the soil, its vulnerability and its interrelationships, both qualitative and quantitative, with other components of the ecosystem. To this extent the present symposium organised by the Commission and the Berlin Senate to establish the present state of knowledge and information gaps and draw up guidelines for future soil research topics meets an urgent environment policy need.

I should also like to mention here a recent discussion between the Commission and the Association of the European Fertilizer Industry which mainly concentrated on the possibility of using economic incentives to bring about a reduction in pollution from nitrates, pesticides and herbicides.

The symposium on 'The environment and chemicals in agriculture' held by the Commission at the end of 1984 in Dublin together with 'An Foras Forbartha' (Irish National Institute for Physical Planning and Construction Research) gave an impressive demonstration of how the use of chemicals and organic fertilisers in agriculture present a serious threat, and not only to groundwater and surface water. A vivid demonstration was given of how the application of pesticides can harm and destroy the very organisms, such as worms and insects, that are indispensable for a healthy soil and its constant regeneration.

With regard to the state of research on integrated plant protection in the Community, the important thing now is to apply existing findings in practice.

It is interesting to note that the different Member States have specialised in certain crops. Germany, for example, has concentrated mainly on wheat and France on apples. However, if existing knowledge is to be applied on a large scale, agricultural advisers must be trained; we are currently organising this at Community level through a network of training centres. It will remain one of the Commission's main objectives to close existing gaps and acquire new knowledge through the research programmes and by coordinating national research, and to find suitable ways of improving access to available soil data.

The Community information system on the state of the environment and natural resources (Corine) and periodic reports on the state of the environment are particularly important initiatives in this connection. The next Community report on the environment, which will also contain a detailed section on soil, will be available in Spring 1987, coinciding with the European Year of the Environment.

The Corine information system, by the way, also includes data on soil quality, soil erosion, land cover and biotopes of Community significance including damage or potential damage.

An important Community legal instrument is the Directive on the assessment of the impact of certain public and private projects on the environment, which was finally adopted in June last year following lengthy negotiations. Environmental impact assessment provides an effective multi-media procedural instrument for systematically integrating environmental aspects into Community structural and sectoral policy measures which should also be of particular value in tackling the complex soil issue.

It goes without saying that within the Commission, too, there must be greater coordination of projects to take account of the impact of integrated development programmes from the point of view of land consumption and soil conservation.

The cause of soil protection has received a boost in connection with agricultural policy, not only as a result of amendments made last year to the 'hill farming Directive', but also from a draft Regulation recently submitted to the Council providing for a Community aid scheme designed to protect certain environmentally important farming areas.

Finally, another Community legal instrument with important consequences for soil protection is the Directive on the application of sewage sludge in agriculture which was recently adopted. It is to be hoped that the standards laid down in this Directive will induce industry and the public and private bodies or water boards responsible for operating sewage plants to develop inexpensive procedures capable, if possible, of preventing heavy metals from getting into sewage sludge.

The problem is that the more heavy metals are removed from effluent in sewage treatment plants, the greater will be the problem of sludge contamination with heavy metals, and, by extension, the waste disposal problems. Consequently, it is essential to stop generation of pollutants at source on the basis of a multimedia environment policy. This can really only be achieved across the board with the aid of integrated techniques.

The 1984 Council Regulation on Community action on the environment has an important role in this respect, as in addition to biotope protection it also contains provisions on integrated and resource-conserving technologies and measuring and monitoring of the quality of the natural environment.

The Commission at any rate believes that the actions now getting speedily under way can serve as a model for comprehensive soil protection in the Community.

4. CLOSING REMARKS

A comprehensive soil protection policy at Community level cannot be restricted to a cumulation of specific individual measures nor neglect the cause-and-effect aspects of spatial planning.

Practically all conceivable packages of measures with a soil protection component are intimately linked with land-use planning, so that this really is the only possible framework for effective action. While the Community has not yet developed a rounded approach based on these premises, a firm beginning has been made.

It will above all be necessary to overcome step-by-step by mutual consent the heavily defended vested interests of traditional planning bodies. This will not least also affect the delicate field of relations between national governments or their responsible administrative departments and all regional and local authorities, as well as the economic agents and interest groups operating in this field.

Our short-term goal therefore has to be to impress on the different decision-making levels the importance of making explicit provision for soil protection in all development projects involving the use of land resources. Thus, together with the principle of preventive action, the principle of cooperation, i.e. coordinated interaction of all competent administrative departments, both in Community Organs and in the Member States, is the second important maxim of an effective Community soil protection policy.

SESSION I

Soil Protection—A Need for a European Programme?

Wasting Europe's Heritage—the Need for Soil Protection

G. H. Moss

*Graham Moss Associates, 12 Eton Street,
Richmond, Surrey, UK*

SUMMARY

*Trends in the use of European Community land over the last two hundred
years suggest three major impacts, each of which has consequences for soil
protection. First is the mass sterilisation of land resources for industrialisation and urbanisation realising economic value with little concern, until
recent times, for environmental effects. Secondly is the growth in airborne
pollution, in which various types of fall-out erode vegetation cover and soil
quality and spreads into the hydrological system and hence to other land
parcels. Thirdly is the less visible but highly pernicious effect of changing
agricultural practices on soil quality which again, until recently, has been
largely neglected in favour of economic gain. This paper seeks, in a
provocative yet constructive way, to analyse these three trends; to assess the
extent to which they are reversible; and to examine means for ameliorating
their effects by reference to examples from throughout the European
Community. Positive contributions by the European Commission are given
particular emphasis.*

1. INTRODUCTION

It is highly appropriate that the European Commission should have
selected soil protection as a major theme for its Year of the Environment.
Growing concern with soil quality has been linked to the coming of age of
environmentalism in the Community as a whole, bringing together many

17

disparate areas of research. Hence in a recent major review of the directions
in which environmental issues are proceeding it was stated: 'Linking a
number of ... problem areas [air pollution, dangerous chemical
production, waste management]—and likely to be of growing importance
in the future—is the question of soil protection' [1].

This concern is indeed long overdue and increasingly doubtful in its
chances of success after many decades of neglect of the land and soil as a
finite natural resource. Nevertheless the Commission is making a major
effort to raise questions about soil protection. This paper will review three
types of highly damaging inputs into the soil system as well as several
damaging outputs which arise from the nature and composition of the soil.
These are illustrated in Fig. 1 which may be taken as the organising

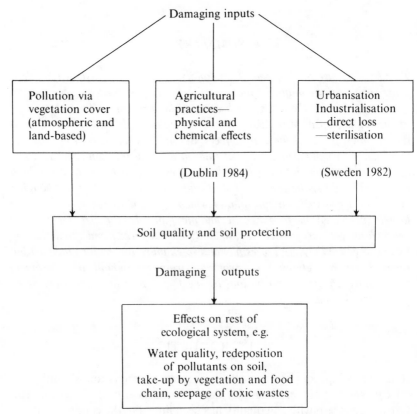

Fig. 1.   Elements of soil quality and soil protection.

framework for the historical review outlined in this contribution. After a brief introduction to each of the three damaging inputs to (and the throughputs of) the soil system, subsequent sections will examine in more detail their effects on soil quality.

The first and most insidious of man's effects upon land lies in the effects of pollution, both air-borne and land-based, which have been reflected by a long-standing concern and in an extensive literature particularly on such subjects as acid rain. Acid rain, through its impact on vegetation cover, is understood to have a detrimental effect upon soil quality. The dramatic and tragic recent incident at Chernobyl has brought home, in an overwhelmingly conclusive manner, the way in which airborne pollution, as direct fallout, can increasingly affect soil quality. Exact links between soil damage and pollution are not yet fully understood.

Secondly are the effects of agricultural practices on soil composition and quality. This area is the most recent concern of the Environmental Directorate and has been late in reaching public awareness probably because the direct impact is not immediately visible. As a result of this lack of early concern, damage done both by physical and chemical means to the soil has tended to continue to the present day largely unabated and unchecked. Increasingly it is being realised that the economic benefits of the Common Agricultural Policy must not outweigh pernicious damage to the heritage of the land itself. Of all the three sets of inputs identified this is probably the most complex; a conference in Dublin in 1984 for the Commission has promoted action in tackling these issues but there is considerably more research needed in this area.

Thirdly are the effects of urbanisation and industrialisation which cause direct loss of soil through sterilisation. There is a long and controversial literature on the study of land loss, particularly the irreversible loss of farmland to urban uses. In 1982 the Commission sponsored a major conference on this theme in Sweden which rightly highlighted these questions. Whatever the controversy over the degree of land loss that has occurred, it is clear that links to soil protection have not been developed as quickly or extensively within the Commission as might have been expected. Equally positive responses to this problem have been missing within Member states.

The effects of soil acting as a throughput to the rest of the ecological system have been investigated in general terms within the circles of soil science, but an understanding of the damage that may be passed on is still at a relatively early stage of enquiry. Obvious cases such as the dumping of pollutants in the Rhine which then find their way through flood-deposition back on to the soil are evident and known, but much work is still needed on areas such as the seepage of toxic wastes underground and their reappearance in potentially harmful locations such as beaches. Again the final resting-point of chemicals used in agricultural production is still poorly understood. Specific areas that the Commission are seeking to ameliorate will be highlighted in the paper.

It can be seen then that a major theme of this paper is that much still needs to be done within the area of environmental protection, and that to date, whilst some very specific areas have received detailed and indeed well-financed investigation, others have been neglected. Interest in soil protection in all its complex forms offers a strong means for the widening of concern throughout Member states for the environment. For example the damage that the removal of hedgerows creates should receive as much attention as the effects of heavy metals in sewage sludge, identified in a recent Commission report [2]. It is important to stress that an integrated approach to the question of soil protection is called for and that this will enable the Community to widen its whole view of environmental protection which to date has been mostly directed towards specific areas of concern in great detail [3]. I will return to this point later in the paper.

2. POLLUTION

The effects of airborne pollutants have figured strongly in much of the Community's programme for improvements to the environment—indeed some areas have now received such coverage that it is being suggested that they are the predominant concerns of the Environment Directorate. Foremost amongst these is acid rain [4]. Although unequal in its effects across the Community with the forests of Germany and Scandinavia most damaged, it has been portrayed as being synonymous with all the land areas and hence potentially as impairing soil quality through damage to the vegetation. This in turn can change the soil composition and encourage physical erosion. Whether this is in fact the most threatening kind of pollution to the soil may be queried, and there is currently considerable debate about the contribution of emissions of nitrous oxides from automobiles and power stations. Clearly then the much earlier concern historically for pollution in the environmental movement has not necessarily guaranteed a successful response when considering an integrated approach to soil degradation.

The three main damaging inputs to soil quality examined here should be the most redeemable. The extent to which direct air- and land-borne pollution is also linked to agricultural practices and their effects is a complication in this issue, but leaving this aside for the moment, most examples suggest that if economic efficiency and the principle that the *polluter pays* are firmly enforced as desired by the Community and repeated in each of its action programmes for the environment, then amelioration of this form of damage is possible.

Clearly there have been scandalous exceptions to the possibility of renewal of the soil where particularly heavy environmental damage has been sustained—toxic metal depositions on abandoned industrial sites is a clear example of this and sterilisation for many decades may result. In other cases there may be effects beyond an acceptable area of sterilisation— recent investigations have suggested for example that the containment of reservoirs of industrial wastes is ignoring seepage that occurs into surrounding land and airborne depositions which pollute farmland. In Scotland this has resulted in the closure of a major chemical processing plant whose airborne emissions were claimed to be killing and deforming cattle, and in Eastern England controversy has surrounded the use of a toxic chemical reservoir, where waste pesticides and industrial chemicals stored in a 'cocktail' reservoir, have been found to be seeping into nearby streams and also moving in the soil substratum towards areas of public use such as parklands and beaches. Similarly pollution to coastal areas by the outfall of sewage and leachage of nitrogen based chemicals into water courses is of increasing concern to the Commission and Member states. However, the treatment of such pollution actually at source has economic consequences that are difficult to ignore.

Derelict and despoiled land can be treated more easily. Even where degradation is quite well advanced programmes for the re-use of waste land have now become an established activity throughout the Community and where such land is returned to agricultural, forest or parkland it can rapidly show improvements to soil quality. In this respect the United Kingdom's Manpower Services Programme has been notable in its successes in linking job creation and environmental improvement and there are other examples such as the Grundstucksfond-Ruhr system for the management of derelict land in Germany and renewal programmes in the Dutch Limburg, both dealing primarily with former mineral sites [5]. This is of course an issue where the Commission can integrate its employment, environmental and economic policy to assist in renewing sterilised or wasteland areas.

However, the majority of interventionary measures that might be applied tend to challenge the economic imperative. There is still some way to go before industry is persuaded of the environmental case. Some areas appear to lack the right approach as yet to this problem. For example in the Mezzogionari it has been found that virtually no urban refuse and waste is recycled leading to considerable loss of good-quality land and soil. Within the Mediterranean Basin as a whole there has been much concern by the Commission for water-based pollution—understandable in an 'internal' sea—but more attention should be focussed on issues of land pollutants. This part of the Community historically has one of the longest records of damage to the land and soil, there being very early examples of erosion from agricultural practices.

Overall the general policy for the treatment of pollutants in soil has been established: 'There is increasing concern about the deterioration of soil quality, the accumulation of pollutants in soils and their transfer to groundwater and the food chain, which justifies a significant research effort in this area, to be complemented by specific research on waste disposal' [6]. It is also acknowledged that there are substantial gaps in the Community's knowledge of the behaviour of pollutants in soil These include the mobility of metal and organic pollutants through soil and into groundwater; the extent to which pollutants are returned from the soil back to plants depending on soil and crop types; and the way in which organic pollutants degrade [8]. Other related areas are examined within the context of agricultural practices.

3. THE EFFECTS OF AGRICULTURAL PRACTICES

'Europe's landscape has been formed and shaped by agriculture over the centuries. However, the development by modern farming practices poses questions which require an urgent response' [9]. This recent policy statement on new directions in environmental protection highlights concisely probably the least understood of the three damaging inputs to soil quality examined, and so far the least researched. It is also the one overall which has the most knock-on effects due to the soil acting as a throughput to the rest of the ecological system and this reinforces the need for urgent attention to be given to damage occurring.

Why has this paradox arisen in which one of the oldest forms of production has only very recently been viewed as threatening to the heritage given by the soil? The reasons are certainly not only technical through the advancement of scientific knowledge about the interaction of the agricultural system and the ecosystem. They arise from a number of complex and interacting forces. Undoubtedly the apparent permanence of agriculture over the last two centuries has probably lulled observers into taking-for-granted its nature as a set of unproblematical practices. The development of large-scale farming methods in recent years has not changed this view dramatically. The impact of landscape changes has only recently become apparent and pollution or physical damage has not been obviously apparent. Most forceful a factor, however, is the elevation of food production both in quantitative and qualitative terms above most environmental considerations. The advent of the Common Agricultural Policy and its absorption of three quarters of the Community Budget has meant generally scant attention to the impact of financial inducements to ever more efficient productivity and the environmental consequences (both to soil and landscape) of the farmer's actions.

To what extent this situation is redeemable depends not only on general policy directions, but also the advancement of our scientific understanding of the problems. The heritage value of soils and environmental damage has only been explored very recently. Until the 1960s soil surveying was a relatively simple affair, and maps produced differentiated only between soil types. Since the 1970s there have been much more marked attempts to move towards land-use ecological mapping with all the interacting elements considered. This approach is typified by its use in examining the Dutch polders [10]. Recently the Commission has made commitments to examine improvements in land suitability in Spain and Portugal [11]. It is also rewarding to see that 80 000 ECU are to be spent on a cartographic study of soil erosion in the Aegean Sea islands of Greece.

Two other major problem areas in the question of soil degradation due to agricultural practices arise. The first of these relates to direct physical changes induced by working the soil and which can be attributed directly to farming methods rather than natural causes. These can include the effects of involving technology and soil compaction caused by heavy machinery [12]. Related to this is the loss of diversity in leys and animals found on arable farms which may have changed or reduced the amount of organic matter found in soil [13]. Further complications are introduced by this: 'Inter-relationships between soil type, slope, rainfall intensity and erodibility index are not sufficiently understood in the Member States and research already indicated should continue' [14]. This kind of problem is further extended by changes induced by the drastic physical effects of farm land used for maximum economic gain. One of the best examples of this found in various Member States (Britain, France and Germany in particular) is the draining of wetland and conversion of apparently fertile pastureland to arable production. Recent investigations have suggested a loss of soil fertility. In cases examined in the South of England crop failure has occurred after several years.

Secondly, and more serious, is the question of chemical changes induced in soils through changes in farming methods, particularly the application of fertilisers and other products some of which may carry pollutants from other parts of the ecosystem. So far a detailed programme of action on this area of concern has focussed on the use of sewage sludge as a fertiliser which may contain trace or heavy metals as well as organic pollutants [15]. Interest here increasingly stresses the take-up of these pollutants from the soil and their transfer to groundwater and vegetation, ultimately to the food chain and then to consumers of agricultural produce. Some very specific programmes for research have already been implemented including a major project on the treatment and use of sewage sludge [16]. Overall there seems to be the beginnings of a concerted policy drive to reduce soil degradation, but it should be noted that this has been viewed as only a partial approach: 'a *limited* effort may be devoted to the effects of agricultural and forestry practice ... on soil and water quality' [17]. This is not really satisfactory if the full heritage given by Europe's soils is to be realised. Member countries must make more effort to respond to these problems. Changes in attitude are now apparent in Germany and the United Kingdom, with the most positive response in the Scandinavian countries. France and the Mediterranean nations could place greater emphasis upon these problems.

4.   URBANISATION AND INDUSTRIALISATION

Loss of soil quality arising from urbanisation and industrialisation has the longest history of concern. Complete sterilisation of land resources in favour of economic determinism is a controversial issue cutting across political and policy boundaries. The development of industrial society over the last two centuries has seen the urbanised landscape develop to cover more than 15% of the total land area in the Netherlands and to nearly 7% in the Community as a whole [18]. Although these totals do not seem high in themselves, it must be remembered that this does not imply that the rest of the land is free from damage. The 64% of the Community's land in direct agricultural production is increasingly damaged by methods used as has been seen above. The 22% in forest and woodland is also similarly affected [19]. Loss of biologically productive land to urban uses between 1961 and

1971 also suggests that this problem is likely to be sustained for many decades to come if action is not taken.

Historically it has been difficult to assess the exact degree of damage to Europe's heritage of land. It is known that one of the most rapid periods of land loss occurred between the two world wars and that this has again accelerated in the post-war period. Serious research investigation of the mix of European land uses did not get under way until the early 1960s when Dudley Stamp was commissioned by the Food and Agriculture Organisation to undertake comparative investigations for up to 20 European nations. Only with the greatest difficulty has any kind of comparative analysis between 1961 and 1971 been possible [20].

Mass sterilisation of land has grown apace and calls for urgent action leading to soil protection. This is undoubtedly the most difficult of the three damaging inputs to examine. It is unlikely that in general terms the loss of the soil heritage is redeemable due to the economic value realised from the change of use of land. Action by the Community will have to be limited to particular forms of intervention which will need to be at least as successful and powerful as the Common Agricultural Policy. Such intervention might take the form of a *Common Environmental Policy* for land use. It will be helpful in emphasising the need to protect soil rather than land and to define urbanised areas much more discretely (a failure so far perhaps in part attributable to Landsat imagery which tends not to give much detail of the built environment). There are pockets of land which can be saved and improved. Encouragement in this field is apparent by the growth of urban ecology which many member countries have boosted by Community Programmes with public money. In the future urban farming and forestry may lead to a reduction in impoverished soils and atmospheres, although this will in part depend on the extent to which pollutants have also affected these. The Commission could well research new areas here in the improvement of derelict land parcels and give further encouragement to the measurement of land loss and quality, identifying those areas at greatest risk from within the general urban landscape.

5. CONCLUSION—TOWARDS A BETTER UNDERSTANDING OF SOIL PROTECTION

It should be a central concern of the Community's Fourth Environmental Action Programme that soil protection receives major attention, a sentiment echoed in recent discussions [21]. Europe's heritage from the soil has been accorded low priority historically and as a result has been wasted. As a consequence substantial damage has occurred that is frequently irredeemable either technically or economically without large scale intervention into market or manufacturing processes.

Already there is an argument propounded by those who assume that it is too late to save much of the land's resources and this has been highlighted most recently in concern with acid rain and the Chernobyl disaster [22]. The same degree of concern is now growing for other previously less obvious forms of degradation, particularly the impact of intensive farming methods and agriculturally created pollutants. These types of degradation can be added to the long-established patterns of land use in the Community which have sterilised much soil through urbanisation and industrialisation. Equally the commercial development lobby suggests that this sterilisation by change of use has only just begun. They argue that there is ample land and with changes in dietary habits and the down-turn in farming, extensive areas of biologically productive land could be brought forward for urban development.

If Europe's heritage is to be saved then the Year of the Environment must see a major leap ahead in thinking, one in which soil protection should play a major role. Thinking needs to be translated into action. In this respect two suggestions have recently been considered. First is the view that new environment-directed funds should be set up under the control of the Environment Directorate—so far one for urban issues has been suggested and there seems to be no reason why the same should not apply to soil protection [23]. Second, and perhaps more important, would be the integration of Environmental Policy through the directorate's work with other areas, particularly agriculture. It seems unfortunate, and short-sighted, that the Agriculture Directorate has generally not sought to incorporate issues of soil quality into its programme whilst emphasising the Common Agricultural Policy as an economic tool of efficiency and gain. Yet the environmental programmes have always considered soil quality to be complementary to efficient production of foodstuffs. This approach should be made more central across both directorates.

The regeneration of derelict and polluted urban and rural land can benefit from joint environmental and employment schemes funded by the Social and Environmental Directorates. The successful examples already evident in certain Member states need to be encouraged more widely.

The Community intervention in the three key areas outlined will take considerable political will and financial commitment if environmental problems are to be tackled at source and ultimately alleviated. Greater integration of policies and funds will be essential if the paradox, for example, of paying twice for agricultural productivity and landscape conservation are not to continue. Environmental issues have assumed considerable importance since the middle of the 1970s and the excellent record of the European Commission in responding to problems could culminate in a major integrated environmental initiative following Environment Year.

6. AN AGENDA FOR ACTION ON SOIL PROTECTION

I would like to thank the Commission for inviting my contribution to this Conference and in order to further the understanding of, and the need for, soil protection I propose that the Community might consider the following six point plan of action:

(i) To integrate more forcefully areas of research into soil quality and the processes leading to its degradation: to date these have been too piecemeal.

(iii) To widen the programme of Research and Development into soil protection to include specific areas such as the nature of sewage sludge. More information gathering and a stronger data base is needed via the existing ecological mapping programme for example. I would wholeheartedly commend the European Commission for its initiative in setting up the Ecological Mapping Programme and in its steadfast move towards Environmental Impact Assessment procedures.

(iii) To create a Community-wide inventory of areas of soil at particular risk, and to map out variations across all member states. Currently certain areas such as the Mediterranean Basin are receiving the most attention, along with, for example, acid rain in Germany and Scandinavia. Other areas of concern should also be examined and this could form the basis of a European Compendium assembled by individual Member States in order to identify trends in land use and change that may lead to degradation or misuse of soil and land.

(iv) To instigate more monitoring in individual Member states of soil quality and degradation and to make an inventory of action already being taken, and proposed.

(v)   To integrate the work of the Directorates better so that soil protection becomes part of an integrated package with complementary economic, physical and social goals.

(vi)   To prepare an overview of Europe's Wasting Heritage to celebrate Environment Year and to enable the Environmental cause to be communicated forcefully throughout Member states as widely and as simply as possible.

REFERENCES

1. Commission of the European Communities; New Directions in Environment Policy, Com. doc. (86) 76 final, Brussels, 1986.
2. Commission of the European Communities—Adopting Multi-annual R & D programmes in the field of the environment (1985–90); Com. doc. (85) 391 final, Brussels, 1985.
3. *Ibid.*
4. Green Europe; Europe's Green Mantle—Heritage and Future of our Forests; Agricultural Information Service No. 204, Brussels, 1984.
5. *The Planner*, September 1986. Articles by C. Crouch, J. Herson and P. Aitken.
6. Com. Doc. (85) 391, *op. cit.*
7. *Ibid.*
8. Green Europe; Agricultural Research—Progress and Prospects; Agricultural Information Service No. 206, Brussels, 1984.
9. Com. Doc. (86) 76, *op. cit.*
10. VINK, A. P. A. (1983). *Landscape Ecology and Land Use*, Longman, Harlow.
11. Green Europe, No. 206, *op. cit.*
12. *Ibid.*; Commission of the European Communities—a Europe-wide Environment Policy; Com. doc. (85) 86 final, Brussels, 1985.
13. Green Europe, No. 206, *op. cit.*
14. *Ibid.*
15. Com. Doc. 85 (391) *op. cit.*
16. *Ibid.*
17. *Ibid.*
18. Best, R. H. (1981). *Land Use and Living Space*, Ch. 9, Methuen, London.
19. *Ibid.*
20. *Ibid.*
21. Com. Doc. (86) 76, *op. cit.*
22. GLOSSOP, C. (1985). Stepping up the pressure for an all European Environmental Policy. *Town and Country Planning*, May.
23. GLOSSOP, C. (1986). UK takes the helm for European Action. *Town and Country Planning*, June.

Land Resources and their Use in the European Communities

J. LEE

*Soil Survey Department, The Agricultural Institute,
Johnstown Castle Research Centre, Wexford, Ireland*

SUMMARY

Between 1969–1971 and 1983 the EEC-10 agricultural area (UAA) declined by 4·5 M ha whereas woodland increased by 1·2 M ha. In the most intensively farmed regions, cattle and pig densities are > 300 per 100 ha UAA (EEC-10 average 80) and > 1000 per 100 ha UAA (EEC-10 average 80), respectively with cereals accounting for > 60% UAA (EEC-10 average 27%). Level of nitrogen (N) use between EEC-12 countries ranges from 238 to 26 kg per ha UAA, with P and K use also showing sizeable variation.

Soil units with Cambisols and Luvisols account for 60% of the total EEC-12 area with Fluvisols and Gleysols accounting for 9%. Podzols, Rankers and Lithosols account for 14%. Although Cambisols and Luvisols of good workability are among the most intensively cultivated they are subject to erosion hazard under certain conditions (e.g. in Spain and France). Soils with gleyic or vertic properties are also intensively cultivated and subject to deterioration (e.g. in England and Italy). In addition, coarse textured Podzols with a high sensitivity to nutrient leaching, ground water pollution and windblow, are among the most intensively farmed soils.

Only 25% of EEC-12 is level (0–8% slope) and 14% is level/sloping (0–15% slope). Almost 26% is moderately steep/steep (15–>25% slope). The comparable moderately steep/steep land figures for the Mediterranean countries, Greece, Italy and Spain are 65, 56 and 42% respectively with countries such as The Netherlands and Denmark having no land in this category.

A preliminary evaluation of the EEC-10 land base for grassland use shows

*that 43·04 M ha (28%) have a high suitability/productivity; 56·33 M ha
(36·5%) are moderate/low–moderate/high and 54·88 M ha (35·5%) are in
the low–very low suitability class. For arable usage, only 30·64 M ha (20%)
are in the highest (Class 1) category, 26·68 M ha (17%) are Class 2, 48·54 M
ha (32%) are Class 3 and 48·14 M ha (31%) are unsuited.*

1. INTRODUCTION

The European Communities have a wide range of soils, climates,
vegetation, topography and a corresponding wide range in land use
suitability. The soil resource is limited and apart from natural limitations, it
may be subject to destruction or deterioration, particularly through erosion
effects, mismanagement and pollution. Intensity of land-use can vary
considerably with land under arable use accounting for 90% of the utilised
agricultural area (UAA) in Denmark and as little as 10% in Ireland. The
agro-ecological environment of North West Europe which is characterised
by high precipitation is conducive to pasture production whereas in the
Mediterranean and much of the Continental zones moisture stress is
limiting and agricultural land-use has a large arable component. Livestock
densities and fertiliser use intensity may also vary considerably. To permit
the most effective land-use planning and management programme, the
properties and use capabilities of the finite land base must be quantified.
Since grassland and arable farming systems comprise the major land-use
types a major objective of this paper is to evaluate the land resource base
for both arable and grassland uses. In addition, attention is focussed on
quantifying land-use structure intensity and trends and the soil
characteristics of the intensively farmed areas are examined.

2. LAND USE STRUCTURE

2.1. Agricultural and forestry use

Agriculture or utilised agricultural area is the largest user of land in the
Member states (Table 1) taking up well over half of the total land area in
most cases. The proportionate range is from 44·3% in Portugal to as much
as 80·8% in Ireland. The high UAA in countries such as UK and Ireland is
explained partly by the inclusion of rough grazing land in the UAA
category.

Table 1

Principal categories of land use in EEC countries

	Utilised agricultural area		Arable land		Permanent grassland		Wooded area	
	000 ha	*% Total area*	*000 ha*	*% Agric. used area*	*000 ha*	*% Agric. used area*	*000 ha*	*% Land area*
Germany	12 137	48·8	7 238	59·6	4 675	38·5	7 328	30
France	31 728	57.8	17 349	54·7	12 734	40·1	14 619	27
Italy	17 836	59·2	9 392	52·7	5 121	28·7	6 079	21
Netherlands	2 012	54·0	830	41·3	1 143	56·8	292	9
Belgium	1 433	47·0	741	51·7	656	45·8	613	20
Luxembourg	127	49·0	55	43·3	70	55·1	90	35
UK	18 814	77·1	6 978	37·1	11 754	62·5	2 141	9
Ireland	5 678	80·8	1 113	19·6	4 562	80·3	437	5
Denmark	2 888	67·0	2 631	91·1	243	8·4	493	12
Greece	9 234	70·0	2 978	32·3	5 271	57·1	2 968	23
Spain	31 530	62·5	15 550	49·3	11 020	35·0	15 270	30·6
Portugal	4 080	44·3	2 965	72·6	530	13·0	3 641	39·5
EEC-12	137 497	60·9	67 820	49·3	57 779	42·0	53 832	24

Source: Eurostat: Crop Production 1983.

Whereas 49·3% of the UAA in EEC-12 is classified in the arable use category the range is from 19·6% in Ireland to 90% in Denmark. Similarly, permanent grassland shows significant variation. Woodland ranges from 39·5% of the land in Portugal (Table 1) to as little as 5% in Ireland compared with an EEC-12 average of 24%.

Germany and France account for 60% of the EEC-10 wooded area. Of the total forest area of 31 M ha in the EEC-9 only 19 M ha were classified as productive in 1978 [1]. Of the remaining 12 M ha at least 4 M ha were considered suited for conversion to productive forest, the remaining 8 M ha being considered less suitable for timber production but serving a vital environmental role in the prevention of erosion etc. They include, for example, certain alpine forests near the upper limit of tree growth, coppice and shrub areas on poor sites, special vegetation types like the Mediterranean maquis as well as potentially productive forest managed as nature reserves. France and Italy between them accounted for 10 M ha of the 12 M ha.

Table 2
Utilised agricultural area in Europe (M ha)

	1969–71	1983	Decrease	Decrease (%)
Europe	233·45	225·37	8·08	3·46
Western Europe	158·11	151·42	6·69	4·23
EEC-10	105·63	101·08	4·55	4·31
EEC-12	142·38	136·35	6·03	4·24

Source: FAO Production Yearbooks 1981 and 1984.

Table 3
Arable land area in Europe (M ha)

	1969–71	1983	Decrease	Decrease (%)
Europe	130·63	125·86	4·77	3·65
Western Europe	79·15	75·71	3·44	4·35
EEC-10	51·59	49·08	2·51	4·86
EEC-12	70·81	67·63	3·18	4·49

Source: FAO Production Yearbooks 1981 and 1984.

Table 4
Permanent pasture in Europe (M ha)

	1969–71	1983	Decrease	Decrease (%)
Europe	88·50	85·55	2·95	3·33
Western Europe	67·57	64·43	3·14	4·65
EEC-10	48·16	46·34	1·82	3·78
EEC-12	60·29	57·56	2·73	4·53

Source: FAO Production Yearbooks 1981 and 1984.

Table 5
Wooded area in Europe (M ha)

	1969–71	1983	Increase	Increase (%)
Europe	148·8	154·0	5·2	3·49
Western Europe	112·5	116·6	4·1	3·64
EEC-10	33·7	34·9	1·2	3·56
EEC-12	51·5	54·1	2·6	5·04

Source: FAO Production Yearbooks 1981 and 1984.

2.2. Changes in land use

Of the 225 M ha of agricultural land (UAA) in Europe (Table 2) 151 million (1983) are in Western Europe of which 101 million are found in EEC-10 and 136 million in EEC-12. Over the period 1969–71 to 1983 the UAA of Europe declined by 8·0 M ha. The comparable declines were 6·7 M ha in Western Europe, 4·5 M ha in EEC-10 and 6·0 M ha in EEC-12. The decline occurred in almost all countries with the highest percentage decline occurring in Italy and Benelux.

Agricultural area comprises arable land and permanent pasture. Tables 3 and 4 indicate that both the arable and grassland components declined by 4·5% in EEC-12. Table 5 indicates that forestry is exerting pressure on

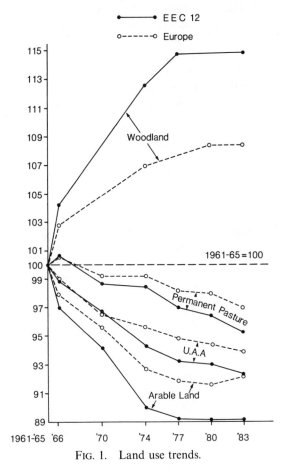

FIG. 1.   Land use trends.

European land showing an increase of 2·6 M ha (5%) in EEC-12 over the 1969–71 period.

Figure 1 depicts the major land-use trends over the 1961–65 to 1983 period.

3.   LAND-USE INTENSITY

Table 6 lists the EEC-10 countries by the percentage contribution their agricultural land and agricultural output made to the total for the whole Community in 1984.

Three countries (France, Italy and Germany) account for 60·5% of the UAA and 63·8% of the agricultural output. When the ratio of output to area is expressed by index the range is from 415 in The Netherlands to 39 in Ireland compared with an EEC-10 average of 100. A scrutiny of output in relation to area indicates that apart from The Netherlands intensity of land-use is also high in Belgium, Germany and Denmark. Global figures mask regional variation and to illustrate regional intensity in land-use three yardsticks are selected:

1.   Cereals as a percent of UAA (Fig. 2).
2.   Total cattle per 100 ha UAA (Fig. 3).
3.   Pigs per 100 ha UAA (Fig. 4).

Table 6

Proportionate contribution of member states to the total agricultural area and agricultural output of the EEC (1984)

	Utilised agricultural area (%)	Agricultural output (%)	Output: area
Germany	11·9	18·4	155
France	31·1	25·8	83
Italy	17·5	19·6	112
Netherlands	2·0	8·3	415
Belgium	1·4	3·1	221
Luxembourg	0·1	0·1	100
UK	18·5	13·4	72
Ireland	5·6	2·2	39
Denmark	2·8	4·0	143
Greece	9·1	5·1	56
EEC-10	100	100	100

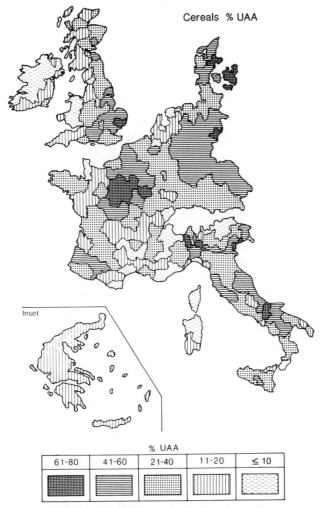

Cereals % UAA

Inset

% UAA				
61-80	41-60	21-40	11-20	≤ 10

Fig. 2.   Distribution of cereals in EEC-10.

Cereals as a percent of UAA range from 60–80% in Paris Basin, parts of South East England, Denmark and Po Valley to <10%. Cereals are very highly concentrated in regions such as Paris Basin, South East England, Denmark and Po Valley (Fig. 2). Table 7 gives the areas of highest cereal concentration.

It is evident (Figs. 3, 4) that highest cattle and pig concentrations in EEC-10 occur in Benelux.

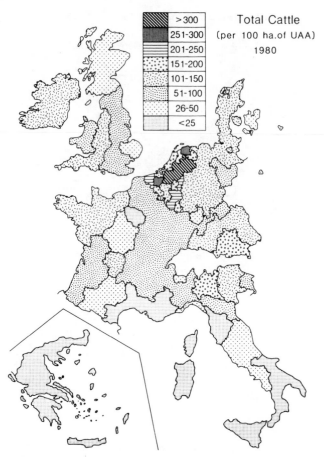

		Total Cattle
	>300	(per 100 ha.of UAA)
	251-300	1980
	201-250	
	151-200	
	101-150	
	51-100	
	26-50	
	<25	

FIG. 3. Total cattle distribution in EEC-10.

Table 7
Areas of highest crop concentration in EEC-10

Crop	Geographic area
Wheat	Paris Basin, Centre/South Italy, SE England, Central Germany
Barley	SE England, Denmark, North and Central Germany, Champagne Ardenne, Nord Pas de Calais Lorraine
Grain Maize	Po Valley, Paris Basin, Sud Ouest France, Rhone Alpes
Potato	The Netherlands, SE England, Germany, Bretagne, Nord Pas de Calais, Paris Basin
Sugar Beet	Benelux, Braunschweig, Ost for Storebaelt (Denmark), Picardie, Paris Basin, E. Anglia (UK), Marche Emilia Romagne, Po Valley

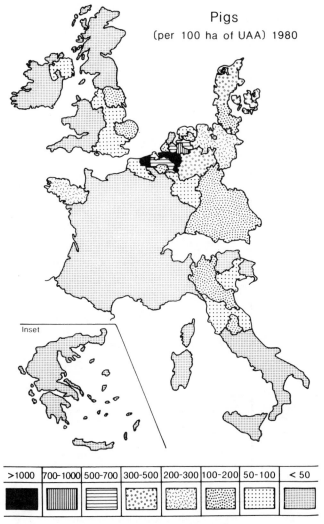

FIG. 4. Distribution of pigs in EEC-10.

3.1. Fertiliser use

Fertiliser consumption per hectare of UAA is shown in Table 8 and Figs. 5, 6 and 7. There is a sizeable disparity between regions in usage of N, P and K. Highest usage occurs in Benelux, Denmark and Germany and lowest usage in the Mediterranean zones. There has been a progressive increase in N

J. Lee

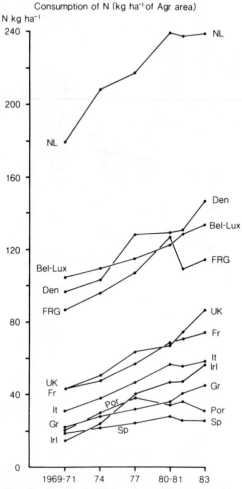

FIG. 5.   Trends in N consumption in EEC-12.

usage (Fig. 5). In contrast a significant decline in usage of P is evident in Benelux with overall usage remaining fairly static at European level and similarly for K.

Nutrient balance studies indicate that for the EEC-10, removals account for only 51, 68 and 32%, respectively of N, P and K applied in fertilisers even when nutrients in purchased feeds are excluded [2]. Figure 8 gives projected levels of fertiliser usage. While N usage is projected to increase, usage of P and K will remain static [3].

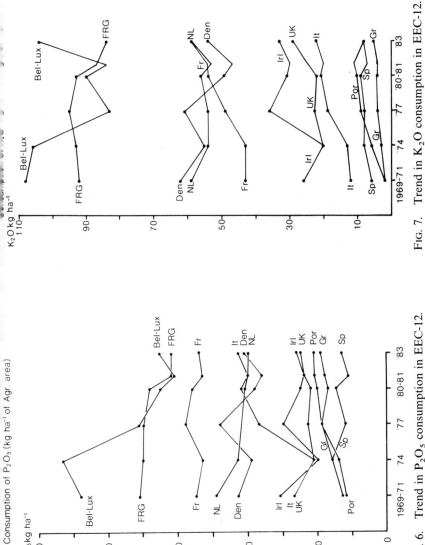

Fig. 7. Trend in K_2O consumption in EEC-12.

Fig. 6. Trend in P_2O_5 consumption in EEC-12.

Table 8

Fertiliser use in European Community as kg of nutrient per ha of UAA

Country	N		P_2O_5		K_2O	
	1969–71	*1983*	*1969–71*	*1983*	*1969–71*	*1983*
Germany	87	114	71	62	92	84
France	43	74	55	54	42	59
Italy	31	58	27	40	12	22
Netherlands	178	238	49	43	59	59
Belgium Luxembourg }	105	133	88	65	108	104
UK	42	86	27	25	26	29
Ireland	15	57	31	26	26	33
Denmark	98	146	43	41	62	54
Greece	22	45	13	19	2	5
Spain	19	26	13	13	6	8
Portugal	21	31	12	18	3	8

Source: 1984 FAO Fertiliser Yearbook Vol. 34.

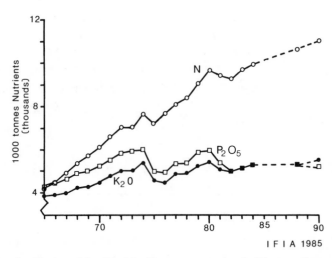

FIG. 8.   Projected levels of fertiliser consumption in Western Europe.

4. CLIMATE AND SOILS

In his classification of world climates Papadakis [4] distinguishes ten major categories. Five of these are represented in Europe (Fig. 9). Each of the zones shown in Fig. 9 is designated by a number representing the climatic type. The dominant European climate is of Marine type which is subdivided into Warm, Cool, Cold Marine, Warm and Cool Temperate climatic units. Southern Europe is characterised mainly by a Temperate Mediterranean climate.

4.1. Soils and land use

Europe includes several major soil regions, i.e. areas dominated by a few kinds of soil [5]. These soil regions are depicted in Fig. 9 which indicates that the EEC-12 is characterised by nine major soil regions. It is evident from Fig. 9 that Europe has a wide range of soils occurring over a wide climatic range. However, soil units with Cambisols and Luvisols dominant account for over 60% of the total area. Their land-use varies according to climate, topography and degree of stoniness. The Eutric Cambisols which are commonly stony or lithic in character are most extensive in the Mediterranean zone whereas the Dystric Cambisols are extensive under Cool Temperate conditions. In the Mediterranean zone the Eutric Cambisols are subject to severe drought limitation. They are devoted to grazing from low yielding pastures with arable cultivation and viticulture being locally important.

The Eutric Cambisols are also highly important arable and grassland soils in the UK and Eastern Denmark and are capable of yielding $7500\,kg$ ha^{-1} from cereals and $10\text{–}12\,000\,kg$ DM ha^{-1} from pastures. The Dystric Cambisols associated with mountainous regions such as Massif Central, Bayerischer Wald, Famenne, the Pyrenees and Italian Alps are largely devoted to forestry or grazing, whereas under lowland conditions in the UK, Ireland and France they are highly important agricultural soils capable of pasture yields of $11\text{–}15\,000\,kg$ DM ha^{-1} and cereal yields of $7500\,kg$ ha^{-1}.

Because of their moderately high nutrient status, the Orthic Luvisols which are most extensive in France and Germany, are highly suited to intensive arable and grassland production where they occur under favourable climatic conditions on level topography. In Ireland pasture yields of $13\,000\,kg$ DM ha^{-1} are attainable and in the Rhine Valley cereal yields of $6000\,kg$ ha^{-1} are achievable. In contrast, Chromic Luvisols are

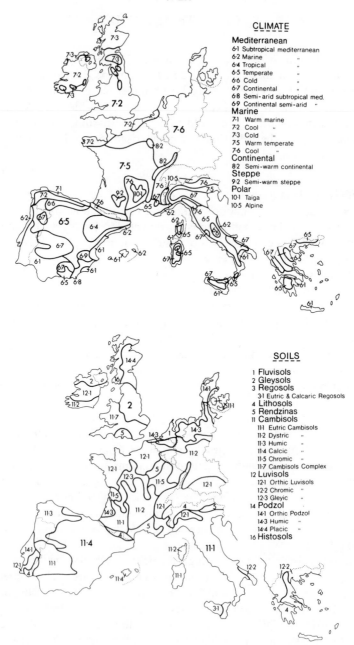

FIG. 9. Climatic and soil regions of EEC-12.

subject to severe drought limitation in Greece, Spain and Italy in addition to susceptibility to erosion.

The Fluvisols, which occupy 5·7% of the land area are among the most intensively farmed soils in Europe. They are particularly significant in the Mediterranean zone where their level topography is well suited to irrigation technology. They are devoted to intensive arable cropping, viticulture and market gardening in the Mediterranean zone whereas under Cool Temperate conditions in The Netherlands they are devoted to intensive grassland and arable cropping. Similarly the Rendzinas which occur on suitable terrain are important arable and grassland soils in France, Germany and the UK. They are devoted to forestry and grassland at high elevation. They also occupy 5% of the land area.

Although the Vertisols which are confined mainly to Spain and Italy are subject to structure and texture limitations they are nevertheless important arable soils. They occur on level terrain suited to irrigation technology and despite their limitations they are intensively cultivated.

The Gleysols which occupy over 3% of the land area are subject to drainage and permeability limitations. They are most extensive in the UK and Ireland where they occur under Cool Marine conditions. They are largely limited to pasture use except in lower-rainfall zones such as Eastern England where they are important arable soils.

The Podzols and Rankers which are commonly subject to depth, drainage, stoniness, elevation, accessibility or climatic constraints occupy over 9% of the land area. In mountainous regions, they are devoted largely to forestry or rough grazing. However, The Podzols on suitable terrain, which occur in West Denmark and North East Scotland constitute important arable and grassland soils in these regions. Similarly, the Lithosols which occupy 5% of the land area are largely associated with mountainous/hilly terrain and are subject to serious limitations of stoniness, rockiness, slope and elevation. They are largely afforested or devoted to rough grazing of garrigue type as in Greece or in the Cevennes. Locally, they may be devoted to vine, olive or almond culture in the Mediterranean zone. Although Histosols in their natural state are subject to severe use constraints, in their reclaimed state in The Netherlands, Germany and Ireland, they are capable of exceptionally high arable, horticultural and grassland yields.

5. EVALUATING LAND BASE FOR AGRICULTURE

The publication of the Soil Map of the European Communities (1:1 000 000) (1985) has enabled an evaluation of land suitability for grassland and arable uses to be carried out at Community level. The results of these exploratory evaluations which were recently carried out by the author [6, 7] are now presented.

5.1. Suitability classification scheme adopted for grassland
Table 9 shows the suitability classification scheme adopted. There are three categories of decreasing generalisation as follows:

   (i)   Suitability Class   —reflecting level of suitability.
  (ii)   Suitability Sub-class—reflecting degree of suitability.
 (iii)   Suitability Unit   —reflecting kinds of limitations.

Five suitability classes are indicated A, B, C, D and E, with degree of limitation increasing from A to E. The classes are subdivided into a number of sub-classes reflecting largely degree of limitation/suitability. The suitability sub-classes are sub-divided on the basis of type of dominant limitation(s) into units.

In the approach adopted, three major land qualities assume major importance:

   (i)   Moisture availability.
  (ii)   Poaching susceptibility.
 (iii)   Accessibility to machinery.

The water balance for a specific area can be expressed as follows:

$$\text{Water balance} = E_p - (P + AWC)$$

where E_p = potential evapotranspiration,
     P = precipitation,
  AWC = available water capacity of soil.

Through use of Turc Formula [8] for computation of E_p climatological precipitation deficits $(E_p - P)$ were computed in respect to approximately 140 representative European meteorological stations, enabling an estimated precipitation deficit figure to be ascribed to each of the 312 soil mapping units depicted on the EEC-10 Soil Map.

To arrive at the water balance for each soil, account was taken of soil water holding capacity (AWC) and the framework illustrated in Table 10 was adopted for this purpose.

5.2. Grassland suitability of each country and EEC-10
Table 11 summarises the percentage distribution pattern by country at suitability class level and Fig. 10 quantifies suitability class distribution at

EEC level. Only 28% of EEC-10 is in the very high and high grassland suitability/productivity Classes A and B capable of achieving pasture DM yields of $10\,000$–$12\,000+$ kg ha^{-1} with most of this area having good trafficability for livestock and being suited to grassland mechanisation. It is noteworthy that 74% of The Netherlands is in this category. Ireland has the highest proportion of land (32%) in the highest suitability Class A compared with a European average of 9%. Twenty six per cent of Europe is moderately suited (Class C) to mechanised grassland having moderately high productivity and having a pasture yield capacity of 8000–$11\,000$ kg DM ha^{-1} under moderately high management. The major limitation in this

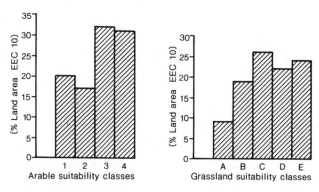

FIG. 10.   Extent of arable and grassland suitability classes in EEC-10.

category is moderate moisture stress (e.g. France) or excessive wetness (e.g. Ireland). Class D land occupies 22% of EEC-10 having a pasture yield potential of 2000–9000 kg DM ha^{-1} depending on degree of moisture stress. Class D land comprises sub-classes D1, D2 and D3 with sub-class D1 land which occupies 10% of EEC-10, having a moderate potential of 6000–9000 kg DM ha^{-1} and should be more appropriately included in Class C category. The remaining 24% of EEC-10 is unsuited to mechanised grassland due to limitations of topography, rockiness, elevation etc. In summary:

43·04 M ha (28%) of EEC-10 has very high–high productivity.
56·33 M ha (36·5%) of EEC-10 has moderate/low–moderate/high productivity.
54·88 M ha (35·5%) of EEC-10 has low–very low productivity.

Table 9

Suitability classification scheme

Suitability class	Suitability sub-class	Suitability unit	Properties	Productivity class
A well suited			High yield potential ($>12\,000$ kg DM ha^{-1}); livestock and machinery trafficability good; slope $<15\%$; moisture regime Class A	Very high
B suited	B_1	t, ts, td, tl	Moderately high yield potential ($10\,000$–$12\,000$ kg DM ha^{-1}), livestock trafficability mostly good; slope up to 25% with limitations for machinery use; moisture regime Class A, B	High
	B_2	s, d, l, ds, dl, c	As above with slight susceptibility to drought (units d, ds) or cattle treading (units s, ds); lateness in spring (unit c); slope $<15\%$; moisture regime Class A, B	
C moderately suited	C_1	s, st, sl	Moderately high yield potential ($9\,000$–$11\,000$ kg DM ha^{-1}), trafficability poor for livestock and machinery due to wetness; restricted grazing season, comparatively long winterage; moisture regime Class A	
	C_2	d, t, dt, ds, dl, tc, tl, ts	Moderately high yield potential ($8\,000$–$11\,000$ kg DM ha^{-1}), livestock and machinery trafficability good (units d, dl); machinery use limitations (unit dt); susceptible to cattle treading (units ds, ts); moisture regime Class C. Growth largely confined to spring/early summer due to drought; drought period May–August. Unit tc subject to topography limitation combined with lateness in spring	Moderate–high

C_3	s, st	Moderate yield potential (8 000–10 000 kg DM ha^{-1}), trafficability very poor for livestock and machinery due to wetness and/or topography; restricted grazing season and long winterage; moisture regime Class A	
D poorly suited			
D_1	d, dt, ds, dl, ts	Moderately low yield potential (6 000–9 000 kg DM ha^{-1}) due to drought; machinery use limitation (unit dt), susceptible to cattle treading (unit ds) when moisture conditions favour growth; moisture regime Class D; drought period May–September; growth largely confined to spring/early summer	Moderate–low
D_2	d, dt	Low yield potential (3 000–6 000 kg DM ha^{-1}) due to severe drought; machinery use limitation (unit dt); moisture regime Class E, F; drought period April–September, growth largely confined to spring	Low
D_3	d, dt, ds	Very low yield potential (2 000–5 000 kg DM ha^{-1}) due to severe drought, machinery use limitation (unit dt); susceptible to cattle treading when moisture conditions favour growth (unit ds); moisture regime Class F; drought period April–October, growth largely confined to winter/spring	Very low
E unsuited	—	Extremely low yield potential due to unfavourable soil/climate, and/or unsuited to mechanisation due to slope/rockiness	—

d = drought limitation; t = topography limitation; s = poaching susceptibility; l = lithic limitation; c = climatic limitation.

Table 10
Moisture regime classes—balance between $E_p - P$ and AWC (mm)

Texture class	50	50–100	100–200	200–300	300–400	400–600	>600
Sandy	B	C	D	E	F	F	F
Loamy	A	B	C	D	E	F	F
Silty	A	A	B	C	D	E	F
Clayey	A	B	C	D	E	F	F
Fine clay	A	B	C	D	E	F	F
Peat	A	A	B	C	D	E	F

A ———————————————→ F
Increasing dryness

5.3. Suitability of land resources for arable use

The methodology applied to assess the suitability of the land resources for cultivated crops assumes a number of principles, namely:

1. Suitability is expressed in qualitative terms without specific estimates of outputs or inputs.
2. The land utilisation type is mechanised arable farming and more specifically suitability for cultivation under mechanisation.
3. The assessment is based on rainfed conditions.

As for the grassland suitability assessment, the EEC-12 1:1 000 000 Soil Map and legend provided the data base for assessing suitability for cultivation. Through superimposing the climatological precipitation deficit map (Turc's formula) on the soil map and taking into account soil AWC derived from texture class, it was possible to assign a moisture regime class to each mapping unit. The map legend defined five texture classes and four slope classes. Degree of wetness could be derived from the classification system adopted, e.g. presence or absence of gleying, and the soil map indicated presence of stony phases, lithic phases, gravelly phases, etc.

5.4. Suitability classification scheme adopted

Soil suitability depends largely on the physical properties of the soil and the environment. These are rarely ideal and the limitations affect productivity and cultural practices. The degree of limitation may be assessed from such land characteristics as wetness (gleying), moisture availability (drought), slope, boulders, rock outcrop (lithic), soil workability or texture and structural properties affecting tilth. Depending on the degree of these

Table 11

Extent of grassland suitability classes in EEC-10 countries (% area provisional)

Suitability class	Germany	France	Italy	Netherlands	Belgium	Luxembourg	England and Wales	Scotland	Denmark	Greece	Ireland	EEC-10
A	13	—	17	22	4	—	12	1	—	11	32	9
B	34	21	6	52	50	12	21	10	38	1	13	19
C	43	29	4	19	45	80	50	27	43	—	21	26
D	5	37	35	4	—	—	3	—	17	19	—	22
E	5	13	38	3	1	8	13	62	3	69	34	24

Irrigation assumed for suitability Class A and B in Italy and Greece which under rainfed conditions would mainly correspond to Class D.

Table 12
Suitability for cultivation classification scheme

Suitability class	Suitability unit	Properties
1 well suited		Well suited to mechanisation, having in general good tilth, favourable moisture regime and favourable slopes (<8%); includes also complexes of soils with <8% and <15% slopes
2 suited	s, d, t, ts, ds, td	Suited to mechanisation but subject to limitations of wetness/tilth, drought or slope (8–15%)
3 moderately/ poorly suited	s, td, tl, ts, d	Moderately/poorly suited to mechanisation due to slope (15–25%), soil wetness/tilth, drought or presence of rock
4 unsuited		Unsuited to mechanisation due to slope (>25%), altitude, rockiness, wetness or climatic constraint

d = drought limitation; t = topography limitation; s = wetness/tilth limitation; l = lithic limitation.

limitations it has been possible to devise a suitability classification (Table 12). The classification scheme includes four suitability classes as follows:

Class 1. Well suited; land with none or slight limitations.
Class 2. Suited; land with slight or slight/moderate limitations.
Class 3. Moderately/ poorly suited; land with moderate or moderate/severe limitations.
Class 4. Unsuited; land with severe or very severe limitations.

Degree of limitation increases from the highest to the lowest category.

5.5. The arable land base of each country and EEC-10

Table 13 quantifies the extent of each suitability class by country. Figures 10 and 11 are graphic illustrations. It may be deduced from Table 13 that 30·64 M ha in EEC-10 are well suited to cultivation (Class 1). This corresponds to 20% of the total land area. On a percentage basis, Germany, France, Belgium and The Netherlands have the greatest proportions of their land areas in this category. However, it is significant that France alone accounts for 50·4% of total EEC-10 Class 1 land. Class 2 arable land (having slight or slight/moderate limitations) occupies 26·68 M ha or 17%

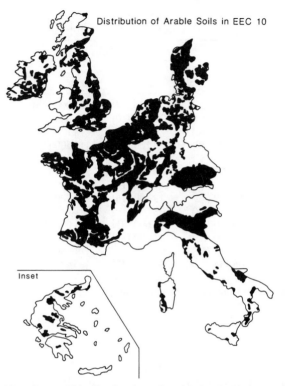

Distribution of Arable Soils in EEC 10

Inset

FIG. 11. Geographic distribution of arable land (Class 1 and 2).

of the land area. France with 7·7 M ha accounts for 29% of the total EEC-10 Class 2 land.

It is notable that 31% of the EEC-10 land area is unsuited to arable cropping and the remaining 32% of land is in the Class 3 category. While cultivation may be feasible on Class 3 land, in practice it is subject to moderate or moderate/severe limitations of soil or topography and must be largely excluded from consideration for mechanised arable systems. Therefore, the land pool having viable arable use options is limited to 57·32 M ha (Class 1 and 2) with France accounting for 40·4% of the total area.

It is pertinent to place these figures in the context of the existing national arable areas. This may be expressed as the proportion of the arable area accounted for by Class 1 and 2 land (Table 14).

Although it is misleading to assume that existing arable areas are

J. Lee

Table 13
Extent of arable land suitability classes in EEC-10[a] (km^2)

	Suitability class			
	1	*2*	*3*	*4*
France	154 400 (29)	77 000 (14)	187 000 (35)	121 000 (22)
Italy	33 500 (13)	26 400 (10)	81 000 (32)	108 700 (45)
Greece	5 500 (4)	9 500 (7)	18 100 (14)	98 100 (75)
Germany	63 000 (30)	28 100 (13)	111 500 (53)	9 300 (4)
England and Wales	11 800 (8)	48 900 (35)	47 800 (34)	33 000 (24)
Scotland	3 300 (5)	7 600 (10)	5 500 (8)	56 500 (77)
Ireland	10 100 (12)	13 800 (16)	11 000 (13)	49 400 (59)
Denmark	7 650 (18)	33 550 (79)	— —	1 300 (3)
Netherlands	9 200 (27)	10 600 (31)	11 000 (33)	3 000 (9)
Belgium	7 600 (25)	11 400 (37)	10 600 (35)	930 (3)
Luxembourg	385 (15)	— —	1 980 (77)	215 (8)
EEC-10	306 435 (20)	266 850 (17)	485 480 (32)	481 445 (31)

[a] Percentage area in brackets. Areas provisional.

Table 14
Proportion of arable area accounted for by Class 1 and 2 land

	Class 1[a]	Class 2[a]	Arable area	Class 1 and 2 land
	(million ha)		*(1983)*	*(% arable area)*
France	15·44	7·70	17·35	134·1
Italy	3·35	2·64	9·39	63·8
Greece	0·55	0·95	2·98	50·3
Germany	6·30	2·81	7·24	125·5
England and Wales	1·18	4·89	5·46	111·1
Scotland	0·33	0·76	1·13	96·5
Ireland	1·01	1·38	1·11	124·1
Denmark	0·77	3·35	2·63	156·6
Netherlands	0·92	1·06	0·83	238·5
Belgium	0·76	1·14	0·80	237·5
Luxembourg	0·04	—	0·06	66·6
EEC-10	30·65	26·68	48·98	108·2

[a] Areas provisional.

synonymous with Class 1 and 2 land, Table 14 nevertheless suggests that the land quality of the arable areas must on average be high. For the EEC-10 area, the extent of Classes 1 and 2 is 8% greater than the existing arable area. The extent of Classes 1 and 2 land as compared with arable area in The Netherlands and Belgium is particularly high and in the Mediterranean zone the reverse is true.

6.   INTENSIVE FARMING AND SOIL CHARACTERISTICS

6.1. Cereal production

Cereals account for 27% of the UAA of EEC-10 extending over an area of about 28 M ha (1983). Table 15 outlines the soil characteristics of the intensive cereal areas (Fig. 2). The most intensive areas (> 60% UAA under cereals) are largely characterised by Luvisols (L) or Cambisols (B). Topography is level/sloping and texture class medium-fine. Workability is good in these soils. However, intensively cultivated Luvisols derived from loess materials are sensitive to water erosion particularly on sloping gradients, e.g. Lo-3b in Germany and France. Areas of high (50–60% UAA) and moderately high (40–50% UAA) concentrations have similar soils.

In England the associated soils may display gleyic properties which would be conducive to machinery compaction or tilth problems, e.g. Soil Units Jeg-4a and Gmf 2/4a. Textures are medium-fine indicating slow permeability which combined with gleyic properties would indicate susceptibility to machinery compaction. However, a comparatively low precipitation and artificial control of water tables would tend to minimise the utilisation hazards. Cereals may be intensively cultivated on Vertisols in Italy, e.g. soil unit Vp-4/5 (Foggia). Their fine—very fine texture and susceptibility to volume changes (cracking) would combine to make tillage operations difficult with an added liability of structural deterioration.

In Germany and Denmark soil textures may be coarse in areas of moderately high/high intensity (40–60% UAA), e.g. Soil Units Phf-1a (Germany), Po-1a and Be-1b (Denmark) and Po-1ab (France). This would suggest a high susceptibility to wind blow in particular.

6.2. Cattle production

The average concentration of cattle in EEC-10 is 80 per 100 ha ranging from < 25 per 100 ha in much of the Mediterranean zone to > 300 per 100 ha in The Netherlands. Particularly high concentrations occur in the Benelux zone, Lombardia in Italy and in the Oberbayern-Schwaben region

Table 15
Soil characteristics of regions of highest cereal concentrations in EEC-10

Country	Cereals (% UAA)					
	40–50		50–60		>60	
	Soil unit	Region	Soil unit	Region	Soil unit	Region
France	Bd-3b Bd-1/2bd Lo-3ab Lo-2/3ab Bk-3/4ab	Paris Basin Picardie Nord Pas de Calais Midi-Pyrennes	Po-1ab Lo-3ab	Aquitaine Paris Basin	Lo-3b Lo-2/3ab Lg-2/3ab	Paris Basin
Denmark	Be-2b Ph-1a	Ribe	Be-1b Be-2a Po-1a	Vibors Nordijylland	Lo-2/3ab Be-1b Be-2b	Rosalde Arhus Sjalland
Germany	Phf-1a Bd-2bc Bd-2b	Luneburg Central	Lo-3ab Lo-2/3a Phf-1a Bd-2b	Hannover Unterfranken, Freiburg	Lo-3ab	Braunschweig

Belgium	Lo-3ab Lo-2/3ab	Brebant	—		—	
Italy	Bef-1/4a Lc-2/4d Vp-4/5ab Bd-2cd Be-2/4cd	Padova Foggia Foggia Potenza Toscana	Bef-1/4a Rc-3/4bc	Padova Marche	Bef-1/4a Gef-2/3a Bd-2cd	Po Valley Molise
England and Wales	Eo-2/3b Ges-4a	Dorset Wiltshire Northamptonshire	Jeg-1a Jeg-4a Gmf-2/4a Lgs-2/4a Lc-1/2a Bgc-4ab	Lincoln Cambridge Norfolk, Suffolk Humberside Essex Bedford, Cambridge	Lgs-2ab	Norfolk Suffolk
Scotland	Be-2b Bd-1/2bc Bd-2bc	Angus Berwick	Be-2bc	East Lothian	—	

For definition of soil units see Soil Map of the European Communities 1:1 000 000—Directorate General for Agriculture Co-ordination of Agricultural Research (1985).

Table 16

Soil characteristics of regions of highest cattle concentrations in EEC-10

Country	Cattle per 100 ha UAA							
	100–150		150–200		200–250		250+	
	Soil unit	Region	Soil unit	Region	Soil unit	Region	Soil unit	Region
Netherlands	Jeg-4a Jeg-2/4a	Groningen	Pg-1a Jeg-2/4a	Drenthe Nord-Holland Zuid Holland	—	—	Jeg-2/4a Pg-1a	Friesland Oost-Nederland West Nederland
Belgium	Lo-3ab	Brabaut Namur	Pg-1a Lo-3ab	Limburg Hainaut	Jeg-2/4a Dd-2b Bd-2bc	West Vlaanderen Liege	Pg-1a Ph-1a	Antwerpen Oost-Vlaanderen
Luxembourg	—	—	Bd-2bc	—	—	—	—	—
Germany	Pg-1a Phf-1a Lo-2/3ab Lo-2/3a Bd-2bc	Nordrhein-West falen Schleswig-Holstein Baden-Württemberg Oberfrankem-Niederbayern	Lo-3ab Lo-2/3a	Oberbayern Schwaben	—	—	—	—

France	Bd-3ab	Normandie								
	Bd-3b	Bretagne								
		Pas de la Loire								
		Limousin								
Italy	Bef-1/4a	Veneto	Lo-2/4ab Lombardia							
			Gef-2/3a							
	Jeg-1/4a	Trentino-								
	Eo-3ad	Alto-Adige								
Denmark	Ph-1a	Vest for								
	Po-1a	Store Baelt								
	Be-2b									
	Be-1b									
England	Bd-1/2bc	South West								
	Eo-2/3b	West Midlands								
	Lo-3b	North West								
	Lgs-2ab									
Ireland	Lo-2ab	Leinster								
	Lo-2/3ab	Munster								
	Bd-2b									
	Bds-1b									

For definition of soil units, see Soil Map of the European Communities 1:1 000 000—Directorate General for Agriculture Co-ordination of Agricultural Research (1985).

of Germany (Fig. 3). These intensive cattle zones are characterised by a high importation of feed concentrates resulting in obvious nutrient build-up.

Table 16 sets out the soil characteristics of the most intensively stocked (>250 LU per 100 ha) cattle areas. Coarse textured Gleyic and Humic Podzols (soil units Pg-1a and Ph-1a) on level topography are associated with the highest stocked regions in The Netherlands. These areas have a high sensitivity to nutrient leaching and ground water pollution. In general high (150–200 cattle per 100 ha) stocking rates are mainly associated with Luvisols (L) and Cambisols (B). In Belgium, Germany and Italy the associated soil units include Orthic Luvisols (soil unit Lo-2/3 ab) on level or sloping topography of texture class medium fine or fine. These soils are well suited to livestock systems with topsoils of good livestock bearing capacity. However, high stocking rates are also associated with Fluvi-Eutric Gleysols (soil unit Gef-2/3a) in Lombardia and with Gleyo-Eutric Fluvisols (soil unit Jeg-2/4a) in Nord and Zuid Holland. Associated topography is level and texture class medium–fine. Gleyic (wetness) characteristics are not conducive to intensive livestock trafficability and have limitations for waste disposal. High stocking rates are also associated with coarse textured Gleyic Podzols (soil unit Pg-1a) in Belgium and The Netherlands.

The regions of moderately high (100–150 cattle per 100 ha) stocking rates are largely characterised by the occurrence of Luvisols (Belgium, Germany, England and Ireland) or Cambisols (Germany, France, Italy, Denmark, England and Ireland). The Luvisols are mainly in the Orthic Great Soil Group (Lo) and the Cambisols in the Dystric (Bd) group. Texture class is mainly medium–medium fine and associated topography level/sloping. These Orthic Luvisols and Dystric Cambisols are characterised by good bearing capacity, drainage status and structural stability. However, in Germany and Denmark moderately high stocking rates are also associated with coarse textured Podzols whereas in The Netherlands (Groningen) and Italy (Trentino-Alto-Adige) the associated soils display hydromorphic characteristics, e.g. Gleyo-Eutric Fluvisols (soil unit Jeg-4a) of fine texture in The Netherlands or coarse–fine texture in Italy (soil unit Jeg-1/4a). All these latter soils display undesirable properties for intensive livestock enterprises.

6.3. Pig production
Over the period 1961–65 to 1984 the pig population of EEC-12 increased by 62%. The comparable increases for The Netherlands and Belgium/ Luxembourg were 252% and 190% respectively. Average pig concentration in EEC-10 is 80 per 100 ha UAA. Highly intensive pig

Table 17

Soil characteristics of regions of highest pig concentration in EEC-10

Country	Pigs per 100 ha UAA							
	300–500		500–700		700–1000		1000+	
	Soil unit	Region	Soil unit	Region	Soil unit	Region	Soil unit	Region
Netherlands	—	—	Jeg-4a Pg-1a	Utrecht Overijssel	Pg-1a	Gelderland	Pg-1a	South Netherlands
Belgium			Pg-1a	Antwerpen Limburg Oost Vlaanderen	—	—	Jeg 2/4a	West Vlaanderen
Denmark	Po-1a Pl-1a Be-1b	Vest for Store Baelt	—	—	—	—	—	—
Germany	Pg-1a	Nordrhein-West falen						

For definition of soil units see Map of the European Communities 1:1 000 000—Directorate General for Agriculture Co-ordination of Agricultural Research (1985).

concentrations (300–1000+ pigs per 100 ha) are largely confined to Benelux, the German hinterland and Denmark (Fig. 4). Pig farming is industrialised, characterised by high importation of feed concentrates with the disposal of animal wastes requiring major attention to minimise pollution. These high pig concentrations are invariably associated with Podzol soils (Gleyic Podzols, Humic Podzols and Orthic Podzols (Table 17) characterised by coarse textures and level topography (soil units Pg-1a, Po-1a, Ph-1a). These soils have a high infiltration capacity. The Dutch and German Podzols in particular are characterised by a shallow depth of groundwater. The combination of coarse texture and shallow groundwater suggests that the sensitivity of these areas to nutrient leaching and groundwater pollution is high. High concentrations are also associated with Gleyo-Eutric Fluvisols of medium–fine texture on level topography (soil unit Jeg-2/4a) in The Netherlands.

7. SLOPE CHARACTERISTICS OF LAND

The slope characteristics of land are significant not only in terms of utilisation but also in terms of susceptibility to water erosion. The legend of the Soil Map of the European Communities (1:1 000 000) (1985) includes four categories of slope as follows:

    a =   0– 8%   Level
    b =   3–15%   Sloping
    c = 15–25%   Moderately steep
    d = > 25%   Steep

The four slope classes indicate the slope which dominates the Soil Association area. Through an analysis of the national legends provided for the Soil Map of Europe project, it is possible to derive a slope inventory of European land. The results are summarised in Table 18. It is significant that only 25% of EEC-12 is categorised as level land with an additional 14% in the level/sloping class. The extent of accessible land from a mechanisation suitability viewpoint ranges from over 90% in The Netherlands to about 16% in Greece. Almost 26% of EEC-12 is moderately steep–steep ranging from about 65% in Greece to zero in The Netherlands and Denmark. The Mediterranean region is characterised by a high proportion of moderately steep/steep land. The percentage areas of Greece, Italy and Spain in this category are >65%, 56% and 42%, respectively. For Portugal about 24% is in this category with an additional 54% in the sloping/moderately steep

Table 18
Percentage distribution of slope classes by country in EEC-12

Country	a	a/b	a/d	b	b/c	b/d	c	c/d	d
Spain	17·9	3·5		14·8	22·2		1·4	40·2	0·8
France	36·6	22·7		13·4	1·3		9·2	16·9	
Italy	15·1	9·6		0·2	16·5	2·9	0·4	44·0	11·2
Greece	12·1			4·1	8·0	10·8		3·3	61·4
Germany	47·2	7·2		16·2	9·2		18·8		1·3
Denmark	45·4			54·6					
Netherlands	90·8			9·2					
Belgium	24·7	1·9		44·7			28·9		
Luxembourg	6·6		16·7				76·7		
England and Wales	23·9	40·2	4·3	3·0	21·6	7·2			1·0
Scotland	1·3	47·2		14·5	20·6			7·3	
N. Ireland	2·1	20·2		67·9	9·8				
Ireland	7·9	25·4		44·0	22·7				
Portugal	2·1	1·6		18·4	29·8	24·2	1·2	22·5	
EEC-12	25·2	13·8	0·3	13·9	13·4	2·5	5·1	20·2	5·5

a = 0–8%; b = 8–15%; c = 15–25%; d = >25%.

and sloping/steep categories. In the Mediterranean zone where sloping land areas are intensively cultivated or where overgrazing of sparsely vegetated areas occurs soil degradation is a major hazard. For instance the intensive cultivation of sloping/moderately steep Chromic Cambisols in Spain (Bk-3bc) under subtropical Mediterranean conditions results in a high vulnerability to erosion [9].

8. OCCURRENCE OF WETLAND

Through an analysis of the EEC-10 1:1 000 000 Soil Map of Europe, it is possible to ascertain the extent of soils having hydromorphic (wetness or gleyic) properties. Such an analysis has been carried out and the results are summarised in Table 19. The table indicates that 17% of EEC-12 is subject to varying degrees of wetness. It is evident that there is a particularly high occurrence of wet soils in UK and Ireland. Hydromorphic soils, especially under the high precipitation conditions of North West Europe, present serious utilisation problems and are subject to structural deterioration from grazing livestock/machinery traffic.

Table 19

Extent of soils (km²) with wetness limitation in EEC-12

	1 Gleysols	2 Fluvisols with gleyic properties	3 Other soils with gleyic properties	4 Eutric Histosols	5 Dystric Histosols	1 + 2 + 3 (%)	4 + 5 (%)	Total (%)
France	1·610 (0·3)	16·540 (3·1)	40·260 (9·0)	0·700 (0·1)	2·330 (0·4)	12·4	0·5	12·9
Italy		8·800 (3·5)		0·300 (0·1)	900 (0·4)	3·5	0·5	4·0
Greece						—	—	—
Germany	7·330 (3·5)	7·700 (3·6)	19·020 (9·0)	2·800 (1·3)	6·200 (2·9)	16·1	4·2	20·3
Denmark	1·170 (2·7)			160 (0·4)	710 (1·7)	2·7	2·1	4·8
Netherlands	1·888 (5·7)	10·030 (30·2)	1·290 (3·9)	3·380 (10·2)	630 (1·9)	39·8	12·1	51·9
Belgium	290 (1·0)	2·200 (7·3)	4·495 (14·8)		150 (0·5)	23·1	0·5	23·6
Luxembourg		120 (0·5)	600 (2·3)			2·8	—	2·8
England and Wales	29·280 (20·6)	8·410 (5·9)	28·720 (20·2)	900 (0·6)	3·300 (2·3)	46·7	2·9	49·6
Scotland	12·700 (17·1)	360 (0·5)	2·320 (3·1)		25·960 (35·0)	20·7	35·0	55·7
N. Ireland	5·566 (48·8)	010 (0·1)		180 (1·5)	1·015 (8·7)	48·9	10·2	59·1
Ireland	18·020 (25·0)		535 (0·7)	2·050 (2·8)	8·380 (11·6)	25·7	14·4	40·1
Spain		27·703 (5·6)	28·138 (5·7)		343 (0·1)	11·3	—	11·3
Portugal		1·070 (1·2)	1·666 (1·9)			3·1	—	3·1
EEC-12	77·945 (5·1)	82·943 (4·0)	135.044 (6·5)	10·470 (0·5)	49·918 (2·4)	14·2	2·9	17·1

Percentage figures in brackets.

9. CONCLUSION

While this paper has of necessity covered a wide perspective, a number of issues emerge which are significant in future soil use/protection policies:

1. UAA is declining and there is a need to quantify the precise allocation of this land.

2. The land resource base of EEC is limited and in view of commodity demand projections farm enterprises will be under serious pressure in the less competitive areas.

3. European agriculture displays a pronounced regional variation in intensity, with intensive farming being carried out in sensitive soil environments.

4. There is a need to establish, on a Community wide basis, the current degradation/pollution status of land areas including natural and man made hazards.

5. The EEC-12 1:1 000 000 Soil Map is a logical framework for the conduct of such a study.

REFERENCES

1. EEC (1978). Forestry policy in the European Community COM (78) 62/final.
2. HEBERT, J. (1984). Levels of fertiliser input and soil nutrient status in European Agriculture. 18th Coll. Int. Potash Institute, pp. 249–72, Berne.
3. GINET, H. (1985). Consumption of $N-P_2O_5-K_2O$. International Fertiliser Industry Association, Paris.
4. PAPADAKIS, J. (1966). *Climates of the World and their Agricultural Potentialities*, Buenos Aires, Argentina.
5. EEC (1985). Directorate General for Agriculture—Co-ordination of Agricultural Research Soil Map of the European Communities (1:1 000 000) (General co-ordinator Professor R. Tavernier).
6. LEE, J. (1984). Suitability and productivity of land resources of EEC-10 for grassland use. A study carried out on behalf of the EC, Johnstown Castle. Wexford.
7. LEE, J. (1985). The impact of technology on the alternative uses for land. A study carried out on behalf of the EC, Johnstown Castle, Wexford.
8. TURC, L. (1954). Le bilan d'eau des sols, relations entre les precipitations, l'evapotranspiration et l'ecoulement. *Annals Agronomique*, **5**, 491–5.
9. FAO/UNESCO (1977). Explanatory Note. UN Conference on Desertification A/CONF. 74/2.

The Production Potential of Soils: Part I— Sensitivity of Principal Soil Types to the Intensive Agriculture of North-Western Europe

N. FEDOROFF

Soils Department, Institut National Agronomique Paris-Grignon, 78850 Thiverval Grignon, France

SUMMARY

A brief review of research dealing with soil degradation under intensive agricultural practices is given. The author defines the sensitivity of soil to agricultural practices as its ability to withstand those negative changes in its properties induced by cultivation. The concept of sensitivity is derived from that of degradation. Water and wind erosion, crusting, translocation of particles, compaction, lowering of the organic matter content and degeneration of the soil ecosystem are all symptoms of such sensitivity. The sensitivity of a soil is assessed by giving a value to each first-order factor (texture, clay mineralogy, pedological fabrics, organic matter, biological activity) and then summing them; these are then compared with the values for second-order factors (aggregation, macroporosity, structural stability, permeability). Soil drainage, slope and rain intensity and distribution are also taken into account.

The sensitivity of the main soil types of North-Western Europe to intensive agriculture is discussed. The sensitivity of luvisols developed on loess—the most productive, but also the most sensitive soils of this part of Europe—is analysed in detail.

This study of the degradation and sensitivity of cultivated soils shows that our knowledge of their behaviour is insufficient, too specific and empirical. The author concludes that priority must be given to the study of alterations in the soil ecosystem of cultivated soils and to the consequences of these alterations for the physico-chemical behaviour of the soil.

65

In the areas of North-Western Europe where intensive agriculture is practised, farmers and local and national authorities are concerned about the extent of certain phenomena such as soil erosion, the fact that crusting is on the increase, the phenomenon of internal compaction and the drop in the proportion of organic matter in the soil. The erosion of silty soils on gentle slopes, which has increased over the years, is causing particular anxiety (De Ploey and van Hecke [14]. This is because, in view of the fertility of the soil and the large areas it covers, it represents the main soil resource of North-Western Europe. In other words, like other parts of the world (FAO [15]), North-Western Europe is suffering from soil degradation apparently aggravated more than anywhere else by intensive farming methods.

Soil degradation is the product of a number of phenomena, including water and wind erosion, salinisation and alkalisation, physical degradation, chemical degradation (acidification, the side effects of fertilisers and plant-health products) and biological degradation (Riquier [30]). Water erosion has been the subject of a great deal of research, particularly by geographers in Belgium (Bollinne [3]; Bollinne *et al.* [4]; Laurant and Bollinne [22]; Pissart and Bollinne, [27]; De Ploey [12]; De Ploey and Mucher [13]), the Netherlands (Mucher [25]) and France (Roux and Fedoroff [31]), and of laboratory tests (Poesen [28]; Savat [32–34]; Poesen and Savat [29]; Savat and Poesen [35]). There has been a symposium on soil erosion due to agricultural activity in temperate climates (Vogt and Vogt [39]). Particular attention has been paid to one aspect of physical degradation, namely surface sealing (Boiffin [1]; Callebaut *et al.* [8]), and its effects have probably been over-estimated. Internal physical changes in the soil, such as compaction and the removal of solid particles are beginning to be relatively well understood (Jongerius [20]; Faure [16]; Grimaldi [17]), although the consequences of these processes as regards the proper functioning of the soil have not been sufficiently emphasised. The steady decrease in the amount of organic matter contained in the soil is one of the main factors on record (Cordonnier *et al.* [10]) but none of the research has given any insight in to the causes. Whilst it is recognised that humus increases structural stability (Monnier [24]) there is still controversy over its role in the soil and no clear relationship between plants and humus has been established. Biological degradation, for instance following impoverishment of the soil ecosystem, has often been suggested as a cause but there is little in the way of scientific data that could be used as a basis for a proper assessment of the consequences. The importance of earthworms in the soil ecosystem is well known (Bouche [6,7]) but there has been no research specifically on the effect of modern agricultural practices on earthworm populations.

One cannot but deplore the compartmentalisation of research by scientists and economists (Cordonnier *et al.* [10]) on the problem of soil degradation in North-Western Europe. Geographers are mostly interested in erosion and its dynamics and consequences in the form of topographical changes whilst others, such as soil scientists, study the processes of degradation in the soil structure. Yet all this specialised work does not help us understand the problem of degradation in agricultural soils, or even that of erosion as a whole. Research has been confined to one or other aspect of the phenomenon. For instance, Boiffin *et al.* [2] stress the role of surface sealing in the process of erosion but neglect the effects of internal compaction. Although pedologists have the conceptual tools necessary for a global approach to the phenomena of degradation, they have not yet taken all the available research and distilled from it a theory on the degradation of the soils in North-Western Europe.

It is generally accepted that soil degradation, and particularly the appearance or aggravation of erosion, is the result of changes in agricultural systems, in particular the disappearance of grasslands, the increase in the size of fields and the use of pesticides and heavier and more powerful agricultural machinery. But this view has not yet been supported by the fragmentary scientific observations available (Boiffin *et al.* [2]).

The concept of the sensitivity of a soil to intensive agricultural practices derives from that of degradation. This sensitivity may be defined as the tendency for the properties of a soil to deteriorate and for the soil itself to be subject to degradation brought on by agricultural practices. Whilst there are many maps indicating the agricultural capability of soils, there are, as far as we know, no maps indicating the sensitivity of soils to intensive agriculture; neither are we aware of any methods having been devised to study such sensitivity.

Because of this gap we cannot immediately proceed to evaluate the sensitivity to intensive agricultural practices of the various types of soil in North-Western Europe. We have therefore tried, firstly, to define the concept of soil sensitivity in respect of agriculture and then to work out a method of estimating this sensitivity from existing research and our own results. Using this method and the European Community's soil map we then go on in the second part of this chapter to estimate the sensitivity of the various types of soils in North-Western Europe to intensive agricultural practices.

SOIL DEGRADATION AND SOIL SENSITIVITY

On a geological timescale, the soil—a complex interface between the atmosphere and lithosphere—is perpetually changing, subject at times to degradation—for instance as a result of lixivation and leaching, whose consequences include a decrease in chemical and physical fertility—and at others to enrichment—as a result of accretions of loess dust which improve fertility. On this scale a soil is never utterly destroyed, the latter concept being an anthropocentric one. Only by restricting our timescale to a period covering a century or a thousand years can we talk of the soils in our temperate Western European regions with their forest cover having achieved relative equilibrium. Before man started clearing the land, North-Western Europe was entirely covered by forest and our reference base is therefore made up entirely of forest soils.

By clearing the land and cultivating it, man broke the 'equilibrium' in the forest soil. We should emphasise that the primitive farmer was far from being a soils expert; historians, geographers and pedologists (e.g. Bork [5]) have shown that there has been considerable soil erosion from Neolithic times to the present. Over the centuries most cultivated soils have suffered erosion, some have turned acid and the most fragile, such as certain sandy soils in the Netherlands which have been transformed into dune fields, have become entirely sterile. All along therefore man has caused soil degradation which has resulted in erosion and biological and chemical impoverishment and has reduced the soil's capacity to produce a harvest.

The modern farmer tends to regard the soil as a mere substrate for his plants for whose every need he provides. He eliminates plant pests by means of pesticides which impoverish, i.e. destroy, the soil fauna. As opposed to the traditional farmer who stimulated biological activity by means of his organic input, the modern farmer tends to destroy this activity by using pesticides. He deals with climatic hazards by:

(1)   draining the soil—a practice which seems to have no negative consequences; and
(2)   irrigating the land—a practice which promotes surface crusting and the collapse of the soil structure.

To counteract meteorological hazards he uses increasingly powerful machines whose weight, together with the tangential forces of the wheels and the vibration of the machinery, cause compaction—particularly under the ploughed horizon—when used on soil that is still wet.

Definitions

The sensitivity of a soil to agriculture is expressed by symptoms of sensitivity such as water and wind erosion, surface sealing and crusting, the translocation of clayey and loamey particles, compaction, a drop in the organic matter content, impoverishment or deterioration of the ecosystem, toxicity due to incomplete elimination of plant products and, although these only affect seaside areas, salinisation and alkalisation.

The sensitivity of a soil to agricultural methods depends on some of the soil's own characteristics which we shall call first-order factors of soil sensitivity and which include the texture of the ploughed horizon and the sub-soil, the mineralogical properties of the clays, the pedological fabric of the sub-soil and the ploughed horizon (where these are clearly different-iated), the organic matter content and the intensity of biological activity of which the number of earthworms is a good indicator. The second-order factors of sensitivity are the characteristics of the soil seen as an aggregation, the stability of its structure, its permeability and its macroporosity—which all depend on first-order factors. The sensitivity of a soil also depends on drainage. Where the soil or one of its horizons is saturated by groundwater or perched water, it becomes more sensitive, particularly if the water rises to a high level in the profile and remains there for a long time. External factors affecting soil sensitivity are the degree of slope and its length, the kinetic energy of raindrops and the duration of a shower of rain, and the distribution and frequency of rain over a given period.

We cannot at present quantify the sensitivity of a soil to agricultural methods. Certain formulae for forecasting the intensity of certain symptoms of soil sensitivity have been proposed, as for instance in the universal soil loss equation of Wischmeier and Smith [40] and the formulae of Skidmore [36] and Riquier [30] for forecasting wind erosion. The validity of these formulae is debatable, as is the erosion map of France drawn up by Pihan [26] using the Wischmeier and Smith [40] R index which indicates a maximum erosion potential (<200) in the Languedoc and a minimum erosion potential (>50) in the Paris basin and the north of France, whereas De Ploey and van Hecke [14] mention major erosion in these areas. The R index calculated from the kinetic energy of raindrops and the duration of an episode of rain is an insufficient basis for estimating the erosion potential of rain in North-Western Europe. The frequency of rainfall and its distribution over a year must also be taken into consideration as rain of average intensity falling on soil which has drained is easily absorbed whereas the same kind of rain falling on saturated soil runs off. In the European North West violent summer storms contribute

Table 1

Sensitivity factor coefficients

	Not at all sensitive (0)	Low sensitivity (1)	Fairly low sensitivity (2)	Average sensitivity (3)	Fairly high sensitivity (4)	High sensitivity (5)
Texture	—	Balanced	Slight imbalance towards clay	Moderate imbalance towards silt	Sandy silt silty sands medium silts	Silt
Mineralogical characteristics of the clays	Abundant swelling clays		Swelling clays present		No swelling clays	
Pedological fabric		Developed calcareous cambic clayey fabrics	Dystro-cambic palaeo-argilaceous fabrics			No fabrics Gley fabrics Hydromorphic clay fabrics
Organic matter		Over 2%		Between 1 and 2%	Less than 1%	
Biological activity		Indicates high level of earthworm activity		Indicates presence of earthworms		Earthworms apparently absent
Aggregation and macroporosity		Very marked crumbly, fluffy, polyhedric and highly macroporous structures		Not very marked to average development of subangular, polyhedric structure. Macroporosity average to low		Massive structure No macroporosity
Structural stability		$\log_{10} I_s < 0{\cdot}8$ and $\log_{10} K > 1{\cdot}8$		$\log_{10} I_s$ between $0{\cdot}8$ and $1{\cdot}2$ $\log_{10} K$ between $1{\cdot}8$ and $1{\cdot}3$		$\log_{10} I_s > 1{\cdot}2$ and $\log_{10} K < 1{\cdot}3$
Permeability (hydraulic conductivity K) (cm × s)		Values of less than 2×10^{-2}		Value of between 2×10^{-2} and 1×10^{-4}		Values of over 1×10^{-4}
	No waterlogging	Sub-soil waterlogged	Exceptional episodes of waterlogging up to the surface of the soil	Occasional waterlogging up to the surface	Waterlogging frequently reaches surfaces	Soil waterlogged up to the surface throughout the winter

towards erosion whilst winter rain has a cumulative effect which is promoted by the low rate of evapotranspiration during this period.

The most we can do is attribute a coefficient of sensitivity to each of the factors of sensitivity (Table 1). We believe that our present knowledge does not allow us to go beyond a scale ranging from 0 to 5 for the coefficients of sensitivity in which 5 = high sensitivity, 4 = fairly high sensitivity, 3 = average sensitivity, 2 = fairly low sensitivity, 1 = low sensitivity and 0 = not at all sensitive. The sensitivities of the first-order factors of sensitivity are additive but not in the mathematical sense. As the second-order factors of sensitivity are linked to the first, their coefficients may not be added on and they therefore do not represent a method of confirming the first. Coefficients of sensitivity must also be attributed to waterlogging, degree and length of slope and the regional characteristics of rain.

There are two ways of determining the sensitivity of soil to intensive agriculture:

(1) Compare the observable original state of the soil under forest cover with the state of cultivated soil. Here a comparison with grassland soil is also helpful. Such comparisons provide information on the state of degradation of a cultivated soil and, from this, it is then possible to extrapolate to the future, i.e. give a verdict as to the sensitivity of the soil.

(2) Evaluate the coefficients of sensitivity of each first-order factor of sensitivity, the sum of which will provide a definition of the intrinsic sensitivity of soil to intensive agricultural methods. In this method one then goes on to compare the intrinsic sensitivity with the sensitivity obtained by means of the second-order factors. The coefficients of sensitivity for waterlogging and external factors are added to those for intrinsic sensitivity.

The second method is based on the land evaluation method advocated by Verheye [38] amongst others and on the soil evaluation method used for the detailed soil map of the United States.

Like the land evaluation map, the map indicating the sensitivity of soils to intensive agricultural methods is produced by determining the sensitivity of each pedological unit, which may be further sub-divided to take account of variations in degree of slope.

SENSITIVITY ATTRIBUTES AND OTHER FACTORS

Texture

Texture is a major, if not the most important, sensitivity factor and, what is more, it is easy to discover. Sensitivity to physical degradation is at its maximum in silty soils and at a minimum in soils with a balanced texture. The role of texture asserts itself in ploughed horizons whereas in the sub-soil the particles and grains are often 'imprisoned' in a pedological organisation which protects them from degradation.

Mineralogical nature of clays

Some clays like smectites and, to a lesser degree, illites swell in the presence of water and shrink when the soil dries out. As a result of this property, soils which are rich in swelling clays are self-structuring and become homogenised in the long run (Vertisols). This property cannot be destroyed by agricultural practices, the exception being irrigation which, when incorrectly carried out, does not allow the soil to dry out and therefore prevents the clays from shrinking. Hence, the soils least vulnerable to physical degradation are those rich in swelling clays whilst the most vulnerable are soils containing only kaolinites.

Pedological fabrics

These fabrics are chiefly characteristic of the sub-soil which agricultural scientists overlook. These fabrics frequently help determine both the sensitivity of a soil to degradation and its fertility. Many pedological fabrics have been described by pedologists. The types most frequently found in North-Western Europe are as follows:

Absence of pedological fabric

On the whole, the fabric of the parent material is intact. Only the least favourable circumstance is taken into consideration here, namely where the mellow sediment is made up of grains and particles simply compressed. Under these circumstances the texture is the dominant, and almost the only, sensitivity factor. This type of material is therefore highly sensitive to physical degradation (compaction). It reacts quickly to saturation by water, external pressure and vibration. Porosity brought about either by the soil fauna or by man (sub-soiling) is unstable (Kooistra [21]). When drains are installed this causes the translocation of silts and clays or even the movement of the whole mass towards the drain when the soil is waterlogged (Sole-Benet [37]).

Clayey fabric

Layers of clay line all the major pores and some clays form part of the parent material. The linings provide the clay horizon with a certain rigidity and help it resist physical degradation, particularly from waterlogging and pressure, even where this is applied to water-saturated clay horizons.

Hydromorphic clay fabric
The typical characteristics of these fabrics are clogging and layering with clay or clayey silt, ferruginous mottling and low or totally non-existent porosity of the type due to man or soil fauna. The ferruginous blotches are the only thing which gives the hydromorphic clay horizons a structure as the textural features are not rigid. The vulnerability of these horizons to physical degradation therefore depends on the amount and hardness of ferruginous mottling.

Palaeo-argilaceous fabric
These fabrics have a clayey content and the clay minerals tend to be organised into tidy sections. The structure is provided by ferruginisation. Generally the fabric is pressure resistant even when the horizon is waterlogged.

Dystro-cambic fabric
This fabric tends to be highly porous with fairly stable fluffy aggregates well able to resist pressure even when the horizon is saturated with water. As opposed to the previous fabric which is found only at depth, the dystro-cambic fabric is also found at the surface.

Gley fabric
The main feature of this fabric is mottling, some of the blotches being poor in iron and pedological fabric whilst others are rich in iron. The stability of this fabric therefore depends on the abundance of ferruginous mottling and the concentration of iron oxides.

Organic matter
Although the evidence is as yet unfortunately rather empirical, it is accepted that organic matter, in the process of forming the humic clay complex—particularly humus— reduces the sensitivity of soils to physical degradation. At best, we can provisionally suggest three coefficients of soil sensitivity in respect of physical degradation, each based on organic matter content, namely: a content of less than 1% = high sensitivity; a content of between 1 and 2% = average sensitivity; a content of over 2% = low sensitivity.

Biological activity

It is difficult to take biological activity into account when estimating soil sensitivity, for two reasons:

1.   Biological activity is both a factor in, and a symptom of, sensitivity.
2.   It is difficult to study.

We propose taking the abundance of earthworms as an indicator of biological activity and using only three soil sensitivity coefficients for degradation each based on the abundance of earthworms, i.e. where earthworms are apparently absent = high sensitivity; earthworms present = average sensitivity; earthworms abundant = low or zero sensitivity.

Type of aggregate and macroporosity

These second-order factors of sensitivity depend on the nature of the clay minerals, the activity of the soil fauna and, for certain types of aggregate, on pedological processes such as the impregnation of mineral matter with pseudo-soluble organic compounds. Absence of aggregates and macroporosity indicates high sensitivity and, at the other end of the scale, a crumbly aggregate in the form of floc, granules or small polyhedrons combined with high macroporosity means absence of sensitivity as long as the activity of the fauna remains high or the soil contains a large proportion of swelling clays.

   The stability of the structure is an attribute which combines all the first-order factors and to which should be added:

(1)   the cation content of the clay and the cations present in solution in the soil;
(2)   the state of the humic clay complex (Guckert [18]);
(3)   the duration and frequence of episodes of water saturation.

Structural stability is measurable and there are various methods (e.g. Henin *et al.* [19]) but unfortunately the results are rather imprecise and, above all, do not clearly show up the causes of instability.

Permeability

This attribute is an indicator of the state of physical degradation of the soil. For instance, a measurement of permeability taken at the surface of the soil reflects the development of surface crusting and the degree of compression in the ploughing horizon, whilst the permeability of the plough sole is a result of its degree of compaction. A permanently permeable soil may be regarded as hardly sensitive whilst by contrast, a soil with little permeability

—whether permanently or as a result of the forming of a crust or a plough sole—is highly sensitive.

Waterlogging

In the absence of the type of pressure caused by heavy machinery soils do not become internally compacted unless they are saturated with water. When a soil is subject to pressure and vibration the effect is aggravated by the presence of free water and is maximised when the soil is in a state of thixotropy. The sensitivity of a soil to physical, chemical and biological degradation depends on the groundwater level and on how long it remains at that level. A soil is in a state of maximum sensitivity when it is saturated up to the surface during the winter months or at the beginning of the spring. Sensitivity decreases considerably when the condition of surface saturation is only episodic and disappears when the soil is well drained. It should be emphasised that waterlogged soils become degraded by natural processes. For instance, in forests highly saturated soils are badly structured, acid and low in organic matter whilst biological activity is reduced and takes place near the surface.

The gradient and length of a slope determine the rate of erosion by run-off. Wischmeier and Smith [40] suggested two coefficients, one for length of slope (1 to 5) and the other for gradient (1 to 10).

Erosive effect of rain

Wischmeier and Smith [40] proposed using an index, R, to represent— for any given type of rain—the total kinetic energy of the rain multiplied by the maximum intensity over 30 min. For North-Western Europe this R index should be modified by an index for the frequence of rain episodes during the period in which evapotranspiration is lower than precipitation.

SENSITIVITY OF THE MAIN TYPES OF SOIL IN NORTH-WESTERN EUROPE TO INTENSIVE AGRICULTURAL PRACTICES

Orthid luvisols on loess

These are the soils of North-Western Europe. They are both the most fertile and the most intensively cultivated of all but are highly sensitive to agricultural practices.

Loess, the basic material from which these soils have developed, is particularly sensitive to erosion because it is neither very well consolidated nor very permeable (De Ploey [11]). The silty loess particles are piled haphazardly one above the other, with hardly any links between them, so that there is a high but fine textural porosity, which is the reason for the fairly low permeability. Their original structural and biological porosity is rather undeveloped. The fact that loess is the product of wind erosion explains these very special characteristics and the high sensitivity of the material to raindrops, to the forming of rills by run-off and even to land slides. The fact that in North-Western Europe the loess was transformed into soil by the biological activities of eluviation and illuviation considerably reduces its sensitivity to erosion because the process resulted in a high degree of biological macroporosity and structuring and formed a clay horizon.

Orthid luvisols formed from loess are potentially highly fertile. There is only one minor constraint, mainly when the texture is too rich in silts and the soil is structurally not very stable at the surface. The texture of the surface horizon depends on particle distribution in the loess and the degree of eluviation. Biological activity, in particular the activities of earthworms and roots prospecting the A and B horizons, has no difficulty in penetrating the loess itself. The clay horizon is not sufficiently thick to stop or slow down infiltrating water so that these soils are well drained. The very open assembly of the loess particles means that a large quantity of capillary water can be retained and the readily available water reserves in these soils are particularly high. Finally, at its lower levels, loess represents a reservoir of nutritive elements, particularly calcium.

These soils began to be cleared and cultivated as early as the Neolithic period since when they have been eroding continuously (Murcher [25]). Bollinne [3] estimates the rate of erosion 1 mm/yr or 15 t/ha/yr. There have been far-reaching changes in the original forest profile. For instance, on mounds the clay horizon now forms an outcrop or is close to the surface whilst in troughs the soil is made up of eluviated material more than 120 cm deep. Between these two extremes there are enormous differences on the theme of horizons since, even on the scale of a single field, but more generally on that of a primary catchment basin, only part of the eroded soil has been carried away whilst the rest has been redistributed to the troughs or the lower edges of fields. When cultivated, and regardless of the farming system, luvisols generated from loess are therefore sensitive to erosion because of their silty texture and lack of structural stability. Soil developed from eluviated material represents both higher constraints (less structural stability and worse drainage) and greater sensitivity to physical degradation.

The regular decrease in organic matter content is the only figure available which indicates how the rate of soil degradation is speeding up under the effect of intensive agricultural methods. It would seem that this decrease affects structural stability which also gradually decreases. There are many consequences to this structural degradation. At the surface, crusting increases, i.e. the aggregates break down faster and as a result the (sedimentary) surface crusts form faster and are thicker, thereby reducing infiltration. Lower down, structural degradation promotes the formation of plough soles but the main reason why these soles form faster is the use on soils that are still saturated with water of heavy, powerful machinery which produces vibrations. Present day plough soles are forming rapidly. For instance Roux and Fedoroff [31] recorded that in a few months, under a store of sugarbeet and in fields ploughed when still in a state of saturation, very dense and almost impermeable plough soles formed within a few months (Fig. 1). We shall call them super plough soles. They promote the formation of perched groundwater within ploughed horizons which, once saturated, set rapidly. All the porosity brought about by farming disappears whilst textural porosity is reduced to a minimum.

Solidified clods from another horizon can resist fragmentation by ploughing implements for several years (Manichon [23]). Compaction is accompanied by the penetration of plugs of a silt and clay mixture which clog this soil making it less porous and therefore less permeable. In horizons saturated with water these plugs take the form of bands of a silt and clay mixture. The material in the plugs originates in the soil surface where the impact of raindrops turns them into a suspension. Taken as a whole, these phenomena (super plough soles, the solidification of the ploughed horizon and surface crushing) help to aggravate erosion. When the ploughed horizons are saturated with water, run-off and erosion are set in train by the slightest amount of rain, even if it is not very heavy. Finally, there are either no earthworms at all, or only very few, to burrow through the plough soles and other compacted areas. Degradation of this type of soil appears to be irreversible.

Our research techniques do not enable us to distinguish between the changes brought about by traditional and modern farming methods in the deeper layers of the sub-soil. Here the main characteristics resulting from cultivation are dark brown clayey linings and plugs rich in small charcoal-like grains of the size of fine silts known as agricutans. It may be assumed that increased surface crusting speeds up the formation of these agricutans and that in the long term sub-soils under luvisols will become increasingly clogged as earthworms, which otherwise continually restore the soil's macroporosity, disappear or almost disappear.

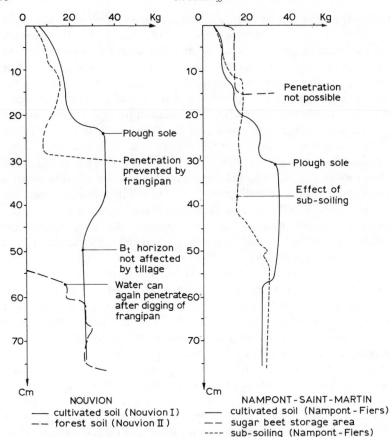

FIG. 1. Examples of plough soles developed in luvisols on sandy loess. Penetrometer graphs. Picardy, France (from Roux and Fedoroff [31]).

The acidification of luvisols is a natural phenomenon which is slowed down under forest cover due to the presence of deep loess. Primitive farmers confronted with this problem quickly found a remedy by adding lime. The danger of acidification therefore does not need to be considered even where rain is acid.

Gleyey and podzolic luvisols over loess-containing silts

These soils, frequently found in association, are a major constraint on soil drainage and their surface horizon has a weak structure. Because of these limitations, this type of soil was cleared later and until recently most of it was still under grass. Farmers are now trying to turn it back into grassland but are having limited success except where the soil is first drained. Under forest cover, sub-soils are already acid, poorly structured (tending to solidification in horizon A and compaction in horizon B) and have a low organic matter content and biological activity (in horizon B this is due to compacting and waterlogging and in horizon A to acidity and water-logging). On the other hand, their sub-soils with their palaeo-argilaceous and, more rarely, hydromorphic clay fabrics are generally fairly resistant to physical stresses. Nevertheless, in the absence of drainage, they are highly vulnerable to degradation—mainly physical, but also chemical and biological. They are therefore highly sensitive to agricultural practices— more so than orthid luvisols. However, the constraints to which these soils are subject guarantee that they are never intensively cultivated and the risk of degradation is therefore limited. When drained, they are comparable in sensitivity to orthid luvisols in an A horizon with a very low clay content.

Dystric cambisols

Before the introduction of liming and mineral manures, these cambisols were poor soils, many of which had been abandoned to heath, and even nowadays very few are intensively cultivated. The fluffy aggregate which is characteristic throughout the profile is relatively stable and not very subject to compacting but on the other hand it is sensitive to wind erosion. It is probable that intensive agriculture destabilises the fluffy aggregates by raising the pH and causing the soil ecosystem to degenerate.

Calcareous cambisols

These soils, which are frequently developed on hard limestone, are often wooded because they are very rocky, have low water reserves and are steeply sloped. When brought under cultivation, extensive systems are used. Their fine matrix, which is rich in clay and generally saturated with calcium, forms stable aggregates easily renewed by the process of swelling and shrinking. They are always highly permeable both at the surface and internally. These soils are therefore not very sensitive to agricultural practices. Sensitivity to erosion is limited even where slopes are steep.

Rendzina

The fertility of rendzinas depends almost entirely on the characteristics of the calcareous rock over which they develop. For instance, friable chalk with its fine pores underlies soils with high reserves of water where roots may penetrate to a great depth but which are poor in fertilising compounds. On the other hand when they overlie dolomite, rendzinas are skeletal and of little agricultural value. The rendzinas overlying the chalk of the Dry Champagne area have been intensively cultivated for several decades and are apparently not deteriorating. Frequently the rendzina matrix contains a relatively high proportion of swelling clays which give permanence to the structure.

Fluvisols

The potential fertility of these soils varies greatly as some of them are subject to frequent swamping and many remain waterlogged until quite late in the year. Nevertheless, these soils may be fertile when not waterlogged or when artificially drained. Because they are found in low-lying ground—generally valley floors—they are not subject to erosion. However, the absence of a pedological fabric makes them particularly sensitive to compaction. Kooistra [21] showed that the interstices created by sub-soiling in the plough sole in fluvisols in the Netherlands could be filled up in a single season.

Gleysols

The common feature of gleysols is bad drainage. This may be due to groundwater in a valley floor, to water welling up to the surface or the result of an impermeable sub-soil, e.g. clay or marl. Where waterlogging has not been counteracted by drainage, it represents a major constraint on agriculture and, unless drained, these soils are frequently left under grass. When drained, their sensitivity to intense agricultural methods depends almost entirely on the type and fabric of the material from which they developed. For instance an alluvial deposit will be more sensitive than a marl or clay.

This list is anything but exhaustive. We have only examined the sensitivity to agricultural practices of a few of the North-Western European soil types most frequently encountered and constituting the larger proportion of the soil resources of this part of our continent. A complete inventory could be made on the basis of the European Community's soils map and a map indicating the sensitivity of soils to intensive agricultural methods, also drawn up on the basis of the soils map [9].

CONCLUSIONS AND PROPOSALS

The sensitivity of a soil to agriculture can only be determined with reference to our knowledge of its degree of degradation as a result of farming practices. The deterioration of cultivated soils is not a new phenomenon. Since man first began cultivating the land, the nature and functions of North-Western Europe's forest soils have been changing. The exposure of bare soil to the impact of raindrops causes crusting and results in:

(1)   the formation of skins and crusts at the surface; and
(2)   the translocation of particles within the soil; the process of eluviation and illuviation is considerably modified.

Centuries of ploughing with primitive and more evolved equipment has created a compacted layer several centimetres deep in the sub-soil—a layer known as the plough sole. North-Western European soils under forest cover or on open plains are not eroded. Where there is no farming, the sensitivity of these soils to erosion depends on the speed with which surface crusting and sealing occurs. Centuries of cultivation have considerably altered the distribution of the surface horizons around the countryside and, in some cases, even that of the lower horizons. This redistribution has generally resulted in decreased fertility. By reducing evapotranspiration, the clearing of hydromorphic soils caused a rise in groundwater levels thereby reducing potential fertility whilst at the same time increasing the sensitivity of the soils.

By adding manure and other organic products such as litter from the neighbouring forests and leaving land fallow, the traditional farmer promoted biological activity. By contrast, the modern farmer with his pesticides and methods of cultivation eliminates a large proportion of the soil fauna, particularly earthworms, and thereby impoverishes and profoundly disturbs the soil ecosystem. This has many direct serious consequences:

(1)   the canalicular macroporosity is not renewed and therefore, sooner or later—depending on the type of soil—sub-soils which do not include swelling clays are tending to become clogged;
(2)   biological aggregation (excreta of the soil fauna) no longer takes place;
(3)   the process of humification is seriously disturbed;
(4)   the soil is no longer homogenised and organic matter is no longer incorporated into the soil matrix as these tasks used to be performed by earthworms.

Absence of biological aggregation, the disturbance of the humification process and unsatisfactory homogenisation lead to less structural stability which promotes crusting, compaction and therefore, erosion. Heavy, powerful machinery used on insufficiently drained soil has its own effect but also amplifies the effects of structural degradation. The sub-soil immediately below the ploughed horizon is rapidly compacted through a far deeper layer than the plough sole produced by traditional methods. This 'super' plough sole is almost impermeable and in its turn causes the ploughed horizon to become waterlogged, causing it to solidify and accelerating the crusting process. All these physical processes of deterioration encourage greater run-off and erosion. On top of run-off from bare soils and the phenomenon of crusting there is the run-off produced by saturation due to the 'super' plough soles.

Soil degradation and therefore the sensitivity of soil to intensive agricultural practices depends on a soil's internal attributes, namely:

(1)   the nature of the parent material, its granulometric and mineralogical composition, and the degree of consolidation;
(2)   the pedological changes the material has undergone; and
(3)   drainage;

and to external factors, such as:

(1)   the degree and length of slope; and
(2)   the rain pattern.

In North-Western Europe, as elsewhere in temperate zones, loess is a material over which highly fertile soils rapidly develop. Unfortunately, because of its wind-borne origin, loess is highly sensitive to erosion and the soils it generates largely inherit this sensitivity. Because of their texture and the way their particles are assembled, loess and the soils to which it gives rise are also particularly sensitive to pressure and the vibrations generated by machinery.

In soils developed on other materials there is less degradation or at least it is less spectacular than that observed in loess soils but calculations may suggest that they are highly sensitive. There are various reasons for the lower rate of degradation. There are major constraints in respect of many of these soils considerably limiting their usefulness in agriculture and thereby protecting them from degradation. Others such as fluvisols are found in valleys, thereby escaping erosion but exposed to physical degradation.

It is therefore possible to assess the sensitivity of a soil to intensive agricultural methods and to draw up soil sensitivity maps using methods similar to those applied to maps assessing the agricultural capability of the different types of soil. But the reliability of these maps is limited. Our method is partly based on hypotheses yet to be verified and, in addition, many of the sensitivity factor coefficients have been established empirically.

Many of the ideas put forward in this chapter are still based on nothing more than assumptions. Many questions are still unanswered. For instance, exactly what is the effect of an intensive farming system on the deterioration of the soil ecosystem? What part do earthworms play in humification? In the past research on the degradation of agricultural soils was compartmentalised and concentrated mainly on erosion. We must go beyond this stage and, in our opinion, priority should be given to studying the biophysical functions of agricultural soils which would put erosion back in context. Research of this type has to be multi-disciplinary as any team studying these functions would have to include biologists (a microbiologist and a zoologist), a biochemist and a soil scientist. Each of these would have to go beyond the limits of his own discipline, keep himself informed of the work of his colleagues and use their results in his own work. The zoologist would determine the precise role of soil fauna—particularly earthworms—in humification, soil homogenisation and aggregation and would study various scenarios representing the degradation of the soil ecosystem whilst the microbiologist would study the microflora in relation to soil fauna. The only way to study organic matter is to use analytical procedures which make it possible to follow the successive transformations discovered by biologists.

The part played in humification by the various components of the ecosystem could thus be determined and, when some of the components disappear, the effects of this disappearance on humification could be described. The soil scientist would study porosity and aggregation and concentrate in particular on the various types of micro-formation (compaction, aggregates, excreta) in order to discover what changes the soil matrix undergoes. This work should include experiments.

REFERENCES

1. BOIFFIN, J. (1984). La dégradation structurale des couches superficielles du sol sous l'action des pluies. Thèse Ing. Doc., INA P-G, 320 pp.
2. BOIFFIN, J., PAPY, F. and PEYRE, Y. (1986). Systèmes de production, systèmes de culture et risques d'érosion dans le Pays de Caux. INA P-G, INRA, Paris, 2 Vols., 154 pp. and 65 pp.
3. BOLLINNE, A. (1977). La vitesse de l'érosion sous culture en région limoneuse. *Pédologie*, **XXVII**, (2), 191–206.

4. BOLLINNE, A., HANOTIAUX, G. and PISSART, A. (1978). L'érosion en milieu agricole. Synthèse et conclusion de la journée d'étude de la société belge de Pédologie du 28 octobre 1977. *Pédologie,* **XXVIII,** 233–245.

5. BORK, H. R. (1983). Die Holozäne Relief- und Bodenentwicklung in Lössgebieten. Beispiele aus dem Südöslichen Niedersachsen. *Catena supplement,* **3,** 1–93.

6. BOUCHE, M. B. (1972). Lombriciens de France. Ecologie et systématique. *Ann. Zoologie-Ecologie animale,* Numéro hors-série, 671 pp.

7. BOUCHE, M. B. (1984). Les vers de terre. *La Recherche,* **156,** 796–804.

8. CALLEBAUT, F., GABRIELS, D. and DE BOODT, M. (Eds) (1985). Assessment of soil surface sealing and crusting. Ghent State University 374 pp.

9. COMMISSION OF THE EUROPEAN COMMUNITIES (1985). Soil map of the European Communities, 1:1 000 000. Office for official publications of the European Communities, Luxembourg, 124 pp. 7 map-sheets.

10. CORDONNIER, P., GUINET, A. and TRICART, J. (1986). Humus et polyculture en Artois-Cambraisis (le modèle Bapomsol). INRA, Economie et Sociologie rurales, Grignon, 40 pp.

11. DE PLOEY, J. (1973). A soil-mechanical approach of the erodibility of loess by solifluxion. *Revue Géomorphologie dynamique,* **22** (2), 61–70.

12. DE PLOEY, J. (1977). Some experimental data on slopewash and wind action with reference to Quaternary morphogenesis in Belgium. *Earth surface processes,* **2** (2–3), 101–116.

13. DE PLOEY, J. and MUCHER, H. J. (1981). A consistency index and rainwash mechanisms on Belgian loam soils. *Earth surface processes and landforms,* **6,** 319–330.

14. DE PLOEY, J. and VAN HECKE, E. (1986). La conservation des sols: une politique anti-surplus. Texte Conférence de presse, Ku Leuven, 5 pp.

15. FAO (1977). Assessing soil degradation. *FAO Soils Bull.,* **34,** 83 pp.

16. FAURE, A. (1978). Comportement des sols au compactage: rôle de l'argile et conséquence sur l'arrangement des grains. Thèse Universitaire Scientifique et Médicale Grenoble, 137 pp.

17. GRIMALDI, M. (1981). Contribution à l'étude du tassement des sols: évolution de la structure d'un matériau limoneux soumis à des contraintes mécanique et hydrique. Thèse Universitaire Rennes, 221 pp.

18. GUCKERT, A. (1973). Contribution à l'étude des polysaccharides dans les sols et de leur rôle dans les mécanismes d'agrégation. Thèse Universitaire Nancy, 138 pp.

19. HENIN, S., GRAS, R. and MONNIER, G. (1969). *Le Profil Cultural,* Masson et Cie, Paris, 332 pp.

20. JONGERIUS, A. (1970). Some morphological aspects of regrouping phenomena in Dutch soils. *Geoderma,* **4,** 311–331.

21. KOOISTRA, M. (1985). The effects of compaction and deep tillage on soil structure in a sandy loam Dutch soil. Abstracts, Soil Micromorphology, Département des Sols. INA P-G, Paris, p. 95.

22. LAURANT, A. and BOLLINNE, A. (1978). Caractérisation des pluies en Belgique du point de vue de leur intensité et de leur érosivité. *Pédologie,* **XXVIII** (2), 214–232.

23. MANICHON, H. (1982). Influence des systèmes de culture sur le profil cultural:

Elaboration d'une méthode de diagnostic basée sur l'observation morphologique. Thèse Ing. Doc., INA P-G, 214 pp.

24. MONNIER, G. (1965). Action des matières organiques sur la stabilité structurale des sols. Thèse Universitaire Paris, 140 pp.

25. MUCHER, H. J. (1986). Aspects of loess and loess-derived slope deposits: an experimental and micromorphological approach. Thesis, University of Amsterdam, 267 pp.

26. PIHAN, J. (1979). Risques climatiques d'érosion hydrique des sols en France. In: *Col. Erosion agricole sols en milieu tempéré non méditerranéen*, H. Vogt and T. Vogt (Eds), University Louis Pasteur, Strasbourg and INRA, Colmar, pp. 13–18.

27. PISSART, A. and BOLLINNE, A. (1978). L'érosion des sols limoneux cultivés de la Hesbaye. Aperçu général. *Pédologie*, **XXVIII** (2), 161–182.

28. POESEN, J. (1981). Rainwash experiments on the erodibility of loose sediments. *Earth surface processes and landforms*, **6**, 285–307.

29. POESEN, J. and SAVAT, J. (1981). Detachment and transportation of loose sediments by raindrop splash. Part II: Detachability and transportability measurements. *Catena*, **8**, 19–41.

30. RIQUIER, J. (1982). Evaluation globale de la dégradation des sols. *Natures et Ressources*, **XVIII** (2), 19–23.

31. ROUX, M. and FEDOROFF, N. (1981). Ruissellement et remembrement. Commune de Nampont-St-Martin (comparaison avec la commune de Nouvion). ADEPRINA, Paris, 37 pp.

32. SAVAT, J. (1977). The hydraulics of sheet flow on a smooth surface and the effect of simulated rainfall. *Earth surface processes*, **2**, 125–140.

33. SAVAT, J. (1980). Resistance to flow in rough supercritical sheet flow. *Earth surface processes*, **5**, 103–122.

34. SAVAT, J. (1981). Work done by splash: laboratory experiments. *Earth surface processes and landforms*, **6**, 275–283.

35. SAVAT, J. and POESEN, J. (1981). Detachment and transportation of loose sediments by raindrop splash. Part I: The calculation of absolute data on detachability and transportability. *Catena*, **8**, 1–17.

36. SKIDMORE, E. L. (1977). Criteria for assessing wind erosion. FAO Soils Bull., **34**, 52–62.

37. SOLE-BENET, A. (1979). Contribution à l'étude du colmatage minéral des drains. Une démarche expérimentale basée sur la micromorphologie pour étudier les transferts solides dans les sols. Thèse 3ème cycle, Universitaire P. et M. Curie, Paris, 207 pp.

38. VERHEYE, W. (1985). Land evaluation. In: *Soil Map of the European Communities* (Edited by Official publications of the European Communities), Luxembourg, pp. 65–66.

39. VOGT, H. and VOGT, T. (Eds) (1979). Colloque sur l'érosion des sols en milieu tempéré non méditerranéen. Universitaire Louis Pasteur, Strasbourg and INRA, Colmar, 264 pp.

40. WISCHMEIER, W. H. and SMITH, D. D. (1958). Rainfall energy and its relationship to soil loss. *Trans. Amer. Geophys. Un.*, **39**, 285–291.

The Production Potential of Soils: Part II— Sensitivity of the Soil Systems in Southern Europe to Degrading Influxes

NICHOLAS J. YASSOGLOU

Athens Faculty of Agriculture, Botanicos 11855, Athens, Greece

SUMMARY

The influxes that threaten the productivity of the soil systems in Southern Europe are discussed and the respective sensitivities of the main soil taxonomic units are qualitatively evaluated. The sensitivities are referred to causes contributing to the degradation of the qualities of the soil peda. Such causes examined were: loss of soil volume and organic matter, physical degradation, chemical degradation and fertility deterioration. Due to the predominantly sloping terrain and the adverse climatic conditions in Southern Europe soil erosion is the dominant deterioration process. It is related to all the causes of productivity degradation, and strongly affects the cultivated soils on the river terraces and on tertiary hills. Erosion also damages the soils on the mountain watersheds that have lost their vegetative cover. The sensitivity of the soil systems to erosion depends on the factors of the USLE and on the depth and morphological configuration of the respective soil peda. Soil taxonomic units having shallow peda and limiting horizons are the most sensitive. Soils of the flood plains are mainly sensitive to physical and chemical degradation and less to fertility deterioration. Their sensitivities depend on the textural and mineralogical compositions, on the acidity of the peda and on the organic matter content of the surface horizon.

1. INTRODUCTION

Soil productivity is usually defined in the physical sense of biomass yield. Inasmuch as soil is a natural resource that contributes to human welfare in many ways, its productivity should be referred to its overall performance and not only to the biomass production.

At this point it would be interesting to recall how the great naturalist, the American Indian, felt about the soil. He loved the earth and all things of the earth, the attachment growing with age. The old people came literally to love the soil and they sat and reclined on the ground.

'It was good for the skin to touch the ground. The great Spirit is our Father but the Earth is our Mother she nourishes us; that which we put on the ground she returns to us, and healing plants she gives us likewise. If we are wounded we go to our mother and seek to lay the wounded part against her, to be healed. Animals too, do thus ... Indians never hurt anything, but white people destroy all ... Everywhere the white man has touched the earth, it is a sore'. (Sayings of Indian Chiefs Compiled by McLuhan [40]).

Soils as dynamic natural systems undergo continuous changes. These changes, however, are slow and do not seriously affect their productivity when regarded in the time scale of human life. Fast rates of degradation can only be caused locally by natural calamities and universally by man's actions. Examples of extensive degradation of soil resources with serious impacts on the welfare of human society are found in Mesopotamia, where man induced erosion of the hilly lands and salinization of the bottomland soils has resulted in the decrease of population from its peak of 25 000 000 in ancient times to its present 4 000 000.

Schumm and Harvey [54] have concluded that agricultural soils are non-renewable resources for the United States, because the average man-induced erosion is 2·0 mm/yr while soils form naturally at rates of 0·5–0·02 mm/yr. In spite of the loss of soil from cultivated lands, crop productivity in the world has been constant or, in some cases, has increased, due to technology inputs [58]. This means that potential losses in productivity from erosion may be compensated by added inputs at a given site. Productivity improvements usually are achieved on high quality land, while erosion takes place mainly on sloping medium quality lands. In some cases, when the soil is deep and uniform, nutrients lost by erosion are replaced by fertilizers. In other cases erosion may bring to the surface soil horizons or layers with higher fertility and more favourable physical properties than the removed sections. In both cases the soil productivity would not deteriorate until the soil depth is critically reduced. From this point on the deterioration would be irreparable. Thus soil loss tolerance is also a function of soil depth and decreases with it [55].

Erosion by truncating the soil pedon removes or damages vital parts of its subsystems and in severe cases irreversibly degrades its functions and productivity. In terminal cases the whole pedon is lost and the unproductive bare rock is surfaced. This is why erosion is considered the most serious threat to soil systems when it acquires an accelerated mode.

Other processes such as those resulting in physical and chemical degradation of the soil are not as destructive as soil erosion, and do not have its global character, but in many cases, they can cause severe damage to soil systems. Cultivated soils and soils indirectly exposed to intensive human activities are those that suffer the greatest damage.

The main influxes that degrade soil productivity in Southern Europe will be analysed in the following sections. The sensitivity of the dominant soil units to these influxes will also be qualitatively assessed. For each soil unit emphasis will be placed on the dominant degrading influxes.

2. A COMPARATIVE VIEW OF SOIL EROSION IN SOUTHERN AND NORTHERN EUROPE

Degradation of soil productivity can be caused by many agents and by several processes. The most important of all, especially in the Mediterranean countries, is accelerated soil erosion, which is advanced on the sloping land by the destruction of protective vegetative cover. It results, in many cases, in irreparable damage by truncating the volume of the soil system and by deteriorating other properties directly and indirectly controlling its productivity.

Accelerated soil erosion has been a problem in the Mediterranean Region for many centuries. Plato [47] described vividly in Critias the destruction of the land of Attica by erosion. He cited that for 9000 years soil had been carried to the bottom of the sea. '... The earthly high mountains that in the past carried tall forests and large pastures have become rocky lands and they look like the bones of a sick body.... In the past the rain water was utilized and it did not run on the barren land to the sea as it does now. It infiltrated into and was stored in the soil and it was distributed in the springs, fountains and river streams of the low lands...'.

The rate of the accelerated erosion and the severeness of its impact on soil productivity varies with the climate, vegetation, topography, geology, soils and with socio-economic conditions. It is greatest where intense rains and vegetative cover are out of phase, as is the case in semiarid and Mediterranean climates [31]. However, it will occur anywhere there is run of water or wind flow on bare soil surface. According to Fournier [18] Europe has the lowest average rate of soil loss of all the continents, but the range is very wide. It varies from $1·2–24·2$ mt/km^2/yr on the gentle slopes of Northwestern Europe to 4000 mt/km^2/yr on the mountains of Southern Europe. The same author reported the following average rates of soil erosion in Europe:

1. Low relief regions in Northwestern Europe 24 mt/km^2/yr.
2. Mountain regions in France 180–250 mt/km^2/yr.
3. Small valleys in the Alps and in the Apennines 2 500 mt/km^2/yr.
4. Albanian mountains 4 000 mt/km^2/yr.

In mountainous watersheds of Greece rates as high as 23 000 mt/km^2/yr have been reported by Margaropoulos [38].

The local factors that affect severe erosion in Europe are:

1. Glaciers in cold regions.
2. Steep slopes in mountainous regions.
3. Adverse climatic conditions in Mediterranean regions.

The climatic conditions of Northwestern Europe are less conducive to accelerated erosion than those of Southern Europe. The higher and more evenly distributed rainfall in the North secures speedy rates of regeneration of the vegetative cover. Thus external influxes causing the removal of natural vegetation are less effective in accelerating soil erosion in Northwestern than in Southern Europe.

The low relief and the weak slope of the terrain over large sections of North Europe also contribute to low average erosion rates.

The removal of vegetation by soil cultivation sharply increases the rate of soil loss even from the gently sloping lands of Northern Europe. Zachar citing respective authors [26, 33] reported rates of soil loss through water erosion ranging from 40 to 450 mt/ha/yr. These rates compared to 2–10 mt/ha/yr soil loss tolerance (T-value) show the gravity of the problem. Hanotriaux [23] reported that soil loss from fallow land on gently undulating land in Belgium with slopes of 5–7% was 82·2 mt/ha/yr. The loss dropped to 30·1 and 4·3 mt/ha/yr under sugar beet and winter wheat respectively.

In lowland England soil erosion occurs in arable bare smooth fields on gently rolling landscape of low relief. Coarse loamy soils are the most vulnerable [14]. Morgan [42] estimated that in England and Wales about 37% of the arable area shows erosion rates above the soil loss tolerance. About 4 000 km^2 of non-arable land associated with blanket peat in the uplands and with coastal sand dunes are at erosion risk. Richter [49] considered as potentially erodible hilly surfaces in Germany with slopes as low as 2–6%.

Belpomm [4] suggests that sound crop rotations, an almost perennial vegetation and large areas of forests ensure an efficient neutralization of erosion factors in France. Among the fragile zones, the most sensitive are the hilly wine areas.

Portier [48] considers a 3–4% slope sufficient to induce soil erosion in Southern France.

The high erosivity of rainfall and the prevailing steep slopes in Mediterranean countries render more than 80% of their lands potentially erodible. The long history of human exploitation of the land in these territories has significantly decreased the effectiveness of the protective vegetative cover. Thus water erosion has irreversibly destroyed the productivity of the soils over large areas and it has significantly depressed it over still larger sections. About 40% of the land of Spain and Greece has its productivity seriously reduced.

The above data show that the greatest portion of the European lands is sensitive to erosion.

3. DEFINING SOIL SENSITIVITY

3.1. Factors involved

Soils are dynamic open physical systems and as such are subjected to continuous changes inflicted upon them by external influxes. Thus soil properties (s) according to Jenny [27] are dependent on five independent variables or otherwise called state factors:

$$s = F(p,c,o,r,t) \qquad (1)$$

The factors climate (c) and organisms (o) represent the influxes (I) on the soil system, while parent material (p) and topography (r) represent the initial state (S_o) of the system. The factor t represents the age of the system. Jenny [28] extended his first formula to include: any properties of the system (l), any vegetation properties (v), any animal properties (a) and any soil properties (s) expressed in eqn. (2).

$$l,v,a,s, = (S_o,I,t) \qquad (2)$$

According to this definition, the value or the magnitude of each property at a particular instant, depends on the initial state (S_o), the influxes (I) and the time elapsed (t). Therefore, the productivity of the soil system, defined as its capability to produce the desirable products or to function in desirable ways, depends on the above three factors.

The sensitivity of the soil system could be defined as the rate of variation of dependent properties caused by given influxes. If we assume that a soil system has reached a slow rate of change under a constant set of state factors, we could also assume that the system is at a steady state for a given time period. A sudden change in the influxes will upset the state of the soil and initiate processes that will lead to changes of soil properties and consequently they will alter the performance of the whole system. Whether these changes would seriously affect soil productivity, depends on their rates. Therefore, the sensitivity of the soil systems could be expressed in terms of reaction kinetics:

$$\frac{\mathrm{d}s}{\mathrm{d}t} = K(I)^n \tag{3}$$

Where the rate constant K is initial state (S_o) of the system in eqn. (2). However, in this case the initial state should be considered as the soil system as a whole (parent material, differentiated soil horizons and topography) at the time, of the change of influx $(t = 0)$. Therefore, the rate constant can express the soil sensitivity only at specific instances. This means that the sensitivity of the soil systems is not a constant parameter but varies with time following the physical, chemical, mineralogical, biological changes in the solum (part of the soil pedon that overlies the parent rock). It is actually a function. Thus it is obvious that we could evaluate the sensitivity of a soil system only at a particular moment. Its evaluation should be based both on Jenny's state factors (parent rock and topography) and on dependent variables (s) that express the composition and dynamics of the system at a specific instance, considering the system at its initial state. Chisci [9] used similar parameters in defining the factors of soil degradation in Italy. FAO [15] also defined soil degradation as a function of similar parameters.

3.2. Soil pedon as reference basis

The pedon is the three-dimensional expression of the soil individual. Specific soil properties affect the sensitivity of the soil systems, but unless they are considered along with the soil, the pedon cannot fully prescribe the results of the degrading influxes on the performance of the soil system.

Soil texture for example is related to soil sensitivity, to compaction and to erosion. But the forces leading to structural deterioration and to loss of soil volume and finally to productivity deterioration depend on the whole soil pedon. Frye *et al.* [17] found that erosion substantially decreased the productivity of strongly developed profiles due to the decrease of available water holding capacity caused by increased clay content in the Ap horizon.

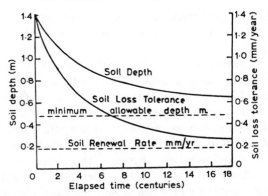

FIG. 1. Relationships between soil depth and soil tolerance. (Skidmore [55]. Reproduced with the permission of the American Society of Agronomy.)

Runge *et al.* [51] found that in the Upper Midwest of US soil vulnerability to erosion calculated from profile characteristics was a more sensitive criterion for productivity losses than soil loss tolerance (*T*) values. Skidmore [55] showed in Fig. 1 that soil loss tolerance decreased with time along with the decrease in the soil depth. Larson *et al.* [35] proposed soil vulnerability curves as an alternative approach to *T* values. Productivity index (*PI*), in Fig. 2 plotted against soil removal, produced curves characteristic of each soil series. The average slope of the curve is the vulnerability of the soil to long term erosion losses. More recent work

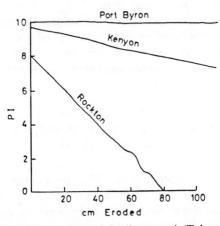

FIG. 2. Productivity Index versus cm of soil removed. (Taken from Larson *et al.* [35]. Reproduced with the permission of the author.)

relating soil loss to soil productivity indexes and to pedon characteristics has been reported by Becher *et al.* [3], Larson *et al.* [36] and Gantzer and McCarty [19].

The above arguments clearly show that the degree of deterioration in soil productivity by degrading influxes depends on the constitution and on the performances of the soil individual and of its sub-systems. Soil pedons should be regarded as entities expressing a particular system. In some ways they resemble and function like organisms. Their pore space along with soil solution and air act like circulatory and respiratory systems. One should not be surprised to find arteriosclerosis in this system. It is actually the cutanic accumulation along the pores of older soils. The high in biological activity of upper sections of the pedon act like digestive systems. The sub-systems are interrelated and operate in a mode that is specific for each pedon and for each section of the pedon. Any degradation or disturbance affects the performance of the whole soil individual. Some of the effects of these so-called *soil qualities* are important for human societies and for preserving the biological balance. Such qualities are: supplying water, oxygen and nutrients to plants and micro-organisms, providing space for the development of root systems and providing a solid basis for construction works.

The sensitivity of the soil systems should be assessed in terms of the damage to soil qualities that the degrading influxes inflict, when they act upon the soil individuals. It also should be assessed in terms of the renewal capacity of the soil system. The Universal Soil Loss Equation (USLE) developed by Wischmeier and Smith [61] is a soil degradation rate determining equation but does not fully evaluate the sensitivity of the soil to erosion unless it is complemented with the soil loss tolerance value (T). The latter should be determined on the basis of the constitution and the performances of each soil pedon and on the respective PI versus soil loss curves.

4. CAUSES AND PROCESSES CONTRIBUTING TO SOIL DEGRADATION

Soils as parts of the ecosystems are usually strong buffer systems, and under natural conditions are stable and resistant to high rates of changes.

Some human actions or natural calamities, however, represent extremely intense influxes that disrupt the ecological equilibrium and cause, in a short time, drastic and in many cases irreversible damage to soil systems.

The main causes contributing to serious deterioration of the soil qualities are:

(a)   Loss of soil volume.
(b)   Degradation of soil structure.
(c)   Loss of organic matter and biological activity.
(d)   Chemical degradation.
(e)   Soil fertility deterioration.

Soil volume is referred in this presentation to that volume of soil pedon that is necessary for securing favourable levels of soil qualities.

The main influxes and processes that result in the above five types of soil degradation are the following:

(a)   Loss of soil volume
       (1) Erosion; (2) compaction; (3) induration; (4) flooding.
(b)   Degradation of structure
       (1) Erosion; (2) mechanical breakdown; (3) alkalinization; (4) flooding; (5) loss of organic matter; (6) raindrop impact; (7) deposition.
(c)   Loss of organic matter
       (1) Erosion; (2) exhaustive soil management; (3) excessive drainage.
(d)   Chemical degradation
       (1) Leaching; (2) acidification; (3) salinization; (4) alkalinization; (5) carbonation; (6) chemical pollution; (7) unbalanced fertilization; (8) erosion; (9) deposition.
(e)   Soil fertility deterioration
       (1) Erosion; (2) leaching; (3) fixation; (4) volatilization; (5) exhaustive soil management.

The above list is indicative of the prime role of erosion in degradation. It contributes to all five causes of soil productivity deterioration and indirectly affects other degrading processes.

5.   ASSESSMENT OF SOIL SENSITIVITY

The assessment of the sensitivity of soil pedon to damaging influxes should be based on its initial states (S_o), described in the previous section, and on environmental parameters that affect its productivity and regenerative capacity. Climatic factors are both influxes causing soil deterioration, such as erosion leaching, and organic matter degradation, and parameters affecting the regenerative capacity of the soil. Therefore, the assessment of the soil sensitivity is a very complex and difficult task. This task could be simplified if the assessment was made for specific soil taxonomic units, because they are descriptive not only of the pedon's initial state but also of the ecological environment.

The taxonomic units of the Map of Europe (1986) in the major landforms will be used in this presentation as a basis of reference for a qualitative and descriptive assessment of soil sensitivity. Before doing that, it would be useful to describe the composition and the functions of some pedons of the main soil units in Southern Europe and try to assess their sensitivities in a general way. More detailed assessment will be made in the following sections.

The pedon (a) in Fig. 3 represents a Fluvisol soil with no profile development. Its performance is mainly controlled by the thickness, the texture, the mineralogy and the relative position of each alluvial stratum in the pedon, and by the amount of organic matter accumulated on the surface.

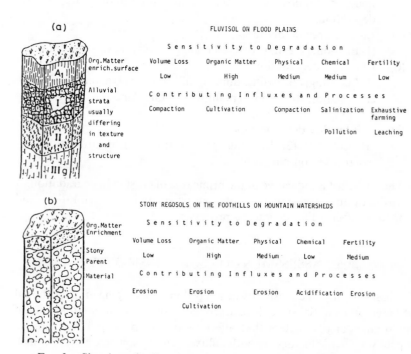

FIG. 3.   Sketches of soil peda and relative sensitivities to degradation.

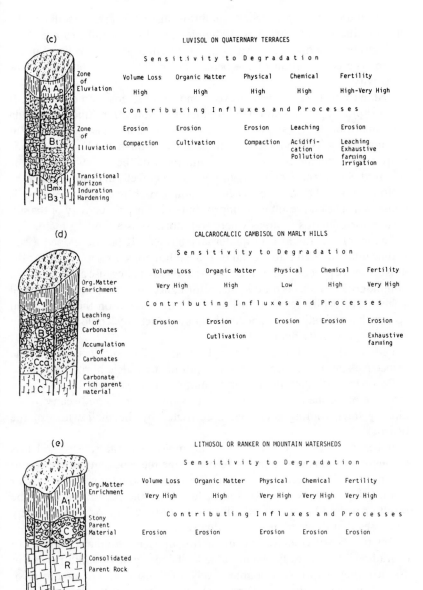

(c) LUVISOL ON QUATERNARY TERRACES

Sensitivity to Degradation

	Volume Loss	Organic Matter	Physical	Chemical	Fertility
Zone of Eluviation	High	High	High	High	High-Very High

Contributing Influxes and Processes

Zone of Illuviation	Erosion	Erosion	Erosion	Leaching	Erosion
	Compaction	Cultivation	Compaction	Acidification Pollution	Leaching Exhaustive farming Irrigation

Transitional Horizon Induration Hardening

(d) CALCAROCALCIC CAMBISOL ON MARLY HILLS

Sensitivity to Degradation

	Volume Loss	Organic Matter	Physical	Chemical	Fertility
Org.Matter Enrichment	Very High	High	Low	High	Very High

Leaching of Carbonates

Contributing Influxes and Processes

Accumulation of Carbonates	Erosion	Erosion	Erosion	Erosion	Erosion
		Cutlivation			Exhaustive farming

Carbonate rich parent material

(e) LITHOSOL OR RANKER ON MOUNTAIN WATERSHEDS

Sensitivity to Degradation

	Volume Loss	Organic Matter	Physical	Chemical	Fertility
Org.Matter Enrichment	Very High	High	Very High	Very High	Very High

Stony Parent Material

Contributing Influxes and Processes

Consolidated Parent Rock	Erosion	Erosion	Erosion	Erosion	Erosion

FIG. 3—*contd.*

Soil volume limiting horizons usually are missing from these soils and the erosion due to flat terrain is negligible. Threatening influxes are those originating from human activities that result in degradations listed in the figure. The composition of the pedon, however, is characteristic of low sensitivity.

Pedon (b) (Fig. 3) is a deep gravelly and stony Regosol on colluvium. It has no horizon differentiation or lithological stratification. Since these peda are usually located on sloping terrain, the rate of soil loss is significant. But erosion affects its productivity only slightly through the loss of organic matter. Soil volume is not affected until most of the colluvial deposit is removed. Surface erosion will continue until a gravelly and stony mulching is formed on the surface. Then erosion ceases and soil moisture is conserved. The end result is an improvement in the productivity of the soil system for orchards in areas where soil moisture is a limiting factor.

Pedon (c) is a Luvisol on quaternary terraces. It has developed a soil profile differentiated into eluvial or illuvial horizons. Erosion is by far the most effective process that threatens its productivity because any loss of soil volume critically affects the overall performance and the qualities of the soil system. The exposure of the fine textured Bt horizon on the surface will reduce the erodibility of the pedon, but its productivity would have been, in most cases, already damaged.

Pedon (d) is a Calcaro-Calcic Cambisol. The productivity of the respective soil unit depends on the depth of the Calcic horizon (Cca).

Pedon (e), a Ranker on the sloping mountainous watersheds, is the most sensitive of all the pedons to volume loss through erosion because of its already small rooting zone that is restricted by the shallowness of the bedrock.

It becomes clear from the above examples that the rate of soil loss through erosion is not sufficient to determine the sensitivity of the soil to this process. It is necessary, in addition to the estimation of the rate of change that influxes cause to soil pedons, that the performances of the resulting pedons be considered. This holds not only for erosion but for any processes that downgrade soil productivity.

Relative sensitivities of the main soil units in Southern Europe to degrading influxes are shown in Table 1. The ratings are qualitative and in essence they express probable levels of sensitivity. A quantitative evaluation needs considerable research effort aiming at the determination of relationships between the external influxes and the performance of each major soil unit. This determination must be based on hard data, which are not available at the present time.

The soil units of the Soil Map of Europe have been grouped into four landforms and descriptive evaluations of their sensitivities are given in the following sections.

5.1. Soils on quaternary formations

This landform includes recent flood plains, older terraces both of holocene and pleistocene age and few pliocene surfaces.

The parent materials of the soils in this landform are deep unconsolidated quaternary deposits characterized by a flat or slightly sloping terrain. These two conditions are responsible for the rather low sensitivity of the soil system to natural influxes, except in cases of natural calamities. On the other hand, these soils are the most intensively used of all other soils in Europe [37]. Therefore, their constitution and their performance is often affected by influxes originating from human activities. The sensitivity of the alluvial peda to these influxes depends on the particular properties of their parent materials.

Soils formed on alluvial deposits are similar the world over in terms of landforms and processes [20]. Soil textures and drainage conditions are in most cases closely related to the alluvial geomorphology. Near the presently active or older inactive stream beds levees are formed, where coarse textured and well drained soils are found. Beyond the levees lower laying soil peda gradually become finer and less water permeable with distance from the stream bed. Swamps, organic soils and poorly drained soils are found in the flood basins and in the deltas of the streams.

Soils on river terraces have ceased to be flooded and silted, and they form chronosequences and chronotoposequences. The sensitivity of their peda to influxes depends on their properties as affected by their age and topography.

Apart from the general description given above, the properties and the sensitivity of the soil peda to influxes vary with their taxonomic classification. The most common taxonomic units (Soil Map of Europe, 1986) of the soils of these landforms in Southern Europe and their respective sensitivity to damaging processes are given below.

Soils of the flood plains

Fluvisols. The main categories of these soils are: *Eutric Fluvisols, Calcaric Fluvisols* and *Dystric Fluvisols.*

Fluvisols have young peda not differentiated into horizons except for the organic matter enriched A1 horizon. The first two categories are the most

Table 1

Relative ratings of soil sensitivity to degradation in Southern Europe. (I = Slight, II = Medium, III = High, IV = Very High)

Soils	Loss of volume	Loss of organic matter	Physical degradation	Chemical degradation	Fertility degradation
a. Flood Plains					
Eutric Fluvisols (Je)	I	II	II	IV	I
Calcaric Fluvisols (Jc)	I	II	II	III	I
Dystric Fluvisols (Jd)	I	II	II	III	III
Vertisols (V)	I	I	III	II	I
Histosols (O)	II	II	III	III	III
b. Quaternary Terraces					
Orthic Luvisols (Lo)	II	III	III	III	III
Chromic Luvisols (Lc)	II	III	III	III	III
Calcic Luvisols (Lk)	III	III	III	IV	IV
Vertic Luvisols (Lv)	II	II	II	I	II
Ferric Luvisols (Lf)	II	III	II	III	IV
Rhod-chrom Luvisols (Lch)	II	II	II	II	II
Plinthic Luvisols (Lx)	IV	II	IV	IV	IV
Eutric Planosols (We)	III	II	III	II	II
Dystric Planosols (Wd)	III	II	III	IV	IV

c. Rolling Tert. Hills

Calcaric Regosols (Re)	I	III	I	I	II
Eutric Cambisols (Be)	III	II	I	I	II
Calc-calcic Cambisols (Bck)	I	II	II	II	III
Calc-vertic Cambisols (Bcr)	I	II	I	I	II
Vertisols (V)	I	I	III	I	I
Chromic Luvisols (Lc)	III	III	III	III	III
Rendzinas (E)	II	III	III	II	IV

d. Mountain Watershed

Lithosols (I)	I	III	I	I	I
Regosols (R)	I	II	I	II	II
Dystric Cambisols (Rd)	III	II	III	IV	IV
Eutric Cambisols (Be)	III	II	III	III	III
Orthic Luvisols (Lo)	III	III	III	III	III
Chromic Luvisols (Lc)	III	III	III	III	III
Acrisols (A)	III	III	III	IV	IV
Rankers	IV	III	IV	III	IV

extensive in Europe. They are rich in nutrients and high in base saturation. They have a mixed soil mineralogy that secures a high level of inherent fertility, and cation exchange capacity (CEC) high enough to make them strong buffer systems.

Loss of soil volume. The above properties in combination with flat topography and great soil depth render the Fluvisols quite resistant to natural influxes. The main natural disturbances that may adversely affect the qualities of these soils are:

(a)   Extensive and frequent flooding.
(b)   Deposition on their surface of thick alluvial material of unfavourable texture and composition.
(c)   Ripening that results in the compaction of subsurface layers.
(d)   Erosion.

The above episodes are not extensive and intensive enough to pose a significant threat to the qualities of Fluvisols in Southern Europe. In some cases flooding is beneficial because it renews soil nutrients and prevents soil ripening.

Erosion can occur even on the smoothest slopes as long as there is movement of a deep enough body of water. Wilkin and Hebel [59] using Cs^{137} techniques in US Midwestern watersheds found that cropped flood plains are the most severely eroded lands in the watershed. Thus both erosion and deposition take place in the flood plains. The net effect of these two processes on the soil qualities, and the sensitivity of the soil peda to them depend on their relative rates in each particular environment. Norton (43) found in an alluvial field of Indiana that erosion and deposition did not agree with USLE predictions. He suggested that the dynamic aspects of the two processes on the landscape are important.

The impacts of water erosion will be more extensively discussed in the analysis of this process for the steeper sloping landforms.

Wind erosion is locally important in drier areas in Southern Europe especially on bare, smooth pulverized soil surfaces. Thus the sensitivity of the soil peda to wind erosion depends on the density of vegetative cover, the roughness, the moisture content, the texture and the aggregation of their surfaces [59]. All these variables are dependent on soil management practices.

Human originated influxes may damage the overall performance of the soil systems through the following processes.

Degradation of soil structure. Soil structure in most of the soils is well developed especially on the surface of the Eutric and Calcaric Fluvisols. However, it may be degraded in the following ways:

(i)   *Compaction* Soil compaction actually acts as a soil volume loss process. Pressures exerted by heavy machinery may drastically reduce the air content of a dry soil. Coarse textured soil layers are more sensitive than fine textured ones. However, at high moisture, fine soil material may lose its aggregation even under low pressure [34].

Compaction reduces the rooting volume in the soil peda and it depresses the transpiration and the yields of the crops [5]. Compaction caused high bulk density in subsurface horizons of Calcaric Fluvisols in the Argos Plain in Greece and it has adversely affected the quality of fruit and the health of orange trees [63].

(ii)   *Pulverization–dispersion* Farm machinery applies forces that lead to soil loosening processes on weakly aggregated soils. Cultivation by reducing organic matter decreases aggregate stability enhancing soil sensitivity to pulverization. Soil structure is often regenerated by natural processes (wetting and drying), especially in fine textured material. Therefore, coarse soil surfaces are more sensitive to pulverization than fine surfaces.

Dispersion of soil aggregates is caused by the increase in the Zeta-potential of the colloidal fraction. Such an increase can be affected by irrigation water high in sodium adsorption ratio (SAR) values. These waters by dispersing the clay particles have long term effects on the hydraulic properties of the soil [45]. Calcaric Fluvisols due to their high carbonate contents are the least sensitive to dispersion by saline-sodic water, whereas Dystric Fluvisols would be the most sensitive.

(iii) *Crusting* Pulverized soil surfaces are sensitive to crust formation caused by rain water and shallow overland flow. Eisenhauer *et al.* (13) reported that the rate of seal development was dependent on the velocity of overland flow and aggregate stability. Data obtained by Onofiok and Singer (44), indicate that soil texture, particle shape and aggregate stability are factors related to crust development. Clay loam and silt loam surfaces were more sensitive than fine loam. Calcium carbonate due to its silt sized particles seems to increase the sensitivity of the soil surface to sealing (22). Biological activity, mainly earthworms, reduce soil crusting (32).

Among the soils of this landform, the weakly structured and medium textured Calcaric Fluvisols would be the most sensitive due to their silt sized carbonates.

Loss of organic matter. Recent alluvial deposits are in many cases rich in organic matter. Furthermore, the high productivity of these materials promotes, under natural conditions, the build-up of organic matter in the soil. Cultivation, crop residue removal, monoculture, unbalanced mineral fertilization and abolition of manure fertilization enhance the degradation of organic matter [12, 30, 52].

Soil properties controlling the rate of organic matter degradation are: internal drainage, texture, acidity, sesquioxide and clay contents [30, 50, 55].

Fluvisols are intensively cultivated and are subjected to human influxes that speed up the rate of organic matter loss. The most sensitive Fluvisols will be those with well drained peda having a coarse textured surface layer.

Surface erosion is a very effective agent of organic matter loss, and it will be discussed in the following sections.

The loss of organic matter indirectly affects the sensitivity of the soil system to other processes that are damaging to their performance. These effects will also be briefly discussed in the following sections.

Chemical degradation. Fluvisols are particularly threatened by chemical degradation, not so much because of their inherent sensitivity, but because of the intensity of human originating influxes they are subjected to. Moreover they are the recipients of surface materials and surface waters, enriched with persistent pollutants and soluble salts. These materials are continuously transferred from the large area of the watershed basin and accumulated on a smaller area of the flood plain.

The enrichment of the Fluvisols with the pollutant carriers depends on the relative rates of erosion and deposition in each particular section of the flood plain.

The main processes leading to the chemical degradation of the soil system are the following:

(i) *Leaching of nutrients* Under normal soil management practices leaching of nutrient out of the rooting volume is not significant in the Eutric and Calcaric Fluvisols. It is a more serious problem for the Dystric Fluvisols since their parent materials are already leached of bases. Also it would be more pronounced under high rather than under low rainfall.

Under irrigation, leaching of nutrients can significantly lower the soil fertility level especially in soil peda with low nutrient retention capacity. Nitrogen is easily leached out. Cations are held by soil minerals and organic colloids and they are less leachable. Therefore, coarse textured highly permeable Fluvisol peda low in CEC would be the most sensitive to leaching.

The irrigation system seems to play a role in the leaching of nutrients. Onken *et al.* (45) reported higher rates of NO_3^- leaching with sprinkler than furrow irrigation.

(ii)   *Salinization–alkalinization* Since ancient times these processes have been a world wide problem for irrigated soils in climatic zones where evapotranspiration exceeds precipitation. Irrigation without adequate drainage and with water applications below the leaching requirement lead to a slow but progressive salinization of the soil peda in the semiarid zones of Southern Europe.

Use of irrigation water with high electrical conductivity and high SAR on soil peda with low permeability and low supply of bivalent cations would speed up their salinization and alkalinization.

The fine textured Eutric Fluvisols in the semiarid zone would have high sensitivities to the above processes.

(iii)   *Heavy metal contamination* Fluvisols due to their convenient landscape and geographical locations have been, in many cases, crowded with man-made heavy metal emitting sources.

The metals reach the soil in many ways, as described by Jones and Jarvis [29]. Once they reach the soil, they are fixed as: organometallic complexes, carbonates, adsorbed on sesquioxides and exchangeable cations. Relatively small amounts pass into solution. Thus the bulk of the contaminants are retained within the upper few centimetres of the soil surface [46]. Their concentrations increase with time as long as the emission sources remain active.

All Fluvisols, except the very coarse ones, have high contents of constituents with heavy metal adsorption and fixing capacities. Therefore, their sensitivity to a toxic build-up is high.

(iv)   *Pesticide contamination* The accumulation of pesticides in the soil peda depends on the transfer processes (evaporation, leaching, erosion, deposition, plant uptake) and on the rate of their degradation in the soil environment.

Transfer processes depend on: (a) the particular physical and chemical properties of the pesticides, (b) on the regional climate and geomorphology and (c) on a number of soil properties. Among the most effective soil properties are: porosity, permeability, drainage, CEC, anion exchange capacity, redox potential, pH, moisture and organic matter.

Micro-organisms are responsible for most of the pesticide degradation in the soils. All the above parameters plus soil fertility indirectly affect the degradation processes by controlling the biological activity of the soil [20]. It seems, therefore, that high rates of degradation would occur in soil peda that have a highly active digestive system. That is peda with well aerated thick surface horizons, which have high levels of organic matter content, fertility and base saturation, and moisture tensions within the available range.

Most of the Calcaric and Eutric Fluvisols have properties that favour both the adsorption and the degradation of pesticides. Degradation, however, and build-up of pesticides, in soils are complex procedures. In some cases the process may act in opposite ways. Adsorption of pesticides on clay surfaces, for example, may catalyse their decomposition. On the other hand adsorption removes them from the soil solution and protects them against microbial attacks [20]. Therefore, it is difficult to estimate the sensitivity of the soil systems to pesticide build-up. It is to be assumed, however, that the properties of the properly managed and well drained Eutric and Calcaric Fluvisols of Southern Europe would place them among soil systems with low sensitivities. Poorly drained Fluvisols in colder regions would be more sensitive.

(v)   *Radioactive contamination* Man-made radionuclides are usually well contained. In case of accidents, they can contaminate the soil but on a localized scale. On the other hand nuclear fission products released in atomic explosions or in accidents involving nuclear reactors contaminate the soil on a large scale.

Two of the fission products have long lasting effects when they fall on the soil surface. These are Sr^{90} and Cs^{137}. Both have considerable long half-lives. As cations they are strongly held in the upper layer of the soil and enter the food chain through plant uptake.

Heavy textured acid soils are the most sensitive to strontium and caesium storage. Plant uptake of caesium would be higher from potassium depleted Dystric Fluvisols than from potassium rich Calcaric and Eutric Fluvisols. All three Fluvisols have high Sr^{90} and Cs^{137} retention power. Thus the radionuclides will remain in the upper layers of the peda for several years.

The rate of their uptake would depend on the kind of plants and the calcium and potassium content of the soil respectively. In Table 2 are estimates of maximum possible plant contamination levels with Cs^{137} taken from the soil.

Soil fertility deterioration. The fertility status of the various soil systems under natural conditions is more or less known and its changes are slow. However, soils intensively cropped under conditions of intensive irrigation, insufficient fertilization and crop residue removal may suffer a significant depletion of their available nutrient supplies. The amount of nutrients removed from the soil through plant uptake, leaching and evaporation have been adequately tabulated by several authors [10, 11, 38].

Soil systems that are sensitive to fertility deterioration are those located in climatic zones with high rainfall, have coarse textural configurations and are depleted of minerals capable of releasing plant nutrients. Thus Calcaric and Eutric Fluvisols are not among the sensitive soil systems. Dystric Fluvisols are more sensitive, due to the advanced stage of leaching that characterizes their peda. It should be pointed out, however, that plant nutrients are transported from the larger watershed to the smaller flood plain area through erosion–deposition processes.

Soil fertilization practised today in the developed countries leads to enrichment rather than to depletion. Heavy fertilization applied to achieve maximum yields is widely recommended. The long term effects of this practice on the overall performance of the soil subsystems has still to be studied and evaluated.

Table 2
Maximum Cs^{137} uptake by plants

Exchangeable soil K (me/kg soil)	K content in plant (% D.W.)	Radioactive contamination (BQ/kg of dry plant tissue)	
		Cs^{137} fallout	(BQ/m²)
		1 000	40 000
	1	1 000	40 000
1	2	2 000	80 000
	3	3 000	120 000
	1	200	8 000
5	2	400	16 000
	3	600	24 000

Nutrient balances, taking into consideration the chemical, physical, biological and mineralogical properties of the soil peda and the interactions of the nutrients in the respective soil systems, seem to be the answer in preserving soil fertility. Nutrient inbalances are practised in developing countries especially in Asia and Africa. These countries are also suffering a 'nutrient mining'. That is nutrients are transported from them to developed countries through agricultural exports.

The increasing Cost/Benefit ratio and the increasing yields due to advances in biotechnology pose a potential threat to the fertility status of sensitive soil systems. This threat will be faced by many developing countries in the near future.

Vertisols. Vertisol peda, due to their high clay contents, high buffer capacities and natural mixing are quite insensitive to influxes. The most significant deterioration of these soil systems is structure degradation. The use of heavy machinery, when the moisture content is high, leads to considerable compaction. Ploughing at low moisture content produces large soil clods. With the first rains, the clods disintegrate into small aggregates and good surface structure is regenerated.

Under irrigation, frequently alternate wetting and drying produces a weak, dense and very firm structure down to a considerable depth below the plough layer. This structure deterioration has been widely observed by the author in the Vertisols of the Thessaly Plain in Greece. Avetjan *et al.* [2] have reported degradation of structure and loss in productivity of heavy montmorillonitic meadow soils within ten years of cultivation and irrigation. They concluded that the compaction of the soil was of physical and not of chemical nature.

Vertisols, due to their high retention capacity and low permeability are sensitive to pollutant concentration on their surfaces.

Histosols. Eutric Histosols are common bottomland soils in Southern Europe. When drained they become highly productive and are intensively cultivated. Their productivity largely depends on the preservation on the organic surface layer.

Under the warm climate of the bottomlands in Southern Europe, Histosols are threatened by high rate of organic matter loss when they are excessively drained and continuously cultivated. Subsidence of soil surface due to speedy loss of organic matter causes serious soil management problems. Intensively cultivated Histosols are sensitive to exhaustive farming, unless balanced fertilization is practised. They are usually more sensitive to minor than to major element depletion. The qualities of the Histosol pedon are largely controlled by the volume of the organic layers and the nature of the underlying mineral substrate. Among the most sensitive to productivity loss are the Histosols with a thin organic layer overlying a marly sediment. These soils suffer a quick loss of their productivity unless organic matter conservation is practised. Examples of this degradation are found in the drained Kopais lakebed in Greece.

Soils of the quaternary terraces

Soil formation on the river terraces is a function of distinctive landform material assemblage and the age of the terrace [20].

The soils of the Mediterranean terraces seem to show standard geomorphic and genetic relationships. Young *Orthic Luvisols* predominate on stabilized holocene terraces. Whereas a series of polygenetic *Chromic Rhodochromic Ultic* and *Ferric Luvisol* profiles have developed on the pleistocene terraces. Rudification decoloration, clay translocation and formation of concretions are prominent features of these soils [1, 24, 50]. Similar patterns have been found by Charters [8] in Central Southern England.

The terraces are characterized by smooth and gentle slopes. There are also flat areas where claypans and fragipans, formed in the subsurface horizons, impede the internal drainage of the soil peda.

With the development of the Bt horizon, the degree of leaching and acidification increase, whereas the fertility level, the organic matter content and the base saturation decrease from the younger to the older terraces. Thus the most productive of these soils are those of the lower terraces, the productivity being reduced with the age of the terrace.

The soils of the river terraces are more sensitive to degradation than the soils of the flood plains. The damaging processes are the same as described for the soils of the flood plains. However, their respective weights are different.

Loss of soil volume. Erosion is the most effective process that results in loss of soil volume. Sheet and rill erosion is the most damaging natural process that threatens the overall performance of the cultivated soil of the terraces. This phenomenon has been extensively studied and there is no need to describe it here. An effort will be made to evaluate only the sensitivities of the soils of this landform to surface erosion and its consequences.

The soils on the river terraces have developed peda showing unique qualities and performances. The combination of coarse to medium textured eluvial surface horizons with the finer texture subsurface illuvial Bt horizons secures favourable thermal, water and plant nutrient storage, irrigation and cultivation conditions. These conditions are particularly favourable on the lower terraces. The coarse to medium textured and low in organic matter soil surface in combination with the low in water permeability subsurface are conditions favouring high soil erodibility values.

Removal of the surface horizons through sheet and rill erosion causes serious disturbances to the performance of the soil peda. The consequences of these disturbance are:

(a)   Loss of organic matter and of soil nutrients contained in the surface horizon.

(b)   The loss of biologically and biochemically active components of the peda. The loss is particularly damaging to the digestive system of these soils because they are low in organic matter and biological activity.

(c)   The decrease in the infiltration and in the water storage capacity within the root zone [52].

(d)   The disruption of the favourable thermal balance that promotes early maturity of the crops in Southern Europe [63].

(e)   The increase of soil sensitivity to compaction and crusting.

(f)   The decrease in the rate of degradation of pesticides in the soil.

The high erodibility of the soil surface and the degrading effects of its removal to soil productivity render these soils highly sensitive to erosion. Irrigation practices usually accentuate erosion risks. Furrow irrigation frequently causes redistribution of top soil material, by eroding the upper parts of the fields and enriching the lower ones. Thus crop productivity sharply decreases in the eroded sections and increases in the silted sections of the fields [7]. This phenomenon has been observed on several occasions by the author even in experimental fields.

Once the surface horizons are removed by erosion, the subsurface Bt horizons become uncovered. These horizons have lower erodibility than the surface horizons due to their fine texture and strong structure. This is why the surface of the cultivated soils of this landform in the Mediterranean countries consists of Bt materials. Gravel often contained in the Bt horizon decreases its erodibility.

Soil peda on the older terraces (late pleistocene early pliocene) are much more sensitive to erosion, because they include limiting horizons such as claypans, fragipans and plinthite. The removal of the surface horizon from these peda critically reduces their rooting volume and their hydraulic properties.

Degradation of structure. Surface structures are generally weak in the peda of this landform mainly due to coarse textures and low organic matter content. Further decrease of organic matter through cultivation renders these soils quite sensitive to compaction and surface crusting.

Chemical degradation. Most of the Orthic Luvisols on the lower river terraces have chemical properties favourable to plant growth, except for the low buffer capacities of the surface horizons. The latter make them quite sensitive to acidification and leaching of bases. Thus acidifying fertilizers and acid rains can lower the pH of the soil to undesirable levels. Exhaustive farming coupled with intensive irrigation can quickly deplete the soil of its available nutrients.

The sensitivity of the soils on the upper terraces (Planoluvisols, Ferric Luvisols and Ultic Luvisols) to acidification is higher than on the lower terraces, because of their advanced stage of development that has already resulted in low pH and extensive losses of bases, low buffer capacities and low organic matter contents.

The above conditions render the soils of the upper terraces highly sensitive to pesticide accumulation.

Erosion, by exposing on the soil surface finer textured and more buffered subsurface horizons, may improve or reduce the productivity of the soil depending on the chemical and mineralogical composition of these horizons. It may also increase the sensitivity of the soil to farm chemical build-up. Frye *et al.* [17] have reported that herbicides often injure crops on severely eroded soils. Erosion may also expose on the surface natric horizons developed under semiarid climates. These horizons have adverse chemical composition and their presence in the rooting zone lowers the suitability of the soil for sensitive crops. Strongly developed Luvisols of the upper terraces may have very acid Bt horizons with toxic concentrations of exchangeable aluminium. The productivity of these soils is drastically reduced if erosion brings the Al-rich horizons into the rooting zones of the plants.

Soil fertility deterioration. The young Orthic Luvisols on the lower terraces usually have medium to high fertility levels. The availability of the nutrient elements is also high due to favourable pH (slightly acid). The relatively high CEC of the subsurface horizons makes them resistant to fertility deterioration.

The soils of the upper terraces are more sensitive due to their advanced stage of nutrient leaching and low buffer capacities. Heavy applications of fertilizers and in some cases lime are necessary for the preservation of soil fertility of these soils.

5.2. Soils of the rolling tertiary hills

This landform is characterized by moderately sloping terrains and unconsolidated deep late tertiary sediments. Land surfaces are more eroded than on the river terraces due to steeper slopes and cultivation. Even though the soils here are not as intensively cultivated as on the latter landform, the steeper slopes make erosion the dominant process that affects both soil development and soil productivity. Soil peda form on young erosional surfaces as eroded phases of older soils developed, in the past, under undisturbed natural vegetation.

The untruncated soil peda that have preserved the undisturbed original profiles are rare. This is particularly true in the drier Mediterranean zones.

The most common soil taxonomic units on the neogene hills are: *Eutric* and *Calcaric Regosols, Orthic Rendzinas, Calcaro-Eutric, Calcic* and *Chromic Cambisols, Calcic* and *Chromic Luvisols* and *Vertisols.*

The parent material of the soils is mostly a deep, friable and carbonate-rich sediment. Therefore, soil depth per se is not a limiting factor for the performances of the peda. However, the depth of the organic matter enriched A1 horizon, the thickness of carbonate leached sub-surface horizons and the depth to Calcic and Petrocalcic horizons are important soil productivity parameters. All these parameters are strongly affected by soil erosion as it is described below.

Loss of soil volume. Erosion is an even more effective agent of soil volume loss on this landform than on the previous ones. Under the given climate of Southern Europe, the sensitivity of the peda of this landform to erosion hazards can be evaluated on the basis of the following factors:

(a)   Slope gradient which usually ranges between 10 and 30%. Steeper slopes are found in Italy.
(b)   Density of the vegetational cover which varies with land utilization type.
(c)   The soil profile and horizon configuration.

The two first factors have been extensively evaluated in recent publications and there is no need to elaborate here. It should only be mentioned that soil peda with fully developed horizon differentiation are found either on plateaus with small slope gradients or on surfaces continuously protected by vegetation. All other surfaces have either truncated or silted peda. The horizons that control the sensitivity of the soil systems to erosion on this landform are the following:

(a) The surface Al horizon usually rich in organic matter, with a strong stable structure and considerable biological activity. When its thickness is greater than the specified limits [57] it is classified as Mollic Epipedon and it characterizes the Rendzina soil peda. Mollic Epipedons are among the least erodible soil surfaces. Al horizons are the first to be removed, but they can regenerate in a relatively short time if protected by vegetation.

(b) The Cambic horizon that underlies the Al and overlies the parent C horizon. It has adequate supplies of nutrients, a fairly strong and stable structure and adequate water holding capacity. It has suffered either a partial (Calcaric Cambisols) or a total loss of carbonates (Calcaro-Eutric Cambisols) but it has retained high base saturation. These properties create a favourable environment for root growth. Therefore, the thickness and the depth of the Cambic horizon in the soil profile has a controlling role on its qualities. The Cambic horizon is the next, after the Al, to be removed by surface erosion. Thus many of the Calcaric and Calcaro-Eutric Cambisols of Southern Europe have their Cambic horizons exposed on the surfaces of the soils.

(c) Argillic horizons are characteristic of the Luvisols. Their presence in the soil peda signifies an advanced stage of leaching and soil development. Their functions in the soil peda are beneficial or damaging depending on their properties and depth in the profile as described in the soils of the river terraces.

(d) Calcic and Petrocalcic horizons are common in the soil peda of the semi-arid zone of Southern Europe. They are located immediately above the parent C horizon and are enriched with carbonates leached from the overlying Cambic, Argillic, and Al horizons. Both horizons constitute an unfavourable chemical (nutritional) environment for the plant roots. The Petrocalcic horizon, due to its induration impedes root penetration. Thus it reduces the rooting volume and the water holding capacity of the soil peda.

The sensitivity of the soil peda of the hilly landform to erosion can be evaluated in relative terms on the basis of the above analysis. The least sensitive would be those with a deep Mollic Epipedon overlying a thick Cambic horizon. Those are well developed uneroded Rendzina peda.

The most sensitive peda would be the truncated Calcaric Cambisols of which the surfaced thin remnants of Cambic horizons overlie a shallow Calcic or Petrocalcic horizon. The latter case being more sensitive.

Intermediate sensitivities would have Cambisols and Luvisols with untruncated Cambic and Argillic horizons respectively. The peda with deep lying Calcic and Petrocalcic horizons belong to the same class.

Regosols present a special case among the soils of this landform. Those on eroded surfaces have lost their productivity to a considerable degree. Further deterioration by erosion depends on the depth to consolidated bedrocks or to strata with unfavourable compositions. These soils have also been deprived, to a large extent, of the properties that reduce soil erodibility (organic matter, structure). Therefore, the rate of soil loss would be high and the danger for accelerated gully erosion considerable for them. Artificial terracing cutting deep into the soil may deteriorate soil productivity and increase erodibility by mixing overlying horizons with the parent material, especially if the latter is of a marly nature. Crop residue burning also increases soil erodibility. However, the unconsolidated parent material has a great depth in most cases and thus their productivity can be preserved through fertilization in spite of the erosion.

Structure and chemical degradation. The main causes of these types of soil degradation are those originating from erosion as described above. The loss of Al and subsurface horizons certainly result in a soil pedon with a less favourable structure and chemical composition.

Extensive land utilization types, applied on these soils, and erosion, reduce the chances for a chemical pollutant build-up.

5.3. Soils of the mountain watersheds

Topography and in particular slope gradient and slope length are dominant state factors affecting erosion, movement and depositional processes on the soils of the mountain watersheds. These processes can selectively add and/or deplete the soil peda of certain chemical and physical characteristics [20].

Fully developed soil peda form in this landform when the sloping land surface is continuously and for a long time period protected by vegetation. Otherwise, truncated or silted soil peda result from the erosion–deposition processes.

In Southern Europe the genetical status and the characteristics of the undisturbed soil peda depend on the type of parent rock, the climate as expressed by the vegetational zone, the composition of the vegetation and the aspect. The main taxonomic units listed in a sequence from the drier and warmer to wetter and colder zones are: *Calcaric* and *Chromic Cambisols, Rhodochromic* and *Chromic Luvisols, Orthic Luvisols* and *Rankers, Dystric Cambisols* and *Alfisols.* Calcareous or acid parent materials may disrupt the above climosequence.

Lithosols and Regosols are extensive in all climatic zones. Lithosols represent the terminal stage of degradation of the mountain soils. They are more frequently found in dry regions on limestone rocks even on hilly landscape. They are also common on the mountain summits.

Mountains are extensively exploited. The main land utilization types are forestry and pasture. Neither of the two types imposes any serious threat to soil qualities and performances, unless the density of the vegetative cover is reduced. In the latter case erosion becomes the dominant threat to the mountain soil systems due to the prevailing steep slopes. The particular weight of each soil degrading process is evaluated below.

Loss of soil volume
Due to the steepness of the landscape, removal of the protecting vegetation leads to accelerated soil erosion. The sensitivity of the soil system to erosion on this landform depends on the parameters of the Universal Soil Loss Equation [61] and on a number of additional state factors and soil properties. Among them are: (a) the type of parent rock, (b) the soil depth and (c) the rate of regeneration of vegetation.

The type of parent rock determines the texture of the soils and consequently its erodibility. It also determines the rate of soil formation and affects the run-off/run-in ratios. The depth of the untruncated soil peda, under given climatic conditions, depends on the relative rates of rock weathering and on the solution loss of the weathering products. These rates produce shallow soils on limestone and deep on flysh. Slope gradient is negatively related to soil depth. Richter [49] reported linear relationships between soil depth and slope inclination for a series of parent rocks shown in Fig. 4.

The depth of the soil peda is generally shallower on the mountain watersheds than on the other landforms. Because of its shallowness, it is the most critical parameter that prescribes the sensitivity of these soil systems to soil loss and determines their productivities.

The regeneration rate of vegetation, after a destructive episode, depends on its composition, on the local climate and on the ability of the soil system to maintain its pre-episode capability to support the vegetation.

Climate determines the erosivity of the rain, the density, the type and the regeneration rate of the vegetative cover. The net effects of the climate

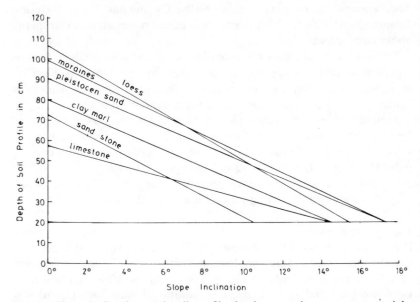

FIG. 4. Slope inclination and soil profile depth on various parent materials. (Richter [49]. Reproduced with the permission of John Wiley and Sons, Ltd, Chichester.)

render the soils in dry climates more sensitive to erosion than in wet climates [31]. The pseudo-alpine mountain summits should be exempted from this rule wherever erosion rates are high.

Episodes that result in total loss of vegetation lead to a rapid degradation of soil productivity and consequently to a suppression of the regenerative capability of the vegetative cover. Finally accelerated erosion follows. Such episodes, that are frequent in mountain watersheds, are the forest fires. Helvey *et al.* [25] have found that the annual sediment yields increased as much as 180 times above the prefire levels after a wild fire had destroyed all vegetation on forested watersheds in Washington. Nitrogen loss was increased by 40 times and available phosphorus by 14 times.

The high evapotranspiration and low rainfall rates prevailing in Southeastern Europe reduce the regeneration ability of vegetation. Thus soil erosion is more effective in denuding the land here than in other European regions. Soils on the southern aspects of limestone mountains and hills are the most sensitive to denudation, because lack of soil moisture over prolonged periods is a severe limiting factor for the growth of vegetation. Soils on limestone are usually fine textured and well structured. Thus their erodibility is low, however, their depth is small due to the high solution loss of the parent material. Therefore, the sensitivity of these soils to erosion is among the highest in spite of their low erodibility. The least sensitive would be soils developed on flysch because of their great depth and high water holding capacity.

Degradation of structure
Soil compaction due to forest harvesting, grazing and recreational activities such as skiing [6] is the major structural deterioration of these soils. It is localized and it affects only small portions of the landform. Erosion, forest fires, and overgrazing are the main processes that indirectly affect degradation of soil structure by reducing the organic content of the top soil. Grazing also causes soil compaction.

Loss of organic matter
The loss of organic matter from the peda of this landform is only second in importance to the loss of soil volume. The surface Al horizons of these peda acquire high contents of organic matter under pasture, forest and shrub vegetation. On the top of the Al horizon a litter floor plays an important protective role against the dispersion of soil aggregates by raindrops.

The combination of a spongy litter with the well structured Al horizon rich in organic matter secures high infiltration, low run-off and low soil loss rates. The productivity of the soil increases with the thickness of the Al horizon because of its high nutrient contents, favourable structure and high available water content.

Removal of vegetative cover followed by loss through erosion and high rate of decomposition are the main processes that deplete the soil peda of its organic matter and result in productivity losses. Shallow soils under forest vegetation in dry climatic zones are the most sensitive to loss of organic matter due to their thin Al horizons and low rate of vegetative regeneration. Deep pasture soils in humid climates are the least sensitive.

Chemical and fertility degradation

In balanced ecological systems such as the forest and pasture lands chemical and fertility degradations have low rates. They are speeded by soil acidification and leaching of nutrient and slowed by the release of bases and nutrients through rock weathering.

Processes that accelerate the rate of degradation are those that enhance loss of the surface horizons through erosion. Among the most effective of these are: overgrazing, overlogging, forest fires and cultivation without erosion control practices.

In recent years acid rain has become a serious threat to the productivity of the affected soils by enhancing their acidification and nutrient depletion. The elaboration of the subject is beyond the scope of this presentation. It can only be stated here that soil peda most sensitive to acid rain impacts are those developed from acid rocks and which have a dystric character.

6.   RECOMMENDATIONS FOR FUTURE RESEARCH

The main theme of this presentation is the importance of evaluating the impact of the degrading influxes not only on single soil properties but also on the performance of the soil systems.

Soil systems and their performances should be the basis of reference for the evaluation and the interpretation of soil degrading processes. Models describing and explaining the degradation processes, based on single soil properties are indispensable. There is also need, however, for the study of the effects of degrading influxes on the performance and the productivity of each major soil unit.

The following information is among the most urgently needed for the protective and sensible exploitation of the soil resources in Southern Europe:

(1)   Rain erosivity assessment under various geographical and climatic conditions.

(2)   Soil volume loss related to potential productivity of representative peda.

(3)   Erodibility functions for the whole soil pedon on sloping lands.

(4)   Cultivation systems and structural and organic matter degradation on major soil units.

(5)   Irrigation effects on salinization structural degradation and erosion.

REFERENCES

1. ANTONAKOPOULOS, G. and YASSOGLOU, N. J. (1986). Distribution patterns of clay and 'free' oxides in polygenetic soil profiles on pleistocene surface in Greece. *Trans. XIII Congr. Int. Soc. Soil.*, pp. 1034–5.
2. AVETJAN, S. S., ROSANOV, B. G. and ZBORISHUK, N. G. (1984). Change of structure and fabrics of clay alluvial soils under agriculture, ILRI Publication No. 37, pp. 82–5.
3. BECHER, H. H., SWERTMANN, U. and STURMER, H. (1985). Crop yield reduction due to reduced plant available water caused by water erosion. In: *Soil Erosion and Conservation* (Eds S. A. El-Swaify, W. C. Moldenhauer and A. Lo). Soil Conservation Society of America, Ankeny, USA, pp. 365–73.
4. BELPOMM, M. (1980). For an application of the soil world charter. In: *Assessment of Erosion* (Eds M. DeBoodt and D. Gabriels), John Wiley, Chichester, pp. 9–22.
5. BOELS, D. (1982). Physical soil degradation in the Netherlands. In: *Soil Degradation* (Eds D. Boels, D. B. Davies and A. E. Johnston), Balkema, Rotterdam, pp. 47–65.
6. BOUNZA, G. (1984). Oberflächenabflüss und Bodenabtrag in alpinen Grasslandökosystemen. *Verhadlungen der Gesellschaft für Ökologie*, **12**, 101–9.
7. CARTER, D. L., BERG, R. D. and SANDERS, B. J. (1985). The effect of furrow irrigation erosion on crop productivity. *Soil Sci. Soc. Amer. J.*, **49**, 207–21.
8. CHARTERS, C. J. (1980). A quaternary soil sequence in the Kennet Valley, Central Southern England. *Geoderma*, **23**, 125–46.
9. CHISCI, G. (1982). Physical soil degradation due to hydrological phenomena in relation to change in agricultural systems in Italy. In: *Soil Degradation* (Eds D. Boels, D. B. Davies and A. E. Johnston), Balkema, Rotterdam, pp. 95–103.
10. COOKE, G. W. (1972). *Fertilizing for Maximum Yield*, Crosby Lockwood, London, 296 pp.
11. COOKE, G. W. (1986). Nutrient balances and the need for potassium in humid tropical regions. In: *Nutrient Balances and the Need for Potassium*. Proceedings 13th Congr. of I.P.I. (to be published).
12. DUTIL, P. (1982). Losses and accumulation of organic matter in French soils. In: *Soil Degradation* (Eds D. Boels, D. B. Davies and A. E. Johnston). Balkema, Rotterdam, pp. 181–5.
13. EISENHAUER, D. E., STIEB, D. J., DUKE, H. R. and HEERMAN, D. F. (1983). Transient surface seal development with shallow overland flow. ASAE Paper No. 2057.
14. EVANS, R. (1980). Mechanics of water erosion and their spatial and temporal controls: an empirical viewpoint. In: *Soil Erosion* (Eds M. J. Kirkby and R. P. C. Morgan), John Wiley, Chichester, pp. 109–28.
15. FAO (1970). *A Provisional Methodology for Soil Degradation Assessment*, Rome.
16. FERERES, E. (1983). Short and long-term effects of irrigation on the fertility and productivity of soils. In: *Nutrient Balances and the Need for Fertilizers in Semiarid and Arid Regions*, IPI, Bern, Switzerland, pp. 283–304.
17. FRYE, W. E., BENNETT, O. L. and BUNTLEY, G. J. (1985). Restoration of crop

productivity on eroded or degraded soils. In: *Soil Erosion and Crop Productivity* (Eds R. F. Follett, B. A. Stewart and I. Y. Ballew). American Society of Agronomy, Madison, Wisconsin, pp. 339–56.

18. FOURNIER, F. (1972). *Conservation des Sols*, Conseil de l'Europe, 206 pp.

19. GANTZER, C. J. and McCARTY, T. R. (1985). Corn yield prediction for a claypan soil using a productivity index. In: *Erosion and Soil Productivity*, American Society of Agricultural Engineers, St Joseph, Michigan, USA, pp. 170–81.

20. GERRARD, A. J. (1981). *Soils and Landforms. An Integration of Geomorphology and Pedology*, George Allen and Unwin, London, 219 pp.

21. GRAHAM-BRYCE, I. J. (1981). The behaviour of pesticides in soil. In: *The Chemistry of Soil Processes*. (Eds D. J. Greenland and M. H. B. Hayes) John Wiley, Chichester, pp. 621–70.

22. HAMDI, H. (1984). Soil of Egypt, problems arising from agriculture. *Fertilizer and Agriculture*, No. 87, 3–7.

23. HANOTRIAUX, G. (1980). Runoff erosion and nutrient losses on loess soil in Belgium. In: *Assessment of Erosion* (Eds M. De Boodt and D. Gabriels), John Wiley, Chichester, pp. 369–77.

24. HEILMANN, P. G. F. (1972). On the formation of red soils in the Lower Crati Basin (S. Italy), Doctoral thesis. State University of Utrecht.

25. HELVEY, J. D., TIEDEMANN, A. R. and ANDERSON, I. D. (1985). Plant nutrient losses by soil erosion and mass movement after wildfire. *J. Soil. Water Cons.*, **40**(1), 168–73.

26. HENIN, S., MICHON, X. GOBILLOR, T. (1954). Etude de l'erosion de vallees de Haute-Durance et du Hautes Drac. Assemblee generale de Rome, publ. 36 de l'Associational d'Hydrologie.

27. JENNY, H. (1941). *Factors of Soil Formation*, McGraw-Hill, New York.

28. JENNY, H. (1980). *The Soil Resource Origin and Behaviour*, Springer-Verlag, New York, 377 pp.

29. JONES, L. H. P. and JARVIS, S. C. (1981). The fate of heavy metals. In: *The Chemistry of Soil Processes* (Eds D. I. Greenland and M. H. B. Hayes), John Wiley, Chichester, pp. 593–620.

30. JOHNSTON, A. E. (1982). The effects of farming system on the amount of organic matter and its effect on yield at Rothamsted and Woburn. In: *Soil Degradation* (Eds D. Boels, D. E. Davies and A. E. Johnston), Balkema, Rotterdam, pp. 187–202.

31. KIRKBY, M. J. (1980). The problem. In: *Soil Erosion* (Eds M. J. Kirkby and R. P.C. Morgan), John Wiley, Chichester, p. 1.

32. KLADIVKO, E. J., MACKEY, A. D. and BRADFORD, J. M. (1986). Earthworms as a factor in the reduction of soil crusting. *Soil Sci. Soc. Amer. J.*, **50** (1), 191–6.

33. KURON, H. (1956). Überblick über die Arbeiten des Unterausschusses für Kulturbauwesen für Bodenerosion im Deutschen Ausschuss für Kulturbauwesen in die Jahren 1938 bis 1945. *Schriftenreihe des Kuratoriums für Kulturebauwesen*, **5**, Hamburg.

34. KUIPERS, H. (1982). Processes in physical soil degradation in mechanized agriculture. In: *Soil Degradation* (Eds D. Boels, D. B. Davies and A. E. Johnston), A. A. Balkema, Rotterdam, pp. 7–18.

35. LARSON, W. E., PIERCE, F. J. and DOWDY, R. H. (1983). The threat of soil erosion to long–term crop production, *Science*, **219** (4584), 458–65.

36. LARSON, W. E., PIERCE, F. J. and DOWDY, R. H. (1985). Loss in long-term productivity from soil erosion in the United States. In: *Soil Erosion and Conservation* (Eds S. A. El-Swaify, W. C. Moldenhauer and A. Lo), Soil Conservation Society of America, Ankey, USA, pp. 262–71.
37. LEE, J. and LOUIS, A. (1986). Land use and soil suitability. In: *Soil Map of Europe* (Eds R. Tavernier and A. Louis), E.E.C.
38. MARGAROPOULOS, P. TH. (1963). *Water Erosion and the Torrent Phenomenon,* Danigeli Press, Athens (in Greek).
39 MENGEL, K. and KIRKBY, E. A. (1982). *Principles of Plant Nutrition,* International Potash Institute, Bern, Switzerland, 655 pp.
40. MCLUHAN, T. C. (1971). *Touch The Earth. A Self-Portrait of Indian Existence,* Simon and Schuster, Rockefeller Center, New York, 185 pp.
41. MILLER, W. P. and MACFEE, W. W. (1983). Distribution of cadmium, zinc, copper and lead in soils of industrial Northwestern Indiana. *J. Environ. Quality,* **12,** 29–33.
42. MORGAN, R. P. C. (1985). Assessment of soil erosion risk in England and Wales. *Soil Use and Management,* **1**(4), 127–31.
43. NORTON, L. D. (1986). Erosion-sedimentation in a closed drainage basin in Northern Indiana. *Soil Sci. Soc. Amer. J.,* **50**(1), 209–13.
44. ONOFIOK, O. and SINGER, M. J. (1984). Scanning electron microscope studies of surface crusts formed by simulated rainfall. *Soil Sci. Soc. Amer. J.,* **48**(5), 1137–43.
45. ONKEN, A. B., WENDT, C. W., WILKE, O. C., HARGROVE, R. S., BAUSH, W. and BARNES, L. (1979). Irrigation system effects on applied fertilizer nitrogen movement in soil. *Soil Sci. Soc. Amer. Proc.,* **43,** 367–72.
46. OSTER, J. D. and SCHOVER, F. W. (1979). Infiltration as influenced by irrigation water quality. *Soil Sci. Soc. Amer. Proc.,* **43,** 444–7.
47. PLATO, *Critias,* The Loed Classical Library, William Heinemann, Harvard University Press, pp. 270–8.
48. PORTIER, J. (1972). Carte pédologique de France au 1/100.000, Toulon, Institute National de la Recherche Agronomique, Versailles, 130 pp.
49. RICHTER, G. (1980). Soil erosion mapping in Germany and in Czechoslovakia. In *Assessment of Erosion* (Eds M. De Boodt and D. Gabriels), John Wiley, Chichester, pp. 29–54.
50. ROQUERO, C. (1979). The potential productivity of Mediterranean Type Climates and their yield potentials. *Proc. 14th Colloquium of the IPI,* Sevilla, Spain, pp. 21–42.
51. RUNGE, C. F., LARSON, W. E. and ROLOFF, G. (1985). Using productivity measures to target conservation programs: a comparative analysis. *J. Soil and Water,* **41**(1), 45–9.
52. SADLER, J. M. (1984). Effects of top soil loss and intensive cropping on soil properties related to crop production potential of a Padzolic Gray Luvisol. *Can. J. Soil Sci.,* **64**(4), 533–43.
53. SAUERBECK, D. R. (1982). Influence of crop rotation, manurial treatment and soil tillage on the organic matter content of German soils. In: *Soil Degradation* (Eds D. Boels, D. B. Davies and A. E. Johnston), Balkema, Rotterdam, pp. 163–77.
54. SCHUMM, S. A. and HARVEY, M. D. (1982). Natural erosion in the USA. In:

Determinants of Soil Loss Tolerance, ASA Publication No. 45, Madison, Wisconsin, pp. 15–21.

55. SKIDMORE, E. L. (1982). Soil loss tolerance. In: *Determinants of Soil Loss Tolerance*, American Society of Agronomy, Spec. Publ. No. 45, pp. 87–93.
56. SÖCHTING, H. G. and SAUERBECK, D. R. (1982). Soil organic matter properties and turnover of plant residues as influenced by soil type climate and farming practice. In: *Soil Degradation* (Eds D. Boels, D. B. Davies and A. E. Johnston), Proc. Land Use Sem. Soil Dey. Sponsored by the C.E.C. Balkema, Rotterdam, pp. 145–62.
57. SOIL SURVEY STAFF (1975). *Soil Taxonomy*, USDA Agr. Handbook No. 436, US Gov. Printing Office, Washington, 754 pp.
58. WATSON, A. (1985). Soil erosion and vegetation damage near ski lifts at Cairn Gorm, Scotland. *Biological Conservation*, **33**(4), 363–81.
59. WILKIN, D. C. and HEBEL, S. J. (1982). Erosion, redeposition, and delivery of sediment to Midwestern streams. *Water Resources Res.*, **18**(4) 1278–82.
60. WILSON, S. J. and COOKE, R. U. (1980). Wind erosion. In: *Soil Erosion* (Eds M. J. Kirkby and R. P. C. Morgan), John Wiley, Chichester, pp. 227–51.
61. WISCHMEIER, W. H. and SMITH, D. D. (1978). Predicting rainfall erosion losses— a guide to conservation planning, USDA /SEA. Agr. Handbook 537.
62. WORTMAN, S. and CUMMINGS, R. W. JR. (1979). *To Feed This World*, John Hopkins University Press, Baltimore, 440 pp.
63. YASSOGLOU, N. J. (1969). Morphological observation on Greek soils producing early maturing fruit and vegetable crops. In: *Value to Agriculture of High-quality Water from Nuclear Desalination*, International Atomic Energy Agency, Vienna, pp. 255–63.
64. YASSOGLOU, N. J., APOSTOLAKIS C., NYCHAS, A. and KOSMAS, C. (1982). *Soil Survey of Argos Plain in Greece*, Ministry of Agriculture and Athens Faculty of Agriculture, Athens (in Greek).
65. YASSOGLOU, N. J., KOSMAS, C., ASSIMAKOPOULOS, J. and KALLIANOU, CH. Heavy metal contamination of roadside soils in the Major Athens Area (to be published.
66. ZACHAR, D. (1982). *Soil Erosion*, Elsevier, Amsterdam, 547 pp.

Sensitivity of European Soils Related to Pollutants

O. Fränzle

Geographical Institute, Kiel University,
Olshausenstr. 40, D-2300 Kiel 1, Federal Republic of Germany

SUMMARY

Sensitivity of soils to pollution is defined as velocity of sequential change in soil properties as related to the impact of pollutants. Its assessment requires a knowledge of the relevant physical and chemical transformation mechanisms in operation and their specific boundary conditions. The internal structures and manifold interrelationships of these sub-systems can be depicted in major synthetic models formulated in matrix or in graph forms, respectively. In view of the numerous feedback mechanisms defined in such comprehensive models a systematic search for the most important variables which thus assume the quality of indicator variables, is necessary. It can be accomplished by a combination of assessment strategies, involving experiments on various levels of complexity and extrapolation techniques based on the areal information of pedological and related maps. The procedure described involves both representative soils as test media and representative reference chemicals as components of test systems which reflect the properties of a major set of 'real-world' cases with measurable accuracy. Comparative leaching experiments in laboratory lysimeters and on experimental plots allow us to define the relative importance of sensitivity-related soil variables. These relationships then permit a relevant interpretation of the areal data contained in large and middle-scale maps of a region relating to its geological, geomorphological, pedological and topoclimatic situation by means of multivariate geostatistical procedures and thus lead to the intended areal assessment of soil sensitivity in relation to specific pollutants such as organics, acid rain or heavy metals.

123

1. EXPERIMENTAL ASSESSMENT OF SENSITIVE SOIL PARAMETERS

Environmental effects of chemicals can only be fully understood and influenced in a lasting manner if studied in their synergetic and systematic relations. Basically this requires a long-term ecosystem research as recommended by Ellenberg *et al.* [1] providing a comprehensive insight into structures, functions and absorptive capacities or stability and resilience of ecosystem compartments, respectively.

In the near future, however, other ecologically-oriented approaches appear necessary in order to tackle these problems in compliance with political and commercial exigencies. In order to avoid unrealistic predictions of chemical distribution this requires an appropriate consideration of the ecological characteristics of target areas in connection with the anticipated use pattern. Ecological and chemical data will become meaningful only if they can be judged in relation to the type of habitat where the compound will eventually be present. A chemical which is unacceptable in a certain application in one place is not necessarily unacceptable in another.

1.1. Ecosystem models

In order to enable realistic hazard predictions to be made, the basic chemical and toxicological data have consequently to be matched to additional data on the properties of the most important types of environments where a substance may ultimately occur. This can be accomplished by defining the fluxes or distribution of chemicals in the main environmental compartments (lower) atmosphere, soil–vegetation complex, and water by means of specific transformation operators or regulatory mechanisms, respectively. The result of such a comprehensive analysis may be summarized in synthetic ecosystem models appropriately formulated in graph form [2]. They describe the input and state variables of the atmospheric system and its linkage with the adjacent terrestrial and aquatic systems, sorption processes in soil as controlled by moisture and microbial activity, and the interactions and cascading of matter and energy in air, water, vegetation and soil in their capacity as essential ecosystem compartments.

Since a reproduction of the complete models is not possible here, in Fig. 1 a representative section is to show one of the most important transformation subsystems, i.e. TSI. This partial model is indicative of the importance of adsorption and desorption which together determine the

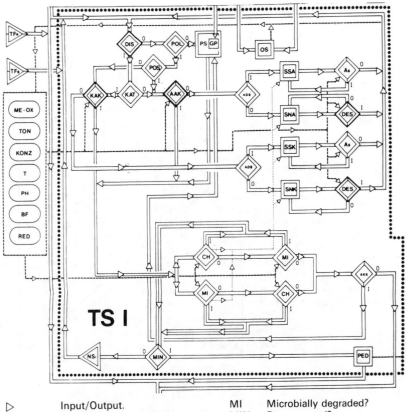

$\triangleright$	Input/Output.	MI	Microbially degraded?
$\diamondsuit$	Regulator .	MIN	Decomposed?
$\square$	Storage element.	NS_I	Newly formed toxic
$\bigcirc$	Phys. and chem. boundary		substances.
	conditions.	OS	Organic matter.
$\diamondsuit\!\!\!\diamondsuit$	OECD Test guideline available?	PED	Pedon.
AAK	Anion exchange capacity	PH	pH value.
	exceeded?	POL	Polarized?
ADS	Adsorbed specifically?	POS	Sorption to positive charges?
AKK	Accumulated?	PS	Porous storage.
A_S	Exchange colloids persistent?	RED	Redox reactions.
BF	Soil humidity.	SNA	Non-specific adsorption in
CH	Abiotically degraded?		anionic form.
DES	Desorbed?	SNK	Non-specific adsorption in
DIS	Dissociated?		cationic form.
GP	Major pores filled with air.	SSA	Specific adsorption in anionic
KAK	Cation exchange capacity		form.
	exceeded?	SSK	Specific adsorption in
KAT	Adsorbable due to cat.		cationic form.
	reactions?	T	Temperature.
KONZ	Concentration of matrix	TF_B	Throughflow loaded.
	solution.	TF_R	Throughflow, unloaded.
ME-OX	Metallic oxides and	TON	Clay mineral content
	hydroxides.		and composition.

FIG. 1. Transformation operator I (TSI).

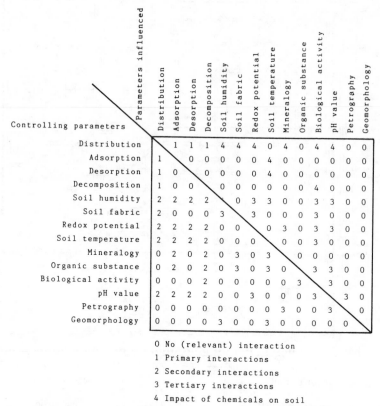

0 No (relevant) interaction
1 Primary interactions
2 Secondary interactions
3 Tertiary interactions
4 Impact of chemicals on soil

FIG. 2. Interrelationships between soil properties and chemical behaviour.

so-called 'buffering capacity' of a soil. Soil constituents with high specific surface and net charge, i.e. primarily organic matter, then clay minerals and metal oxides and hydroxides, largely determine adsorption and desorption. The relevant boundary conditions are concentration and dissociation or polarity of the chemical on the one hand, soil moisture, temperature, pH value and oxidation and reduction potentials on the other. The connectivity matrix in Fig. 2 which is mathematically equivalent to a graph, illustrates the essential interrelationships in a concise manner.

1.2. Test systems

In view of the numerous feedback mechanisms operative among the variables defined in the above models a systematic search for determining the most important ones and establishing a list of relative priority becomes imperative. Appropriate approaches to this end follow the hypothesis that the highest priority chemicals are those with the least margin of safety, defined as the gap between the non-observable effect concentration and the ambient exposure concentration. Since the latter is largely controlled by adsorption/desorption processes in soils the respective pedochemical and physical qualities of soils together with their microbial activity must be determined with a sufficient amount of accuracy.

This, however, implies that soils and chemicals whose interactions are to be tested, are representative in the sense that their relevant physical and chemical properties reflect those of a major set of cases reliably. In this connection the term 'representative' (i) means reproducing faithfully the properties of various phenomena in the light of characteristic frequency distributions, and (ii) relates to specific spatial patterns. In fact soil variability is such that only a careful and systematic primary study of the particular distribution functions or associations, respectively, can ascertain that a sample taken is also representative from the regional point of view.

1.2.1. Selection of representative soils

Soils, like many other things, are not discrete, independent and unambiguously identifiable objects or entities; the habitual and well-known statistical procedures cannot be applied to them as a consequence. The particular problems relating to areal data such as mapping units 'concern (i) the arbitrariness involved in defining a geographical individual, (ii) the effects of variation in size and shape of the individual areal units, (iii) the nature and measurement of location' [3]. Difficulties encountered in separating individual areal units from a continuum are most frequently, and at least partially, overcome by the selection of grid squares as the basic units, geographical characteristics being averaged out for each grid square. Since grid squares are all of the same shape and size their use eliminates variability in these properties and thus solves the second problem. The most common solution of the third problem, which is peculiar to geography, is to make relative location as measured by spatial contiguity the dominant variable of analysis. It can be accomplished by means of special diversity analyses or regionalization procedures which are based on comprehensive geographical data matrices.

Their elements are derived from the digital evaluation of soil maps. In the present instance the digitalization of the 1:2·5 mio. FAO Soil Map of Europe [4] furnished some 26 600 binary data for the territory of EC Member States. The analogous evaluation of the 1:1 mio. Soil Map of the Federal Republic of Germany [5] provided some 2500 nominal soil data, supplemented by about 10 000 data relating to geology, physical soil properties and land use characteristics.

The first step in selecting representative test soils is the determination of their acreages, i.e. a simple frequency analysis in terms of descriptive statistics. The next step is to define the characteristic spatial patterns of soil distribution by means of neighbourhood analysis. The methodology (which is described in mathematical detail elsewhere) basically consists in determining the individual nearest-neighbourhood relationships of each grid point, i.e. the positive or negative spatial autocorrelation which is a distance-weighted measure for each point in relation to 80 neighbours. The resulting data matrix allow definition of average association frequencies of all the soil units of the original maps, which, in turn, is the basis for a comparison of each individual grid point as defined by positive or negative autocorrelation with these average frequencies. The vectorial distance of each grid point from the corresponding soil average is a measure of similarity or representation. In terms of spatial structure it ensues that those soil units are most representative which differ least in their neighbourhood relationships from the average association pattern of the respective soil type.

The sampling sites of these representative soil units are more precisely determined by subsequent application of the above analytical techniques to large-scale maps, the results of which are eventually corroborated by visual inspection in the field.

Within the EC Member States the most common soils (in the sense of mapping units of the FAO Soil Map) are the Grey Brown Podzolic Soils, covering 12.8% of the whole area with a distinct maximum in the northern parts. The regionally representative soil unit is located near Caen (Normandie, France). The Brown Mediterranean Soils are typical representatives of the soil cover in southern regions, reflecting 6.8% of the European Community. The corresponding typical locality is on Sicily. In the light of frequency statistics the Acid Brown Forest Soils cover nearly the same area (6.5%) as the Brown Mediterranean Soils, but they differ in the physical–chemical properties as well as in their specific distribution pattern. While the Brown Mediterranean Soils are concentrated mostly in coastal areas of the EC's southern parts, the Acid Brown Forest Soils are typical for higher latitudes and altitudes. The typical locality is near Cardiff (Great Britain). Brown Forest Soils and Rendzinas—distributed over 6.1% of the area—reflect the group of moderately developed soils on calcareous material with the representative sampling site in Greece near Korinthos. Podzols have an acreage of 4.2% of the EC territories, occurring mostly in the northern parts. The sampling site for the ideal Podzol appears to be located in the Federal Republic of Germany near Lauenburg (Schleswig-Holstein).

The five soil groups have an explicit chorological representation of more than one third (36.4%) of the whole EC soil inventory and reflect the wide variability of parameters responsible for the sorption behaviour of chemicals in soils. In the light of pedological considerations related to the legend of the FAO Soil Map and the results of spatial statistics an implicit chorological representation of 62.0% is given, because the legend indicates similarities and/or transitions between mapping units.

An analogous analysis of German soils [6] proved the following soils, summarized in Table 1, to be representative. In terms of acreage these soils represent more than 80% of the total German soil inventory with a correspondingly wide span of pedophysical and chemical properties.

Table 1

Locations of regionally representative soils in the Federal Republic of Germany, arranged in order of decreasing acreages

Soil type	Geographical coordinates	
Albic Luvisol	13°10′ E,	48°23′ N
Dystric Cambisol	9°10′ E,	50°19′ N
Orthic Podzol	8°13′ E,	51°54′ N
Rendzina	12°00′ E,	49°07′ N
Dystric Histosol	8°09′ E,	53°07′ N
Eutric Fluvisol	9°04′ E,	54°24′ N

On the regional level, finally, i.e. related to the Federal Land Schleswig-Holstein, a combination of following soils exhibits a maximum spatial representativity: Gleyic Luvisol, Ferric-humic Podzol, Dystric Cambisol, Histic Gleysol. They represent about 80% of the Schleswig-Holsatian soil cover in terms of acreage, and no less than 70% of the German soil inventory.

It follows from the above selection methodology that ecologically oriented experiments with these regionally representative soil specimens are likely to provide data which are best suited for spatially valid extrapolations. This, however, implies a correspondingly careful selection of test chemicals.

1.2.2. Selection of representative test chemicals

Given the large number (about 100 000) and variety of chemicals produced in commercial quantities a rational selection of test chemicals involves a classification of existing and new chemicals in terms of environmental properties. These relevant physical and chemical properties which can only in minor part be deduced from chemical structure by means of structure–activity relationships [7] comprise: density, vapour pressure, molar mass, melting and boiling points, water solubility, n-octanol/water partition coefficient, dissociation constant, polarity, particle size, tendency to complexation, viscosity, volatility.

These m variables define the environmental behaviour of the above n chemicals with respect to soils in the form of a $n \times m$ matrix, or a geometrically equivalent point distribution in a m-dimensional space, respectively. This is due to the fact that any matrix with $n \times m$ elements may be represented by a vector for a row and another vector for each column such that the elements of the matrix are the inner products of the vectors. When the matrix is of rank 2 or 3, or can be closely approximated by a matrix of such rank, the vectors may be plotted and the resulting matrix representation, i.e. the so-called biplot [8, 9] inspected visually, which is of considerable practical interest for the analysis of large matrices.

A biplot can be made unique by introducing a particular metric for either row or column comparisons. To approximate the original $n \times m$ matrix of rank r by a matrix of rank 2 or 3 the singular value decomposition is used. This approximate biplot not only permits viewing of the individual data and their differences, but further allows scanning of the standardized differences between units and to inspect the variances, covariances and correlations of the variables. In the present context the biplot is a most useful graphical aid in interpreting the multivariate matrix of chemicals as defined by the above characteristic physical and chemical properties, and to test the mathematical validity of subsequent numerical classification strategies. In the form of cluster algorithms they aim at defining homogeneous classes of chemicals in the precise operational sense of the term.

The exemplary analysis of 34 organic compounds taken from the OECD 'Minimum Pre-Marketing Set of Data' as listed in Table 2 therefore started with a three-dimensional biplot analysis, then used the WARD algorithm for clustering purposes, and eventually optimized the result by means of the iterative procedure RELOCATE [10].

The result of the complex classification strategy is summarized in Fig. 3. A graphic display of the classes defined can be given by means of their T-

Table 2

List of substances grouped by means of biplot analysis and multivariate
aggregation algorithms

1	Benzene	25	γ-Hexachlorocyclohexane
2	Toluene	27	n-Octane
3	Phenol	28	Styrene oxide
4	4-Nitrophenol	30	Thiourea
5	PCP	31	2,4,6-Trichlorophenol
6	Aniline	32	Tris-(2,3-dibromopropyl)-phosphate
7	4-Chloroaniline	33	DDT
8	2,4-D	36	DOP
11	Methanol	39	Ethene
13	Trichloroethene	43	N-Ethylaniline
14	1,2,4-Trichlorobenzene	44	2-Chloroaniline
15	Hexachlorobenzene	45	3-Chloroaniline
16	Fluoroanthene	47	N,N-Dimethylaniline
17	2-Nitrophenol	49	1,2,3-Trichlorobenzene
18	Perylene	51	2-Nitroanisole
22	1,1-Dichloroethene	52	Nitrobenzene
23	Dichlorodifluoromethane	53	1,4-Dichlorobenzene

Variable:		
1	Molar mass	—
2	Melting point	—
3	Boiling point	—
4	Density	—
5	Vapour pressure	log
6	Water solubility	log
7	Partition coefficient P_{ow}	log
8	K_{oc}	log

ratios, i.e. the standardized deviations of the defining variables of each class
from the global mean values. They illustrate that homogeneous classes of
chemicals were formed, which means that each member of a class is class-
specifically representative in the operational sense of the term, and in terms
of the above environmentally relevant parameters.

In conclusion it may be said that only the systematic combination of
regionally representative soil samples with class-specific chemicals is likely
to yield realistic interaction or impact data. Yet considerable effort has
been, and partly continues to be devoted to trying to produce some simple,
globally-acceptable standard tests. It is hoped this will produce accurate
and useful information about the sorption of chemicals in spite of the
increasing recognition of the low value of the information obtained from
such tests outside of very special fields. There is ample evidence in the

O. Fränzle

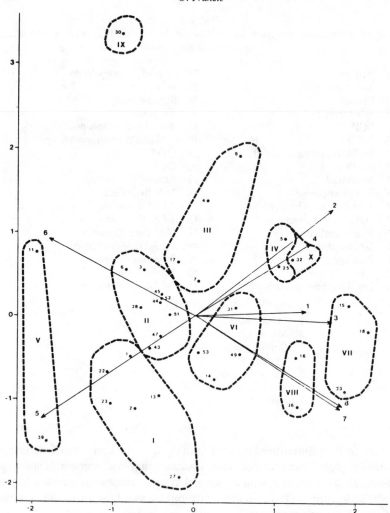

FIG. 3.   X-Y biplot of 34 chemicals of the OECD 'Minimum Pre-Marketing Set
of Data" (Arrows: indicator variables as in Table 2).

literature, however, and also in related fields of research such as
ecotoxicology there are quite a few authoritative voices on both the
national and international level to support this, that premature 'rigidly
formulated regulations specifying in detail test procedures may be
inadequate and misleading' [11].

1.3. Soil leaching experiments

Based on the selection methodology described, exemplary leaching experiments with representative soils and chemicals are appropriate tools to assess interactions and to more precisely define sensitivities. To this end a four-level approach may be used starting with simple stirring experiments and small-scale laboratory lysimeter experiments. Since these basic experiments involve an excess of solution or, in the second case, small saturated soil columns, they are simplifications or idealizations from the realworld situation. Therefore they must be complemented by larger-scale monolithic lysimeter leaching experiments in both laboratory and field, and finally by still larger-scale field experiments on selected sites comprising various polypedons.

1.3.1. The test systems 2,4-D-podzol-luvisol

The present chapter is put forward as a methodological one. It is not the purpose to compare in detail the variety of field situations with the derived theoretical relations. Rather, it is hoped merely to provide an exemplary basis for some broader generalizations about the physical and chemical principles controlling soil sensitivity in the field. Therefore the presentation is deliberately limited to the analysis of 2,4-D impact on the above-mentioned Holsatian Gleyic Luvisol and the Ferric-humic Podzol developed from Weichselian till or Weichselian cover sand on Saalian moraine, respectively.

The relevant physical and chemical properties of these experimental soils are summarized in Table 3.

The test substance used for the comprehensive leaching experiments was 2,4-dichlorophenoxyethane acid. 2,4-D with a molar mass of 221·04 and an (estimated) world production of 100 000 t/a is used as a herbicide; its partition coefficient n-octanol/water has a log P_{ow} value of 1·57, the dissociation constant pK_K being 2·73 at 298 K. The analytical detection of 2,4-D in both soil matrix and solution was made by gas chromatography, using a Perkin-Elmer SIGMA 2 B (technical details in Fränzle *et al.* [12]); the preparation of samples followed Horner *et al.* [13].

The evaluation of the numerous data obtained from three years' comparative experiments on the four levels of differentiation described above enable us to define the relative importance of the variables involved in sorption or buffering processes by means of correlation and regression analyses. While field and large-scale lysimeter experiments allow definition of the influence of the locally varying water balance on migration and transformation of chemicals on and in soils, small-scale lysimeters and in

Table 3

Granulometry and chemical characteristics of a representative Holsatian Gleyic Luvisol (Boksee) and a Podzol (Bucken)

(a) Plot: Boksee

Horizon	Depth	Granulometry (%) at (μm):							fs: mS	Org. matter (%)	pH-value (0·1 N KCl)	CEC (mval/100 g)	% Base saturation	mval/100 g				
		<2	2–6	6–20	20–63	63–200	200–630	630–2000						Ca	Mg	K	Na	H^+Alg
A_h	0–6	10·5	4·7	8·1	16·6	34·1	20·8	5·2	1·64	9·0	3·3	12·9	11	0·9	0·3	0·1	0·1	11·6
A_hB_v	–14	10·2	4·2	10·0	15·1	34·6	21·7	4·2	1·59	4·2	3·4	8·3	6	0·3	0·1	tr.	0·1	7·8
E_tB_v	–33	8·7	4·8	11·6	12·4	35·7	21·9	4·9	1·63	1·8	3·7	5·7	7	0·2	0·1	—	0·1	5·3
SE_t	–60	12·3	4·2	9·6	16·7	33·6	19·2	4·4	1·75	0·6	3·6	5·7	7	0·2	0·1	tr.	0·1	5·3
$(S)B_t$	–110	21·0	5·3	10·9	14·8	29·6	14·6	3·8	2·03	0·4	3·2	9·1	20	1·0	0·6	0·1	0·1	8·2
B_{vt}	>110	19·8	4·9	11·5	14·7	29·7	15·6	3·8	1·90	0·2	3·4	8·2	66	3·2	1·9	0·2	0·1	2·8

Horizon	Depth	Exchangeable cations (%)					Fe_o (‰)	Fe_d (‰)	$\dfrac{Fe_o}{Fe_d}$	Mn_d (‰)	Si_i (‰)	Al_o (‰)	Al_i (‰)	$\dfrac{Fe_d}{Clay} \times 100$	$\dfrac{Mn_d}{Clay} \times 100$	$\dfrac{Si_i}{Clay} \times 10$	$\dfrac{Al_i}{Clay} \times 10$
		Ca	Mg	K	Na	H^+Alg											
A_h	0–6	7	2	1	1	90	2·02	3·46	0·58	0·05	5·68	0·78	1·69	3·30	0·48	5·41	1·61
A_hB_v	–14	4	1	tr.	1	94	2·16	3·48	0·62	0·06	4·18	0·77	1·70	3·41	0·59	4·10	1·67
E_tB_v	–33	4	2	—	2	92	1·93	3·39	0·57	0·11	3·69	1·04	2·77	3·89	1·26	4·24	3·18
SE_t	–60	4	2	—	2	92	1·19	3·49	0·34	0·13	5·58	0·87	3·98	2·84	1·06	4·54	3·24
$(S)B_t$	–110	11	7	1	1	80	1·16	4·85	0·24	0·08	6·46	0·84	3·98	2·30	0·38	3·08	1·90
B_{vt}	>110	39	23	2	1	35	0·80	4·58	0·17	0·10	5·53	0·42	2·36	2·31	0·50	2·79	1·19

(b) Plot: Bucken

Horizon	Depth	Granulometry (%) at (μm):							fs: mS	Org. matter (%)	pH-value (0·1 N KCl)	CEC (mval/100 g)	% Base saturation	mval/100 g				
		<2	2-6	6-20	20-63	63-200	200-630	630-2000						Ca	Mg	K	Na	H+Alg
$A_h E$	0-6	—	—	2·5	2·3	38·5	50·9	5·8	0·76	6·6	3·2	4·0	20	0·5	0·1	0·1	0·1	3·2
$E_{h,Fe}$	-12	0·4	0·3	1·4	3·8	41·1	48·4	4·6	0·85	1·8	3·4	2·9	17	0·3	0·1	tr.	0·1	2·4
$B_{h,Fe}$	-17	2·4	—	3·2	—	39·1	48·1	7·2	0·81	15·4	3·3	17·2	17	0·7	0·1	0·1	0·1	16·2
$B_{Fe,Al}$	-30	—	2·1	1·3	4·9	43·3	43·6	4·8	0·99	1·8	4·6	6·8	7	0·3	0·1	0	0·1	6·3
B_{Fe}	-47	—	—	1·4	1·3	33·1	55·5	8·7	0·60	0·4	4·8	1·8	17	0·2	tr.	0	0·1	1·5
BC	>47	—	1·3	1·1	1·6	44·7	46·2	5·1	0·97	0·2	4·7	1·8	17	0·2	tr.	tr.	0·1	1·5

Horizon	Depth	Exchangeable cations (%)					Fe_o (‰)	Fe_d (‰)	$\dfrac{Fe_o}{Fe_d}$	Mn_d (‰)	Si_1 (‰)	Al_o (‰)	Al_1 (‰)
		Ca	Mg	K	Na	H+Alg							
$A_h E$	0-6	13	3	3	3	78	0·10	0·18	0·55	tr.	1·38	0·17	0·33
$E_{h,Fe}$	-12	10	3	tr.	3	84	0·06	0·15	0·41	tr.	1·34	0·09	0·34
$B_{h,Fe}$	-17	4	1	1	1	93	4·64	4·92	0·95	0·04	1·20	3·49	4·42
$B_{Fe,Al}$	-30	3	1	0	1	95	2·82	3·89	0·72	0·04	3·19	5·58	7·38
B_{Fe}	-47	11	tr.	0	6	83	0·29	1·84	0·16	0·01	2·02	1·22	2·69
BC	>47	11	tr.	tr.	6	83	0·08	1·07	0·07	0·01	2·18	0·67	2·04

particular stirring experiments are indicated to rapidly obtain large sets of comparable data on relevant soil properties.

Therefore the following presentation is confined to the determination of soil properties which proved to be of prime importance for 2,4-D sorption. The adsorption isotherms show that for minor concentrations (i.e. 0·1, 0·2, 0·5, 1·0, 2·0 g 2,4-D/litre) the Freundlich theory yields an appropriate description of the relevant processes, while in the following range of concentrations (3–4 g 2,4-D/litre) the interpretation is rendered difficult by irregular fluctuations and feedback phenomena. Consequently the systematic search for the most sensitive soil properties involves a two-step exclusion procedure, the first of which is a rank correlation of every soil parameter of Table 3 with all the sorption data measured. In the next step multiple linear regressions are computed with those parameters which had the highest Spearman correlation coefficients on the 95% level of significance. The principal results are summarized in Table 4.

It follows from Table 4 that granulometry (as an indicator variable of specific surface), organic matter, and (frequently more important) the lye-, oxalate- and dithionite-soluble Fe, Mn, Al and Si-fractions (which are indicative of pedogenic clay minerals and oxides) and potential acidity (H^+Alg) account for more than 90% of the observed variance of sorption rates. This involves a strong pH control of the non-specific charge of the sorption-relevant soil colloids [14]. If, for instance, due to the acid character of a dissociated compound applied or because of another type of acidification, the pH of the soil solution falls below the point of zero charge (IEP) a positive net charge of the sorption complex results, which favours the reversible fixation of anions while cation sorption is proportionally diminished. In a more general context these and related reactions will be considered with regard to soil sensitivity in Section 2, while a more specific interpretation of geo-scientific maps in the light of the above findings will be given in the following sub-section.

1.3.2. Combined geostatistical interpretation of geological, pedological and topographic maps

The relationships summarized in Table 4 permit, to a certain extent, a sorption- and buffering-oriented interpretation of the areal data contained in large and middle scale maps of a region relating to the geological, geomorphological, pedological and topographic situation. The 1:25 000 geological, petrographic, pedological and topographic maps of a Saalian moraine complex with Weichselian cover sand SW of Kiel (Schleswig-Holstein) may serve as an example.

Table 4

Multiple regressions of sorption rates as a function of 2,4-D concentration and soil properties: A. Gleyic Luvisol (Boksee); B. Humic-ferric Podzol (Bucken)

A

$V\ 49 = -0.083 + 0.1\ Al_0 + 0.01\ Si_1$ (Mult R = 0.95)

$V\ 50 = 0.13 - 0.01\ OS - 0.01\ Al_1$ (Mult R = 0.80)

$V\ 51 = 0.33 - 0.06\ Si_1 + 0.18\ Al_0$ (Mult R = 0.74)

$V\ 52 = 0.24 + 0.05\ Al_1 - 0.03\ Fe_d$ (Mult R = 0.82)

$V\ 53 = 0.54 - 0.009\ (H^+Alg) + 0.9\ Al_0$ (Mult R = 0.88)

$V\ 54 = 2.27 - 2.06\ Al_0 + 0.013\ (H^+Alg)$ (Mult R = 0.71)

B

$V\ 49 = 0.01 - 0.008\ (<2\,\mu m) + 0.003\ (<63\,\mu m)$ (Mult R = 0.98)

$V\ 50 = 0.06 + 0.03\ (<2\,\mu m) - 0.006\ (<63\,\mu m)$ (Mult R = 0.98)

$V\ 51 = 0.13 + 0.13\ (<2\,\mu m) - 0.02\ OS$ (Mult R = 0.98)

$V\ 52 = 0.01 + 0.04\ (<63\,\mu m) + 0.04\ (<2\,\mu m)$ (Mult R = 0.71)

$V\ 53 = 0.14 + 0.06\ Fe_d + 0.008\ (<63\,\mu m)$ (Mult R = 0.98)

$V\ 54 = 1.54 + 0.4\ (<2\,\mu m) - 0.1\ Fe_d$ (Mult R = 0.74)

V 49	Amount of 2,4-D adsorbed at concentration 0.1 g/litre
V 50	Amount of 2,4-D adsorbed at concentration 0.2 g/litre
V 51	Amount of 2,4-D adsorbed at concentraiton 0.5 g/litre
V 52	Amount of 2,4-D adsorbed at concentration 1.0 g/litre
V 53	Amount of 2,4-D adsorbed at concentraiton 2.0 g/litre
V 54	Amount of 2,4-D adsorbed at concentraiton 4.0 g/litre

Al_1	Al in soil extract
Al_0	Oxalate-soluble Al
Fe_d	Dithionite-soluble Fe
Fe_0	Oxalate-soluble Fe
(H^+Alg)	Sum of exchangeable cations in %
Mn_d	Dithionite soluble Mn
OS	Organic matter in %
Si_1	Si in soil extract

Mult R: Multiple regression coefficient.

By means of digitalization (3249 grid points) the relevant cartographic information at each point and its immediate neighbourhood relating to sedimentology (e.g. till, coversand, peat), substrate (sand, silt, clay, etc.), soils (category label: soil group), relief (i.e. exposition and slope angle classes), and land use is transformed into nominal data. After an appropriate reduction of the prohibitive number of primary data thus obtained by using representative neighbourhood relationships, cluster analysis in the form of entropy analysis [15, 16] is indicated as a method of evaluation. The latter is based on the information theoretic concept of

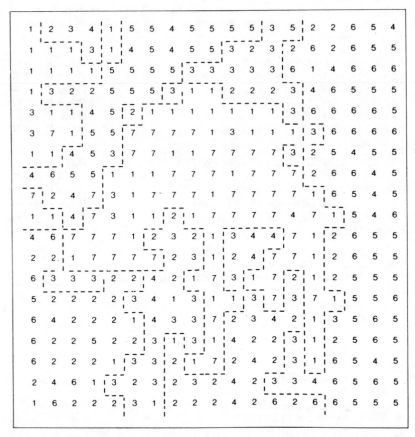

FIG. 4.   Computer map of 7 geopedological units in the Hennstedt area (Schleswig-Holstein) resulting from entropy analysis. 1, orthic and Humic Podzols from sandy till on marked to steep slopes facing W and SE; arable land and broad-leafed forest; 2, orthic and Humic Podzols from sandy till and fluvioglacial deposits on moderate slopes facing NE to N and on hilltops; arable land; 3, dystric Gleysols and Podzols (pp. Luvisols) from clayey to sandy till on moderate to steep slopes facing NW and SW; arable land and grassland; 4, histosols in valley bottoms; grassland and arable land; 5, humic Podzols and Histosols on level fluvioglacial sands; grassland and arable land; 6, humic (pp. Gleyic) Podzols from fluvioglacial sands in valley bottoms and on subordinate slopes; arable land and grassland; 7, orthic Podzols from sandy till on hilltops, on steep slopes and in valleys of highly variable exposition; arable land and broad-leafed forest.

entropy [17] and defines measures of association between nominal or binary variables such that the information loss suffered in conversion to binary data is somewhat compensated by additional advantages. The aggregation of the binary data is accomplished by a centroid sorting procedure [15]. This clustering method employs heuristic devices for adjusting the number of clusters to conform to the apparent natural structure of the data set [18].

Three of the most prominent motivations for allowing the number of clusters to vary are (i) entropy increase, (ii) spatial diversity of the area depicted, and (iii) specific interactions of a selected chemical with the environment. In the present framework sorting the data set into 7 clusters (i.e. geopedological units) appears optimal, 10 introducing an element of useless complexity on the one hand, and 5 yielding poorly defined clusters on the other, which would not permit a sufficiently precise chemical hazard assessment [19].

The result of entropy analysis is summarized in Fig. 4. The broken lines indicate associations of geopedological units, in order to make the spatial structure more easily discernible, while the legend explains the units in terms of environmental criteria according to Figs. 1 and 2.

1.3.3. Assessment of buffering capacities and sensitivities

Geopedological units of the above type are homogeneous entities in the precise operational sense of cluster algorithms or entropy analysis, respectively. Hence each unit may be expected to display specific reactions to defined chemical impacts. In addition to Fig. 2, showing the essential relationships between physico-chemical characteristics of a compound and its distribution properties, Table 5 summarizes in an analogous way the relevant systemary relations between pedological parameters of a potentially affected soil and the environmental chemical characteristics of a compound released.

The information contained in Table 4 and Fig. 4 forms the basis of the concluding exemplary assessment of the different buffering capacities of the above geopedological units in regard to the agrochemical 2,4-D. In Table 6 the possible interactions of each unit are labelled positive ($+$), negative ($-$) or indifferent (0) in order to indicate their relative potential to reduce or enhance the chemical concentration in the soil solution. The sum of these positive and negative effects is deemed indicative of the buffering capacity or resilience of each unit which is thus defined on the relative level of an ordinal scale. In a few critical cases additional information was drawn from the geomorphological situation.

Table 5
Relationships between pedological parameters and
distribution characteristics

	V	S	D[a]
Fabric	×	0	0
Mineralogical composition	0	×	×
Temperature	×	×	×
Aeration/redox potential	×	×	×
Microbial activity	0	0	×
Organic matter	0	×	×
Soil moisture	×	×	×
pH-value	×	×	×

× = yes (positive or negative); 0 = no.
[a] V = distribution; S = sorption; D = decomposition.

The buffering capacities of the 7 geopedological units with regard to 2,4-D, decrease as in the following sequence:

$$1 > 2 > 7 > 3 > 6 > 5 > 4$$

Assuming that an evaluation of the buffering capacities of the same geopedological units in regard to groundwater pollution is intended, an inversion of the V signs in rows 'fabric' and 'soil moisture' would be necessary. The sequence would then be:

$$3 > 1 > 2 > 7 > 6 > 5 > 4$$

A transformation of these ordinal sequences to the level of metric scales (i.e. kg/ha) appears possible by means of adsorption isotherms of the Freundlich or Langmuir types. Their determination must be based on a comprehensive set of samples such that the essential requirements of a spatially representative distribution are met [20].

2. SOIL SENSITIVITY RELATED TO ACIDIFICATION

Soil acidification is a natural process in areas with a humid climate, and many of the compounds deposited from the polluted atmosphere occur naturally in soils. The problem of assessing their effects is therefore one of comparison of the amount of acid deposited and the amount produced in soil in relation to pedogenic sinks and buffering systems.

Table 6

Relationships between pedological parameters of 7 geopedological units in the Hennstedt area (Schleswig-Holstein) and physical-chemical properties of 2,4-D

Geopedological units	1			2			3			4			5			6			7		
	V	S	D[a]	V	S	D	V	S	D	V	S	D	V	S	D	V	S	D	V	S	D
Fabric	+[b]												+			+					
Mineralogical composition		+	−		+	−		+	+		−	−		+	−		+	−		+	−
Temperature	0	+	+		+	+		+	−		−	−	0	−	−	0	−	−	0	−	+
Aeration/redox potential	0	+	+		+	+		+	+		−	−	0	−	−	0	+	−	0	+	+
Microbial activity			−			−			−			−			−						−
Organic matter	+	+	−	+	+	−	+	+	−		+	−		+	−		+	−		+	−
Soil moisture	+	−	−		−	−		−	−		+	−	+	−	−	+	−	−	+	−	−
pH-value	+	+	−	+	+	−	+	+	−	0	+	−	+	+	−	+	+	−	+	+	+
Geomorphological situation	+			−			−						−			+					
Total positive	10(11)			10			9			5			6			8(9)			9		
Total 0	2			2			2			3			2			2			2		
Total negative	6			6(7)			7			11			10(11)			8			7		
Relative buffering capacity	7			6			4			1			2			3			5		

[a] V = distribution; S = sorption; D = decomposition.

[b] − = negative; + = positive; 0 = indifferent.

Natural H-ion production, best illustrated by its effects on fossil interglacial soils with very low pH value [21, 22], occurs in different parts and due to several processes in an ecosystem. It may enter the system by precipitation in the form of carbonic acid, and also pedogenic CO_2 production by soil organisms and plant roots is of major importance. H ions are furthermore produced when N and S compounds are mineralized. The main part of H-ion production in the soil can be estimated by the accumulation of soil organic matter and by the accumulation of excess cations in the biomass which is equated by a corresponding release of H ions in the soil [23]. Consequently the natural H-ion production in forest ecosystems varies considerably with soil properties and productivity, tree species, climate, forest management, etc.

Some of the more important soil properties to be considered in the present context are: fabric, amount and composition of organic matter, mineral composition, cation and anion exchange capacities, base saturation and soil depth [2]. Soil acidity increases when H ions exchange with base cations adsorbed to soil particles. Thus weathering reactions are the main sink of H ions in soils. The intensity and efficiency of these exchange processes is dependent on the base saturation, soil pH and the relative proportions of permanent and (pH-controlled) variable charges in the soil. These pedochemical properties have to be matched to the pattern of infiltration and flow through the soil. This, in turn, is governed by soil structure, pore size spectrum and the quantity of water flowing through the soil. In a concise way the interrelationships and feedback mechanisms operative can be depicted in a model graph only [2].

Thus general knowledge of soil chemistry supplemented by acidification experiments [24–28] and systematic comparisons of soil chemistry over time [29–34] can infer that acidification of soil and water depends on the balance between the sum of internal H-ion production and atmospheric deposition, and the consumption of H ions in weathering processes. Consequently the following set of buffering systems appears to play an essential role in regulatory mechanisms:

(i)   carbonic acid/carbonates (8.6 > pH > 6.2);
(ii)   carbonic acid/silicates (6.2 > pH > 5.0);
(iii) mineral acids/clay (5.0 > pH > 4.2);
(iv)   mineral acids/Al, Fe oxides (pH ∼ 3.0).

In terms of sensitivity to acidification this means, other things being equal, that pH decrease is likely to be largest for a soil horizon of but marginal buffering capacity. It should be mentioned, however, that much of the sesquioxide aluminium is associated with organic matter, and much of the aluminium in soil solutions and seeps appears to be complexed to organic solutes [35]. On the other hand, when no significant increase in exchangeable acidity is found, this is equivalent to a very marked decrease in the contents of exchangeable Ca, Mg and K [30, 33].

It follows from the relationships enumerated that the effects of acids on soil acid/base properties vary enormously with climatical conditions and soil properties. This is reflected on soil maps. In Scandinavia and northern USSR the Orthic and Gelic Podzols dominate; the natural vegetation is characterized by conifers and birch. At the other extreme, in the Mediterranean climatic belt with severe summer drought, the soils are mainly Calcic and Eutric Cambisols, Chromic Luvisols and Xerosols. Between these extremes many other soils are represented, ranging from Calcaric Phaeozems, Calcic Chernozems and Gleyic Luvisols rich in bases, to thin Lithosols and extremely acid Cambisols or Podzols. The regionalized analysis of their buffering systems in relation to element mobility, chemical transport, soil microfauna and flora are challenging tasks for the future.

ACKNOWLEDGEMENTS

I am particularly indebted to the Federal Environmental Agency (Umweltbundesamt)/Bundesministerium des Innern and Bundesministerium für Umwelt, Naturschutz und Reaktorsicherheit of the Federal Republic of Germany for valuable grants which permitted the implementation of the comprehensive research schemes upon which the present chapter is based. Mrs. Weller prepared the manuscript.

REFERENCES

1. ELLENBERG, H., FRÄNZLE, O. and MÜLLER, P. (1978. Ökosystemforschung im Hinblick auf Umweltpolitik und Etnwicklungsplanung. Umweltforschungsplan des Bundesministers des Innern—Ökologie—Forschungsbericht 78–101 04 005, pp. 1–144, Bonn.
2. FRÄNZLE, O. (1982). Erfassung von Ökosystemparametern zur Vorhersage der Verteilung von neuen Chemikalien in der Umwelt. Schriftenreihe 'Texte' des Umweltbundesamtes Berlin, pp. 1–347, Berlin.
3. MATHER, P. M. (1972). Areal classification in geomorphology. In: *Spatial Analysis in Geomorphology* (Ed. R. J. Chorley), London, pp. 305–22.
4. FAO. (1963). Soil Map of Europe, Bruxelles.

5. HOLLSTEIN, W. (1963). Bodenkarte der Bundesrepublik Deutschland, Maßstab 1:1 Mio., Bundesanstalt für Bodenforschung, Hannover.

6. FRÄNZLE, O. and KUHNT, G. (1983). Regional repräsentative Auswahl der Böden für eine Umweltprobenbank–Exemplarische Untersuchung am Beispiel der Bundesrepublik Deutschland. Forschungsbericht 106 05 028 im Umweltforschungsplan des Bundesministers des Innern/Umweltbundesamtes, pp. 1–72, Berlin.

7. MILL, T. (1981). Minimum data needed to estimate environmental fate and effects for hazard classification of synthetic chemicals. Proceedings of the Workshop on the Control of Existing Chemicals Under the Patronage of the Organisation for Economic Cooperation and Development, pp. 207–227, Berlin.

8. GABRIEL, K. R. (1971). The biplot graphic display of matrices with application to principal component analysis. *Biometrika*, **58**, 453–67.

9. FRÄNZLE, O., KILLISCH, W., INGENPASS, A. and MICH, N. (1980). Die Klassifizierung von Bodenprofilen als Grundlage agrarer Standortplanung in Entwicklungsländern. Ein Beispiel aus dem Savannengebiet Nordostghanas. *Catena*, **7**, 353–81.

10. WISHART, D. (1984). *Clustan 1 C User Manual*, London.

11. BROWN, V. M. (1978). Assessment of the toxicity of chemicals to fishes. Umweltbundesamt Report ¡10/78, pp. 75–86, Berlin.

12. FRÄNZLE, O., KUHNT, D. and KUHNT, G. (1985). Die ökosystemare Erfassung von Bodenparametern zur Vorhersage der potentiellen Schadwirkung von Umweltchemikalien. *Verh.d.Ges.f. Ökologie*, **XIII**, 323–40.

13. HORNER, J., QUE HEE, S. S. and SUTHERLAND, R. G. (1974). Esterification of 2,4-dichlorphenoxyacetic acid—a quantitative comparison of esterification techniques. *Anal. Chem.*, **46**, 110–12.

14. FRÄNZLE, O. (1982). Modellversuche über die Passage von Umweltchemikalien und ihrer Metaboliten durch die ungesättigte Zone natürlicher Bodenprofile sowie durch Bodenschlämme in Laborlysimetern und im Freiland. Umweltforschungsplan des Bundesministers des Innern, Forschungsbericht 106 02 005/02, pp. 1–315, Berlin.

15. ANDERBERG, M. R. (1973). *Cluster Analysis for Applications*, New York, London.

16. WILLIAMS, W. T. and LANCE, G. N. (1966). Multivariate methods in plant ecology. *J. Ecol.*, **54**, 427–45.

17. SHANNON, C. E. and WEAVER, W. (1949). *The Mathematical Theory of Communication*, Urbana, Ill.

18. VOGEL, F. (1975). *Probleme und Verfahren der numerischen Klassifikation*, Göttingen.

19. HOFFMANN, D. (1982). Pufferungsvermögen und Belastbarkeit schleswigholsteinischer Böden—eine exemplarische Untersuchung am Beispiel des Blattes Hennstedt 1.25.000. *Schr. Naturwiss. Ver. Schlesw.-Holst.*, **52**, 1–28.

20. FRÄNZLE, O. (1984). Regionally representative sampling. In: *Environmental Specimen Banking and Monitoring also Related to Banking* (Eds A. Lewis, C. Lewis and N. Stein), *Proc. Int. Workshop Saarbrükken*, May 10–15, 1982, 164–79, Boston/The Hague/Dordrecht/Lancaster.

21. FRÄNZLE, O. (1965). Die pleistozäne Klima- und Landschaftsentwicklung der

nördlichen Po-Ebene im Lichte bodengeographischer Untersuchungen. Abhdl. Akad. Wissensch. und der Literatur in Mainz, *Math.-Nat. Klasse, Jahrgang*, **8**, 1–144.

22. STREMME, H. E., FELIX-HENNINGSEN, P., WEINHOLD, H. and CHRISTENSEN, S. (1982). Paläoböden in Schleswig-Holstein. *Geol. Jb.*, **F14**, 311–61.

23. ULRICH, B. and MATZNER, E. (1983). Raten der ökosystem-internen H^+-Produktion und der sauren Deposition und ihre Wirkung auf Stabilität, Elastizität von Waldökosystemen. VDI-Berichte Nr. 500, 289–300, Düsseldorf.

24. WOOD, T. and BORMANN, F. H. (1977). Short-term effects of a simulated acid rain upon the growth and nutrient relations of *Pinus strobus L.*, *Water, Air, Soil Pollution*, **7**, 479–88.

25. OGNER, G. and TEIGEN, O. (1980). Effects of acid irrigation and liming on two clones of Norway spruce. Expanded version with basic data included. The Norwegian Forest Research Institute, 1432 Aas-NLH, Norway, Report, pp. 1–38.

26. BJOR, K. and TEIGEN, O. (1980). Effects of acid precipitation on soil and forest. 6. Lysimeter experiment in greenhouse. In: *Proc. Int. Conf. Ecol. Impact Acid Precip.* (Eds D. Drabløs and A. Tollan), SNSF-Project, pp. 200–1, Norway.

27. LEE J. J. and WEBER, D. E. (1982). Effects of sulfuric acid rain on major cation and sulfate concentrations of water percolating through two model hardwood forests. *J. Environ. Qual.*, **11**, 57–64.

28. FARRELL, E. P., NILSSON, J., TAMM, C. O. and WIKLANDER, G. (1980). Effects of artificial acidification with sulphuric acid on soil chemistry in a Scots pine forest. In: *Proc. Int. Conf. Ecol. Impact Acid Precip.* (Eds D. Drabløs and A. Tollan), SNSF-Project, pp. 186–7, Norway.

29. WIKLANDER, L. (1973/74). The acidification of soil by acid precipitation. *Grundförbättring*, **26**, 155–64.

30. TROEDSSON, T. (1980). Long-term changes of forest soils. *Ann. Agr. Fenniae*, **19**, 81–4.

31. LINZON, S. N. and TEMPLE, P. J. (1980). Soil resampling and pH measurements after an 18-year period in Ontario. In: *Proc. Int. Conf. Ecol. Impact Acid. Precip.* (Eds D. Drabløs and A. Tollan), SNSF-Project, pp. 176–7, Norway.

32. BLUME, H. P. (1981). Alarmierende Versauerung Berliner Forsten. *Berliner Naturschutzblatt*, 1981, 713–15.

33. BUTZKE, H. (1981). Versauern unsere Wälder? Erste Ergebnisse der Überprüfung 20 Jahre alter pH-Wert-Messungen in Waldböden Nordrhein-Westfalens. *Der Forst- und Holzwirt*, **36**, 542–8.

34. ABRAHAMSEN, G. and STUANES, A. O. (1983). Effects of acid deposition on soil: An overview. VDI-Berichte Nr. 500, pp. 279–87, Düsseldorf.

35. DRISCOLL, C. T., VAN BREEMEN, N., MULDER, J. and VAN DER POL, M. (1983). Dissolution of soil bound aluminium from the Hubbard Brook Experimental Forest, New Hampshire. VDI-Berichte Nr. 500, 349–61, Düsseldorf.

Sensitivity of European Soils to Ultimate Physical Degradation

R. P. C. MORGAN

Cranfield Institute of Technology, Silsoe College,
Cranfield, Bedford, MK43 0RL, UK

SUMMARY

The sensitivity of European soils to ultimate physical degradation is assessed by using the model of Stocking and Pain [35] to calculate their life span for commercial agricultural production. Assuming a soil loss tolerance of 1 t/ha, soils shallower than 0·30 m for sandy loams and 0·25 m for clay loams will reach ultimate physical degradation within 50–75 years wherever the mean annual erosion rate exceeds 20 t/ha. Data on present-day erosion rates reveal that there are certainly parts of the Mediterranean countries and the sandy and loamy belt of northern Europe in danger of serious degradation. The data undoubtedly understate the severity of the problem in the Mediterranean where millions of hectares of land are badly eroded and have been abandoned. The extent of the area at risk of or already affected by ultimate physical degradation in the European Community as a whole is not known. A suitable methodology, combining erosion modelling and assessment techniques, needs to be devised to determine this.

1. INTRODUCTION

Ultimate physical degradation is defined here as the state when soil has lost its value as a growing medium and has to be taken out of production. Since about 60% of the area of the European Community is devoted to agriculture, this chapter will be concerned solely with the agricultural aspects of degradation. Soil degradation as a result of modern agricultural practices was recognised as a problem for the European Community during

147

a Workshop held at Wageningen in October 1980 (Boels *et al.* [4]). Attention was drawn specifically to the effects on soil structural stability of compaction under heavier machinery and to the loss of organic matter as a result of continuous use of soils for arable farming. More important to ultimate physical degradation, however, are the effects of decreasing shear strength of the soil, decreasing infiltration rate and increasing run-off that follow loss of soil structure, a combination which invariably results in soil erosion. Since soil erosion represents the most extreme form of degradation, the effects of erosion on the productivity of the soil must form the basis of any assessment of ultimate physical degradation.

An awareness of soil erosion as a problem in the European Community is quite recent but scientists have now presented data at two Workshops, Firenze, October 1982 (Prendergast [26]), and Cesena, October 1985 (Chisci and Morgan [9]), which give cause for concern. Following the recommendations of the Cesena Workshop, the Commission of the European Communities has set up a Steering Group in the Land and Water Use and Management Programme of Directorate-General VI/F to advise on what initiatives should be taken in the Community on soil erosion and conservation.

Thus increasing recognition is being given to the problem of accelerating soil erosion as a result of changes in agricultural practices over the last 20–30 years. Crop rotations based on grass leys to restore soil structure and organic matter, together with the regular application of farmyard manure to the soil, have declined as, throughout the European Community, mixed farming has given way to all-arable rotations of cereals and root crops. Where, as in the hilly cultivated areas of Mediterranean Europe, mechanical measures for water control and soil conservation were practised, increased mechanisation, greater crop specialisation and rural depopulation have combined to bring about their abandonment. Improvements to pastures in the hill farming areas of Northwest Europe have created greater livestock pressures on the common rough grazing and encouraged the ploughing up of moorland for seeding to grass.

This chapter places the high present-day erosion rates associated with these practices in perspective by examining their effects on the life span of European soils to support sustained agricultural production.

2. CONCEPT OF SOIL LIFE

Most attempts at assessing the long-term implications of erosion on soil productivity in Europe have been based on comparing the mean annual rate of erosion, estimated or measured, with a maximum acceptable rate known as the soil loss tolerance. This rate is meant to reflect the rate of new soil formation as a result of natural pedogenetic processes, the addition of manures and fertilisers and the mixing of top soil and sub-soil during tillage. Where the erosion rate exceeds the tolerance level, the difference between the two, after allowing for the bulk density of the soil, can be converted into a decrease in soil depth. Through studies of the relationship between crop yield and depth of soil, this decrease can be expressed in terms of a lower crop yield.

Stocking and Pain [35] set these changes in soil depth against a finite point, dependent upon the prevailing economic conditions and levels of technology, at which the soil no longer yields a productive crop. This is taken for commercial farming as the point at which yields fall to 75% below the maximum possible. The finite point is defined as a minimum soil depth and is determined by the available water capacity for plant growth which, in turn, is linked to rates of evapotranspiration and to water stress on the crop. Although this approach ignores other factors that affect the fertility of the soil, it may be justified since, without water, plants cannot make use of the nutrients in the soil. Becher *et al.* [1] provide support from within Europe for the approach through their studies in Niederbayern. These show that where the mean annual erosion rate is 30 t/ha, maize yields on loess soils will decline by 3 t/ha over a ten-year period as a result of a decrease in plant-available water capacity.

According to Stocking and Pain [35] the life-span (L) of a soil, that is the number of years it can remain productive at a given erosion rate, can be calculated from:

$$L = \frac{(D_z - D_0)M}{(Z - Z_f)}$$

where D_z is the existing depth of soil (m), D_0 is the minimum depth of soil required to support production (m), M is the bulk density of the soil (t/ha-m or Mg/m^3 × 1000), Z is the mean annual rate of soil erosion (t/ha) and Z_f is the soil loss tolerance (t/ha).

3. APPLICATION

Table 1 shows the calculated life spans for sandy loam and clay loam soils assuming minimum soil depths for productive yields of 0·20 and 0·15 m respectively and a mean annual soil loss tolerance of 1 t/ha. The values for

Table 1
Calculated life spans of soils

Soil type	Sandy loam			Clay loam		
Bulk density (t/ha-m)	14 000			13 000		
Soil loss tolerance (t/ha)	1			1		
Minimum soil depth (m)	0·20			0·15		
Available soil depths (m)	0·50	0·30	0·25	0·30	0·25	0·20
Life span (years) at mean annual erosion rates of (t/ha):						
2	2 800	1 400	700	1 950	1 300	650
5	700	350	175	488	325	163
10	311	156	78	217	144	72
20	147	74	37	103	68	34
50	57	29	14	40	27	13
100	28	14	7	20	13	7

minimum soil depths are based on data given by Stocking and Pain [35] for a range of crops and are meant as guide values only. They are reasonable for cereals but rather low for root crops. They are also close to the minimum depth of soil that is feasible to plough. The soil loss tolerance level is a best estimate taking account of the shallowness of many European soils and the rather slow rate of new soil formation (Kirkby [20]). It has been recommended as a suitable level for England (Evans [14]).

From Table 1 it can be seen that soils shallower than 0·30 m for sandy loams and 0·25 m for clay loams will reach ultimate physical degradation within 50–75 years wherever the mean annual erosion rate exceeds 20 t/ha. Where existing soil depths are only 0·05 m above the minimum, a mean annual erosion rate of 10 t/ha will exhaust the soil in about 75 years. With a mean annual erosion rate of 2 t/ha, such soils will have a life span of just over 500 years. Since sandy loam soils have a higher minimum depth value than clay loam soils, they will exhaust earlier. With a mean annual erosion rate of 10 t/ha and a depth of 0·30 m, sandy loam soils will have a life span of about 150 years whereas a clay loam soil will last more than 200 years.

4.   COMPARISON WITH EROSION RATES

Most measurements of the rates of erosion in the countries of the European Community are either very long period averages based on interpretation of alluvial and colluvial fills in valley bottoms or very short period, often single storm, values determined from erosion plots or calculated from the volumes of deposited material. Data on present-day erosion rates measured over periods of five years or more are few.

Hence the information available to assess life spans of European soils is rather limited. However, the data base can be extended by including the short period measurements as indicators of annual rates of erosion. This is permissible because, in much of Europe, most of the annual erosion takes place in only two or three events. Tropeano [37] found that four storms contributed 58% of the erosion recorded over a two-year period under vineyards at two sites in central Piemonte, Italy. Marques and Roca [22] recorded 92% of the erosion in three storms during one year of observation under peach trees near Barcelona, Spain. Morgan [23] measured 88% of the erosion in 12 storms over a seven-year period on bare sandy loam soils near Silsoe, Bedfordshire, England. Further supporting evidence for the importance of events of this frequency comes from studies by Sfalanga and Franchi [34], Raglione *et al.* [27] and Zanchi [40] in Italy; Richter [30] in Germany; and Fullen and Reed [16] in England.

Table 2 summarises published data on measured erosion rates in the European Community which exceed 10 t/ha. Clearly the information cannot be treated as an unbiased statement on the situation in the Community generally. It is dependent on where researchers have chosen to measure and on published material known to the author. As a basis for assessing the life spans of soils the data are also inadequate because few researchers indicate the depth in their descriptions of the soil. Nevertheless, some conclusions can be drawn.

First, there are certainly some areas where erosion rates are high enough to cause serious physical degradation in the first part of the next century. Second, these areas are concentrated in the Mediterranean countries and in the sandy and loam belt of northern Europe. Third, surprisingly, the data suggest that the situation is worse in the north than in the south of Europe. The main reason for this is that almost all the measurements made in northern Europe are on erodible sandy, sandy loam and loam soils devoted to arable farming whereas the measurements made in the Mediterranean countries are more often on clay and clay loam soils and under a wide variety of land uses. Fourth, almost no data exist for those parts of the Mediterranean which have long been deforested, are deeply gullied and have reached ultimate physical degradation. Mean annual erosion rates, either measured or predicted, exceed 150 t/ha in these areas, sufficient to cause ultimate physical degradation of a soil within 0·05 m of its minimum depth in just 5 years. Thus, whilst the information available clearly

Table 2
Measured rates of erosion in the EEC

Location	Rate (t/ha)	Conditions
Oudenaarde, Belgium	10	Ploughed loamy sand, 3° slope, winter wheat following 213 mm rain Nov.–Dec. 1974 (Gabriels *et al.* [17])
Hesbaye, Belgium	82	Bare fallow on loess soils, 4° slope, rate for 1974 (Bollinne [5])
	30	Sugar beet on loess soils, 4° slope, rate for 1974 (Bollinne [5])
	15	Annual average rate for agricultural land over the last 150 years based on colluvial infill (Bollinne [5])
Leuven, Belgium	74	Sandy loam soil, 6° slope, winter wheat, annual average rate 1981/82–1984/85 (Govers [18])
Silsoe, Bedfordshire, England	45	Bare sandy loam soil, 11° slope, annual average 1973–79 (Morgan [24])
Woburn, Bedfordshire, England	24	Sandy loam soil under winter cereals, 7° slope, annual average 1976–79 (Morgan [24])
Ashwell, Hertfordshire, England	21	Chalky soils under winter wheat, 10° slope, annual average 1976–79 (Morgan [24])
Hilton, Shropshire, England	44	Bare sandy soils, 8–15° slope, annual average 1982/83–1983/84 (Fullen and Reed [16])
	11–18	Same site, annual averages for different plots, 1976–83 (Reed [29])
East Shropshire, England	156	Bare sandy soil following 83 mm rain 24–26 Sept. 1976 (Reed [28])
Cromer, Norfolk, England	195	Gullies on sandy loam soil, 7 May–15 June 1975 (Evans and Nortcliffe [15])
Beavendean, Sussex England	100	Chalky soil, ploughed and recently sown with grass, 20 Sept.–25 Oct. 1982 (Boardman and Robinson [3])
Albourne, Sussex, England	181	Fine loamy soil, 2–9° slope, under strawberries, rate for 9 months, 1979/80 (Boardman [2])
Vale of York, England	21–44	Wind erosion on sandy soils in spring and early summer, 1974 and 1975 (Wilson and Cooke [39])
Hey Clough, Derbyshire, England	34	Annual average rate around sheep scrapes on Agrostis-Festuca moorland, 1966–68 (Evans [13])
Yeovil, Somerset, England	14–27	Fine sandy soils under winter cereals, rates for 1982–83 winter (Colborne and Staines [11])

Table 2—*contd.*

Location	Rate (t/ha)	Conditions
Albugnano, Piemonte, Italy	47	Clay loam soil, 20° slope under new vineyard, soil repeatedly worked, annual average 1981–82 (Tropeano [37])
	20	Silt loam soil, 17° slope, 30-year old vineyard, no ground cover, annual average 1981–82 (Tropeano [37])
Fagna, near Firenze, Italy	31	Clay soil under maize, 8° slope, annual average 1979–82 (Zanchi [40])
	10	Clay soil under wheat, 8° slope, annual average 1979–82 (Zanchi [40])
Savoi watershed, near Cesena, Italy	17	Annual average suspended sediment load 1950–67 (Chisci [7])
Tacina Valley, Calabria, Italy	165	Severely eroded bare clay soils, annual average 1979/80–1980/81 (Raglione *et al.* [27])
Alt Penedes, near Barcelona, Spain	24	Bare loam soil under peach trees, 5° slope, rate for 1983 (Marques and Roca [22])
Fuirosos Basin, near Barcelona, Spain	105	Sandy soil, woodland, 27–29° slopes, annual average 1977–79 (Sala [32])
	116	Sandy soil, shrub vegetation, 23–25° slopes, annual average 1977–79 (Sala [32])
Jucar Basin, near Alcira, Spain	26	Sandy soils, 14° slopes, shrub vegetation following forest burning a few years before, rate for October 1982 following 700 mm rain in 3 days (Calvo and Fumanal [6])

demonstrates how erosion is threatening the productivity of land in northern Europe, it undoubtedly underestimates the severity of the problem in the Mediterranean countries.

5. EXTREME EVENTS

Table 2 includes several examples of individual storms resulting in erosion rates of 100 t/ha or more. Such extreme events can be very important in their effects. For instance, 100 t/ha of lost soil is equivalent to a decrease of 7 mm in the depth of a sandy loam soil. This loss can be critical for a soil close to its minimum depth for agricultural production. A sandy loam soil of only 0·21 m depth will have a life of 140 years if erosion takes place at a mean annual rate of 2 t/ha. If one event reduces the soil depth by 7 mm and erosion then continues at 2 t/ha, the soil life span will be only 42 years. The dramatic nature of extreme events is illustrated by the effects of a rainfall of 175 mm on 18 October 1973 in the Ugijar area of Murcia, Spain. This caused scarring of hillsides by a dense network of rills and gullies, 0·3–0·7 m deep, and completely changed the morphology of the river channels. Soil depths in the Alpujarras were reduced in this event by between 56 and 420 mm (Thornes [36]). The very rapid degradation that can be produced means that the sensitivity of soils to extreme events should be considered when assessing ultimate physical degradation.

6.   EXTENT OF AREA AT RISK

It is not possible at present to state the area of the European Community countries that is at risk of or has already reached ultimate physical degradation. Several surveys of erosion risk have, however, been made at a local level.

Between 4 and 5 million hectares in France are subject to soil erosion (Henin and Gobillot [19]); about 3·5 million hectares in Greece have a high risk of erosion and it is predicted that croplands in many hilly and mountainous regions will have to be abandoned within 5–10 years (Vousaros et al. [38]); about 2 million hectares of arable land in England and Wales have a risk of erosion by water, wind or both (Morgan [25]); and about 0·5 million hectares in Denmark suffers from wind erosion with a further 1 million hectares having the potential to erode but protected by the network of shelterbelts (Kuhlman [21]).

Richter et al. [31] surveyed an area of 30 km² near Trier, Germany, and found that 12% had soil depths less than 0·20 m and a further 23% had depths of between 0·20 and 0·40 m. Sanroque et al. [33] investigated two regions near Valencia, Spain, using the Universal Soil Loss Equation as a method of erosion assessment. They found that 93% of the Tuejar and 66% of the Higueralas study areas had mean annual erosion rates greater than 20 t/ha and that 21 and 12% respectively had rates over 300 t/ha. Chisci [8], also using the Universal Soil Loss Equation, records mean annual erosion rates greater than 20 t/ha in the Diano d'Alba region of Piemonte and rates over 10 t/ha in parts of Toscana and Emilia-Romagna. Although these data relate only to 250 hectares, the areas involved were selected as representative of the situation in the north-central Apennines of Italy.

Unfortunately, these surveys have been carried out with different techniques and different standards of what constitutes a high erosion risk.

So the results are not necessarily compatible. Since few of the surveys include information on soil depth, they cannot be used directly to assess sensitivity to ultimate physical degradation. Surveys based on the Universal Soil Loss Equation may not be entirely reliable because of the need to modify the rainfall (R) and soil (K) factor values for European conditions (Chisci and Zanchi [10]; De Ploey [12]). A more effective method of survey is required which combines the techniques used in erosion assessment with those of erosion rate prediction so that, given local conditions of climate, soils, topography and land cover, the effects on soil depth of both dominant and extreme erosion events can be evaluated.

Some progress to providing better quality information is being made. A Mediterranean-wide assessment of erosion risk at a reconnaissance level is in hand under the present CORINE Project of Directorate-General XI of the Commission of the European Communities. Larger-scale and more detailed surveys are required for planning purposes, however, and the whole exercise needs extending to the non-Mediterranean countries. The Commission has also responded to the advice of the Steering Group by arranging a Workshop on Erosion Assessment for the EEC: Methods and Models to be held in Brussels, 2–3 December 1986, under the Land and Water Use and Management Programme of Directorate-General VI.

7. CONCLUSIONS

Simple calculations of the life span of soils reveal that with present-day rates of erosion certain areas of the European Community, in both northern and Mediterranean countries, are in danger of ultimate physical degradation within 50–75 years. Other areas, particularly in the Mediterranean, have already reached this state and have been abandoned. Erosion is thus a serious threat to the quality of the EEC's soil resources. There is an urgent need to make a proper assessment of the extent of the problem.

REFERENCES

1. BECHER, H. H., SCHWERTMANN, U. and STURMER, H. (1985). Crop yield reduction due to reduced plant available water caused by water erosion. In: *Soil Erosion and Conservation* (Eds S. A. El-Swaify, W. C. Moldenhauer and A. Lo), Soil Conservation Society of America, Ankeney, IA., pp. 365–73.
2. BOARDMAN, J. (1983). Soil erosion at Albourne, West Sussex, England. *Appl. Geog.*, **3**, 317–29.

3. BOARDMAN, J. and ROBINSON, D. A. (1985). Soil erosion, climatic vagary and agricultural change on the downs around Lewes and Brighton, autumn 1982. *Appl. Geog.*, **5**, 243–58.
4. BOELS, D., DAVIES, D. B. and JOHNSTON, A. E. (1982). *Soil Degradation*, Balkema, Rotterdam.
5. BOLLINNE, A. (1978). Study of the importance of splash and wash on cultivated loamy soils of Hesbaye (Belgium). *Earth Surf. Proc. Landf.*, **3**, 71–84.
6. CALVO, A. and FUMANAL, M. P. (1983). Repercusiones geomorfologicas de las lluvias torrenciales de octubre de 1982 en la cuenca media del rio Jucar. *Cuad. de Geogr.*, **32–33**, 101–20.
7. CHISCI, G. (1982). Il dissesto idrogeologico della collina italiana con particolare riferimento all collina cesenate: situazione e prospettive di intervento. *Annali Ist. Sper. Studio e Difesa Suolo*, **13**, 31–51.
8. CHISCI, G. (1986). Influence of change in land use and management on the acceleration of land degradation phenomena in Apennines hilly areas. In: *Land Degradation due to Hydrological Phenomena in Hilly Areas* (Eds G. Chisci and R. P. C. Morgan), Balkema, Rotterdam.
9. CHISCI, G. and MORGAN, R. P. C. (1986). *Land Degradation due to Hydrological Phenomena in Hilly Areas*, Balkema, Rotterdam.
10. CHISCI, G. and ZANCHI, C. (1981). The influence of different tillage systems and different crops on soil losses on hilly silty-clayey soil. In: *Soil Conservation Problems and Prospects* (Ed. R. P. C. Morgan), John Wiley, Chichester, pp. 211–17.
11. COLBORNE, G. J. N. and STAINES, S. J. (1985). Soil erosion in south Somerset. *J. Agric. Sci. Camb.*, **104**, 107–12.
12. DE PLOEY, J. (1986). Soil erosion and possible conservation measures in loess loamy areas. In: *Land Degradation due to Hydrological Phenomena in Hilly Areas* (Eds G. Chisci and R. P. C. Morgan), Balkema, Rotterdam.
13. EVANS, R. (1977). Overgrazing and soil erosion on hill pastures with particular reference to the Peak District. *J. Br. Grassld. Soc.*, **32**, 65–76.
14. EVANS, R. (1981). Potential soil and crop losses by erosion. *Proc. Conf. Soil and Crop Loss: Developments in Erosion Control*. Soil and Water Management Association, Stoneleigh.
15. EVANS, R. and NORTCLIFF, S. (1978). Soil erosion in north Norfolk. *J. Agric. Sci. Camb.*, **90**, 185–92.
16. FULLEN, M. A. and REED, A. H. (1986). Rainfall, runoff and erosion on bare arable soils in east Shropshire, England. *Earth Surf. Proc. Landf.*, **11**, 413–25.
17. GABRIELS, D., PAUWELS, J. M. and DE BOODT, M. (1977). A quantitative rill erosion study on a loamy sand in the hilly region of Flanders. *Earth Surf. Proc.*, **2**, 257–9.
18. GOVERS, G. (1985). Rill erosion on a winter wheat field in Bertem. The Huldenberg Experimental Field: excursion guide. Workshop, Soil erosion and hillslope hydrology with emphasis on higher magnitude events, Leuven.
19. HENIN, S. and GOBILLOT, T. (1950). L'erosion par l'eau en France metropolitaine. *C.R. Acad. Sci. Paris*, **230**, 128–30.
20. KIRKBY, M. J. (1980). The problem. In: *Soil Erosion* (Eds M. J. Kirkby and R. P. C. Morgan), John Wiley, Chichester, pp. 1–16.
21. KUHLMAN, H. (1986). Vinden og landbruget. In: *Landbrugsatlas Danmark* (Eds

K. M. Jensen and A. Reenberg), Kong. Dansk. Geogr. Selskab, Kobenhavn, pp. 17–23.

22. MARQUES, M. A. and ROCA, J. (1986). Soil loss measurements in an agricultural area of northeast Spain. *Proc. 1st Int. Conf. Geomorphology*, John Wiley, Chichester.

23. MORGAN, R. P. C. (1980). Soil erosion and conservation in Britain. *Prog. Phys. Geog.*, **4**, 24–47.

24. MORGAN, R. P. C. (1985). Soil erosion measurement and soil conservation research in cultivated areas of the UK. *Geogr. J.*, **151**, 11–20.

25. MORGAN, R. P. C. (1985). Assessment of soil erosion in England and Wales. *Soil Use and Management*, **1**, 127–31.

26. PRENDERGAST, A. G. (1983). Soil erosion. Commission of the European Communities Report No. EUR 8427 EN.

27. RAGLIONE, M., SFALANGA, M. and TORRI, D. (1980). Misura dell'erosione in un ambiente argilloso della Calabria. *Annali 1st. Sper. Studio e Difesa Suolo*, **11**, 159–81.

28. REED, A. H. (1979). Accelerated erosion of arable soils in the United Kingdom by rainfall and runoff. *Outlook on Agriculture*, **10**, 41–8.

29. REED, A. H. (1983). The risk of compaction. *Soil and Water*, **11**(2), 29–33.

30. RICHTER, G. (1980). On the soil erosion problem in the temperate humid area of central Europe. *GeoJournal*, **4**, 279–87.

31. RICHTER, G., MULLER, M. J. and NEGENDANK, J. F. W. (1977). Landschaftsokologische Untersuchungen zwischen Mosel und unterer Ruwer. *Forsch. z. dt. Landeskunds*, 214.

32. SALA, M. (1983). Fluvial and slope processes in the Fuirosas basin, Catalan Ranges, north east Iberian coast. *Z. Geomorph.*, **27**, 393–411.

33. SANROQUE, P., RUBIO, J. L., SANCHEZ, J., SALVADOR, P. and ARNAL, S. (1984). Evaluacion del riesgo de erosion hidrica en la provincia de Valencia. Zonas piloto de Higueruelas y Tuejar. *Rev. Agroquim. Tecnol. Aliment.*, **24**, 134–50.

34. SFALANGA, M. and FRANCHI, R. (1978). Relazioni tra carica solida in sospensione, caratteristiche fisiografiche e sollecitazioni energetiche in due piccolo bacini idrografici (Botro dell' Alpino Valdera). *Annali Ist. Sper. Studio e Difesa Suolo*, **9**, 183–201.

35. STOCKING, M. and PAIN, A. (1983). Soil life and the minimum sol depth for productive yields: developing a new concept. University of East Anglia, School of Development Studies Discussion Paper No. 150.

36. THORNES, J. B. (1976). Semi-arid erosion systems: case studies from Spain. London School of Economics Geogr. Papers No. 7.

37. TROPEANO, D. (1984). Rate of soil erosion processes on vineyards in Central Piedmont (NW Italy). *Earth Surf. Proc. Landf.*, **9**, 253–66.

38. VOUSAROS, A., ROMANOS, L. and KOROXENIDIS, N. (1983). The state of research into soil erosion in Greece. In: *Soil Erosion* (Ed. A. G. Prendergast), Commission of the European Communities Report No. EUR|8427|EN, pp. 42–4.

39. WILSON, S. J. and COOKE, R. U. (1980). Wind erosion. In: *Soil Erosion* (Eds M. J. Kirkby and R. P. C. Morgan), John Wiley, Chichester, pp. 217–51.

40. ZANCHI, C. (1983). Primi risultati sperimentali sull'influenza di differenti colture (frumento, mais, prato) nei confronti del ruscellamento superficiale e dell'erosione. *Annali Ist. Sper. Studio e Difesa Suolo*, **14**, 277–88.

SESSION II

Assessment of Impacts on the Soil Environment

Effects of Agricultural Practices on the Physical, Chemical and Biological Properties of Soils: Part I— Effect of some Agricultural Practices on the Biological Soil Fertility

W. BALLONI and F. FAVILLI

Istituto di Microbiologia Agraria e Tecnica, Università di Firenze, Ple delle Cascine 18, 50144 Florence, Italy

SUMMARY

The microbiological aspects of the positive effects of some agricultural practices on soil fertility and crop production has been recently revised.

Tillages, crop rotation and fertilization improve the physical condition of soil; they provide a more favourable reaction for the activities of the numerous soil bacteria and admit large quantities of oxygen which are necessary for the growth of the aerobic organisms. These treatments are of great importance in soil fertility, not only because they produce in the soil favourable physical and chemical conditions for plant growth, but also create more favourable conditions for the activity of the microbes which effect more rapid liberation of the soil nutrients.

However, recent researches showed that soil tillages affect in a favourable way mainly the important groups of micro-organisms involved in the soil organic matter transformation, mineralization and nitrification; but affect unfavourably the most important process bound to the biological soil fertility as the nitrogen fixation by free micro-organisms.

The biological soil fertility is also more stimulated in soils submitted to crop rotation than in soils with monocultures, since the succession of different rhizospheric microbial populations exert a positive effect not only on the organic matter transformation and on biological nitrogen fixation but also against the pathogenic plant microbes.

161

1. INTRODUCTION

The knowledge of the distribution of micro-organisms in the soil and the role they carry out in the genesis, stabilization and degradation of its structure, are indispensable in order to understand, from the microbiological point of view, the effects of tillage and crop rotations on the biological fertility of the soil.

2. STRUCTURE AND MICROHABITAT OF THE SOIL MICROBIOLOGICAL RESPONSES

When we speak about the soil structure as an aggregation of crumbs or aggregates varying in dimensions from 0·05 to 10 mm and consisting of heterogeneous mineral particles (sand, lime, clay) cemented by mineral cements (amorphous SiO_2, clay, Al and Fe oxides) or organic cements polysaccharides, proteins, fats, waxes, humic compounds), we often forget that microbial cells are also present in these aggregates (Fig. 1).

The formation and the stability of the soil aggregates are correlated to the fertility, often defined as the physical fertility of the soil. Well aggregated soils drain well, favour the diffusion of gases and solutes and favour the fixation and the growth of roots and an active microbial population.

On the other hand, poorly aggregated soils, during degradations, are subject to erosion; when waterlogged they become anaerobic and inhibit microbial proliferation. Factors that facilitate the development and stabilization of aggregates are: coagulation of the clay particles, due to the attraction of many cations by the electronegativity of clay colloids, the physical penetration of roots and of fungal hyphae and the activity of microflora. The latter participates in the formation of aggregates both

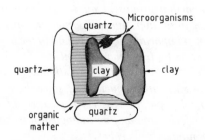

FIG. 1.   Greenwood's model aggregate of soil particles [6].

directly (for instance fungal or actinomycetal hyphae that hold together the dispersed particles, or capsulate microbial cells that fix mineral particles in their mucous layer) and indirectly through the formation of extracellular products (polysaccharides, polyuronic acids, etc.) that act as 'soft' organic cements or with aromatic polymers of the humus responsible for more 'hard' cements. Given the multiple contribution of micro-organisms to the formation of aggregates, their organic composition is different from that of a non-aggregated soil. This difference can be further increased by the subsequent metabolism of micro-organisms situated in the aggregates.

As a result, the biochemical activity of aggregates is qualitatively different from that of degraded particles. Many microsites exist within the aggregate itself; each having a distinct niche of microbial colonization. These niches were called 'elementary microbial microecosystems' by Nikitin [1]. In this author's conception, microecosystems are populated not by single cells, but by microaggregates of microbial cells in the shape of rosettes, spirals and stars and equipped with fimbriae, slime and stalks. Elementary microcolonies do not exceed 5–10 μm in diameter and multiply through repeated, multiple aggregation, until they form aggregates, called 'population granules' whose dimensions vary from 300 to 500 μm. The further association of some 'population granules' together with the mineral soil particles gives rise to the formation of a microplot 0·5–2·5 mm in diameter. This formation and working mechanism of microaggregates on a microscale, postulated by Nikitin [1], shows how microbial cells can be found in the soil in closely-packed colonies or in pluricellular filamentous masses. On this micromorphological and structural aspect of the soil characteristic of the habitat, depends the composition, the density and the activity of the microflora, which in turn participates not only in the genesis of the structure and micro-morphology of the soil, but also in the continuous variations from one microecological area to another or also in the same area in time. According to Allison [2], the majority of micro-organisms found within the aggregates have been trapped during their formation (Fig. 2).

2.1. Micro-organisms responsible for the formation of aggregates
The tendency of soil micro-organisms to form aggregates both directly and indirectly is widespread among the various groups.

Fungi. Species of *Cladosporium*, *Trichoderma*, *Mucor*, *Sclerotium* and *Fusarium* are recognized to be particularly efficient in the formation of aggregates.

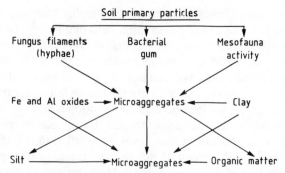

FIG. 2. Formation of aggregates by the activity of the microflora and microfauna [6].

Yeasts. The species *Lipomyces starkeyi* and *Cryptococcus terreus*, which
are frequent in the soil, are characterized by the extracellular production of
polysaccharides.

Actinomycetes. Many species of the genus *Streptomyces* and *Nocardia*
have been recognized as responsible for the formation of aggregates [3].

Bacteria. Some of the most efficient species are *Bacillus mycoides*,
B. polymyxa, *B. subtilis*, *B. circulans*, *Rhizobium japonicum*, *Rh. legumino-
sarum*, *Beijerinckia indica*, *Chromobacterium violaceum* and *Pseudomonas*
spp.

Cyanobacteria. The role carried out by this group of oxygenic,
photosynthetic prokaryotes in the formation of aggregates is of
considerable importance especially in tropical soils. The species involved
belong mainly to the genera *Nostoc*, *Anabaena* and *Scytonema* [4, 5].
   Bacteria, fungi, actinomycetes and cyanobacteria give rise to different
types of aggregates. For instance, bacteria play an important role in the
initial stabilization of small aggregates (0·5–2 mm), whereas fungi favour
the formation of aggregates with a diameter greater than 2 mm, such
mesofauna (for example earthworms) give rise to granules and aggregates
which differ from microbial ones. Consequently, the microbial discontinu-
ity is more the rule than the exception in the soil, in which aerobic and
anaerobic environments are present in proportions and sites determined by
various factors, such as the water content, pore dimensions, oxygen, etc.
Generally speaking, anaerobic bacteria are concentrated inside aggregates
whereas fungi and other obligate aerobes reside in the peripheral region.

The humidity gradient, especially during drought, affects the micro-organisms situated in the surface substrates; they will be the first damaged by dehydration and to feel the positive effects of rehumidification. Micro-organisms living inside the aggregate can be protected from predators and can be physically preserved by extended colonization of their microenvironment.

2.2. Factors that affect the genesis and microbial degradation of the soil structure

Various factors affect the genesis and the microbial degradation of the soil structure, such as organic matter, liming, aeration and water regime of the soil, roots, tillages and crop rotations.

2.2.1. Organic substances

The influence of microbial activities on the soil structure differs according to the origin of the organic substances (decomposition of fresh vegetal residues of humic compounds). In this way, for example, polysaccharides and polyuronides that are synthesized by the microbial activities degrading the fresh organic matter (litter, green manure, crop residues, etc.), have an immediate effect on the soil structure which is detected a few months after the introduction of fresh organic matter.

They are, however, 'soft' cements which, once degraded, lead to a parallel degradation of the structure. However, when organic compounds of the soil are formed (humic ones), they bind themselves permanently to the scattered mineral particles together with amino acids, fats and compounds of various organic nature and act for a long period of time on the formation and stability of the aggregates, unlike polyelectrolytes such as Krilium, which are biologically inactive [6, 7]. On the other hand the stability of the structure depends on the complexity of the substance cementing the soil particles.

In fact, until the organic materials are not degraded by the microflora, they have no favourable influence on the structure, and this also applies to humic compounds formed elsewhere and brought to the soil (such as compost).

2.2.2. Liming (fertilization with lime)

Liming generally brings about a net improvement in the structure's stability. Its effect is not only physicochemical, increasing the colloid flocculation processes, but also microbiological because following on the rise of the pH, micro-organisms producing aggregating substances are stimulated.

2.2.3. *Influence of aeration and water regime of the soil*

Laboratory and field experiments show that biological aggregation is favoured by microaerophilic conditions often related to high humidity. Harris *et al.* [8], on comparing the influence of aerobiosis and anaerobiosis on sucrose enriched soils, showed that in aerobiosis, the initial structure improvement phase is followed by a phase of slow degradation; on the other hand, in anaerobiosis, the structure improvement phase is followed by a high level stabilization phase (Fig. 3).

2.2.4. *Influence of the roots and rhizosphere*

Vegetation causes a direct and indirect effect on the structure depending on the vegetal species and how long it remains in the soil.

The direct effect concerns the mechanical action of the roots in creating spaces and pores in the aggregates, but it is commonly believed that it is the indirect effect due to the activity of micro-organisms of the rhizosphere that have a greater effect on the soil structure. In fact, as it is well known, close to the plant roots the microflora is richer than in the surrounding area, stimulated in its development by organic compounds exuded by the roots.

In the rhizosphere there are many capsulated microorganisms or producers of mucilaginous substances aggregating the soil particles. As a result, the soil as a substrate for microbial life is a basically discontinuous and heterogeneous environment in which the main supports of life (water, oxygen and nutrients) are unequally divided around the interior of soil

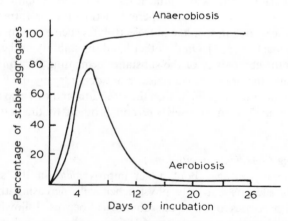

FIG. 3.   Influence of aerobic and anaerobic conditions on the formation of stable soil aggregates [8].

structural elements. To this situation corresponds an aggregation and microdistribution of the soil that is equally discontinuous and heterogeneous. Hence the notion of microhabitat understood as a microenvironment of dimensions proportioned to the microbial scale in which microcoenosis takes place, and where the soil microflora carries out its activities.

Therefore, the overall biological soil activity is the summation of the activities of a large number of microhabitats. In this conception of the soil structure, it is clear that any change to the chemical and physical soil conditions is accompanied by changes in the biological state of the soil expressed by the nature, abundance and activities of the soil microflora.

3. INTERACTIONS BETWEEN SOIL MANAGEMENT AND MICROBIAL ACTIVITIES

The biological soil fertility is the result of many processes in which micro-organisms and agricultural practices take part. The growth of higher plants and also of animals including man depends upon the activity of the soil micro-organisms.

Consequently there is little likelihood of overestimating the importance of microbes in soil processes. The growth of soil microbes is intimately concerned with the availability of nutrients which determine the fertility of the soil.

Anything that hastens the decomposition of the soil organic matter also favours an increase in soil fertility.

A fertile soil is distinguished from an infertile soil not by the fact that it contains more nitrogen, phosphorus and potassium, but by the fact that the nutrients present in the soil are liberated with greater rapidity in the fertile than in the infertile soil.

There are numerous important processes in which micro-organisms take an active part, among these are: rock disintegration, soil structure formation and degradation, decomposition of organic matter in soil, N_2-fixation, nitrogen transformation, liberation of mineral elements, improvement of plant growth by synthesis of growth factors, plant and animal parasites, antagonisms of plant pathogenic microbes.

It is clear that any change in the chemical and physical conditions is accompanied by changes in the biological state of the soil expressed by the nature, abundance and activity of the soil microflora. Some treatments result in profound alterations of the biological activities which persist for considerable periods of time; other treatments exert only slight effects of short duration.

Among the important treatments of the soil are the following:

(1)   The addition to soil organic matter in the various forms of manure (stable, green manure, artificial manure, plant stubble, or other plant and animal residues) as well as the organic fertilizers of commerce.

(2)   Physical modification of the soil by ploughing, cultivation and similar mechanical processes.

(3)   The addition of different inorganic fertilizers or lime in its various forms.

(4)   Cropping the soil according to the various systems.

(5)   Partial sterilization of the soil by heat or antiseptics.

(6)   Air drying of the soil.

The frequent cultivation of the soil considerably modifies the number and the activities of soil microbes in various ways. The processes of ploughing, harrowing and cultivating the soil accomplish several distinct purposes:

(1)   They mix the plant residues with the soil itself, thus leading to better conditions for the decomposition of the various residues by micro-organisms.

(2)   They favour aeration of the soil, thus accelerating the exchange of carbon dioxide of the soil for the oxygen of the atmosphere: the oxygen is required for the various oxidation processes in the soil, especially nitrate formation, decomposition of organic compounds and the oxidation of the reduced inorganic substances.

(3)   Finally, soil cultivation brings about a more uniform distribution of the moisture necessary for the activity of the soil microbes.

Tillage, crop rotations and fertilizations are of great importance in soil fertility not only because they create more favourable conditions for the activity of micro-organisms which effect more rapid liberation of the soil nutrients, but also they modify the chemical–physical aspect of the soil from the point of view of its structure, permeability and water retention. The structure is connected to the soil's natural filtration capacity, which allows the micro-organisms to go across or to be retained by a given soil. This is related to the colloids' power to absorb cells of micro-organisms.

The structure of the soil also affects the water circulation, enabling a uniform, continuous diffusion of nutritional solutions and aeration.

The degree of humidity in the soil exerts considerable influence on the number of micro-organisms and on their activities; a humidity of 60–70% of the soil's water capacity is an ideal percentage, but there can be variations in the physiological groups. Repeated changes in the water state damage the microbiological equilibrium. Long periods of dryness lead to the selection of species with resistance organs (spores, cysts, capsules, etc.). The actinomycetes for example are favoured, among the microbial soil groups, by low humidity in the soil. It is, however, certain that the resistant species undergo a slowing down of their vital activity. Aeration generally favours all aerobic species and their biochemical processes (nitrification, mineralization of the organic matter, oxidation of sulphur, etc.); on the other hand, a scarcely aerated or reducing environment favours anaerobic activities (denitrification, reduction of sulphates, anaerobic transformation processes of the organic matter).

The temperature of the soil influences the microbial activities in the sense that a slight increase favours respiration (production of CO_2), the mineralization of nitrogen and nitrification. Sharp rises that often take place during the summer, specially in the surface horizon, can stop biological activities altogether. The same occurs when the temperature drops to less than $5°C$.

By modifying the soil physicochemical properties, estimate tillage of the soil affects the microbial activities; but in order to estimate the true effect of this influence it is necessary to examine not only the total microflora or enzymatic activities, but also the variations that have occurred in the soil's microbial functional groups.

The many contributions of the effects of tillage on the soil's fertility mainly concerned the soil's physicochemical properties, neglecting its microbiological aspects and the few data available on the effect on the total microflora and some physiological groups often have contrasting results.

According to Verona [9] all types of tillage have a favourable influence on the increase of the microbial population and on the equilibrium between the various microbial groups. The increase of the microbial population can be seen by the increase in the emission of CO_2, following the activation of the oxidative processes on the organic matter. There is an increase in the processes of nitrogen fixation and nitrification and a decrease in the denitrification process. Any change in the microbial world of the soil by cultivation is in proportion to the conditions of aerobiosis and anaerobiosis which regulate mainly the humification and mineralization processes of the organic substance.

In fact, as the soil is a complex mosaic of microsystems in which various elements (air, water, aggregates) come into play, every action that tends to modify one of the above-mentioned elements brings about modifications in the soil structure and consequently in the microbial life of the soil itself.

Part of the soil's microbial population consists of anaerobes (obligate and facultative), of which certain groups (sulphate reducers, cellulolytics, methanogenics) present an interesting feature of soil microbiology, whereas other anaerobes operate important transformations of nutritional elements in the soil; however, very little is known about the influence of the latter on the soil fertility and structure.

In fact, important microbial activities such as nitrogen fixation and denitrification can take place in anaerobic microaggregates. Denitrification is not considered negative in relation to the various reduction levels; moreover in anaerobiosis, there is greater stability of the structure with a slower degradation of the organic substance cementing the scattered particles.

Anaerobiosis is not the only factor that can determine the development of specific groups of anaerobes in the soil because their activity is connected either to the redox values of the environments or to the capacity of assimilating nutritional elements (in particular the presence of nitrates prevents the anaerobic decomposition of cellulose, reduction of sulphates and methanogenesis) and also the physical factors (temperature and humidity).

Important factors for anaerobic conditions are on the one hand the physical movement of the air in the soil caused by water, through which the oxygen diffusion is very slow and on the other hand, the utilization of oxygen in the air through microbial respiration which occurs at a faster rate than that of the diffusion of the same gas at the sites of metabolic activities · of the soil.

The concept that aerobiosis favours nitrogen fixers, based on the diffusion of *Azotobacter*, has been revised by knowledge that nitrogenase is inhibited by oxygen. As a result, at low oxygen tensions many micro-organisms in the soil fix nitrogen. Therefore with our present knowledge, it is hard to assert that physical and mechanical treatments of the soil must necessarily be done to stimulate aerobes, since some anaerobic or microaerophilic conditions enhance important processes in the soil such as nitrogen fixation.

According to Verona [9] the overturning of the layers would damage both the nitrification and humification processes of the organic substances as the surface layer of the soil brought to the bottom would allow a large part of its assimilable nutrients to be dispersed. Other harmful consequences of the inversion of the layers would be the interruption of oxidation and decomposition of toxins formed by the microbial activities present in the surface layers of the soil and the bringing to the surface of toxic substances formed deep down with harmful effects on the microflora and vegetation. The same applies to ploughing beyond a depth of 40 cm, such as subsoiling.

However, it must be pointed out that these conditions of microbiological unbalance change in time, and after a period of reduced biochemical activity a certain equilibrium is regained in the microflora.

On the contrary, surface or shallow tillage, by removing the surface layers of the soil, change the aeration and the water conditions of the soil with stimulating effects on activities and numbers of micro-organisms [6].

3.1. Effect of tillage on microbial activities

Verona [9] investigating the effect of repeated harrowing on the water content of the soil and on the number of micro-organisms showed that harrowing improves the moisture content of the soil and causes an increase in the total number of micro-organisms from 6 to $18 \times 10^6/g$. Furthermore, an enhancement of the nitrification process and of the nitrogen fixing bacteria belonging to *Azotobacter* was detected. Waksman and Starkey [10] found that surface tillage of the soil favours a high increase of the bacterial population that passes from 2×10^6 cells/g, before tillage, to over 12×10^6 cells/g 24 days after the treatment.

However, these works do not mention the variations of the soil's microbial functional groups.

More information on the influence of tilling on the soil microflora can be obtained from a report of Giacomelli *et al.* [11]. In a series of researches conducted on soil left uncultivated for 10 years and then submitted to a 3-year series of different mechanical processes (hoeing, ploughing, turning over and ripping), they discovered that surface tillage and ploughing had a very marked effect on the soil's microbial activities, whereas the ripping practice was not at all efficient. On the other hand, the microbial activities carried out at low oxygen tensions or in anaerobiosis were depressed by tillage. In fact, data on the anaerobic bacteria count and the overall nitrogenase activity were lower in tilled soils than in the others.

The unfavourable effect of tillage on nitrogenase activity disappeared 2 years from the mechanical intervention after soil settling.

The inverse correlation between nitrogenase activity and aeration was shown by the effect carried out by the ripping. By producing a lesser aeration of the soil as compared to that produced by other tillages, it had lesser dampening effect on the nitrogen fixation activity.

The development of the nitrogen fixing population was in complete agreement with the course of the nitrogenase activity.

Lynch and Panting [12] also found significant variations in the microbial biomass in the surface layer of soils submitted to surface tilling and ploughing. The biomass was much greater in soil submitted to surface tilling than in the ploughed soil. Higher respiratory activity and microbial population were also discovered by Arcara and Tellini [13] in cultivated but not tilled soil as compared to soils submitted to ploughing.

3.2. Agricultural practices and enzymatic activities of the soil

In recent years a great deal of attention has been paid to the study of the enzymes of the soil. It is known that many transformations of the soil's organic matter are catalysed by enzymes and that enzymatic activities are often used as an index of the soil fertility [6]. Furthermore it has been shown that many agricultural practices such as tilling, fertilization and crop rotations, in which leguminose plants take part, have more or less favourable effects on the soil enzymatic activities [14, 15].

The cultivation of the soil significantly alters the enzymatic activities; Lupinovich and Yakushava [16] noted an increase in the catalase, phosphatase and saccharase in a virgin soil put into cultivation. Khan [17] observed significant increases in enzymatic activities in soil submitted to 5 years rotations with wheat and legumes as compared to soil kept on wheat and fallow.

Some researches on enzymatic activities in soil seeded directly without tillage and in soils tilled conventionally indicate that crops in untilled soils also have favourable effects on other soil properties such as humidity, organic substance content and total nitrogen [18, 19].

Similar research was conducted from 1964 to 1974 by Vrestraete and Voets [20] on the correlations between agricultural practices and biochemical and microbiological characteristics of the soil.

They showed the enzymatic activities, phosphatase, saccharase, urease and β-glucosidase are directly correlated with the soil biological fertility and can be used to characterize the various treatments of the soil.

Speir *et al.* [21] noted that the enzymatic activities (sulphatase, urease and protease) of two soils, one left fallow and the other cultivated with ryegrass, considerably decreased in the fallow soil as compared to the cultivated one, where slight increases in activity were observed. However, they recorded that the decrease in enzymatic activities in the fallow soil was due to the denaturation of enzymes caused by the temperature.

Research carried out by Klein and Koths [22] on untilled soil seeded with wheat and forage maize and on ploughed soil to evaluate the influence of no-tillage on some enzymatic activities of the soil (urease, protease and phosphatase) showed that both the humidity and organic substance content of the soil as well as the enzymatic activities decrease in tilled soil seeded with wheat and forage maize as compared to that in which the crop is produced without tillage. They showed a correlation between enzymatic activities and organic substance content and humidity of the soil examined.

4.   INTERACTION BETWEEN CROP ROTATION AND MICROBIAL ACTIVITIES OF THE SOIL

The type of plant grown in a soil will also considerably influence the microbial population. The influence of the plant is due to several factors.

Considerable material becomes added to soils from plant roots, either as excretion or as part of the roots themselves.

The influence of plants on soil moisture, salt concentration and physical conditions is largely responsible for the change in the microbial population.

Subsequent to the death of the plant, appreciable amounts of organic residues may find their way into the soil and serve as food for the microbial inhabitants.

The activities of soil microbial population and plants are so closely connected as to ensure living conditions for all the organisms living in the soil.

Apart from these general relationships, there are other more specific ones, especially with regard to roots, as a result of plant and microorganisms living in the same habitat: the soil.

The importance of these particular rhizospheric relationships is emphasized by the fact that considerable differences are encountered in the number and activity of microbes in uncultivated and cultivated soils also in fields submitted to different types of crops.

The rhizosphere is the soil biotope, with the typical nature of the microhabitat in which more intense biological processes are carried out, albeit indirectly, and biochemical activities of the microflora–plant–soil complex are developed to the maximum [23].

The plant–micro-organisms relations carried out in the area of ground concerning the plant root system influence the vegetal development since the complex of microbial activities regulates the availability of nutrient elements. The plant's roots, in turn, influence the soil microflora through the emission of organic, energetic and stimulating substances.

All these actions are more marked when they take place in the area closer to the roots. From the agronomical and silvicultural point of view, the rhizosphere differs according to whether the roots grow in cultivated or in natural soils. In forage crop and intensive crop farming soils, the greater part of the layer of soil must be considered rhizospheric.

The aforementioned structural heterogeneity of soil diminishes in the conditions is largely responsible for the change in the microbial population. colonized, creates highly active microbial niches that are continuously supplied with carbon sources and nutrient through root exudation, vegetal and microbial metabolites.

The rhizosphere continuously supplies the rhizospheric microflora with organic material in contrast to the discontinuous, episodic and irregular availability of energy sources for the telluric microflora.

Root excretions are important in the microbiological equilibrium in the rhizosphere and the rhizospheric microflora constitutes a biological barrier in defence of the plant's health. By modifying the exudate composition, the plant has a considerable possibility of influencing the rhizospheric microflora. Some vegetal species were able to create an efficient defence against the parasitic microbes thanks to the selection and processing of appropriate substances contained in the exudates and the stimulation of the antagonistic microflora around the roots.

The rotations and crop consociations with the succession of selected rhizospheric microflora represent an efficient control method for the rhizospheric effect capable of acting on the crop yields, soil productivity and defence from plant diseases.

Unfortunately, there is not a great deal of research that indicates that the beneficial effects of crop rotations on soil productivity are due to the increased biological fertility following the stimulation of the biochemical activity of the microbial biomass of the rhizosphere. Also because it is difficult to correlate the increase in biological activity of the microflora with the increase of soil productivity.

A general positive effect exerted by agricultural practices, by crop rotations and by organic amendments, on the soil productivity has been known for some time [24, 9, 25], but research in recent years has mainly shown the stimulating effect of various rotation systems on several activities of particular groups of micro-organisms. Florenzano and Favilli [26] on 2 fields cultivated for 12 years with 4-year rotations and wheat monoculture found no significant microbiological differences in the total microflora between the two systems whereas significant differences were discovered in the physiological groups of cellulolytics, nitrifying and nitrogen fixing micro-organisms compared to the continuous single wheat crop.

Analogous results were detected by Martyniuk and Wagner [27] with significant differences in the bacterial, fungal and actinomycetal populations between soils submitted to single crops (maize, wheat) and 4-year rotations (maize–oat–wheat and clover).

They showed that the bacterial and actinomycetal population was higher in rotation combined with fertilizing treatment as compared to single crop growing submitted to the usual fertilizing treatments.

On the other hand, the fungi were increased more by single crop mono-culture than by rotation.

Arcara [28, 29] found significant differences in the enzymatic and microbiological activities on 3 fields submitted for 10 years to deep ploughing, minimum tillage with continuous wheat and permanent grassland. In the latter two crop systems, the values of enzymatic and microbiological activities were higher than in the deep ploughing [30–32].

In two systems of different rotations: winter wheat–spring peas and winter peas (used as green manure) and winter wheat–spring peas, Bolton *et al.* [33] discovered a higher number of micro-organisms (bacteria, fungi and actinomycetes) and a greater microbial and enzymatic activity in the 3 crop rotation as compared to the 2 crop rotation, whereas in the latter a higher number of nitrifying and denitrifying bacteria was found.

Increases in the bacterial and actinomycetal populations and in the enzymatic activities in soils submitted to long rotations as compared to soil growing wheat alone were detected by Biederbeck *et al.* [34]. They found that these extensive, persistent effects in the soil surface layer were reduced deep down and under adverse weather conditions.

One aspect, of no lesser importance than the influence of rotations and crop succession on the biological fertility and productivity of the soil, concerns the control exerted by these crop practices on phytopathogenic micro-organisms [35]. Some typical examples of these interactions suffice to highlight the effect of rotations on the phytosanitary condition of the soil [36, 37]: the introduction of Lucerne of *Arachis* in the crop system can, for example, prevent the expansion of cotton diseases since the rhizosphere of the *Arachis*, a development of mycolytical micro-organisms that are antagonists of *Verticillium dahliae* and *Fusarium vasinfectum*, pathogenic for cotton, seems to occur.

The introduction of soybeans in rotations has a regressive effect on strawberry rot. On the other hand, certain crops can cause on subsequent cultivations a recreduscence of the attack of pathogens. In strawberry cultivated after red clover or maize or after *Phleum pratense* an increased virulence of the root rot agent was detected [38]. In many Canadian soils, the black rot of cereals due to various species of *Pythium* is shown with particular virulence when the cereals are grown after fallow as this practice causes a phosphate deficiency. To avoid this trouble, it is advisable to grow cereals after another crop or to correct the phosphorus deficiency [39].

These brief examples show that the repeated cultivation of the same plant in the same soil considerably favours the spread of diseases of the root system and therefore it is particularly important to carefully choose the rotations and crop sequences to apply to a certain soil.

5. CONCLUDING REMARKS

With their multiple activities, micro-organisms influence the chemical–physical state of the soil, making it suitable for the development of cultivated plants; in fact a microbial biomass of 10 t (d.w.)/ha, which characterizes a biologically fertile soil, is capable of ensuring the circulation of nutrients, the turnover of the organic substances and the biological equilibrium necessary for plant growth. To take into account these basic principles means to apply chemical–agricultural, agronomical practices to conservation and to the increase of soil fertility.

The foregoing quite clearly shows how the increase and preservation of soil biological fertility can be obtained with the use of tillages, rotations, suitable crop consociations and rational mineral and organic fertilization that takes into account both the soil microbiology and microbial ecology as well as the adaptation and genetic characteristic of agricultural and forest plants.

Promising methods, which according to Garrett [40] were 'killed' by the aforementioned agronomical applications, are sometimes made quite useless by current trends in agriculture and by new problems: pollution, single crop growing, ever increasingly intensive and specialized crops, progressive lack of suitable organic matter and, lastly, the defence of the soil fertility.

REFERENCES

1. Nikitin, D. J. (1973). The microstructure of the elementary microbial ecosystems. *Bull. Ecol. Comun.*, **17**, 357–66.

2. ALLISON, F. F. (1973). *Soil Organic Matter and its Role in Crop Production*, Elsevier, Amsterdam.
3. JONES, D. and GRIFFITHS, E. (1964). The use of soil sections for the study of soil microorganisms. *Plant and Soil*, **20**, 232–40.
4. BALLONI, W. and FAVILLI, F. (1973). Il ruolo delle microalghe nella fertilità del suolo. *L'Agricoltura Italiana*, **73**, 267–74.
5. FAVILLI, F., MARGHERI, M. C. and TOMASELLI, L. (1984). I cianobatteri in agricoltura. In: *Impiego degli azotofissatori in agricoltura (a cura di M.P.Nuti)*, Consiglio Nazionale delle Ricerche P.F.IPRA Monografia n.7, pp. 71–84.
6. FLORENZANO, G. (1983). *Fondamenti di Microbiologia del terreno*, Edizioni Reda.
7. MALQUORI, A. (1979). *Lineamenti di chimica del terreno*, Edizioni Scuola Universitaria Firenze.
8. HARRIS, R., ALLEN, O., CHESTERS, G. and ATTOE, O. (1963). Evaluation of microbial activity in soil aggregate stabilization and degradation by the use of artificial aggregates. *Soil Sci. Soc. Amer. Proc.*, **27**(5), 542–6.
9. VERONA, O. (1977). *Microbiologia Agraria*, Edizioni UTET.
10. WAKSMAN, S. A. and STARKEY, R. L. (1947). *The Soil and the Microbe*, John Wiley, New York, London.
11. GIACOMELLI, E., CAROPPO, S. and MICLAUS, N. (1976). Recherches preliminaires sur les interactions entre de systèmes differents de travail du soil et ses proprietes microbiologique. 9 Conferencia International de mecanizacion Agraria. *Zaragoza (Espana)*, **1**, 49–56.
12. LYNCH, J. M. and PANTING, L. M. (1980). Cultivation and the soil biomass. *Soil Biol. Biochem.*, **12**, 29–33.
13. ARCARA, P. G. and TELLINI, M. (1971). Primi rilievi sulle caratteristiche biologiche di suoli argillosi soggetti a diverse tecniche colturali. *Ann. Inst. Sperim. Studio e Difesa del Suolo, Firenze*, **II**, 167–77.
14. HOFMAN, E. and BRÄUNLICH, K. (1955). The saccharase content of soils as affected by various factors of soil fertility. *Z. Pflernähr, Düng Bodenk.*, **70**, 114–23.
15. TOOGOOD, J. A., BENTLEY, C. F., WEBSTER, G. R. and MOORE, A. W. (1962). Gray wooded soils and their management. University of Alberta, Edmonton, Canada Bull. SM 1, n.21.
16. LUPINOVICH, I. S. and YAKUSHAVA, V. I. (1965). Change of the fermentative activity of dernogley soils in connection with their cultivation. *Vestri Akademii Navuk Belaruskai SSR. Seryya Sel's Kagaspadarchykh Navuk*, **3**, 23–5.
17. KHAN, S. U. (1970). Enzymatic activity in a gray wooded soil as influenced by cropping systems and fertilizers. *Soil Biol. Biochem.*, **2**, 137–9.
18. BLEVINS, R. L., COOK, D., PHILIPPIS, S. H. and PHILIPPIS, R. E. (1971). Influence of no-tillage on soil moisture. *Agronomy Journal*, **63**, 593–6.
19. LAL, R. (1975). No-tillage effects on soil conditions and crop response on an alfisol in southern Nigeria. American Society of Agronomy, Abstracts, p. 36.
20. VERSTRAETE, W. and VOETS, J. P. (1977). Soil microbial and biochemical characteristic in relation to soil management and fertility. *Soil Biol. Biochem.*, **9**, 253–8.
21. SPEIR, T. W., LEE, R., PANSIER, E. A. and CAIRUS, A. (1980). A comparison of sulphatase, urease and protease activities in planted and in fallow soils. *Soil Biol. Biochem.*, **12**, 281–91.

22. KLEIN, T. M. and KOTHS, J. S. (1980). Urease, protease and phosphatase in soil continuously cropped to corn by conventional or no-tillage methods. *Soil Biol. Biochem.*, **12**, 293–4.

23. FLORENZANO, G. (1972). Fondamenti bioecologici dell'habitat radicale e prospettive di controllo dell'effetto rizosfera. Colloquio: Rapporti piante-microrganismi. Atti Soc. Ital. Scienze del Suolo. Pisa, pp. 99–130.

24. SMITH, G. E. (1942). Sanborn Fifty: Fifty years of field experiments with crop rotations, manures and fertilizers. Missouri Agricultural Experiment Station, Bulletin 458, Columbia.

25. LARSON, W. E., WALSH, L. M., STEWART, B. A. and BOELTER, D. H. (1981). Soil and water resources: Research Priorities for the nation. *Proc. Soil Sci. Soc. Amer.*, Madison.

26. FLORENZANO, G. and FAVILLI, F. (1973). Confronto tra rotazione e concimazione in ambiente mediterraneo. I Influenza di rotazioni e concimazioni sulla microflora del terreno. *Riv. Agron.*, **2–3**, 105–14.

27. MARTYNIUK, S. and WAGNER, G. H. (1978). Quantitative and qualitative examination of soil microflora associated with different management systems. *Soil Sci.*, **125**, 343–50.

28. ARCARA, P. G. (1972). Revisione e completamento di un'indagine sulla microflora dei terreni argillosi soggetti a differenti coltivazioni. *Ann. Inst. Sperim. Studio e Difesa del Suolo, Firenze*, **III**, 73–86.

29. ARCARA, P. G. (1983). Rapporto tra attività microbiologia e tecniche di coltivazione nei prati pascoli delle aree marginali. In: *La problematica delle terre marginali*, Collana P. F. 'Promozione della Qualità dell'Ambiente', CNR AC/4/151, Vol. IV, pp. 191–6.

30. ARCARA, P. G. and SPARVOLI, E. (1981). Fixation de l'azote et biomasse microbienne dans le sol de prairie soumis à trois engrais differents. *Colloque Humus-Azote*, **1**, 177–82.

31. ARCARA, P. G., SPARVOLI, E. and PIOVANELLI, C. (1982). Influence of agronomic techniques on soil biological activity in meadows of the north Apennines. Evolution du fertilite des soils dans differents systemes de culture. Criteres pour mesurer cette fertilite. Seminaire CEE. Agrimed, Bari, p. 357–66.

32. GREGORI, E. and MICLAUS, N. (1980). Effetti a medio termine di colture e tecniche di coltivazione diverse su alcune caratteristiche biologiche degli orizzonti superficiali di un suolo limoso-argilloso della Valdera, Pisa. *Ann. Inst. Sperim. Studio e Difesa del Suolo, Firenze*, **IX**, 183–204.

33. BOLTON, H. JR, ELLIOT, L. F., PAPENDICK, R. J. and BEZDICEK, D. F. (1985). Soil microbial biomass selected soil enzyme activities: effect of fertilization and cropping practices. *Soil Biol. Biochem.*, **17**, 297–302.

34. BIEDERBECK, V. O., CAMPBELL, C. A. and SCHNITZER, M. (1986). Effect of wheat rotations and fertilization on microorganisms and biochemical properties of a brown loam in Saskatchewan. *Transactions of XIII Congress of the International Society of Soil Science*, Vol. II, pp. 552–3.

35. VIENNOT-BOURGIN, G. (1964). Interactions entre les champignons telluriques et les autres organismes composant de la rhizosphere. *Ann. Inst. Pasteur, suppl.*, **107**, 21–62.

36. KRASILNIKOV, N. A. (1958). Soil microorganisms and higher plants. Office of technical services, US dept. of Commerce, Washington.

37. KRASILNIKOV, N. A. (1968). Sanitation of soil by microorganisms. In: *Ecology of Soil Bacteria* (Eds Gray and Parkinson), Liverpool University Press, pp. 422–38.
38. POCHON, J. and DE BARJAC, H. (1958). *Traité de microbiologie du sol*, Dunod, Paris.
39. GARRETT, S. D. (1965). Towards biological control of soil-borne plant pathogens. In: *Ecology of Soil-borne Plant Pathogenes* (Eds Baker and Snyder), University of California Press, pp. 4–17.
40. GARRETT, S. D. (1981). *Soil Fungi and Fertility*, 2nd edn, Pergamon Press, Oxford.

Effects of Agricultural Practices on the Physical, Chemical and Biological Properties of Soils: Part II—Use of Sewage Sludge and Agricultural Wastes

D. SAUERBECK

*Plant Nutrition and Soil Science Institute,
Federal Agricultural Research Centre, Bundesallée 50,
Braunschweig-Völkenrode, Federal Republic of Germany*

SUMMARY

In future the recirculation of organic wastes into soils must be considered, not only in the light of their possible toxic pollutant content but also with respect to the amount of plant nutrients acceptable for maintaining the nutrient balance and to avoid eutrophication of soils and waters by sewage sludge and agricultural wastes alike. In this connection, the very high concentration of livestock in some regions poses serious soil protection problems.

Water authorities and farmers must therefore work together to find an acceptable compromise between the most economic and the most environmentally sustainable level of farming and fertilizer use. This may entail production cuts, but it should also put an end to excessive demands which agriculture cannot meet.

In the long run, there is no unequivocally safe lower limit for inputs and accumulation of inorganic and organic pollutants in farmland. This implies that instead of consciously exploiting the existing threshold or limit values, lower levels still should be maintained wherever feasible. The only reliable way to achieve this goal is effectively to reduce usage of all potentially harmful substances and meticulously to prevent them from entering waste waters and sewage sludges.

181

1. INTRODUCTION

Returning organic wastes from agriculture or urban origin to the soil
wherever possible has always been regarded as theoretically desirable, for all
the reasons mentioned in the previous chapter and recapitulated in Table 1
[8, 65]. Just twenty years ago this sort of recycling met with virtually
universal approval. In the meantime, however, this has given way to a far
more discriminating approach, following the relatively belated discovery of
the pollutant content in urban wastes [21, 24, 55, 66]. By contrast, the main

Table 1
Influences of sewage sludges and agricultural wastes on soil

1. Physical:	Soil texture
	Soil structure
	— pore volume
	— pore size distribution
	Soil colour
2. Chemical:	pH value
	Redox potential
	Carbon content
	Exchange capacity (cations and anions)
	Colloidal and ionic precipitation and bonding reactions (e.g. complexation, etc.)
	Nutrient content
	Pollutant content
3. Biological:	Biological activity
	Microflora (bacteria, fungi and algae)
	Fauna (microfauna, mesofauna and macrofauna)

source of organic residues from agriculture is intensive livestock farming,
which inevitably poses a risk of over-abundant inputs of, in essence, useful
fertilizer [22, 73, 94, 95]. Accordingly, like the central subject addressed by
this symposium, this paper will put its emphasis less on the positive and far
more on the negative impacts of agricultural and non-rural wastes.

Heavy metals and persistent organic pollutants are the foremost
potentially harmful components in municipal wastes in general and sewage
sludges in particular. By contrast, the main problem with liquid manures
and other agricultural residues lies in their plant nutrient content, which
can also have an adverse impact if excessive quantities are applied.

2. HEAVY METALS

Domestic sewage and wastes each contain considerable amounts of organic matter and plant nutrients (primarily nitrogen and phosphorus) which are suitable for spreading as fertilizers on farmland, just like farmyard manure (see Table 2; [21]). Already in 1980 the Federal Republic of Germany produced 36 million tonnes of sewage sludge containing around 0·9 million tonnes of organic matter, around 65 000 tonnes of nitrogen and about 75 000 tonnes of phosphates (P_2O_5) [66]. Since then, extension of sewerage facilities and improved sewage treatment methods have raised the figure to between 42 and 47 million tonnes of sludge in Germany containing over 1 million tonnes of organic matter [13].

Theoretically, this would be enough to replace over 5% of the nitrogen and around 10% of the phosphates spread in mineral fertilizers in Germany. In practice, however, hitherto at most 40% of Germany's sewage sludges have been used as fertilizers or soil improvers. Recently this figure has even been falling. In the European Community as a whole, the figures vary widely from one country to another, depending on the capacity of the sewage network and existing sewage treatment facilities (Table 3; [21, 22, 73]). Overall, however, the Member States produce very large amounts of such sewage sludge. Spreading this material on farmland would undoubtedly not only be beneficial but also be far simpler for the local authorities than any other method of disposal [66].

But the other side of the picture is that sewage sludges contain undesirable and often downright harmful substances (Table 4; [33, 68, 92]). Heavy metals, e.g. cadmium, chromium, copper, mercury, nickel, lead and

Table 2

Composition of sewage sludge, compost from town refuse and sewage sludge, and farmyard manure

(Source: Dam Kofoed [21])

	% dry matter	% organic matter	% in the dry matter of		
			N	P_2O_5	K_2O
Dewatered sewage sludge	25	15	2·0	3·7	0·1
Compost	60	20	0·8	1·1	0·2
Farmyard manure	27	16	2·2	1·7	1·8
Liquid cattle manure	8	6	5·0	2·9	6·0
Liquid pig manure	6	6	8·3	3·8	4·0
Liquid poultry manure	15	15	7·3	6·1	3·2

Table 3
Use of sludge in various countries as a percentage of total amount
(Source: Dam Kofoed [21])

Country	Final disposal (%)		
	Agriculture	Controlled dumping	Discharged to the sea
Sweden	41	48	—
Norway	18	82	—
Finland	21	51	—
Denmark	48	52	—
Federal Republic of Germany	39	57	2
France	33	50	—
Belgium	15	85	—
The Netherlands	34	25	13
United Kingdom	40	31	29
Ireland	4	52	44
Switzerland	61	—	—
Italy	20	60	—

zinc, are the main culprits, either because they are particularly prevalent in sewage sludges or because even very small doses can already be toxic to plants or animals [49, 69]. In a considerable number of places large-scale sewage sludge spreading has in fact produced a perturbing or even unequivocally harmful accumulation of such heavy metals [11, 70, 71].

As all too often in such cases, many people immediately leap to the conclusion that all sewage sludges are dangerous and, therefore, unsuitable for spreading on farmland. Naturally, this only holds true for sludges containing significant amounts of heavy metals, i.e. if these heavy metals are not carefully and strictly enough kept out of the sewage. Admittedly, however, in modern-day society it can never fully be ruled out that some heavy metals enter sewage and, hence, sewage sludges, despite the careful controls [82].

Accordingly, over the last 10–15 years every Community Member State has been seriously considering the maximum heavy metal content tolerable in farmland or which amounts may be added to it in sewage sludges without further concern [33, 92]. Without going into the wide variety of different ideas in individual countries, this is an appropriate place simply to mention the recently adopted EEC Directive, which is a first attempt to harmonize the existing regulations in this area to some extent [18]. According to this Directive, apart from local exceptions, sewage sludges may no longer be

Table 4

Comparison of heavy metal concentrations (mg/kg dry weight) in United Kingdom
sewage sludges and agricultural soils
(Source: Webber *et al.* [92])

Element	Sludge		Soil		Ratio[a] sludge/soil
	Range	Common value	Range	Common value	
Cd	2–1 500	20	0·01–2·4	1[b]	20
Cr	40–14 000	400	5–1 000	100	4
Cu	200–8 000	650	2–100	20	32
Hg	0·2–18	5	0·01–0·3	0·03	167
Ni	20–5 300	100	10–1 000	50[b]	2
Pb	50–3 600	400	2–200	20	20
Zn	600–20 000	1 500	10–300	50	50
As	3–30	20	1–50	6	3
B	15–1 000	30	2–100	10	3
Co	2–260	15	1–40	10	1·5
F	60–40 000	250	30–300	150	2
Mo	1–40	6	0·2–5	2	3
Se	1–10	3	0·01–2	0·2	15

[a] Based on common values.
[b] Common values for Danish soils; Cd 0·2 and Ni 10.

applied to soils exceeding the heavy metal concentrations set out in Table 5,
whatever the reason might be.

But to prevent future sludge applications from increasing the heavy
metal content in soils to near the maximum limits, or at least to defer this
event for as long as possible, the Community Member States have also laid
down limit values for the heavy metal content in the sewage sludges
themselves. Sludges exceeding those threshold values may no longer be
used on farmland (Table 6; [73, 92]). The Joint EEC Directive takes a
different approach; instead of setting strictly defined maxima, it lays down
maximum ranges which may be exceeded only under exceptionally
favourable conditions but should otherwise not even be reached at all [18].

Most of the Community Member States which have been working
together for years on COST Project 68(1) on processing and use of sewage
sludge have initially concentrated on just the seven heavy metals listed in
the Directive since, as can be seen from Table 4, these are the elements
usually found in far higher concentrations in sludges than in most soils
[23]. Obviously, this is the reason why sewage sludge spreading can lead to

D. Sauerbeck

Table 5

Normal background (bkgd) and maximum permissible (perm) heavy metal concentrations (mg/kg dry wt) in agricultural soils
(Source: Webber et al. [92]; CEC [18])

Element	Germany		France perm	United Kingdom			CEC Directive perm
	bkgd	perm		bkgd	perm		
					Non-calcareous	Cal-careous	
Cd	0·2	3	2	1	3·5	3·5	1–3
Cr	30	100	150	100	600	600	—
Cu	30	100	100	5	140	280	50–140
Hg	0·1	2	1	<0·1	1	1	1–1·5
Ni	30	50	50	1	35	70	30–75
Pb	30	100	100	50	550	550	50–300
Zn	50	300	300	2·5	280	560	150–300
As				5	10	10	
B				1	3·25	3·25	
F				200	500	500	
Mo				2	4	4	
Se			10	0·5	3	3	

Values are total concentrations in soil except United Kingdom. Zn, Cu and Ni extracted by EDTA and B extracted by hot water.

a worrying accumulation of heavy metals in the soil at all. Of course, at least theoretically it is not impossible that elements other than the seven listed could also reach critical concentrations (Table 4). Up to now, however, it would nonetheless be reasonable to confine the statutory routine monitoring to these seven elements, since they are considered the most important or most hazardous (Table 7).

On the heavy metal limit values or maximum loads in soils and sewage sludges, the EEC Directive deliberately departs from the approach adopted by most national regulations in that it did not yet set a limit value for chromium. All past studies on heavy metal absorption by plants agree that chromium is particularly firmly bound in the soil and, therefore, all the less available to plants [12, 48, 49].

The same also applies to lead. Accordingly, the EEC Directive allows a soil lead content up to three times higher than the German Sewage Sludge Order (Table 5). In principle, mercury too is very firmly bound in the soil. However, since this element is particularly zootoxic there is little prospect

Table 6

Maximum heavy metal concentrations (mg/kg dry wt) in sludges considered
acceptable on agricultural lands
(Source: Webber *et al.* [92]; CEC [18]; Scheltinga and Candinas [73])

Element	B	D	DK	F	NL	Range	CEC Directive 1986
Cd	10	20	8	20	5	5–20	20–40
Cr	500	1 200		1 000	500	500–1 200	—
Cu	500	1 200		1 000	600	500–1 200	1 000–1 750
Hg	10	25	6	10	5	5–25	16–25
Ni	100	200	30	200	100	30–200	300–400
Pb	300	1 200	400	800	500	300–1 200	750–1 200
Zn	2 000	3 000		3 000	2 000	2 000–3 000	2 500–4 000
As	10				10		
Co	20						
Mn	500						
Se	25						

United Kingdom—sludge for public distribution should not contain more than
20 mg Cd/kg dry wt. Sludge applied to pasture land should not contain more than
3500 mg F/kg dry wt, and sludge applied to pasture land, gardens and recreational
areas should not contain more than 2000 mg Pb/kg dry wt.

of any further relaxation of the current, undeniably very cautious, limit
values [49, 70].

On the other hand, the bioavailability of copper and nickel depends far
more on soil acidity. There is, therefore, every reason not only for lower
limit values than for chromium and lead but also for setting the tolerance
ranges depending on the soil properties (Table 5). To date none of the
numerous plant species tested has absorbed large enough amounts of
copper or nickel from such contaminated soils as to demand limit values
stricter than the ones set out in the EEC Directive [70].

In marked contrast, with zinc and cadmium at least the vegetative plant
parts must be assumed often to contain almost as much if not more of these
heavy metals as the soil on which they are grown. Accordingly, to protect
plant and animal health, the European Community has set maximum
ranges for sandy and acidic soils in particular which are more cautious than
the corresponding limit values in the German Sewage Sludge Order (Table
5). This appears advisable for these particular elements and soils, not only
for the sake of plants but also because of the impairment of biological soil
functions otherwise possible [19, 28, 36].

Table 7

CEC limit values for heavy metals added annually to
agricultural land (kg/ha/yr)
(Source: CEC [18])

Parameters	Limit values[a]
Cadmium	0·15
Copper	12
Nickel	3
Lead	15
Zinc	30
Mercury	0·1
Chromium[b]	—

Based on a 10-year average.
[a] Member States may permit these limit values to be exceeded
in the case of the use of sludge on land which at the time of
notification of this Directive is dedicated to the disposal of
sludge but on which commercial food crops are being grown
exclusively for animal consumption. Member States must
inform the Commission of the number and type of sites
concerned. They must also ensure that there is no resulting
hazard to human health or the environment.
[b] It is not possible at this stage to fix limit values for
chromium. The Council will fix these limit values later on the
basis of proposals to be submitted by the Commission within
1 year following notification of this Directive.

There has in fact been a considerable reduction in microbial biomass and
in its functions on several soils treated with sewage sludges containing
heavy metals over many years [15, 56]. Although it is not always possible to
distinguish unequivocally between the primary and secondary effects, an
overview of the available data in Fig. 1 indicates that the heavy metal
damage thresholds for key biological processes and capacities in the soil are
of a similar magnitude as the values set in the EEC Directive [27].

Under these circumstances, it would certainly be wrong for the official
sewage sludge controls to be satisfied simply by keeping to the maximum
permissible heavy metal limit values. It may be theoretically correct that
even if these maxima were fully exploited it would still take between 100
and 250 years before today's normal soils gradually reach the heavy metal
limit value (Table 8). In practice, however, heavy metals accumulate far
faster since they also enter the soil from other sources, not least from the
contaminated air [68].

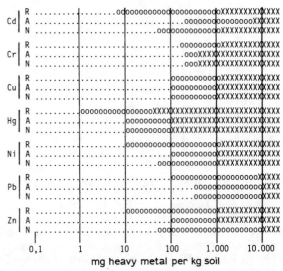

FIG. 1. Overview of available data on the impact of heavy metals on microbial respiration (*R*), nitrogen mineralization (*A*) and nitrification in the soil (*N*) (Source: Doelman [27]). ···· = no inhibiting effect observed; ○ ○ ○ ○ = inhibiting effect occasionally observed; × × × × = inhibiting effect always observed.

Table 8

Maximum permissible heavy metal content in German sewage sludges and in the soils on which they are spread, with the correspondingly calculated accumulation periods

(Source: Sauerbeck [68])

Element	Sewage sludge		Soil		Accumulation period (years)
	Limit value (ppm)	Maximum load (g/ha.a)	Limit value (ppm)	Average content[a] (ppm)	
Cd	20	33	3	0·2	250
Cr	1·200	2·000	100	26	110
Cu	1·200	2·000	100	11	130
Hg	25	42	2	0·1	140
Ni	200	333	50	30	180
Pb	1·200	2·000	100	22	120
Zn	3·000	5·000	300	50	150

5 t sewage sludge (dry matter) in 3 years.

3 kg heavy metal/ha = 1 ppm.

[a] Figures for arable land in Hesser (Source: Brüne and Ellinghaus [16a]).

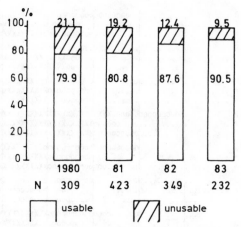

FIG. 2.   Percentage of usable and unusable sewage sludges, as defined in the German Sewage Sludge Order, in the samples surveyed at the Augustenberg Agricultural Test and Research Centre (Source: Hoffmann [43]).

Fortunately, however, the controls made over the last four to six years have shown relatively few cases where sewage sludges intended for use in agriculture have attained or even exceeded the set limit values (see Fig. 2). Moreover, in many places heavy metal contents have even fallen sharply following the introduction of improved waste water regulations [43, 54, 76]. Therefore, in future every means available must be used to promote this trend towards eliminating heavy metals as effectively as possible directly at their source [2, 76].

Table 9

Results of sewage sludge surveys in the North Rhine area between 1 April 1983 and 31 March 1986

(Source: Linssen and Rieß [54])

Pollutant content category	Limit value utilization rate (%)	Relative share for						
		Pb	Cd	Cr	Cu	Ni	Hg	Zn
1	0–25	83	80	91	59	74	93	24
2	>25–50	15	18	7	31	23	5	43
3	>50–75	1	2	2	7	2	1	26
4	>75–100	1	0	0	2	1	1	7
5	>100–125	0	0	0	1	0	0	0

Number of samples: 927 or 928.

For example, one effective incentive to achieve this would be to classify all the sewage sludges tested into quality categories, as several agricultural testing centres in Germany have done already (Table 9; [54, 60]). Leaving aside zinc—which, unfortunately, enters effluent mainly from municipal sources rather than from industry or commerce—Table 9 shows that over 90% of all the sewage sludges tested contained less than half the maximum permissible heavy metal concentrations.

In future such a quality classification could probably result in an even sharper reduction in the heavy metal content in municipal sewage sludges, whereas tightening up the limit values would only prompt more local authorities to resort to landfilling as an alternative means of disposing of sewage sludge.

3. ORGANIC POLLUTANTS

All the foregoing suggests that it is difficult enough to ascertain the true situation with heavy metals in sewage sludges, even after narrowing down the list to just seven elements. By the same token, at present it is entirely impossible to assess the potential hazard posed by the 20 000 or so organic pollutants known [10, 30, 52, 68]. In the specific context of waste water and sewage sludge, this enormous number of organic chemicals at large in the environment can be sub-divided into eight families, some large, some small, as shown in Table 10. Of these, the polychlorinated biphenyls (PCBs) and polyaromatic hydrocarbons (PAHs) are generally regarded as the most critical ones [14, 25, 42, 53].

Table 10
Main families of organic micropollutants in sewage sludges
(Source: Tarradellas *et al.* [80])

Halogenated aromatics	Polyaromatic and heteroaromatic
polychlorinated biphenyls (PCBs)	hydrocarbons
polychlorinated terphenyls (PCTs)	Halogenated aliphatics
polychlorinated naphthalenes (PCNs)	Aliphatic and aromatic
polychlorobenzenes	hydrocarbons
Aromatic amines and nitrosoamines	Phthalate esters
Halogenated aromatics containing oxygen	Pesticides
phenols	Lindane
chlorophenols	Dieldrin
polychlorodiphenyl ethers	DDE
polychlorodibenzofuranes	organophosphorus
polychlorodibenzo-*p*-dioxins	

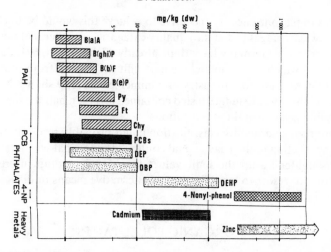

FIG. 3. Micropollutant concentration in sewage sludges from Switzerland
(1983–1984) (Source: Tarradellas *et al.* [80]).

The survey of representative samples from Switzerland (Fig. 3) gives an
idea of the ranges of organic pollutants to be expected in sewage sludge
[81]. It shows that European sewage sludges contain between 0·5 and about
10 mg PAH/kg, depending on the precise substance concerned, most of
them obviously arising from road and surface runoff [3, 16, 52, 68].
Comparable contents of PCBs are also found, though in exceptional cases
of industrial pollution even readings of 1000 ppm or more have been
recorded [26, 32, 34, 43, 51, 76].

In the past, abundant quantities of organochlorine pesticides were also
common in sewage sludge. Today only traces well under 1 ppm remain [45,
57]. Only more recently have phthalates received the attention they
deserve, now that levels of up to several hundred ppm of these plasticizers
have been detected in sewage sludge [29, 74, 80]. Finally, levels of over 1 g/
kg dry matter have been recorded for nonylphenol, a toxic decomposition
product of many detergents present in effluents [1, 72].

Analyses have also detected PAHs and PCBs in the soils treated with
sewage sludge, but mostly in rather small quantities. Although both these
groups are renowned for their persistence, this nonetheless suggests that
they are gradually broken down or degraded in the soil [75], unless these
decreases can simply be attributed to the relatively high volatility of such
substances, as e.g. in the case of many chlorinated hydrocarbons [31a]. The
latest surveys, carried out on long-term German sewage sludge field

Table 11
PCBs in soils fertilized with municipal waste for many years
(Six German institutes. Source: Kampe [45])

Substance	Control		Treated with sludge		
	Minimum	*Maximum*	*Minimum*	*Maximum*	*Average*
K28	—	—	0·001	0·011	0·005
K52	0·001	0·002	0·001	0·123	0·017
K101	0·001	0·008	0·001	0·299	0·036
K138	0·001	0·009	0·001	0·339	0·049
K153	0·001	0·008	0·001	0·303	0·044
K180	0·001	0·006	0·001	0·197	0·031

All figures in mg/kg dry soil.

experiments show no appreciable accumulation of organochlorine pesticides in such soils [45, 68], and even the more highly chlorinated PCBs were found only in quantities well below 1 ppm (Table 11; [45, 57, 59]).

Nevertheless, a large number of chlorinated hydrocarbons are known to be broken down by microbes in soil and sediments only under anaerobic conditions [37, 38, 51]. The opposite is obviously true of the phthalates, which, in spite of relatively high inputs, are not likely to survive for long under aerobic soil conditions (Table 12; [74]). Presumably the same should apply to nonylphenol, even though specific figures do not yet exist for this particular substance.

Table 12
Breakdown of phthalates in the soil (0·5 mg DEHP/kg garden soil)
(Source: Shanker *et al.* [74])

Incubation period (d)	Residual concentration (μg/kg garden soil)			
	Aerobic		Anaerobic	
	DEHP	*Phthalic acid*	*DEHP*	*Phthalic acid*
0	480 ± 9	0	478 ± 9	0
5	430 ± 8	8 ± 1·1	460 ± 8	traces
10	320 ± 11	7 ± 1·1	439 ± 6	2 ± 0
20	120 ± 4	11 ± 0·6	389 ± 5	8 ± 1·1
30	40 ± 8	5 ± 0·6	318 ± 7	11 ± 0·6
Sterile control	471 ± 4	0	478 ± 7	0

Earlier studies on assimilation by plants have been confined so far to just a few organochlorine pesticides and selected PAHs and PCBs. They have revealed hardly any translocation of both PAHs into the aerial parts of the plant. Small amounts of PAHs and PCBs, although well below the corresponding level in the soil itself, were detected only in the roots of carrots, for example (Table 13). Accordingly, there is definitely no fear of any marked accumulation of these substances in terrestrial food plants [41, 68, 75, 79, 90].

Swiss studies have, however, shown that at least PCBs too could well accumulate in soil organisms, with a corresponding biomagnification in their predators (Fig. 4; [81]). As the soil concentration figures show, these are not even soils suffering from exceptionally high PCB contamination from urban wastes. Nonetheless, there was a clear correlation with the PCB content in the biota investigated. Consequently, in the future not only the soil–plant–animal–human food chain must be taken into account when assessing organic pollutants but also soil fauna and their predators.

Table 13
Uptake of HCB and PCBs by carrots and radishes
(Source: Wallnöfer *et al.* [90])

Substance	Concentration (ppm)			
	In the soil		In the plants	
	Start of experiment	End of experiment	Radish	Carrot
HCB	0·05	0·028	n.d.	0·011
2-PCB	0·05	n.d.	n.d.	n.d.
4-PCB	0·05	0·029	n.d.	n.d.
6-PCB	0·05	0·032	n.d.	n.d.
HCB	0·5	0·30	n.d.	0·05
2-PCB	0·5	0·59	n.d.	n.d.
4-PCB	0·5	0·30	n.d.	0·022
6-PCB	0·5	0·44	n.d.	0·008
Arochlor 1254	0·5	0·53	n.d.	n.d.
HCB	5·0	2·20	0·025	0·50
2-PCB	5·0	4·20	n.d.	0·81
4-PCB	5·0	3·30	0·025	0·47
6-PCB	5·0	3·30	0·025	0·48

n.d. = not detectable.

This applies not least because of the recent finding that some earlier theories on the degradability of organic chemicals in soil and water need to be revised. First, this is the case with the older ideas about the stability of such substances fixed in soil organic matter [9, 17, 61, 68]. Alongside irreversible polymerization into stable humic matter, some organic pollutants are only bound relatively loosely by radicals, which may then be re-released and made bioavailable once more at a later time, albeit in most limited quantities [46, 47, 61, 78].

Secondly, recently there has been a growing awareness that many of these substances are only partly broken down by the soil microflora—in

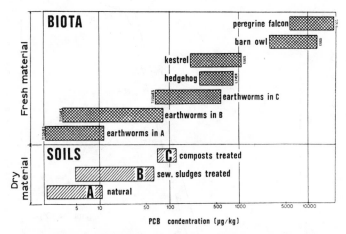

FIG. 4.   Magnification of PCB concentrations in Swiss agro-ecosystems (Source: Tarradellas *et al.* [80]).

other words, the breakdown products formed can be even more persistent or more toxic than the original substance was before [67]. Under these circumstances, chlorinated or nitrated compounds in general, and aromatic organic compounds in particular, could very well pose a potential hazard in the soil environment which should not be underestimated and which at any rate has not yet been studied in sufficient depth [40]. Not least, this holds true because it has been found that the microbial degradation often depends on cometabolism and on concentration levels. In practice, this means that some of these substances can be biodegraded only at higher but not in the lower concentrations expected in soil and water (Fig. 5). This phenomenon is one possible explanation for the recent upswing in

detections of such substances in groundwater [58, 91]. It therefore follows
that it is undoubtedly desirable to work out suitable limit values for the
organic pollutant contents in soils too. Naturally, bearing in mind the
multitude of substances and turnover products involved, this would be an
enormous task. For the time being, therefore, other countries are happy to

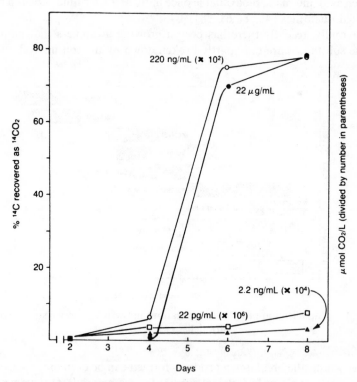

FIG. 5. Formation of CO_2 from 2,4-D added to stream water at four initial
concentrations (Source: Alexander [7]).

draw on the corresponding values for reference, monitoring and clean-up
requirements published by The Netherlands, Leidraad Bodemsanering
(Table 14; [50]). Just as in the case of heavy metals, however, as a general
rule the accent must be put on keeping inputs of persistent organic
pollutants into sewage sludges and soils to the minimum wherever possible
[68].

Table 14

Guide values for assessing soil contamination by organic substances (in mg/kg dry matter)
(Source: Leidraad Bodemsanering [50])

Category	A	B	C
Polycyclic aromatic hydrocarbons			
Naphthalene	0·1	5	50
Anthracene	0·1	10	100
Phenanthrene	0·1	10	100
Fluoranthene	0·1	10	100
Pyrene	0·1	10	100
Benzo(a)pyrene	0·05	1	10
PAHs—total	1	20	200
Chlorinated hydrocarbons (CH)			
Aliphatic CHs ind.	0·1	5	50
Aliphatic CHs total	0·1	7	70
Chlorobenzenes ind.	0·05	1	10
Chlorobenzenes total	0·05	2	20
Chlorophenols ind.	0·01	0·5	5
Chlorophenols total	0·01	1	10
CH—total	0·05	1	10
PCBs	0·05	1	10
EOCl—total	0·1	8	80

Category	A	B	C
Pesticides			
Organochlorine ind.	0·1	0·5	5
Organochlorine total	0·1	1	10
Pesticides—total	0·1	2	20
Aromatic compounds			
Benzene	0·01	0·5	5
Ethyl benzene	0·05	5	50
Toluol	0·05	3	30
Xylol	0·05	5	50
Phenol	0·02	1	10
Aromatics—total	0·1	7	70
Other organic compounds			
Tetrahydrofuran	0·1	4	40
Pyridine	0·1	2	20
Tetrahydrothiophen	0·1	5	50
Cyclohexanone	0·1	6	60
Styrol	0·1	5	50
Gasoline	20	100	800
Mineral oil	100	1 000	5 000

A = Reference value. B = Test requirements. C = Clean-up limit.
Rough guide only; the precise values will depend on the use made of the soil and on the specific conditions.

4. EXCESSIVE PLANT NUTRIENTS

Many problems posed by sewage sludge spreading on farmland, and the associated pollution loads, are likely to be cut down in future, since the EEC Directive explicitly recommends that the quantities of sludge applied should be kept in line with the amount of nutrients required by the plants. This takes account of the fact that it is impossible simply to increase the nitrogen stock, and hence the nitrogen turnover in the soil, limitlessly without increasing the risk of N losses into the groundwater and the environment at the same time (Fig. 6; [62, 67, 84, 95]).

Another problem is the phosphate eutrophication of land on which organic waste is frequently spread as a fertilizer. Soil runoff can then pollute surface waters or even lead to the phosphates penetrating deeper strata in light soils with high humus levels (Fig. 7; [44a, 88]). Consequently, if the doses spread were tailored more strictly to the amounts of nutrients required by the plants, in future sewage sludges would actually have to be classified as phosphate fertilizers, since under balance considerations phosphates are the nutrient which limit the appropriate amounts to be applied [35, 39, 83].

Quite apart from the considerably lower pollution levels, in this case sewage sludge spreading would actually be almost better for the environment than the current practice in some livestock farming regions. In recent years, imported animal feed has pushed cattle and, even more so, pig rearing into hitherto unknown intensities, concentrating on very high animal numbers per farm or per hectare. Within the European

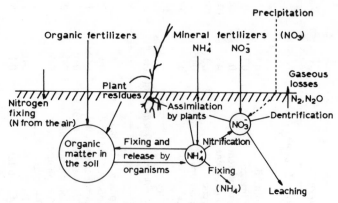

FIG. 6.   Nitrogen cycle in farmland.

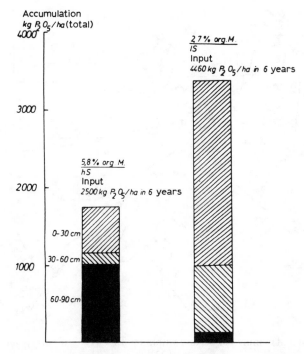

FIG. 7. Phosphorus accumulation at different depths in humic sand and sandy clay soils after 6 years' spreading of 90 m^3 of liquid pig manure per hectare per year (Source: Vetter and Steffens [88]).

Community, this holds particularly true of the Benelux countries, North West Germany, Denmark and the United Kingdom (Fig. 8; [22, 73]). But the accompanying livestock droppings inevitably add large amounts of plant nutrients to the soil, the regional average standing at between 80 and 180 kg of nitrogen per hectare per year and around 40 kg of P or 90 kg of P_2O_5 per hectare per year [22]. It is not even unusual for individual farms to record readings far above twice these averages.

Yet there is no denying that in order to maintain the nutrient balance one and a half livestock units per hectare would be ample to provide all the phosphates needed by the crops. Although more nitrogen is required, this can be provided by farmyard manure alone only at a considerable loss, because of the seasonal spreading limitations and the corresponding leaching and runoff hazard. Consequently, from both the environmental and economic points of view, 1·5 livestock units per hectare would be the

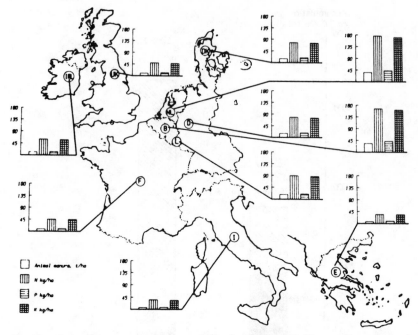

FIG. 8.   Animal manure per hectare within the EEC (1982) (Source: Dam Kofoed [22]).

desirable optimum, since it allows (a) fullest possible utilization of farmyard manure and (b) supplementary spreading of the exact amount of mineral fertilizer required by the particular crops at the right time (Table 15).

However, current farming practices are frequently far from achieving this, since labour-saving liquid manure techniques often make it impossible to spread this manure at the right time, for shortage of storage space [63].

Table 15
Nutrient supply with liquid manure (in kg/year)

Unit	m^3/a	% dm	N	P_2O_5	K_2O	MgO
1 Dairy cow (5·000 litre)	20	7·5	80	40	130	18
1 Livestock unit, cattle	16	10	80	50	70	19
7 Slaughter pigs (equivalent to 3 sows plus piglets)	14	7·5	80	55	45	17
100 Hens, 300 chicks	9	12	80	70	40	14

Table 16
Nitrogen fractions and their bioavailability in organic fertilizers
(Source: Sluijsmans and Kolenbrander [77])

| Fertilizer | min. N | org. N | |
	immediately	available rapidly	slowly
Solid manure	10	46	44
Cattle slurry	50	25	25
Pig slurry	51	34	15
Chicken slurry	54	32	14
Slurry	94	3	3

All figures expressed as a percentage of total N content.

This problem of spreading only as much as is needed is particularly acute with animal slurry, since it contains a far higher proportion of mineral nitrogen, which is easily lost, than the solid farmyard manure commonly used in the past (Table 16; [77, 88, 95]).

Admittedly, in practice the danger of overfertilization is somewhat reduced by the considerable ammonia losses almost always associated with animal manure spreading—in other words, not all the nitrogen in the manure actually enters the soil (Table 17; [64, 68]).

Leaving aside the environmental concern about these gaseous ammonia losses, in many cases this still leaves too much readily convertible nitrogen in the soil which can then be leached out during the non-growing season [93]. Accordingly, many studies have confirmed that in autumn three to four times more liquid manure is needed to attain the same fertilizer effect as in the spring (Table 18; [4, 77, 86, 87, 95]).

The often-mentioned danger of copper accumulation from liquid pig manure is a relatively secondary problem, well behind that of the nutrient

Table 17
Nitrogen losses from farmyard manures

| Process | Storage | Spreading | In the soil |
	Preparation Stabilization	Distribution	Leaching Denitrification
% loss	30–90	0–50	0–50
Form	NH_3, N_2	NH_3	NO_3^-, N_2, N_2O

Table 18
Amounts of animal slurry required to obtain 80 kg available nitrogen
per hectare in autumn (15% utilization rate) and spring (60%)
(Source: Vetter [86])

		m^3	kg N	Utilization rate
Cattle slurry	Autumn	120	500	15
(7·5% DM)	Spring	30	120	60
Pig slurry	Autumn	90	500	15
(7·5% DM)	Spring	22·5	120	60
Chicken slurry	Autumn	50	500	15
(15% DM)	Spring	12	120	60

balance and ammonia losses. True, for years many feedstuffs on the Community market contained very high Cu levels, most of which then entered the soil via the livestock droppings [44]. But this is far more a long-term problem than the nitrogen and phosphorus enrichments. The existing calculations suggest that it is unlikely to reach critical proportions in the next few decades except on land with a particularly high livestock headage

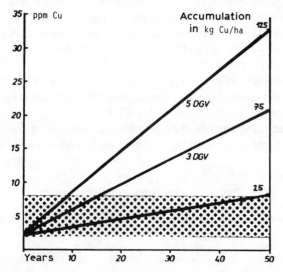

FIG. 9.   Copper accumulation in the soil after 50 years' application of liquid pig manure with an average copper content (Source: Vetter and Steffens [89]). (0·04 kg Cu/m³ liquid manure; shaded area = desirable level of Cu.)

(Fig. 9; [20, 89]). Fortunately, recently this danger too has been taken into account by introducing regulations limiting the copper content in animal feed [31]. On the other hand, however, still too little is known at the moment to say whether or not other livestock feed additives also harbour long-term soil contamination problems [85].

5. CONCLUSIONS

In short, the following conclusions can be drawn from the foregoing considerations for future agricultural and environment policy:

1. In principle, it is still desirable to return organic wastes to the soil. In future, however, not only the potential content of dangerous substances must be taken into account but also, to a greater extent, the exact quantity needed for plant nutrition.
2. Concerning the nutrient balance and eutrophication, this applies to the impact of sewage sludges and agricultural wastes alike, though the problem is particularly acute in regions with a very high livestock density.
3. Water authorities and farmers must therefore work together to find an acceptable compromise between the most economic and most environmentally sustainable level of farming and fertilizer use. This may entail production cuts, but it should also put an end to unrealistic environmental demands which agriculture cannot meet.
4. In the long run, there is no unequivocally safe lower limit for inputs and accumulations of pollutants in farmland. This implies that instead of consciously exploiting the existing threshold or limit values, lower levels still should be attained wherever feasible.
5. The only reliable way to achieve this goal is effectively to reduce the usage of all potentially harmful substances and meticulously to prevent them from entering waste waters and sewage sludges.

REFERENCES

1. AHEL, M., GIGER, W. and KOCH, M. (1986). Behaviour of nonionic surfactants in biological waste water treatment. In: *Organic Micropollutants in the Aquatic Environment*, CEC-Symposium Vienna, Austria, October 22–24, 1985 (Eds A. Bjørseth and G. Angeletti) (EUR 10388), Reidel Publ., Dordrecht, pp. 414–28.
2. AICHBERGER, K., MAYR, E. and SCHMOIGL, K. (1984). Systematische

Klärschlammkontrolle und landwirtschaftliche Verwertung in Oberösterreich. In: *Processing and Use of Sewage Sludge*, CEC-Symposium Brighton, UK, September 27–30, 1983 (Eds P. L'Hermite and H. Ott) (EUR 9129), Reidel Publ., Dordrecht, pp. 158–60.

3. AICHBERGER, K. and REIFENAUER, D. (1984). Determination of polycyclic aromatic hydrocarbons in sewage sludge. In: *Processing and Use of Sewage Sludge*, CEC-Symposium Brighton, UK, September 27–30, 1983 (Eds P. L'Hermite and H. Ott) (EUR 9129), Reidel Publ., Dordrecht, pp. 161–3.

4. AIGNER, H. (1978). In: *Faustzahlen für Landwirtschaft und Gartenbau*, Landwirtschafts-Verlag GmbH, Münster-Hiltrup.

5. ALEXANDER, M. (1981). Biodegradation of chemicals of environmental concern. *Science*, **211**, 132–8.

6. ALEXANDER, M. (1983). Ecologically significant microbial transformations of synthetic chemicals. In: *Environmental Biogeochemistry, Ecol. Bull. Stockholm* (Ed. R. Fallberg), Vol. 35, pp. 503–10.

7. ALEXANDER, M. (1985). Biodegradation of organic chemicals. *Environ. Sci. Technol.*, **18**, 106–11.

8. BALLONI, W. and FAVILLI, F. (1986). Effect of some agricultural practices on the biological soil fertility. Manuscr. CEC-Symposium *Scientific Basis for Soil Protection in the European Community*, Berlin, October 6–8, 1986 (in press).

9. BARTHA, R., YOU, I.-S. and SAXENA, A. (1983). Humus-bound residues of phenylamide herbicides: their nature, persistence and monitoring, IUPAC, Pergamon Press, Oxford, pp. 345–50.

10. BERROW, M. L. (1986). An overview of soil contamination problems. In: *Proc. Int. Conf. Chemicals in the Environment* (Eds J. N. Lester, R. Perry and R. M. Sterritt), Selper Ltd, London, pp. 543–52.

11. BERROW, M. L. and BURRIDGE, J. C. (1984). Persistence of metals in sewage sludge treated soils. In: *Processing and Use of Sewage Sludge*, CEC-Symposium Brighton, UK, September 27–30, 1983 (Eds P. L'Hermite and H. Ott) (EUR 9129), Reidel Publ., Dordrecht, pp. 418–22.

12. BERROW, M. L. and REAVES, G. A. (1984). Background levels of trace elements in soils. In: *Proc. 1st Int. Conf. on Environmental Contamination*, London, July 1984, pp. 333–40.

13. BLICKWEDEL, P. T. and SCHENKEL, W. (1986). Klärschlamm-Menge und Anfall in der Bundesrepublik Deutschland. *Korrespondenz Abwasser*, **33**, 680–5.

14. BRIDLE, T. R. and WEBBER, M. D. (1983). A Canadian perspective on toxic organics in sewage sludge. In: *Environmental Effects of Organic and Inorganic Contaminants in Sewage Sludge*, CEC-Symposium Stevenage, UK, May 25–26, 1982 (Eds R. D. Davis, G. Hucker and P. L'Hermite) (EUR 8022), Reidel Publ., Dordrecht, pp. 27–37.

15. BROOKES, P. C., McGRATH, S. P., KLEIN, D. A. and ELLIOTT, E. T. (1984). Effects of heavy metals on microbial activity and biomass in field soils treated with sewage sludge. *Proc. 1st Int. Conf. on Environmental Contamination*, London, July 1984, pp. 574–83.

16. BRÜNE, H. (1986). Schadstoffeintrag in Böden durch Industrie, Besiedlung, Verkehr und Landbewirtschaftung. In: VDLUFA-Kongreßband 1985 Gießen 'Bodenbewirtschaftung, Bodenfruchtbarkeit, Bodenschutz', VDLUFA-Verlag, Darmstadt, pp. 85–102.

16a. BRÜNE, H. and ELLINGHAUS, R. (1982). Schwermetallgehalte in landwirt-
schaftlich genutzten Ackerböden Hessens. *Landwirtsch. Forsch. Sonderheft*, **38**, 338–49.
17. CAPRIEL, P., HAISCH, A. and KHAN, S. U. (1985). Distribution and nature of bound (nonextractable) residues of atrazine in a mineral soil nine years after the herbicide application. *J. Agric. Food Chem.*, **33**, 567–9.
18. CEC DIRECTIVE (1986). Council Directive of 12 June 1986 on the protection of the environment, and in particular of the soil, when sewage sludge is used in agriculture. Official Journal of the European Communities No. L 181/6-12, 4 July 1986.
19. COPPOLA, S. (1986). Summary of investigations in Italy and effects of sewage sludge on soil microorganisms. In: *Factors Influencing Sludge Utilisation Practices in Europe*, CEC-Seminar Liebefeld, Switzerland, May 8–10, 1985 (Eds R. D. Davis, H. Häni and P. L'Hermite) (EUR 10360), Elsevier Applied Science, London, pp. 72–9.
20. DAM KOFOED, A. (1980). Copper and its utilization in Danish agriculture. *Fertilizer Research*, **1**, 63–71.
21. DAM KOFOED, A. (1984). Optimum use of sludge in agriculture. In: *Utilisation of Sewage Sludge on Land: Rates of Application and Long-term Effects of Metals*, CEC-Seminar Uppsala, Sweden, June 7–9, 1983 (Eds. S. Berglund, R. D. Davis and P. L'Hermite) (EUR 8822), Reidel Publ., Dordrecht, pp. 2–20.
22. DAM KOFOED, A. (1985). Pathways of nitrate and phosphate to ground and surface waters. In: *Environment and Chemicals in Agriculture*, CEC-Symposium Dublin, Ireland, October 15-17, 1984 (Ed. F. P. W. Winteringham) (EUR 10050), Elsevier, London, New York, pp. 27–69.
23. DAVIS, R. D. (1980). Control of contamination problems in the treatment and disposal of sewage sludge. Water Research Centre, Stevenage, UK, Technical Report TR 156.
24. DAVIS, R. D. (1984). Crop uptake of metals (cadmium, lead, mercury, copper, nickel, zinc and chromium) from sludge-treated soil and its implications for soil fertility and for the human diet. In: *Processing and Use of Sewage Sludge*, CEC-Symposium Brighton, UK, September 27–30, 1983 (Eds P. L'Hermite and H. Ott) (EUR 9129), Reidel Publ., Dordrecht, pp. 349–57.
25. DAVIS, R. D., HOWELL, K., OAKE, R. J. and WILCOX, P. (1984). Significance of organic contaminants in sewage sludges used on agricultural land. In: *Proc. 1st Int. Conf. on Environmental Contamination*, London, July 1984, pp. 73–9.
26. DIERCXSENS, P. and TARRADELLAS, J. (1983). Presentation of the analytical and sampling methods and of results on organo-chlorines in soils improved with sewage sludges and compost. In: *Environmental Effects of Organic and Inorganic Contaminants in Sewage Sludge*, CEC-Workshop Stevenage, UK, May 25–26, 1982 (Eds R. D. Davis, G. Hucker and P. L'Hermite), Reidel Publ., Dordrecht, pp. 59–68.
27. DOELMAN, P. (1986). Resistance of soil microbial communities to heavy metals. In: *Proc. Europ. Microbiol. Soc. Symp. Copenhagen, Denmark*, August 4–8, 1985 (Eds V. Jensen, A. Kjøller and L. H. Sørensen), Elsevier, London, New York, pp. 369–84.
28. DOMSCH, K. (1985). Funktionen und Belastbarkeit des Bodens aus der Sicht der Bodenmikrobiologie. Stuttgart, Mainz: Kohlhammer, 67 S.

29. ECE (1985). Technical Report No. 19. An assessment of the occurrence and effects of dialkyl ortho-phthalates in the environment. Brussels, 22 May 1985.

30. EEC (1976). European Economic Community Council Directive on pollution caused by certain dangerous substances discharged into the aquatic environment of the community. EEC-OJL 76/464 Brussels.

31. ENTEL, H.-J., FÖRSTER, N. and HINCKERS, E. (Hrsg.) (1970–1985) Futtermittelrecht. Textsammlung mit Begründungen und Erläuterungen. Loseblattsammlung, Parey, Berlin, Hamburg.

31a. FAIRBANKS, B. C. and O'CONNOR, G. A. (1984). Toxic organic behavior in sludge-amended soils. In: *Proc. 1st Int. Conf. on Environmental Contamination*, London, July 1984, pp. 80–3.

32. FIEGGEN, W. (1981). Pesticides and PCBs in sewage sludge. Commission of the European Communities. COST Project 68 Bis 'Treatment and use of sewage sludge', Final Report, III. Technical Annexes, pp. 73–100.

33. FLEMING, G. A. and DAVIS, R. D. (1986). Contamination problems in relation to land use. 4th International Symposium 'Processing and use of organic sludge and liquid agricultural wastes' (CEC), Rome, Italy, October 8–11, 1985 (in press).

34. FUJIWARA, K. (1975). Environmental and food contamination with PCBs in Japan. *Sci. Total Environ.*, **4**, 219–47.

35. FURRER, O. J., GUPTA, S. K. and STAUFFER, W. (1984). Sewage sludge as a source of phosphorus and consequences of phosphorus accumulation in soils. In: *Processing and Use of Sewage Sludge*, CEC-Symposium Brighton, UK, September 27–30, 1983 (Eds P. L'Hermite and H. Ott) (EUR 9129), Reidel Publ., Dordrecht, pp. 279–94.

36. GUPTA, S. K. and STADELMANN, F. X. (1984). Effect of sewage sludge on the biorelevant cadmium concentration. In: *Processing and Use of Sewage Sludge*, CEC-Symposium Brighton, UK, September 27–30, 1983 (Eds P. L'Hermite and H. Ott) (EUR 9129), Reidel Publ., Dordrecht, pp. 435–45.

37. HAIDER, K. (1983). Abbau und Umwandlung von γ-HCH und anderen HCH-Isomeren durch Bodenmikroorganismen. In: *Hexachlorcyclohexan als Schadstoff in Lebensmitteln. Materialien zur DFG-Veranstaltung 28./29.11.1979 und 06.03.1980*, Verlag Chemie, Weinheim, pp. 73–8.

38. HAIDER, K. (1983). Anaerobic microsites in soils and their possible effect of pesticide degradation. In: *IUPAC—Pesticide Chemistry. Human Welfare and the Environment*, Pergamon Press, Oxford, pp. 351–6.

39. HALL, J. E. and WILLIAMS, J. H. (1985). Nitrogen and phosphorus value of sewage sludges. Concerted action treatment and use of sewage sludge. Revision of Document Nr. SL/82/82-XII/ENV/35/82 by R. Chaussod *et al.*, Commission of the European Communities, Brussels, Belgium.

40. HALLAS, L. E. and ALEXANDER, M. (1983). Microbial transformation of nitroaromatic compounds in sewage effluent. *Appl. Environ. Microbiol.*, **45**, 1234–41.

41. HARMS, H. and SAUERBECK, D. (1983). Toxic organic compounds in town waste materials: their origin, concentration and turnover in waste composts, soils and plants. In: *Environmental Effects of Organic and Inorganic Contaminants in Sewage Sludge*, CEC-Symposium Stevenage, UK, May 25–26 (Eds R. D. Davis, G. Hucker and P. L'Hermite) (EUR 8022), Reidel Publ., Dordrecht, pp. 38–51.

42. HARMS, H. and SAUERBECK, D. (1984). Organische Schadstoffe in Siedlungs-abfällen: Herkunft, Gehalt und Umsetzung in Böden und Pflanzen. *Angew. Botanik*, **58**, 97–108.

43. HOFFMANN, G. (1984). Bodenkundliche und pflanzenbauliche Aspekte beim Einsatz von Siedlungsabfällen in der Landwirtschaft. Veröffentlichungen der Landwirtschaftlich-technischen Bundesanstalt. *Linz/Donau, Österreich*, **17**, 17–64.

44. HOVMAND, M. F. (1984). Cycling of Pb, Cd, Cu, Zn and Ni in Danish agriculture. In: *Utilization of Sewage Sludge on Land: Rates of Application and Long-term Effects of Metals*, ECE-Seminar Uppsala, Sweden, June 7–9, 1983 (Eds S. Berglund, P. D. Davis and P. L'Hermite) (EUR 8822), Reidel Publ., Dordrecht, pp. 166–85.

44a. HUCKER, T. W. G. and CATROUX, G. (Eds) (1981). Phosphorus in sewage sludge and animal waste slurries. Proc. CEC-Seminar Groningen, Netherlands, June 12–13, 1980, Reidel Publ., Dordrecht, 443 p.

45. KAMPE, W. (1986). Potentielle organische Schadstoffe in Böden und Pflanzen nach langjähriger Anwendung von Klärschlammen (unveröffentl.).

46. KHAN, S. U. (1982). Bound pesticide residues in soil and plants. *Residue Review*, **84**, 1–25.

47. KHAN, S. U. and HAMILTON, H. A. (1980). Extractable and bound (non-extractable) residues of prometryn and its metabolites in organic soil. *J. Agric. Food Chem.*, **28**, 126–32.

48. KICK, H. and BRAUN, B. (1977). Wirkung von chromhaltigen Gerbereischläm-men auf Wachstum und Chromaufnahme bei verschiedenen Nutzpflanzen. *Landwirtsch. Forsch.*, **33**, 160–73.

49. KLOKE, A., SAUERBECK, D. and VETTER, H. (1984). The contamination of plants and soils with heavy metals and the transport of metals in terrestrial food chains. In: *Changing Metal Cycles and Human Health* (Ed. J. O. Nriagu), Springer, Berlin, pp. 113–41.

50. LEIDRAAD BODEMSANERING (1983). Leitfaden Bodensanierung. Hrsg. vom Niederländischen Ministerium für Wohnungswesen, Raumordnung und Umwelt (Ministerie van Volkshuisvesting, Ruintelijke Ordening en Milieu-beheer. Staatsdruckerei (Staatsdrukkerij, Uitgeverij).

51. LESTER, J. N. (1983). Occurrence, behaviour and fate of organic micropollutants during waste water and sludge treatment processes. In: *Environmental Effects of Organic and Inorganic Contaminants in Sewage Sludge*, CEC-Symposium Stevenage, UK, May 25–26 (Eds R. D. Davis, G. Hucker and P. L'Hermite) (EUR 8022), Reidel Publ., Dordrecht, pp. 3–18.

52. LESTER, J. N. (1984). Presence of organic micropollutants in sewage sludge. In: *Processing and Use of Sewage Sludge*, CEC-Symposium Brighton, UK, September 27–30, 1983 (Eds P. L'Hermite and H. Ott) (EUR 9129), Reidel Publ., Dordrecht, pp. 150–7.

53. LINDGAARD-JØRGENSEN, P. and NEERGAARD JACOBSEN, B. (1986). A data base on behaviour and effects of organic pollutants in waste water treatment processes. In: *Organic Micropollutants in the Aquatic Environment*, CEC-Symposium Vienna, Austria, October 22–24, 1985 (Eds A. Bjørseth and G. Angeletti) (EUR 10388), Reidel Publ., Dordrecht, pp. 429–39.

54. LINSSEN, K. and RIESS, P. (1986). Ein Beitrag zur Klassifizierung von Böden und Klärschlämmen. *Korrespondenz Abwasser*, **33**, 686–95.

55. MATTIGOD, S. V. and PAGE, A. L. (1983). Assessment of metal pollution in soils. In: *Applied Environmental Geochemistry* (Ed. I. Thornton), Academic Press, London, pp. 355–94.

56. MCGRATH, S. P. and BROOKES, P. C. (1986). Effects of long-term sludge additions on microbial biomass and microbial processes in soil. In: *Factors Influencing Sludge Utilisation Practices in Europe*, CEC-Seminar Liebefeld, Switzerland, May 8–10, 1985 (Eds R. D. Davis, H. Häni and P. L'Hermite) (EUR 10360), Elsevier, London, New York, pp. 80–9.

57. MCINTYRE, A. E. and LESTER, J. N. (1984). Analysis and incidence of persistent organochlorine micropollutants in sewage sludge. In: *Proc. 1st Int. Conf. on Environmental Contamination*, London, July 1984, pp. 600–5.

58. NOVICK, N. J. and ALEXANDER, M. (1985). Cometabolism of low concentrations of propachlor, alachlor, and cycloate in sewage and lake water. *Appl. Environ. Microbiol.*, **49**, 737–43.

59. PAL, D., WEBER, J. B. and OVERCASH, M. R. (1980). Fate of polychlorinated biphenyls (PCBs) in soil–plant systems. *Residue Review*, **74**, 45–98.

60. POLETSCHNY, H. (1986). Folgekosten der Klärschlammverordnung aus der Sicht des Bodenanalytikers. *Korrespondenz Abwasser*, **33**, 721–3.

61. RACKE, K. D. and LICHTENSTEIN, E. P. (1985). Effects of soil microorganisms on the release of bound ^{14}C residues from soils previously treated with ^{14}C Parathion. *J. Agric. Food Chem.*, **33**, 938–43.

62. RUDAZ, A. and GUPTA, S. K. (1986). Predictability of estimated mobilizable N pool in sludge and soils. In: *Factors Influencing Sludge Utilisation Practices in Europe*, CEC-Seminar Liebefeld, Switzerland, May 8–10, 1985 (Eds R. D. Davis, H. Häni and P. L'Hermite) (EUR 10360), Elsevier, London, New York, pp. 64–71.

63. SACHVERSTÄNDIGENRAT FÜR UMWELTFRAGEN (1985). Umweltprobleme der Landwirtschaft. Sondergutachten. Kohlhammer, Stuttgart, Mainz, 423 S.

64. SAUERBECK, D. (1979). Der Stickstoffkreislauf in Agrarökosystemen. Land-bauforsch. *Völkenrode Sonderheft*, **47**, 44–62.

65. SAUERBECK, D. (1981). Einfluß der Humusversorgung und Düngung auf Bodenleben und Bodenstruktur. *Landwirtsch. Forsch. Sonderheft*, **37**, 146–56.

66. SAUERBECK, D. (1981). Möglichkeiten und Grenzen der Ausbringung von Siedlungsabfällen auf pflanzenbaulich genutzte Flächen. *Berichte über Landwirtsch*, **197**, Sonderheft, 104–13.

67. SAUERBECK, D. (1984). Die land- und wasserwirtschaftliche Bedeutung des Stickstoffkreislaufes in Böden. *Gewässerschutz-Wasser-Abwasser*, **65**, 627–61.

68. SAUERBECK, D. (1985). *Funktionen, Güte und Belastbarkeit des Bodens aus agrikulturchemischer Sicht*, Kohlhammer, Stuttgart, Mainz, 259 S.

69. SAUERBECK, D. (1986). Schadstoffeinträge in den Boden durch Industrie, Besiedlung, Verkehr und Landbewirtschaftung (anorganische Stoffe). VDLUFA-Kongreßband 1985 Gießen 'Bodenbewirtschaftung, Bodenfrucht-barkeit, Bodenschutz', VDLUFA-Verlag, Darmstadt, pp. 59–72.

70. SAUERBECK, D. and STYPEREK, P. (1986). Long-term effect of contaminants. 4th International Symposium Processing and use of organic sludge and liquid agricultural wastes (CEC), Rome, Italy, October 8–11, 1985 (in press).

71. SAUERBECK, D. and DIEZ, T. (1984). Heavy metal contents of soils and plants from field experiments receiving large amounts of sludge or town waste

composts in the FRG. Newsletter from the FAO European Cooperative Network on Trace Elements. 3rd issue (4th Consultation Meeting of the Network, held in Aarhus, Denmark 31 May–3 June 1983). Coordination Centre at the State University Ghent/Belgium, pp. 19–32.

72. SCHAFFNER, C., BRUNNER, P. H. and GIGER, W. (1984). 4-Nonylphenol, a highly concentrated degradation product of nonionic surfactants in sewage sludge. In: *Processing and Use of Sewage Sludge*, CEC-Symposium Brighton, UK, September 27–30, 1983 (Eds P. L'Hermite and H. Ott) (EUR 9129), Reidel Publ., Dordrecht, pp. 168–71.

73. SCHELTINGA, H. M. J. and CANDINAS, T. (1986). Political and administrative considerations in the formulation of guidance for sludge utilization. In: *Factors Influencing Sludge Utilisation Practices in Europe*, CEC-Seminar Liebefeld, Switzerland, May 8–10, 1985 (Eds R. D. Davis, H. Häni and P. L'Hermite) (EUR 10360), Elsevier, London, New York, pp. 90–102.

74. SHANKER, R., RAMAKRISHNA, C. and SETH, P. K. (1985). Degradation of some phthalic acid esters in soil. *Environ. Pollution (Series A)*, **39**, 1–7.

75. SIMS, R. C. and OVERCASH, M. R. (1983). Fate of polynuclear aromatic compounds (PNAs) in soil–plant systems. *Residue Review*, **88**, 1–68.

76. SJÖQVIST, T. (1984). Trends in heavy metal; PCB and DDT contents of sludges in Sweden. In: *Utilisation of Sewage Sludge on Land: Rates of Application and Long-term Effects of Metals*, CEC-Seminar Uppsala, Sweden, June 7–9, 1983 (Eds S. Berglund, R. D. Davis and P. L'Hermite) (EUR 8822), Reidel Publ., Dordrecht, pp. 194–7.

77. SLUIJSMANS, C. M. J. and KOLENBRANDER, G. J. (1977). The significance of animal manure as a source of nitrogen in soils. *Proc. Int. Seminar on Soil Environment and Fertility Management in Intensive Agriculture (SEFMIA)*, Tokyo, Japan, pp. 403–11.

78. STEVENSON, F. J. (1982). *Humus—Chemistry, Genesis, Composition, Reactions*, Wiley, New York.

79. STREK, H. J. and WEBER, J. B. (1982). Behaviour of polychlorinated biphenyls (PCBs) in soils and plants. *Environ. Pollution*, **28**, 291–312.

80. TARRADELLAS, J., MUNTAU, H. and BECK, H. (1985). Abundance and analysis of PCBs in sewage sludge. In: *Polychlorinated Biphenyls (PCB)* (Eds R. Leschber, J. Tarradellas and P. L'Hermite), Commission of the European Communities, Brussels, pp. 11–42.

81. TARRADELLAS, J., MUNTAU, H. and BECK, H. (1986). Abundance and analysis of PCBs in sewage sludge. 4th Int. Symposium Processing and use of organic sludge and liquid agricultural wastes (CEC), Rome, Italy, October 8–11, 1985 (in press).

82. THORMANN, A. (1986). Eintrag von Schadstoffen über Klärschlamm und Abfälle. In: *Dachverband Agrarforschung (Hrsg.) Belastungen der Land- und Forstwirtschaft durch äußere Einflüsse*, DLG-Verlag, Frankfurt, pp. 31–44.

83. TIMMERMANN, F., CERVENKA, L. and BARAN, E. (1980). Phosphatverfügbarkeit von Klärschlämmen aus der dritten Reinigungsstufe. Second EEC Symposium on Characterisation, Treatment, and Use of Sewage Sludge, Vienna, Austria, October 20–24, 1980, pp. 511–20.

84. TIMMERMANN, F. (1982). Verlagerung und Verlust von Nährstoffen. In: *Schriftenreihe des Bundesministers für Ernährung, Landwirtschaft und Forsten. Reihe A 'Angewandte Wissenschaft'*, Heft 263, pp. 107–20.

85. VDLUFA (1986). Bodenbewirtschaftung, Bodenfruchtbarkeit. Bodenschutz. Kongreßband 1985 Gießen. VDLUFA, Darmstadt, 642 S.
86. VETTER, H. (1973). *Mist und Gülle*, DLG-Verlag, Frankfurt.
87. VETTER, H. (1977). *Wieviel düngen?*, DLG-Verlag, Frankfurt.
88. VETTER, H. and STEFFENS, G. (1981). Nährstoffverlagerung und Nährstoffeintrag in das oberflächennahe Grundwasser nach Gülledüngung. *Z. Kulturtechnik und Flurbereinigung*, **22**, 159–72.
89. VETTER, H. and STEFFENS, G. (1982). Düngen wir zuviel? *Landwirtschaftsblatt Weser-Ems*, **129** (2), 18–22, and (3), 7–9.
90. WALLNÖFER, P., KONIGER, M. and ENGELHARDT, G. (1975). The behaviour of xenobiotic chlorinated hydrocarbons (HCB and PCBs) in plants and soils. *Z. Pflanzenkrankheiten u. Pflanzenschutz*, **82**, 91–100.
91. WANG, Y.-S., MADSEN, E. L. and ALEXANDER, M. (1985). Microbial degradation by mineralization or cometabolism determined by chemical concentration and environment. *J. Agric. Food Chem.*, **33**, 495–9.
92. WEBBER, M. D., KLOKE, A. and TJELL, J. C. (1984). A review of current sludge use guidelines for the control of heavy metal contamination in soils. In: *Processing and Use of Sewage Sludge*, CEC-Symposium Brighton, UK, September 27–30, 1983 (Eds P. L'Hermite and H. Ott) (EUR 9129), Reidel Publ., Dordrecht, pp. 371–86.
93. WELTE, E. and TIMMERMANN, F. (1982). Über den Nährstoffeintrag in Grundwasser und Oberflächengewässer aus Boden und Düngung. VDLUFA-Schriftenreihe 5, VDLUFA, Darmstadt, 236 S.
94. WELTE, E. and TIMMERMANN, F. (1985). *Düngung und Umwelt*, Kohlhammer, Stuttgart, Mainz, 93 S.
95. WILLIAMS, J. H. and HALL, J. E. (1986). Efficiency of utilisation of nitrogen in sludge and slurries. 4th International Symposium Processing and use of organic sludge and liquid agricultural wastes (CEC), Rome, Italy, October 8–11, 1985 (in press).

To complete this reference list, the papers on this subject in the following compendium should also be mentioned, even though, unfortunately, the book was not brought to the author's attention until after completion of this manuscript.

PAGE, A. L., GLEASON, T. L., SMITH, J. E., ISKANDAR, I. K. and SOMMERS, L. E. (Eds) (1983). *Proceedings of the 1983 workshop on Utilization of Municipal Wastewater and Sludge on Land*, University of California, Riverside, CA, USA, 480 p.

Effects of Agricultural Practices on the Physical, Chemical and Biological Properties of Soils: Part III—Chemical Degradation of Soil as the Result of the Use of Mineral Fertilizers and Pesticides: Aspects of Soil Quality Evaluation

F. A. M. DE HAAN

Department of Soil Science and Plant Nutrition,
Agricultural University De Dreijen 3, 6703 BC Wageningen,
The Netherlands

SUMMARY

Agricultural crop production is dependent on application of compounds to soil. The purpose may be improvement of the nutrient supply as with fertilization practices, or crop protection and disease control as in the case of pesticide use. Such additions may induce chemical degradation of soil because of compound presence at undesirable level. This may specifically exert itself in long-term effects as the result of compound accumulation. For mineral fertilizers this is amongst other things caused by imbalance between supply and plant uptake or requirements, resulting for example from non-optimal relative presence of elements in the fertilizers. For pesticides the compound degradability plays a major role, in addition to its toxicity and mobility in soil.

A prerequisite for the establishment of rules and measures for soil protection is a means for soil quality evaluation. With respect to contamination or pollution of soil, behaviour of the compound of interest in the soil system then takes a key position. This behaviour is governed by a large number of variables. As most important in this respect the buffering capacity of soil, compound speciation, heterogeneity of soil systems and bioavailability of compounds are elaborated somewhat further. According to the author's opinion these present at the same time the problem areas which deserve major attention in research programs directed towards the scientific foundation of soil protection.

1. INTRODUCTION

Soil as plant growth medium constitutes a most essential production factor in agriculture and horticulture. The development of techniques and means for the improvement of physical and chemical growth conditions by tillage and fertilization, respectively, has made agriculture a very important modifier of soil in most countries of the European Community. With respect to the surface area in use agriculture is to be considered as the prime conservator of soil quality.

Large scale application of the results of scientific agricultural research and the development of technologies induced agricultural production systems of very high intensity over extended areas of the EC. This intensity is primarily characterized by high yields and intensive land use. Specific developments with respect to, for example, intensive animal production systems and their consequences for soil are left out of consideration here, as they will be given attention elsewhere in this symposium.

Addition of compounds to soil has become common practice in present agriculture. Main purposes are the improvement of the crop nutrient supply as in the case of fertilizer applications; or crop protection and disease control as in the case of the use of pesticides. The practice of such compound additions may appreciably differ from one region of the EC to another, mainly in line with the intensity of the local agricultural production system. They all have in common, however, that they may cause chemical degradation of soil as the result of accumulation of compounds or compound constituents at undesirable levels. Such accumulation may on short or on long term adversely affect the quality of soil. This may be so either with respect to its inherent crop production function or with respect to other soil functions as for example the protection of the quality of groundwater and surface water, or soil's role in element cycling which is mainly biologically controlled. It is self-evident that the time period involved for the appearance of such effects is, at comparable other conditions, strongly related to the rate of application.

In this contribution some attention is focussed on the role of mineral fertilizers and pesticides in this respect. Additional consideration is given to the foundation of soil quality evaluation.

2. MAJOR ADVERSE EFFECTS OF MINERAL FERTILIZER AND PESTICIDE USE

2.1. Mineral fertilizers

Crop production with high yields requires sufficient availability of plant nutrients. In accordance with the relative amounts of elements required, usually a distinction is made between so-called macronutrients and micronutrients. Important representatives of the first group are N, P and K, of the latter Cu, Mo, Mn and Zn. As indicated in this list, in many cases micronutrients are heavy metals.

The natural availability of plant nutrients in soil is usually insufficient for continuous or repeated crop production. Enhancement of this availability by fertilization has thus become a long existing intervention of mankind in plant growth and hence one of the earliest agricultural activities. From the very beginning agriculture was to this purpose dependent on waste products from the then existing human society. Therefore organic materials mainly constituted the first fertilizers used, at a time when the actual action of the soil additive was not fully understood.

The epoch-making work of Justus von Liebig, which in essence showed that plant growth directly responds to the availability of mineral nutrients in the plant root environment and to the resulting plant uptake, was the instigation for the production of mineral fertilizers on an industrial scale. Modern agriculture thus commonly leans upon mineral fertilizer supply. It is very well realized that this bird's eye view of the historical development of soil fertilization, compressed into one or two sentences, does not by far do any justice to all the efforts involved, neither to the benefits thereof for mankind with respect to food production, nor to the complexity of soil fertility problems. For the present purpose, however, it suffices to mention that in addition to the above a soil quality problem may result from the use of mineral fertilizers. The main reasons therefore are threefold, namely:

(a) The supply of fertilizer may be so high that it adversely affects other soil functions, although still being economically feasible from a plant production point of view.

(b) The soil conditions are such that an overdose of certain nutrients as compared to plant uptake is required for proper plant production; this overdose may in the long term result in undesirable effects.

(c) The imbalance between the relative elements' presence in mineral fertilizers and plant requirements may make that adjustment of fertilizer application on the basis of one element lead to an overdose of other constituents; again in the long term possibly inducing negative consequences.

These three cases will, although by no means exhaustively, be elaborated further and illustrated with an example.

2.1.1. Excessive mineral fertilizer supply; nitrogen as an example

First the term 'excessive' deserves some elucidation. Present intensive animal husbandry systems have resulted in overdosing of plant nutrient elements in the soil in the form of animal manures. Then the nutrient application is evaluated as excessive because it far exceeds plant requirements. It is undeniable that in a number of regions of several EC countries such practices are causing severe environmental problems. Excessive doses of plant nutrient elements in those cases, however, are the direct result of the abundant availability of manure. This means that they are considered more as waste products than as highly valuable fertilizers. Tietjen [23] sharply characterized them as 'resources out of place'. For the farmer, disposal on soil is usually the cheapest way to get rid of them. Rates of application under such circumstances may also considerably exceed the ones that would be used if the nutrient element would have to be bought in the form of mineral fertilizer. It is commonly agreed that such overdosing of soil is unacceptable because of damage to interests such as water quality and soil fertility, and that regulations and measures are to be developed in order to combat such practices.

However, not only with waste products but also with mineral fertilizers practical application rates may be so high that they are negatively affecting other interests that should be safeguarded by soil. It apparently pays in those cases to apply the commercial fertilizer beyond the level of adverse effects. This thus presents a problem area where different soil functions are conflicting.

With respect to nitrogen it then concerns nitrate leaching to groundwater at such fluxes that the standard for drinking water use cannot be met anymore. This EC standard amounts to $11.3 \, g/m^3 \, N$ (corresponding to $50 \, g/m^3 \, NO_3^-$), whereas the value preferably should not exceed $5.6 \, g/m^3 \, N$ ($25 \, g/m^3 \, NO_3^-$).

In a schematic approach to the nitrate leaching problem as influenced by agriculture, the following factors are of prime consideration:

—nitrogen application rate (also type: mineral, organic);
—land use (arable land, pasture);
—soil type (sand, clay, peat);
—water management (groundwater level, irrigation).

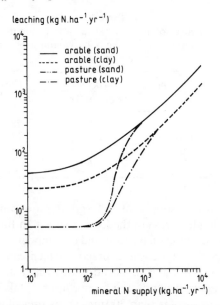

FIG. 1. Leaching of nitrogen as a function of mineral nitrogen supply to soil, for two different soil types and land uses [12].

Kolenbrander [12] gives a compilation of research data which cover these factors (Fig. 1).

It is thus apparent that in the region up to about 200 kg mineral N/ha/year, nitrate leaching on arable land considerably exceeds that on pasture. This is because mineralization proceeds after harvest in the absence of a crop which can take up the mineralized nitrogen. The larger denitrification in clay soil as compared to sandy soil results in a decrease of leaching for arable land. Over the application region mentioned there is no difference for pasture with respect to soil type. At considerably higher application rates the leaching lines are coincident for different land use on the same soil type. From that point on, leaching on sandy soil stays above that of clay soil as the result of lower denitrification.

These experimentally derived data allow the calculation, or at least an estimation, of the nitrate content of upper groundwater as the result of mineral nitrogen supply in agricultural practice. Table 1 shows the outcome of this exercise for an application rate of 200 kg mineral N/ha/yr at a supposed precipitation surplus of 300 mm/yr. In the case of arable land the crop rotation scheme is taken as potatoes, wheat (2 ×) and sugarbeets, for each four years.

Table 1

Nitrogen leaching and corresponding nitrate concentration in upper groundwater at a mineral nitrogen supply of 2000 kg N/ha/yr and a precipitation surplus of 300 mm/yr

Land use	Soil type	Leaching (kg N/ha/yr)	Conc. groundwater ($g\,N/m^3$)	($g\,NO_3^-/m^3$)
Arable	sand	85	28	124
Arable	clay	40	13	58
Pasture	sand	8	3	13
Pasture	clay	6	2	9

A mineral nitrogen supply of 200 kg/ha/yr on sandy soil is fairly common over extended areas in the EC. As shown in Fig. 1, lower supply does not decrease leaching on pasture, and only moderately on arable land. This then implies that agricultural practice which can be considered as common in the EC is not able to meet drinking water standards over large areas, when the upper groundwater composition is taken as reference. A lower precipitation surplus than 300 mm/yr, as prevailing in parts of the EC, would lead to a corresponding increase of concentration.

Groundwater as a source of drinking water is usually pumped from much larger depths than the upper groundwater layer. This implies that denitrification may substantially contribute to the improvement of groundwater quality on its way to the aquifer involved. Quantitative information about this contribution, however, is still fairly scarce although the required conditions for denitrification are pretty well known at present. Many calculations and guesses in this respect take an assumed denitrification of roughly 50% as a point of departure [19]. This assumption is also based on experimental observations [21].

Taking this value of 50% denitrification and using the information in Fig. 1, one may calculate the maximum mineral nitrogen supply to soil which would not exceed the EC standard of 11·3 g/m^3 N. This would imply an acceptable net nitrogen leaching of 33 kg N/ha/yr without denitrification and 66 kg N/ha/yr with 50% denitrification. The data are given in Table 2.

These data show that use of soil for arable crop production would only allow nitrogen supply on sandy soil which stays far below the optimal dose, even if 50% of the nitrate leached were denitrified. At zero denitrification no nitrogen fertilization at all would be allowed.

When the drinking water standard is hard to meet, it is evident that the

Table 2

Maximum mineral nitrogen supply for different land use and soil type in order to meet the EC drinking water nitrate standard, for 0% and 50% denitrification.

Land use	Soil type	Max. N supply, kg/ha/yr	
		0% denitrification	50% denitrification
Arable	sand	0	70
Arable	clay	100	360
Pasture	sand	320	380
Pasture	clay	500	725

'target value', i.e. the nitrate concentration which should preferably not be exceeded and which is half of the drinking water value, is really in conflict with contemporary agricultural management.

Although the above presentation is necessarily an oversimplification of the actual problem, it nevertheless shows that common agricultural practice may easily lead to undesirable side effects. It remains an intriguing question how the different interests involved should be weighed and which adjustments could be considered as reasonably acceptable for practical agriculture.

2.1.2. Overdose with respect to plant uptake; phosphate as an example

Basically the nitrogen leaching as described above is caused by incomplete use or uptake of nitrogen present in the rootzone in soil solution. Reasons for this incomplete element utilization may be different, e.g. the absence of roots at certain times of mineralization of soil organic matter as mentioned earlier. Because nitrogen in the form of nitrate is not hampered in its downward movement by bonding on soil constituents, this nitrate surplus almost immediately becomes a burden on groundwater.

For phosphate the situation is comparable with respect to certain aspects, but completely different with respect to others. Different in as far as the mobility in the soil system is concerned, which in many soils is very low for phosphate. Comparable with respect to incomplete use of phosphate present in the rootzone; this time, however, as the result of partial unavailability due to strong bonding on soil constituents. Thus the same soil properties which prevent phosphate leaching are responsible for the requirement of a relative overdose of P. It will be clear that this overdose has to be larger the stronger the interaction of phosphate with soil (so-called 'P-fixing' soils). This situation will remain until a certain level of saturation of the phosphate bonding capacity is reached. From that moment on all P supplied is in principle available for plant growth.

This soil property of effectively bonding phosphate ions has been applied in practice for removal of phosphate from wastewater [3,4]. At the same time it is the reason that an excessive supply of phosphate in manure slurries in certain areas has not led so far to large-scale leaching of phosphate from soil, which would damage surface water bodies by eutrophication. In the long term, however, any significant overdose as compared to plant uptake will inevitably induce complete saturation of the bonding capacity. Again of course, the time scale involved is dependent on the magnitude of the remaining bonding capacity on the one side and of the overdose on the other.

The behaviour of phosphate in the soil thus has received considerable attention in research programs. Initially mainly with respect to plant nutrition and soil fertility aspects, presently also in relation to possible environmental damage. The main soil compounds responsible for phosphate bonding are organic matter, clay minerals and (hydr)oxides of Al and Fe. Beek and Van Riemsdijk [5] gave an extensive review on phosphate bonding in soil. A model which satisfactorily describes the bonding mechanism in sandy soils was developed by Van Riemsdijk et al. [17, 25]. A formula of general validity for homogeneous soil describing the transport of the phosphate saturation front in soil was presented by Lexmond et al. [13]. As long as the shape of the saturation front in penetrating the soil does not alter, the rate of movement of this saturation front may be described by the following simple relationship between the proceeding saturation, phosphate load and bonding capacity:

$$y = \frac{(A - U)}{BS} 141 \tag{1}$$

in which:

y = velocity of saturation front, in cm/yr
A = phosphate supply to soil, in kg P_2O_5/ha/yr
U = uptake and removal by crop, in kg P_2O_5/ha/yr
B = dry bulk density of soil, in kg/m^3
S = remaining phosphate sorption capacity, in mmol P/kg.

The value of S thus constitutes a crucial point. It may be determined by the use of sophisticated techniques as developed by Van Riemsdijk and van der Linden [18]. Fortunately it can also be reasonably well approached by analytical techniques which are more easily accessible [16]. As shown by

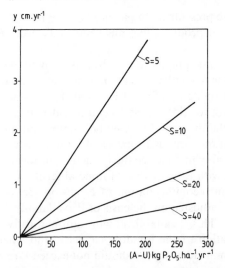

FIG. 2. Velocity of phosphate saturation front in soil as a function of net P addition, for a number of values of remaining phosphate bonding capacity; soil bulk density $= 1500 \, kg/m^3$.

Lexmond *et al.* [13] and Van der Zee and Van Riemsdijk [26] for many soils a good correlation exists between the total amount of P that can be bound and the amount of reactive Fe and Al as determined by oxalate extraction. However, it is the presently remaining bonding capacity, S, that counts. S thus is the main parameter for vulnerability of soil for phosphate leaching.

The relationship of eqn. (1) is graphically shown in Fig. 2.

The value of U is related to the type of crop and to crop yield. In Table 3

Table 3

P removal from soil, in kg P_2O_5/ha/yr for a number of crops at assumed dry matter production corresponding to good yield

Crop	Dry matter production (kg/ha)	P removal (kg P_2O_5/ha)
Wheat	9·000	57
Rye	7·200	32
Corn (grain)	4·700	32
Silage corn	12·500	82
Potatoes	9·200	55
Sugar beets	19·600	102
Hay	4·000	32

some examples are presented. On pasture farms P removal is usually very low because animal products like meat and milk are the only ways of discharge there.

The value of A (P supply with fertilizer) is strongly dependent on the fertility status of the soil with respect to phosphate, which again is inversely related to the remaining bonding of P in soil. There is not much sense in giving average data for this supply as can be calculated from P fertilizer consumption in the different EC Member States and agricultural areas involved. Agricultural soils with extremely low capacity for phosphate bonding will probably first induce the P leaching problem. Probability for saturation of soil with P by the use of mineral fertilizers is low when the official fertilization recommendation of extension services are followed. Then the supply is adjusted on the basis of P availability according to soil sample analysis. This availability will automatically increase with proceeding saturation of the bonding capacity, and when saturation is reached, the recommended supply should not exceed P removal by crops. There are several reasons in practice, however, for deviations from these official recommendations. One would be non-regular soil sampling and analysis, and supply on the basis of the farmers' experience. Another one would be the use of, for example, 'starting' doses of mineral P in silage corn, even when the P fertility status of the soil is very high. As compared to mineral P, however, P overdosing as the result of manure surpluses seems far more important since in that case values of $(A - U)$ may easily amount to several hundreds of kg P_2O_5/ha/yr.

The approach presented above also allows a risk evaluation for P leaching. Figure 3, taken from Van Riemsdijk *et al.* [16] presents an

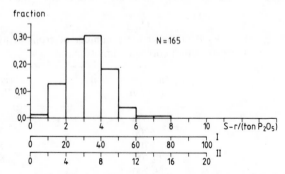

FIG. 3.   Frequency distribution of remaining P sorption capacity for topsoil of 165 different parcels in a certain region in The Netherlands [16]. Axes I and II refer to time of saturation (years) for P supply rates of 160 and 560 kg P_2O_5/ha/yr.

example. Here the frequency distribution for the value of the remaining P sorption capacity, expressed in metric ton P_2O_5/ha, has been given for the top layer of 165 agricultural parcels in a certain region in The Netherlands. By translating the capacity axis for given supply rates (160 and 560 kg P_2O_5/ha/yr, in this case), a transformation is obtained to the number of years it will last before this layer will be saturated with P.

Saturation of soil profiles should be avoided because the combined effect of desorption and continued supply will harm surface water quality. These effects are most severe with animal manures, because of the high P concentration involved.

2.1.3. Imbalance of elements in mineral fertilizers; cadmium as an example

Compounds that are used for fertilization purposes usually contain a number of different constituents. This means that application of a certain element necessarily implies the addition of others. Striking examples in this respect are sewage sludge and copper-containing pig manure. If these are added to soil on, for example, nitrogen or phosphorus basis, a strong overdose of heavy metals in general and copper specifically will be applied.

However, mineral fertilizers also contain different elements which would in general just by chance exactly meet the plant requirements. The development of so-called compound mineral fertilizers, in which the macro-nutrients N, P and K prevail at varying ratios, provided a way to adjust the supply at least more closely to the requirement than is usually possible with organic fertilizers. But they still may contain constituents at amounts which are not in accordance with demand for plant nutrition. A good example in this respect is given by cadmium, which may occur as an impurity in mineral phosphate fertilizers. Because cadmium is not an essential element for plant growth any addition of it to soil is redundant from that point of view. Such additions are undesirable because cadmium, which can easily be taken up by plants, may form a health-hazard after human consumption of agricultural products containing Cd. Cadmium contamination of soil may thus damage its plant production function.

Cadmium is present in nature in many compounds albeit at low concentration levels. Because animal feed and fodder do contain cadmium it has to be expected in the excreta (next to its occurrence in specific animal organs) and consequently in animal manure. Its natural presence also induces its appearance in fertilizers derived from fossil products as in the case of phosphate, lime and to a certain degree potassium fertilizers. As shown by Henkens [9] the contribution of mineral fertilizers, which in turn is just one part of the total soil cadmium burden, is for Dutch conditions by far (i.e. more than 90%) caused by phosphate fertilizers. Although a similar analysis is not available in detail for all EC states, it may be assumed that P fertilizers are in a comparable way the main contributors as a combined result of fertilization practice and Cd content in different fertilizers. Therefore the attention is limited here to P fertilizers.

The cadmium content of P fertilizers in the first place depends on the origin of the ore used as raw material. Because the Cd from the ore passes into the P fertilizer, differences in contents of different products are mainly caused by the ore used.

Raw phosphate may vary from as low as 5 mg Cd/kg (Kolaphosphate from USSR) to as high as 100–300 mg Cd/kg (Idaho BPL 70/72 from USA). Many raw phosphates from Morocco and Tunisia are somewhere around 15–30 mg Cd/kg of product.

Table 4, taken from the compilation by Henkens [9], presents for a number of different phosphate fertilizers the Cd content expressed as mg Cd/kg P_2O_5. These data show that there are considerable differences for different P fertilizers and brands. This at the same time provides the main way for protecting soil against cadmium pollution along the pathway of P fertilization. Use of low Cd phosphate fertilizers should be stimulated, and possibly even regulated.

The above information also allows a simple estimate of the quantitative

Table 4

Cd content of a number of phosphate fertilizers used in three European countries; (within brackets other brands of fertilizer)

Fertilizer *(N + P + K)*	*Cd (mg/kg P_2O_2)*		
	Netherlands	*Belgium*	*Sweden*
23 + 23 + 0	39 (74)		
12 + 10 + 18	50 (90, 100)	25	
17 + 17 + 17	106		
15 + 15 + 15		20	
10 + 10 + 10		22	
9 + 9 + 12		28	
20 + 11 + 9			122
14 + 14 + 17			29
20 + 14 + 6			47
16 + 16 + 3			112
26 + 14 + 0			54
Triplesuperphosphate		61 (100)	
Superphosphate	39	22	112

contribution to Cd in soil. At a Cd content of 60 mg/kg of P_2O_5 and a P supply of 100 kg P_2O_5/ha, the yearly increase of the Cd content of the 20 cm topsoil at a dry bulk density of 1500 kg/m^3 would amount to 0.002 mg/kg, if all cadmium were to stay in this layer. Since Cd is one of the most mobile heavy metals in many soils [6], the above increase would be a maximum for the example used. Dissipation to other environmental components like water and plants may, although lowering the build-up in soil, be undesirable for other reasons. The above accumulation rate seems very low at first sight. Nevertheless, unchanged practice of P fertilizer use will in the long term induce undesirable Cd levels in soil. Moreover, it is not so much the average situation as used in the sample calculation that cause the most severe problems, but much more the extremes (high application rates as in vegetable growing, sensitive crops, etc.). Therefore also this pathway of soil contamination deserves attention, at the same time making it necessary to reduce all other sources of Cd emissions to the environment.

In conclusion it can be said that many compounds are used in agricultural practices in order to supply the crop with nutrients. As far as mineral fertilizers are concerned the side effects of such additions with respect to chemical degradation of soil are relatively low because of the relative purity of the compounds involved. Nevertheless some problems may result. For mineral fertilizer the use of nitrogen already has an effect on water quality. Phosphate overdosing and impurity of Cd in mineral P fertilizers will in the long term adversely affect soil functions.

2.2. Pesticides

In the same way as modern agricultural production systems rely on fertilizer use with respect to plant nutrition, they are dependent on the application of compounds to soil for crop protection and disease control purposes. These compounds are commonly given the collective name of pesticides. Specific designations point in more detail to intended effects and/ or target organisms, e.g. herbicide, fungicide, insecticide, nematicide, etc.

The nature of the active component of pesticides may be organic or inorganic. Both may cause chemical degradation of soil as undesirable side effects. Prototypes of inorganic pesticides are, e.g. the Cu-containing Bordeaux mixture and Pb-arsenates. In these cases soil degradation is the result of heavy metal and arsenic accumulation and their effects can be treated in a comparable way as from other sources, e.g. sewage sludges. Since this is given consideration elsewhere in this symposium, and because organic pesticides have become far more important than the inorganic ones, attention is limited here to organic pesticides.

Because of the large number of organic chemicals involved (several ten thousands of active compounds are known of which roughly 300 are applied on large scale in EC Member States), a further limitation to model treatment is required. In line with the purpose of this contribution the main factors governing soil degradation resulting from pesticide use is given consideration, whereas reference to specific chemicals is given merely by way of illustration.

The awareness, as gradually arose some decades ago, of hazards involved in the repeated application of low degradable organic chemicals has led to the development in most industrialized countries of regulations which strictly control the use of such compounds and especially the introduction of new products. Nevertheless, still a number of problems remain.

Pesticides are compounds which exert biological activity. The degree of selectivity determines whether the target organism only is attacked, or non-target organisms are also influenced. The latter then in many cases is an undesirable side effect of pesticide use. Low selectivity is indicated as the 'broad working spectrum' of a compound.

Because of the toxicity of pesticides, which is an inherent property of their application, undesirable side effects may occur with respect to soil organisms and to human beings. These effects may be direct as in the case of exposure of non-target soil organisms to low selective soil fumigants or the occupational exposure of man to highly volatile compounds. Indirect effects occur, e.g. by leaching to groundwater or surface water, uptake by plants and contamination of foodchains.

While toxicity constitutes the basic problem, the way in which undesirable effects are brought about largely depends on the compound behaviour in the soil system. In this respect degradability, usually expressed as transformation rate, is one of the most important characteristics because it governs the persistency of the compound. In addition, behaviour in soil is controlled by volatility, solubility and bonding onto soil constituents. Bonding also for pesticides, in the same way as for all compounds, substantially governs the mobility in the soil system. In this way it mainly controls emissions to the other environmental systems: air and water. The above mentioned factors will in different combinations shortly be treated and illustrated again with examples. These examples concern pesticides which are used in soil applications.

2.2.1. Accumulation as the result of strong bonding; diquat and paraquat as examples

Diquat and paraquat are very reactive herbicides with a broad working spectrum for plants. These compounds have now been used for about 20 years.

The two main reasons for relatively high accumulation in soil are their low degradability in combination with strong bonding. The transformation rate is very debatable and estimated somewhere between 0 and 10% per year. This means that the half-life could probably be around 10–20 years or longer.

The strong bonding in soil results from the ionic character of these compounds. They are actually divalent cations as schematically illustrated in Fig. 4, for the bonding of diquat on soil organic matter [22].

The most important bonding constituents in soil in this respect are organic matter and clay particles. The latter even exhibit a fixation-like bonding of these two compounds. The bonding takes care of an inactivation which makes it possible that a gradual build-up in soil can take place which is not perceived.

Because of the low selectivity involved, such a build-up could possibly induce long term damage to the soil productivity for plant growth. This is so because of the gradual increase of compound activity and its prevalence. It is therefore that approval of continued use of these compounds is a point of discussion at present. Some extension services already recommend termination of this use, especially on sandy soils which because of their low bonding capacity would be most vulnerable.

2.2.2. *High volatility, resulting in emission to the air; methylbromide as an example*

Application of volatile pesticides has become a practical technique for disinfestation of soil. This is commonly known as soil fumigation. Examples of large scale applied soil fumigants are methylbromide, 1-3

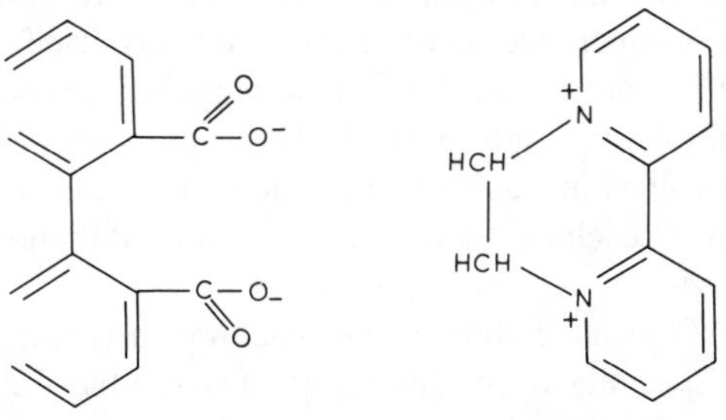

diquat

FIG. 4.   Schematic representation of diquat bonding on soil organic matter constituents [22].

dichloropropene and methylisothiocyanate. An important side effect of these applications is emission into the air by volatilization. Circumstantial factors that strongly affect the volatilization are temperature and moisture content of soil; high temperature and dry soil usually enhance air emission. The technique of application may also play a considerable role.

Methylbromide has been used extensively in preplanting treatment of glasshouse soil in order to control fungi and nematodes. Besides several accidents due to occupational exposure, mainly resulting from insufficient care during the execution of the treatment, considerable public concern has also arisen for the health conditions of people living in the environment. This was initiated following the measurement of a methylbromide concentration up to $10 \, mg/m^3$ in the outside air at a distance of a few metres from the wall of a glasshouse [10].

De Heer *et al.* [8] showed by use of mass balance approaches that volatilization indeed may be appreciable; several tens of per cents up to more than 50%. This research at the same time studied the effectiveness of the use of different types of soil covers during treatment.

Since 1981 the use of methylbromide in The Netherlands is permitted only under license. License is given under a number of severe conditions with respect to distance to inhabitation, covering of soil, sealing of glasshouse, etc.

2.2.3. Leaching to surface water and groundwater; aldicarb as an example

Aldicarb is an insecticide and nematicide. It has widespread application during springtime in pretreatment of soil for potato growing. Also soils in glasshouses are regularly treated with aldicarb, e.g. in ornamental crop production.

At normal soil conditions aldicarb is readily oxidized to aldicarb sulphoxide; in turn this is in part further oxidized to aldicarb sulphone. Both oxidation products are very toxic as is aldicarb itself. This introduces severe hazards in case of leaching. Leaching can easily occur because aldicarb as well as its oxidation products are only very weakly bound in soil. The risk involved in leaching to surface water concerns the effects on water organisms. Leaching to groundwater may damage its quality for drinking purposes.

Leistra *et al.* [15] have shown that leaching to surface water may be considerable. Especially if conditions are favourable for leaching as in many glasshouse practices. Here, a relatively large precipitation surplus as the result of irrigation practice, in combination with intensive drainage systems enables an easy contact between soil and surface water bodies.

Leistra and Smelt [14] showed that under conditions of low transformation rates of the oxidation products of aldicarb, severe contamination of groundwater in aquifers might be possible. Because of the hazards involved, aldicarb is one of the compounds that has been put on the so-called 'black list' in The Netherlands.

Table 5, taken from Anon. [1], presents the pesticides that have been put on this list. This list provides a typical example of restrictions to pesticide use in agriculture because of priority given to other soil functions, in this case conservation of groundwater quality. The compounds mentioned are not allowed to be used in the protection zones for drinking water supply. This is prescriptively indicated on the labels and packings of these compounds.

Table 5
'Black list' of pesticides, for protection zones of drinking water
supply areas in The Netherlands

Alachlore	Dichloropropene	Sodium arsenite
Aldicarb	Dikegulac Na	Oxycarboxin
Alloxydim-Na	Endothal-Na	Propachlore
Asulam	Ethiofencarb	Propoxur
Benazolin	Glufosinate	Pyridate
Bentazon	Hexazinone	TCA
Borates	Lenacil	Thifanox
Bromacil	Metalaxyl	Triclopyr
Chloro-methyl	Metazachlore	Vamidothion
Chloralhydrate	Methomyl	
Dicamba	Metalachlore	

Next to the black list a white list has been defined, containing at present 206 compounds. These may be used in drinking water areas when the directions on the label are strictly followed. Some of these compounds are subject to specific additional restrictions, e.g. with respect to time of the year and defined soil properties.

World-wide regulations, as formulated in the so-called Codex Alimentarius, have been developed in order to prevent health hazards via residual effects in food. These aspects therefore seem pretty well covered on a basis of broad international agreement.

Between different EC Member States common agreement on the importance of harmonization with respect to approval of new products and continued use of existing compounds is growing. Since also here different practices and interests are sometimes conflicting, adverse environmental effects of pesticide use as illustrated above will probably continue for some time to come.

3. SOIL QUALITY EVALUATION

The principal aim of soil protection is conservation (or restoration) of desirable soil quality. Hence a quantitative evaluation of soil quality constitutes a prerequisite for the development of soil protection measures. Such an evaluation must take the effects to be expected as a starting point. In case of soil pollution these effects directly or indirectly result from the abundant presence of compounds in soil. Because all compounds may negatively influence soil functioning, no further distinction with respect to chemical nature will be made here. Suffice to say that risk assessment of pollution is a prime consideration in soil quality evaluation. In turn, a basic problem in risk assessment is the establishment of the quantitative relationship between exposure and effect. As mentioned in the introduction, effects of interest may be manifold, e.g. on human and animal health, on soil organisms, on plant growth and plant quality, on surface water and groundwater composition.

From the above it will be apparent that the behaviour of a compound of interest in the soil system constitutes the central point in chemical soil quality evaluation. As recently discussed, although in a somewhat different context by De Haan et al. [7], this behaviour is controlled by a large number of variables which may be compound related and soil related as well. Immediately arising topics then are soil buffering capacity, compound speciation, soil heterogeneity and bio-availability of the compound. These are given some further attention.

3.1. Soil buffering capacity

The buffering capacity of soil with respect to soil contamination may be defined as its capacity to delay the negative effects of the contaminant's presence because of inactivation. Inactivation is mainly achieved by effective bonding onto soil constituents or sometimes conversion into insoluble compounds. A schematic presentation for an essential element (full line) and a non-essential element (broken line) is given in Fig. 5.

Considering an essential element, e.g. a micronutrient which is at the same time a heavy metal, with insufficient content in the soil a positive effect will result from an increase as obtained by fertilizer supply. Then a level is

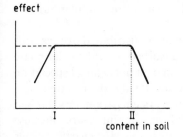

effect

I   II

content in soil

FIG. 5. Schematic relationship between compound content in soil and effect, for an essential (full-line) and a non-essential (broken line) element.

reached where further increase does not have any influence. Beyond a certain point, further increase induces adverse effects. The region up to point II on the content axis in Fig. 5 is indicated as the buffering capacity. Beyond this content the soil must be considered as polluted. While agricultural production will pursue value I, soil protection should be aimed at not exceeding value II.

The extent of the soil buffering capacity varies widely for different compounds and different soils. It reflects the vulnerability of soil to contamination and pollution. The actual content as compared to point II provides information about the proximity of the hazardous or threshold value and its determination thus may constitute an important part of the risk assessment procedure. It is at the same time indicative of the need of measurement in order to prevent any further increase of the compound in soil.

Because of the huge variety of soil types and soil properties, a quantitative soil quality evaluation cannot be based on single values. Since the content value should be interpreted with respect to effects, preferably sliding scales should be considered. This would at least introduce an opportunity to give justice to specific soil properties which govern compound behaviour. One of these is the chemical form of the compound in the soil system.

3.2. Compound speciation
Usually it is not the total content of a contaminant which governs its effect, but much more its active form, amount or concentration. Moreover, it depends on the soil phase via which the effect of interest predominantly exerts itself whether the contaminant's presence or activity in either the gaseous, the liquid or the solid phase of soil should deserve major attention. Sometimes all three have to be taken into account, especially when volatile and soluble compounds are involved. In many cases, however, the presence

in the soil solution is of major concern, either with respect to plant uptake or with respect to leaching.

The total amount of contaminant in soil may be distributed over various chemical forms: precipitated as a mineral, adsorbed onto the solid phase and dissolved in the soil solution. The development of chemical speciation models, which can be used in order to calculate the distribution of a compound over different forms as a function of system parameters, has had considerable attention in soil chemistry during recent years. Because usually many species and system parameters are involved, such calculations are in general fairly complex, thus requiring the use of electronic computers. Widely applied computer codes in this respect are MINEQL [24] and GEOCHEM [20]. The practical validity of the derived results is dependent on the correctness and reliability of input data. For that purpose, a good description of the system and also the selection of proper formation constants for all species involved is required.

Important reactions governing the chemical speciation of, for example, heavy metals in soil solution are hydrolysis, formation of chloride complexes and complexation with organic compounds. With respect to the latter, specifically synthetic organic compounds like, e.g. EDTA, DTPA and NTA have been given comprehensive attention. Interest in these compounds primarily originated from their ability to improve heavy metal availability in soil with respect to plant uptake and growth. Recently this interest increased again, this time as result of their possible applicability for the displacement or removal of heavy metals from polluted soil. Figure 6, taken from Keizer *et al.* [11], presents an example of the results of heavy metal speciation calculations when using EDTA for soil extraction.

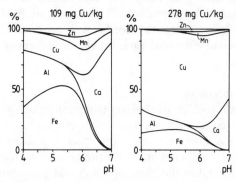

FIG. 6.   Calculated distribution (%) of EDTA over 6 metals as a function of pH, for heavy metal extraction of soil at two different total Cu contents [11].

In this case the sorption process of soil constituents was also incorporated in the speciation calculation by treating the sorbed amount as a species present in solution. Heavy metal adsorption was then described with a pH-dependent Freundlich sorption equation. The data refer to calculations for copper contaminated soils, where the EDTA distribution over six different elements as a function of pH was computed. The incorporation of the adsorbed amount as a species leads to different results for different copper contents of soil (109 and 278 mg Cu/kg dry soil, respectively). For more detailed information and comparison with experimental data the reader is referred to Keizer *et al.* [11].

The importance of compound speciation in relation to soil quality evaluation is derived from the fact that contaminants present in different forms will exhibit different mobility and reactivity in soil, different availability for plant uptake and toxicity for organisms. In this respect again the variability within and between soil systems constitutes a complicating factor.

3.3. Soil heterogeneity

The most pronounced characteristic of soil heterogeneity is its occurrence on a large number of different scales. Several of these are briefly mentioned here by way of example. Distinction between the scales is not based on sharp boundaries but are merely arbitrarily chosen. Scales involved may vary from molecular to global.

Heterogeneity plays a role in the interaction of compounds within the system and thus greatly influences compound behaviour in soil. Therefore it must be taken into account for the understanding and prediction of effects of the compounds' presence in soil. As such it deserves major attention in risk assessment and consequently in soil quality evaluation.

The smallest scale that lends itself for discussion is the molecular scale. One of the important processes in soil governing compound concentration in solution is adsorption. In many cases this is the result of the presence of reactive sites on soil constituents, e.g. in the form of electric charges. Distribution patterns of such reactive sites over the constituents may be homogeneous as well as heterogeneous. Heterogeneity at this scale may then be caused by, for example, surface imperfections and different crystal faces. Although sometimes a distinction between both is not a necessary condition for proper evaluation of adsorption data this is not always so.

Differences in size of soil particles provide a second example of heterogeneity, at another scale. Although this type of heterogeneity is of a physical nature, its effect may well be chemical. This may be caused by an increase of specific surface area for smaller particles, resulting in increased reactivity. Clear examples are clay particles with respect to ionic bonding and oxide particles with respect to dissolution.

Heterogeneity at the scale of soil samples has required the development of special methods for soil sampling and analysis, in order to obtain sufficient reliability for interpretation of the results. Sometimes the methods employed may be adjusted to specific purposes for which the collected data are used. Interpretation of the final results, however, must always take the variability and uncertainties involved into account.

At the scale of a soil profile, heterogeneities are readily visible in many cases. Figure 7 gives an illustration, indicating that in horizontal and vertical directions important differences may occur.

Finally, heterogeneity on a regional scale is reflected on soil maps where soil types are grouped according to specific similarities and differences. The dependency of heterogeneity on the scale of observation is illustrated well by differences between maps of the same region but of different scales.

It is thus apparent that many properties of soil which may govern compound behaviour are spatially variable. This implies that also a resulting composite property such as the buffering capacity with respect to contamination shows spatial variability and should be considered

FIG. 7.   Illustration of soil heterogeneity at the scale of a soil profile.

stochastic. This is in many cases also true for the dosage of contaminants given to soil. Risk assessment and soil quality evaluation should thus have a probabilistic basis. How this can be worked out for leaching of compounds at undesirable concentrations, has been described in Van Der Zee and Van Riemsdijk [26]. Risk evaluation of soil pollution for contamination of crops that are to be used for human and animal consumption can probably be approached in a comparable manner.

3.4. Bio-availability

As mentioned before usually no simple relationship exists between the total content of a compound in soil and its biological action. This action may reveal itself in growth and development of plants but also in the form of reactions of soil (micro-)organisms.

The absence of such a distinct relationship is caused by the fact that apparently only part of the total amount of the compound is available to induce a reaction; at least in a direct way. Interpretation of content values with respect to effects is a principal part of soil quality evaluation. This interpretation thus requires insight into the availability of compounds involved. Because of the interest in the effects upon biological constituents of the system, this problem area is commonly known as bio-availability.

The concept of bio-availability has played an important part from the very beginning in the field of plant nutrition problems in relation to crop production. In that case it usually concerns elements or compounds which are present at insufficient availability, whereas in soil pollution mainly the excess of compounds constitutes the problem. However, the same approaches are probably applicable in both areas. This then points to the necessity, or at least the evident advantage, to rely in soil protection considerations on experiences derived from conventional soil fertility.

In the initial approach the problem of plant nutrient availability gained practical application in the search for extraction procedures. These were aimed at a good correlation between contents as determined by a given extraction method and the corresponding crop response. Many analytical procedures applied in the field of fertilizer supply recommendations are still based more or less on these empirically derived relationships.

A more theoretical approach to bio-availability has evolved from concepts put forward by Bray and Schofield as concisely discussed by Bache [2]. In this respect the concepts of quantity and intensity were given consideration, combined with compound mobility as a determining factor.

Quantity, Q, then is considered as a so-called extensive variable referring to amount of mass of a compound; it is expressed in mole or kg.

Intensity, I, on the other hand is an intensive variable, referring to the (electro)chemical potential of the compound of interest; it is expressed in joules per mole as it reflects the amount of free energy per unit amount of compound. With respect to compounds in solution, the activity or concentration provides a good practical means to express the value of I.

The relationship between Q and I is indicated as the Q–I curve. It presents important information via its slope, which is referred to as the buffering intensity for the compound at the conditions prevailing in the system under consideration, i.e. for which the Q–I curve is given. This buffering intensity may vary widely with soil properties and system parameters, as well as with the position on the Q–I curve.

Detailed information on the Q–I relationship of contaminating compounds may greatly contribute to the understanding and evaluation of contaminant availability for organisms. This is so even when the picture is strongly complicated by the fact that in addition to quantity and intensity also mobility may play a role, and that there are mutual influences between compound behaviour and organism behaviour. Mobility then refers to the rate at which a contaminant can move or be displaced through the soil from the point of its prevalence to the point of its biological action, e.g. association with plant roots. While compound behaviour thus may greatly affect organisms via uptake or other means of exposure, there is also an influence of organisms on compound behaviour. This is so because many conditions in soil which govern compound behaviour (e.g. pH, redox status) are at least in part biologically controlled, particularly in the rhizosphere.

4.   SOME CONCLUDING REMARKS

Agricultural practices related to the use of mineral fertilizers and pesticides probably have at present relatively small contributions only to the chemical degradation of soil. This means: relative to other burdens on soil such as those caused by industrial and agricultural waste disposal, and diffuse contamination from the air. Nitrogen and pesticide leaching are apparently the most severe problems at present. In the long term, however, additional undesirable effects may result from ongoing accumulation of compounds in soil. Because agriculture controls quality of soil over extended surface areas, these aspects, although not yet alarming, should also be given serious consideration.

Soil protection should take the conservation or restoration of desirable soil quality as the main aim. If soil protection is considered as urgent then there consequently exists an urgent need for ways and means to arrive at a quantitative evaluation of soil quality. In this context compound behaviour in the soil system with respect to presumed effects constitutes the central theme.

If this can be commonly agreed upon then there would also be agreement on the research areas which deserve major attention in programs directed towards the scientific foundation of soil protection. In this respect soil buffering capacity, compound speciation, soil heterogeneity and bio-availability are to be considered as priority areas. At the same time they constitute subject areas with equal validity for all EC Member States, and even past the boundaries thereof. This thus justifies concerted efforts in these research fields for the common interest of soil protection.

REFERENCES

1. ANON. (1986). Bericht 3, Bestrijdingsmiddelen en Waterwingebieden, Plantenziektenkundige Dienst—CAD Gewasbescherming, Wageningen, 15 April.
2. BACHE, B. W. (1977). Practical implications of quantity-intensity relationships. Proc. SEFMIA, Tokyo, 777.
3. BEEK, J. and DE HAAN, F. A. M. (1973). Phosphate removal by soil in relation to waste disposal. *Proc. Int. Conf. Land for Waste Management, Ottawa, Canada,* p. 77.
4. BEEK, J. (1979). Phosphate retention by soil in relation to waste disposal, Ph.D. thesis, Agricultural University, Wageningen.
5. BEEK, J. and VAN RIEMSDIJK, W. H. (1979). Interaction of orthophosphate ions with soil. *Soil Chemistry B, Physico-chemical Models* (Ed. G. H. Bolt), Chap. 8, Elsevier Applied Science, Amsterdam, p. 259.
6. CHARDON, W. J. (1984). Mobiliteit van cadmium in de bodem. Ph.D. thesis, Agricultural University, Wageningen.
7. HAAN, F. A. M. DE, KEIZER, M. G., LEXMOND, TH.M., VAN RIEMSDIJK, W. H. and VAN DER ZEE, S. E. A. T. M. (1986). Some recent developments and approaches in soil protection research. *Neth. J. Agri. Sci.,* 34.
8. HEER, H. DE, HAMAKER, PH., TUINSTRA, L. G. M. TH. and VAN DER BURG, A. A. M. (1983). Use of gas-tight plastic films during fumigation of glasshouse soils with methylbromide. *Acta Horticulturae,* 152, 103.
9. HENKENS, CH. H. (1983). Cadmium in meststoffen. *Bedrijfsontwikkeling,* 14, 484.
10. HUYGEN, C. and VAN IJSSEL, F. W. (1981). Luchtverontreiniging bij kassen door bodemontsmetting met methylbromide. IMG-TNO, Delft, Report no. G 1048.
11. KEIZER, M. G., VAN RIEMSDIJK, W. H. and DE HAAN, F. A. M. The extraction of heavy metals from contaminated soils by organic chelates: a model description. Submitted for publication in *J. Environ. Qual.*
12. KOLENBRANDER, G. J. (1981) Leaching of nitrogen in agriculture. In: *Nitrogen Losses and Surface Run-off from Landspreading of Manures, Developments in Plant and Soil Sciences, Vol.* 2 (Ed. J. C. Brogan), Martinus Nijhoff/Dr. W. Junk Publishers, p. 199.

13. LEXMOND, TH. M., VAN RIEMSDIJK, W. H. and DE HAAN, F. A. M. (1982). Onderzoek naar fosfaat en koper in de bodem, in het bijzonder in gebieden met intensieve veehouderij. Reeks Bodembescherming no. 9, Staatsuitgeverij, 's-Gravenhage.

14. LEISTRA, M. and SMELT, J. H. (1983). Evaluation of leaching of aldicarb residues from soil to groundwater. 10th Int. Congress of Plant Protection, Vol. 2, p. 729.

15. LEISTRA, M., DEKKER, A. and VAN DER BURG, A. M. M. (1984). Leaching of oxidation products of aldicarb from greenhouse soils to water courses. Arch. Environ. Contam. Toxicol. 13, 327.

16. RIEMSDIJK, W. H. VAN, LEXMOND, TH. M. and DE HAAN, F. A. M. (1983). Fosfaat-en kopertoestand van de cultuurgrond in de Provincie Gelderland; Report Dept. Soil Science and Plant Nutrition, Agricultural University, Wageningen.

17. RIEMSDIJK, W. H. VAN, BOUMANS, L. J. M. and DE HAAN, F. A. M. (1984). Phosphate sorption by soils. I. A diffusion-precipitation model for the reaction of phosphate with metal oxides in soil. Soil Science Society of America Journal, 48, 537.

18. RIEMSDIJK, W. H. VAN and VAN DER LINDEN, A. M. A. (1984). Phosphate sorption by soils. II. Sorption measurement technique. Soil Science Society of America Journal, 48, 541.

19. RIJTEMA, P. E. (1985). Uitspoeling van stikstofmeststoffen. PAO-Waterkwaliteit Landelijk gebied, Agricultural University, Wageningen.

20. SPOSITO, G. and MATTIGOD, S. V. (1980). GEOCHEM: a computer program for the calculation of chemical equilibria in soil solutions and other natural water systems. University of California, Riverside, USA.

21. STEENVOORDEN, J. H. A. M. (1983). Nitraatbelasting van het grondwater in zandgebieden; denitrificatie in de ondergrond; Nota 1435, ICW, Wageningen.

22. STEVENSON, F. J. (1972). Organic matter reactions involving herbicides in soil. J. Environ. Qual., 1, 333.

23. TIETJEN, C. (1977). The admissible rate of waste (residue) application on land with regard to high efficiency in crop production and soil pollution abatement. In: Land as a Waste Management Alternative (Ed. C. Loehr), Ann Arbor Science, p. 63.

24. WESTALL, J., ZACHARY, J. and MOREL, F. (1976). MINEQL—a computer program for the calculation of chemical equilibria composition of aqueous systems. Techn. Note no. 18. Massachusetts Institute of Technology, Cambridge, Mass., USA.

25. ZEE, S. E. A. T. M. VAN DER and VAN RIEMSDIJK, W. H. (1986). Sorption kinetics and transport of phosphate in sandy soil. Geoderma, 38, 293.

26. ZEE, S. E. A. T. M. VAN DER and VAN RIEMSDIJK, W. H. (1987). Transport of phosphate in a heterogeneous field. Transport in Porous Media, 1, 339.

Effects of Forestry Practices on the Chemical, Biological and Physical Properties of Soils

HUGH G. MILLER

*Department of Forestry, University of Aberdeen,
St Machar Drive, Aberdeen AB9 2UU, UK*

SUMMARY

Concern over the effects of forest management on soils relate to one or more of (a) afforestation of previously bare ground, (b) a change from uneven- to even-aged forests, or (c) a change in species. Soils are dynamic entities, the features of which change both with changing forest practice and stage of development of the crop. Thus, in an even-aged forest there is a reduction in soil exchangeable cations and an increase in accumulated humus during the early stages of the rotation but both processes subsequently reverse and ultimately there may be no net change by the end of the rotation. The ability to resist an overall chemical change, however, depends on the presence of an adequate supply of readily weatherable minerals in the surface soil, the supply of which can be increased through soil mixing as a result of either windblow or ploughing. Physical factors of the soil may be damaged through compaction by machinery, but this should be avoidable. Also theoretically avoidable is the real danger of initiating erosion. Improperly designed road systems not infrequently initiate erosion with, in Mediterranean regions, the consequences of fire being a further worrying feature in this regard.

1. INTRODUCTION

The forests of Europe are subject to a great range of management intensities, varying from haphazardly managed selection forests to the intensive cultivation of plantations. Even the latter, however, represents a very low input system in comparison to agriculture. Thus, forest ground is unlikely ever to be completely ploughed and would only be strip ploughed or scarified at most twice a century. Generally no fertilizers are applied but where these are necessary four or five applications a century is the maximum needed. The requirements for herbicides are similarly low and most will never receive insecticides. Despite these very low inputs there are often repeated fears that forestry practices may in some unspecified way damage the environment, in particular the soil.

In examining this matter the first difficulty is defining the base level against which comparisons properly should be made. The heathlands of Europe are impoverished, species-poor man-maintained ecosystems. Their afforestation with spruce involves changes including soil changes; are these to be compared to the preceding conditions under ericaceous vegetation or to some notional condition that may have pertained under indigenous pine or oak? The work of Page [1] and more recently that of Miles [2] has drawn attention to the plastic nature of soils and the extent to which developments manifest early in a forest rotation are subsequently reversed so that the net long-term effect may be one of little or no change. Thus, pH at 15 cm depth in the soil declined from 4·5 to 4·2 over the first half of the rotation of coniferous trees only to return to 4·5 by the time of final felling [1]. Clearly, comparisons made early or midway through the rotation would give very misleading indications of long-term trends. To a certain extent, therefore, discussion of the likely effects of forest operations on soils has to involve reasonable deductions based upon knowledge of soil and plant processes rather than on extrapolation from the few available studies comparing soils beneath different species or ages of trees.

2.   FOREST DEVELOPMENT

Most concern relates to the management of even-aged forest plantations that have involved one or more of (a) a change from previously unforested ground, (b) a change from uneven-aged forests, or (c) a change of species. A change in structure or species of forest is usually brought about to ensure increased yield in total or of a particularly desired product. There is also the factor that because all the trees in an even-aged forest pass through each development stage simultaneously management is simplified and lends itself to large-scale operations.

It is as well to examine the stages through which an even-aged forest passes before discussing the effects these, and associated management operations, may have on the soil. At planting, or during early seedling growth in a naturally regenerated forest, the trees represent a tiny proportion of the total biomass on the site and can have little or no effect. Over the ensuing 20 years or so growth increases more or less exponentially, with the dominant feature being the build-up of the photosynthetic display [3]. Leaves are nutrient-rich organs, particularly in comparison to stem wood. Once canopy has closed, i.e. maximum tree leaf area has developed, further net growth is concentrated on the accumulation of stem and structural branches while the total weight of leaf declines slowly. Leaves have a limited life of from less than 1 year (deciduous trees) up to 7 years, but prior to their being shed about two-thirds of the nutrients they contain are recovered by the tree (retranslocation) and are then available for the formation of their replacements.

Thus, the first period in the life of an even-aged forest is characterized by high nutrient demands necessary for the establishment of the leaf biomass (Stage I in the system proposed by Miller [4]). There follows a period (Stage II) when efficient recycling of nutrients within the tree and through the litter system mean that, despite a high net primary production in terms of stem wood, nutrient demands are much reduced. Indeed, for potassium the order of reduction in terms of the demands made on soil supplies is from 12 kg/ha/yr in a 2-m tall crop to 1 kg in an 11-m tall crop [5].

There follows a prolonged stage when nutrient demands on the soil are low and for some elements may even be less than the amount introduced in rainfall. Whereas prior to canopy closure growth may need to be assisted by the application of fertilizers, thereafter nutrient deficiencies are unusual unless some event, such as an epidemic of defoliating insects, seriously interrupts the normal cycle of nutrients. An important exception, however, is the development of late rotation nitrogen deficiency, or sometimes phosphorus deficiency, that can occur in coniferous forests of northern regions. This results from the immobilization of excessive amounts of nitrogen or phosphorus in the mor humus layer. As such the problem is limited to coniferous species in cold northern regions where mor humus development is a pronounced feature, and even there will only occur on soils with a very low nutrient capital.

The final feature of an even-aged forest is that harvesting is invariably by clear felling. This entails the removal of the vegetation cover so exposing the humus to the warmth of the sun, stimulating rate of decomposition, and the soil to the direct impact of rain and wind, with possible risk of erosion.

By contrast, an uneven aged forest exhibits all development stages in intimate mixture over a small area. Thus, at no stage are all the trees placing a heavy nutrient demand upon the soil nor are exposed areas ever more than a few tree crowns in extent. All extremes, therefore, are greatly reduced provided the forest never suffers catastrophic destruction. However, in the natural state it is probable that such catastrophic destruction, from wind, fire or avalanche, were a regular, if widely spaced, feature.

3.   CHEMICAL FACTORS

The afforestation of bare ground, conversion of an uneven-aged to even-aged forest or conversion to a faster growing species would each accentuate the demand placed upon the soil during the period of canopy formation. There is thus a period characterized by a massive shift of all nutrients from the soil to the tree. In an acid-dominated soil, where nitrogen supply will be largely as ammonium, there is an excess uptake of cations over anions and in consequence hydrogen ions pass from the tree to the soil to maintain electrical neutrality. The consequences for the soil is a rapid decrease in exchangeable cations, primarily potassium, calcium and magnesium, and an increase in the hydrogen ion saturation of the soil exchange surfaces. It should be emphasized that this is a transient phase, as already pointed out, once the canopy has closed the transfer of ions from soil reserves to tree falls dramatically. In a soil with a reasonable supply of soil minerals weathering release of cations will now exceed the rate of uptake by tree roots and the soil base status should slowly recover to pre-planting levels, the reverse pattern pertaining for soil acidity. The lack of appreciation that there is such a recovery stage has led some workers to suggest absurd extrapolations on the basis of soil changes measured early in the forest rotation.

Whether or not the soil fully returns to its original state depends on the amounts and nature of the readily weatherable minerals in the surface soil horizons. If the soil has long remained undisturbed and is heavily leached, the lack of availability of relatively unweathered minerals may become a limiting factor. In this context the mixing of the soil, to bring up unweathered subsoil, that accompanies windblow is a beneficial factor, and a factor that the manager can mimic by use of suitable ploughing techniques. Paradoxically, it can be argued that even intermittent erosion, to reduce the depth to the mineral-rich subsoil, may be beneficial provided that there is sufficient depth of soil (although there will then be a period of slow growth while stocks of organic nitrogen recover). Of course, should there be no removal of timber from the site all cations taken up by the trees will eventually be returned to the soil.

Theoretically, timber removal from uneven-aged and even-aged forests should be about the same if species of similar growth rate are involved. Over an extended period, therefore, the acidification stress will be unchanged by conversion to even-aged forest although in the latter it will proceed somewhat stepwise because of the simultaneous canopy development. In the uneven-aged forest, however, the forest manager is denied the opportunity to use ploughs to bring up mineral rich subsoil. The creation of plantations on unproductive heath and moor, or the conversion of forest to a more productive species, will, by consequence of the increased net primary production, increase the rate of transfer cations and hence potential acidification of the soil. Here the availability of weatherable minerals is of prime importance and again the opportunity to plough or otherwise cultivate should be valued.

A change in species or forest type may also influence the soil through a change in the nature and pattern of accumulation of organic matter. The amount of organic matter in an uneven-aged forest should remain fairly stable with time over any extended area. In an even-aged forest, by contrast, the humus layer (if it accumulates at all) will rise to a maximum around the middle years of the rotation and then decline, with particularly accelerated decomposition occurring following clear-felling [1]. There is, however, a species factor to be taken into consideration, for the fast-growing conifers, that must necessarily dominate in commercial forestry, not only produce more litter than do the slow-growing broadleaved species but also produce a litter than both decomposes slowly and is acid in nature (although the distinction between species is not always clear and there can be a species × soil interaction). This may lead to accelerated surface soil leaching and podsolization, particularly if there is nitrate leaching following clear-felling, but the significance of any such change is still open to debate. As mentioned above, soils are dynamic entities that reflect their vegetation cover. Many effects are known to be reversible but our appreciation of the speed, extent and permanence of any change is still very far from complete. What is clear is that change should not be regarded as permanent, far less deleterious, unless there is clear supporting evidence.

One change that is at least very slow to reverse is that associated with the destruction of the organic layers by fire, the first adverse consequence of which is the development of nitrogen deficiency. Fortunately, the nitrogen capital in most European forests, but by no means all, is sufficiently high to withstand the losses associated with an occasional fire. Furthermore, fire has not been used as a management tool on any extensive scale in European forests for more than 50 years.

4. BIOLOGICAL PROPERTIES

The biological properties of the soil reflect its chemical and physical characteristics. Thus, a change from a broadleaved species with characteristic mull humus to a mor-forming coniferous species is accompanied by a marked shift in the soil microbial population from bacteria-dominated to fungi-dominated. To soil scientists brought up on agricultural criteria this is regarded as an adverse change, a highly opinionated view that derives from the use of an inappropriate yardstick. Again, it is quite clear that microbial changes can be induced and reversed by simply changing the tree species forming the forest.

A change from uneven- to even-aged forest should have little or no effect on soil microbial populations provided that there is no change in species. However, either such a change or the afforestation of open land can have a profound effect on the range and types of ground and shrub vegetation found within the forest. Again these will shift to the communities characteristics of the forest type now created and as such can only be considered harmful in relation to the value of the vegetation types they replace. If the latter have conservation importance, either because they are rare or because they represent a valued local amenity, then creation of a forest, or changes in species or structure of the forest, will have deleterious effects [6]. With the exception of situations where the survival of a species is at stake, conservation values are essentially subjective and hence not open to rational analysis. It is the mark of civilized man that he tries to conserve things of beauty; that he is able to do so is firstly dependent upon him creating the wealth that permits of such behaviour and secondly requires some agreement as to what is worthy of conservation, or rather agreement on priorities in a situation of limited resource and conflicting interests.

A related aspect is the conservation of a forest tree gene resource in face of a trend, not only towards selection of species but of selection of provenance within species and the anticipated selection of particular clones within provenances. Foresters are well aware of the potential dangers and no breeding programme in Europe envisages forests of severely limited numbers of clones (usually afforestation would be with mixes of several tens of clones) nor of a stage where new clones are not being continually introduced from breeding programmes. Indeed, the present inability to maintain juvenility (avoid maturation) will ensure that there is a finite limit to the number of individuals that can be bulked up from a single seed. The prospects in Europe still fall well short of the step the Japanese took with *Cryptomeria japonica* several centuries ago and it should be pointed out that they have neither suffered any biotic or abiotic disasters in their clonal forests nor do the soils beneath their clonal forests show any abnormal feature.

5. PHYSICAL PROPERTIES

The forest manager interacts most markedly with the physical factors of the soil. Frequently when he plants, either to create a new forest or improve the species or genotype of an existing forest, he will endeavour to overcome growth limiting factors in the site by drainage and cultivation. A properly designed drainage system is one of the few means of permanently improving site quality. By contrast, the immediate effects of cultivation are usually short-lived with the exception, as mentioned above, of the value of mixing into the surface soil fresh minerals from the relatively unweathered subsoil. The drainage and afforestation of peat bogs affects a change that can never be reversed, a change that is essentially in the physical nature of the peat (water content, pore size, bulk density, etc.) but which leads to chemical changes (e.g. increased CEC and exchangeable acidity) and to biological changes (increased soil respiration and reduced anaerobes). Drainage of very wet mineral soils may have a similar, if less pronounced, effect. Less extreme soils, however, are little altered, if at all, in their physical characteristics by the actions of forest cultural operations. The best soils would seldom be ploughed, although the planting spots may be scarified free of vegetation and humus (a technique that has been used for centuries), and mediocre sites would usually only be striped ploughed to a fairly shallow depth. Sometimes, however, complete ploughing or rotovation has been used (although because the advantages are evident only over a few years after planting such treatment is now seldom favoured) and where there are restrictions to profile drainage deep ploughs, or more usually ploughs fitted with deep tines, are used to disrupt the impediment. The effect of the latter is probably one of drainage rather than of cultivation in the strict sense. At all events such treatments are only given every 60–100 years and there are still very few sites that have been deep drained or ploughed more than once.

Most concern over the effects of forest practices on the soil must relate to the inadvertent consequences of using machines in the wood or of building roads. Timber is bulky and heavy and has to be moved to an extraction road either by dragging along the ground (skidding) or by placing on large and powerful all-terrain machines (forwarders). The skidded tree stem inevitably gouges a channel in the soil, pushing aside the organic layer and upper soil, and the wheels of the tractor used to skid the logs can cause rutting and compaction. Similarly, forwarders are wheeled vehicles that may rut and compact. Root growth is much reduced in compacted soils [7] and the growth of both naturally regenerated and planted trees, and of other vegetation, may be retarded as a result. In addition, the displacement of the organic layers can lead to nutritional problems, particularly nitrogen deficiency in the wheel tracks. Damage can be reduced by limiting the use of skidding to soils that resist compaction, by using machines with low ground pressure, by prohibiting operations in unsuitable weather, and, through proper design of extraction routes, ensuring minimum occurrences of repeated passes over the same soil. Where compaction has been a problem some forest departments prescribe subsequent ploughing or subsoiling of the worst affected soil, which, including the landings on which timber is stacked, would seldom account for more than 2–3% of the area felled. Compaction is a problem that good management should be able to avoid or minimize.

Erosion, similarly, should not occur in a properly managed forest. Some forest operations, however, do carry with them the real possibility of initiating erosion. Essentially any action that exposes the soil to wind or rain must be managed with care. Clear-felling itself is a major factor and in areas of high erosion risks, such as some of the unconsolidated sediments in the Alps, must simply be avoided. In such areas selection or group selection systems, in which only single trees or small groups of trees are felled in any one area at any time, have to be adopted. In less fragile areas clear-felling should carry no risk provided that the soil surface has not been disturbed or potential erosion channels created. Such can well occur if extraction is not planned and carried out with due care. The creation of deep ruts across the contour has to be avoided and if necessary properly angled drains should be dug to harmlessly discharge any channelled water.

Poorly designed drainage systems, particularly if the drains are at too steep an angle (a slope of no more than 2° on any significant drain), can initiate erosion but the danger is so obvious that few foresters commit such a crime. One can be less sanguine, however, about the problems associated with new road systems. Roads must always be regarded as potential erosion channels and steps taken to ensure that drainage water is properly channelled alongside and beneath the road. Because many roads have a slope of more than 2° the roadside drains must not be allowed to run for any distance before discharging onto flat ground or into a suitable drainage system. This will entail the installation of frequent culverts of sufficient size to cope with all storms, because once erosion has been initiated it is extremely difficult to arrest.

6. CONCLUSIONS AND NEED FOR RESEARCH

It is only very rarely that the stability, in the widest sense, of the forest soils of Europe are threatened by the action of forest management or mismanagement. Soils change under forests but many of these changes reverse with development of the crop or change in forest species. Mineral poor soils, however, may have limited ability to recover in all aspects. Research is clearly required on the relation between rate of recharge of exchangeable cations in the soil and soil mineralogy with the aim of seeking suitable ameliorative techniques such as deep ploughing and soil mixing.

Soil organic matter accumulation also fluctuates with species and stage of development, although there is also clear evidence for a species–site interaction. The variations in soil organic matter are such that it is difficult to believe that species do not leave some long-term imprint on this aspect of the soil. The ease with which soil can be converted to different uses suggests that any such aspect has little practical significance but research is clearly indicated.

The most worrying operations are those that expose the soil to the risk of erosion. Poorly designed drainage and roading schemes can be very damaging but the problem is less one of research than of proper supervision and professional competence. Fire is also an initiator of erosion and has the further disadvantage of loss of volatile nutrients, notably nitrogen. The causes of fire are easily established and need little research. The prevention and control of fires, however, is a pressing problem in Mediterranean countries that calls for an urgent and massive increase in research funding. A useful start could be made by examining the appropriateness of the techniques, including fuel reduction burning, developed in Australia.

Finally, much has been made about the plasticity of soil and the extent to which its characteristics reflect transient features of the vegetation cover rather than any progressive and irreversible intrinsic change with time. Such an idea accords well with casual observation in many parts of the world but there are relatively few well-designed studies to support the contention. Further detailed investigation is urgently required but any such work must cover, by the sampling of different stands, all stages in the life of a forest. Ideally any programme should also initiate a long-term programme of monitoring changes at particular sites.

REFERENCES

1. PAGE, G. (1968). Some effects of conifer crops on soil properties. *Commonwealth Forestry Review*, **47**, 52–62.
2. MILES, J. (1985). The pedogenic effects of different species and vegetation types and the implications of succession. *J. Soil Sci.*, **36**, 571–84.
3. ATTIWILL, P. M. (1979). Nutrient cycling in a *Eucalyptus obliqua* (L'Herit) forest. III. Growth, biomass and net primary production. *Aus. J. Bot.*, **27**, 439–58.
4. MILLER, H. G. (1981). Forest fertilization: some guiding concepts. *Forestry*, **54**, 157–67.
5. MILLER, H. G. (1984). Dynamics of nutrient cycling in plantation ecosystems. In: *Nutrition of Plantation Forests* (Eds G. D. Bowen and E. K. S. Nambiar), Academic Press, London, pp. 53–78.
6. NATURE CONSERVANCY COUNCIL (1986). *Afforestation and Nature Conservation in Britain*, NCC, Northminster House, Peterborough, UK.
7. SANDS, R. and BOWEN, G. D. (1978). Compaction of sandy soils in Radiata pine forests. II. Effects of compaction on root configuration and growth of Radiata pine seedlings. *Aust. For. Res.*, **8**, 163–70.

Impacts on Soils Related to Industrial Activities: Part I—Effects of Abandoned Waste Disposal Sites and Industrial Sites on the Soil: Possible Remedial Measures

V. Franzius

Federal Environment Agency, Bismarckplatz 1, 1000 Berlin 33, Federal Republic of Germany

SUMMARY

Contaminated sites, such as areas which have been used for the disposal of solid or liquid waste (abandoned waste disposal sites) and former industrial areas where environmentally harmful substances have been in use are termed abandoned hazardous sites (or inherited pollution sites—Altlasten) if it is proved that they are potentially hazardous. The potential risk posed by these sites mainly concerns soil and groundwater, which are protected assets (Schutzgüter). Because of the cases which have occurred of severe damage, and the continual addition of new contaminated sites, the problem of abandoned hazardous sites (Altlasten) is among the environmental topics which receive the highest priority in the Federal Republic of Germany. Even if the precise extent of the Altlasten problem is not yet known—estimates place the number of sites requiring rehabilitation at around 5000, with the cost of rehabilitation amounting to some DM 20 000 million—the consequences for soil and groundwater are already evident. Strategies for dealing with the problem generally comprise the stages of: listing; investigation and evaluation; rehabilitation and monitoring. Responsibility for implementing these stages devolves on the respective Länder. While progress has been made in the listing of contaminated sites, especially abandoned waste disposal sites, and the process has begun of adding abandoned industrial sites to the list, systematic investigation and evaluation (hazard assessment) is only in its preparatory phase. There is no uniform method of assessing potential risk and determining priorities as there is, for example, in the United States. Some Länder have developed formal procedures of their own. At the same time as

247

these activities, rehabilitation has been undertaken in extreme cases of soil or groundwater contamination. A basic distinction must be made between techniques which temporarily render a site safe and permanent rehabilitation. Abandoned waste disposal sites which pose particular problems have to be rendered safe by the use of techniques to prevent the contamination from spreading, such as containment or stabilization, owing to the lack of suitable technology for their treatment and of technical and economic resources for their disposal. These methods, together with techniques for the treatment of contaminated soils which are used for the rehabilitation of abandoned industrial sites, are being developed and applied on a large scale with the support of the Federal Minister of Research and Technology.

1. INTRODUCTION

Owing to the many spectacular cases in which damage has occurred, such as the seepage of oils containing dioxin from the landfill site at Hamburg-Georgswerder and the former sludge landfill at Bielefeld-Brake, on which housing had been built and in which toxic production residues had been dumped, and the premises of the defunct firm Pintsch Öl in Berlin-Britz, which was contaminated with chemicals which endanger water supplies, contaminated sites (*Altlasten*) have increasingly become the subject of environmental discussions. To be sure, these examples of environmental damage are not representative, but they illustrate the range of possible soil and water pollution caused by former waste disposal sites and the use of substances which endanger water supplies on industrial premises, which must now be rehabilitated at great expense.

Contaminated sites may be defined in accordance with the Federal Government's Soil Protection Plan (*Bodenschutzkonzeption*). According to this, closed and abandoned dumping grounds for local authority and industrial refuse (abandoned waste disposal sites), uncontrolled dumping, tips and cases of backfilling with environmentally hazardous production residues (whether or not combined with mining material and building waste), former industrial sites (*Altstandorte*), corrosion of piping systems, defective sewers, war material dumps, inappropriate storage of materials which endanger water supplies and other soil pollution, are all causes of so-called *Altlasten*. Agricultural pollution, which also contributes to soil and groundwater pollution, caused for example by the over-application of fertilizers or by remnants of pesticides, are not dealt with under this heading, being cases of pollution covering wide areas.

In general, the problem with *Altlasten* is that there is as a rule insufficient information about them. The special problem with abandoned waste disposal sites is that they may contain hazardous waste which, at the present state of the art, must be disposed of in special waste treatment plants. On the other hand, the identification of pollutants in contaminated industrial premises is generally easier, as it is often possible to identify firms and the particular materials used in their production, even after long periods.

The Soil Protection Plan outlines the problem of contaminated sites as follows: such pollution endangers soil, ground and surface water and limits their subsequent use; it may also constitute an immediate health risk. Even the further use of such sites for industrial or commercial purposes is environmentally hazardous, because, for example, foundations of buildings or supply pipes and sewers may be destroyed.

2. PROCEDURE

There is no uniform procedure throughout the Federal Republic of Germany for solving the *Altlasten* problem. The general procedure essentially consists of the following three stages: listing; investigation and evaluation hazard assessment; monitoring and rehabilitation. In Stage 1, all known and newly identified sites are listed in data files and where possible identified and pinpointed in atlases (*Altlasten* registers). The numerous results obtained to date from such activities, which have been initiated in a number of *Länder* on the basis of official listings issued by the *Land* authorities which have been charged with this task, in conjunction with the competent local government bodies, are reflected in numerical form in Table 1.

In Stage 2, the suspected sites listed undergo a hazard assessment. A decision as to whether or not a particular suspected site is a contaminated site and thus an *Altlast* is generally taken by the competent authorities in an initial assessment based on the current dossier. Suspected sites which are clearly shown by the initial assessment to present no actual or likely danger may be exempted from regular monitoring. However, if the initial assessment provides sufficient evidence of a latent or present hazard, the suspected area must be listed as a contaminated site or *Altlast*, for which, depending on the nature and extent of the hazard, either emergency measures (e.g. where the danger is acute) or rehabilitation measures must be initiated.

Table 1

Number of estimated or listed suspected sites in the Federal Republic of Germany

Land (State)	Number	Observations	Situation up to:
Schleswig-Holstein	Over 2000	*Altablagerungen* (closed or abandoned waste-disposal sites)	Dec. 1984
Free Hanseatic City of Hamburg	About 2400	Sites suspected of being *Altlasten* (1831 locations are recorded in the *Altlasten* notification register, of which 162 have been rehabilitated, partially rehabilitated or otherwise disposed of)	May 1985
Hanseatic City of Bremen	45	Public or private dumps mainly containing building waste, household refuse, garden waste and workshop waste similar to household refuse	April 1985
Lower Saxony	3500	Abandoned waste disposal sites of all kinds listed. According to an initial hazard assessment, there are about 100 inherited pollution sites which could be prejudicial to the public good	1984
North Rhine-Westphalia	About 8000	Abandoned waste disposal sites and industrial sites listed	May 1985
Hesse	5000	Abandoned hazardous sites (*Altlasten*) is the estimated number. Over 3200 waste-polluted locations have been registered	March 1985
Rhineland-Pfalz	About 5000–6000	Estimated number of waste disposal sites	Jan. 1985
Baden-Württemberg	Over 4600	Waste disposal sites listed	May 1984
Bavaria	About 5000	Local authority refuse dumps up to 1982, of which about 4300 had been closed and recultivated by 1984	May 1984
Saarland	738	Abandoned landfill sites	Dec. 1984
Berlin	178	Inherited pollution sites (waste disposal sites, contaminated industrial land and abandoned sites polluted with war material)	Jan. 1985

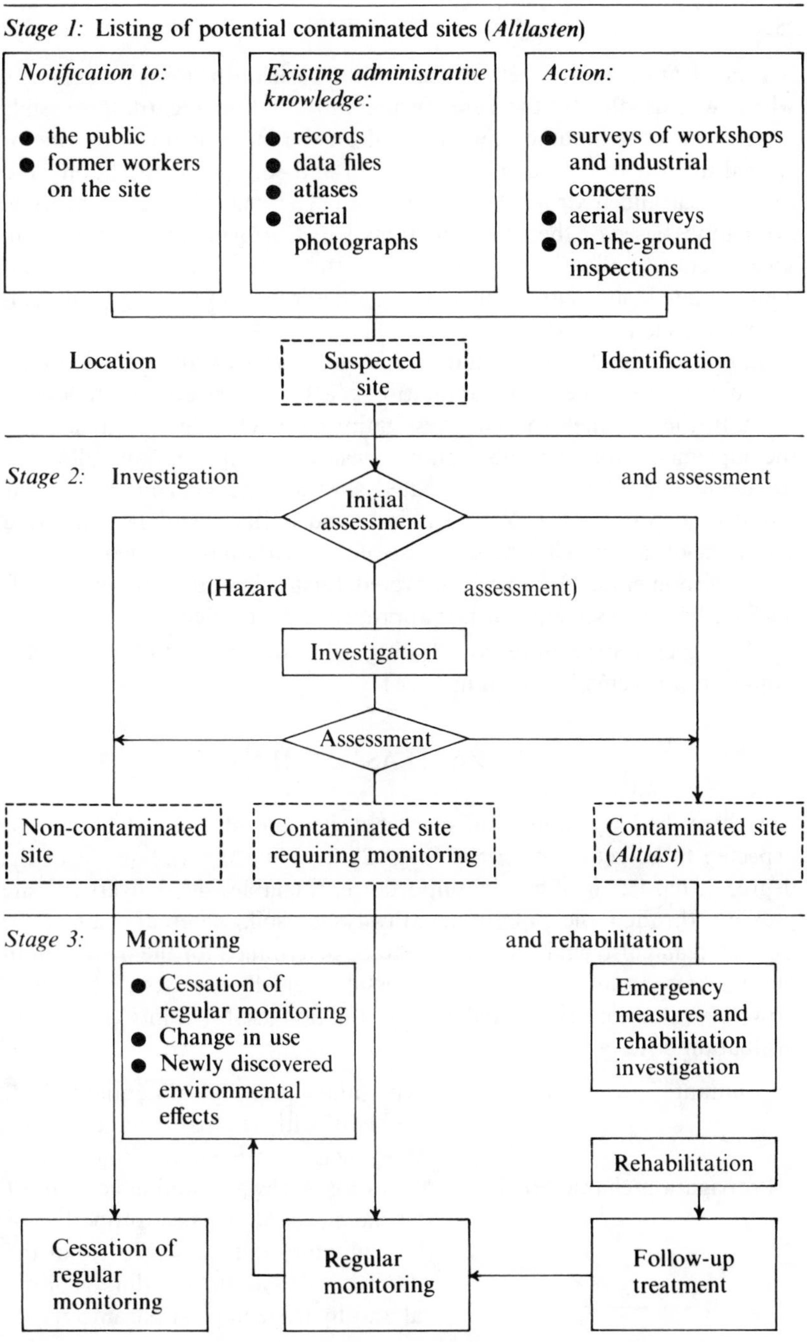

Stage 1: Listing of potential contaminated sites (*Altlasten*)

Notification to:	*Existing administrative knowledge:*	*Action:*
● the public ● former workers on the site	● records ● data files ● atlases ● aerial photographs	● surveys of workshops and industrial concerns ● aerial surveys ● on-the-ground inspections

Location   [Suspected site]   Identification

Stage 2: Investigation   and assessment

◇ Initial assessment

(Hazard   assessment)

[Investigation]

◇ Assessment

Non-contaminated site	Contaminated site requiring monitoring	Contaminated site (*Altlast*)

Stage 3: Monitoring   and rehabilitation

● Cessation of regular monitoring
● Change in use
● Newly discovered environmental effects

Emergency measures and rehabilitation investigation

Rehabilitation

Cessation of regular monitoring	Regular monitoring	Follow-up treatment

FIG. 1. Flow diagram of the general procedure when rehabilitating inherited pollution sites in the Federal Republic of Germany.

If no clear decision is yet possible on the basis of the initial assessment, which will usually be the case, further information regarding possible hazards must be obtained by means of on-site investigations (routine or special investigations, such as water, soil and soil-air investigations and geophysical and hydrogeological surveys). The hazard assessment (final evaluation) based on these investigations, which are generally carried out in several stages, allows in principle for three possible decisions: non-contaminated site, contaminated site requiring monitoring and contaminated site (*Altlast*).

Stage 3 includes the rehabilitation and monitoring of contaminated sites. In order to determine the rehabilitation goals and measures to implement them, further detailed special investigations are generally required. After the implementation of rehabilitation measures and appropriate follow-up treatment, rehabilitated contaminated sites are included in the regular monitoring programme. Where it is intended that regular monitoring should be discontinued, where the use of the location is changed or where new environmental effects are discovered, further investigations culminating in a final assessment must, if appropriate, be carried out.

The general procedure for dealing with the problem of inherited pollution problems is shown in Fig. 1.

3. HAZARD ASSESSMENT

In view of the large number of suspected sites and of those which may be expected to be actual *Altlasten*, Stage 2, involving the assessment of the degree of hazard, is of crucial importance. It enables the dangers the site presents through the pollution pathways of soil, water and air to be estimated and also enables priorities to be determined for the treatment of many *Altlasten* cases. The hazard assessment chiefly consists of the laying down of evaluation criteria and of the evaluation itself. The most important evaluation criteria are:

Pollutant: Type and quantity, potential effect, especially with regard to toxicity, carcinogenicity, persistency, cumulability.

Propagation characteristics: Migration pathways, taking account of the chemical and physical properties of the pollutants and the sensitivity of the site, in particular surrounding geological and hydrogeological parameters.

Protected assets: Likelihood of damage to assets which are protected for public safety and use, in particular human health, water, animals, etc.

The accuracy of the assessment is largely dependent on the data available. Before the concluding evaluation is made, a number of stages of investigation and evaluation must generally be completed. An exact, scientifically corroborated assessment is not possible, as the parameters and interactions to be considered are too complex. Case-by-case consideration will continue to be decisive. However, in view of the large number of suspected sites, there is a need for a comparable, objective and repeatable assessment. As there is no unified Federal evaluation scheme, some *Länder* have developed their own assessment models.

Most of these evaluation models consist of formalized evaluation procedures for determining priorities for further investigation and/or rehabilitation procedures. Examples of these are the evaluation procedures of the *Länder* of Schleswig-Holstein (Classification of Abandoned Refuse Dumps), Hamburg (AGAPE—Assessment of the Potential Hazard of Abandoned Waste Disposal Sites in order to Determine Priorities (Environmental Authority) and Determination of the Potential Hazard to Groundwater—the Building Authority's initial evaluation procedure), Baden-Württemberg (evaluation scheme in the Soil and Subsoil Rehabilitation Manual Requirements for Survey, Rendering Safe and Rehabilitation of Polluted Sites) and Saarland (Evaluation of Abandoned Waste Disposal Sites for Determination of Priorities).

Further approaches to the assessment of hazards are under development. The conference of Environment Ministers has commissioned a working party of the *Länder* Waste Disposal Study Group to investigate, among other things, the assessment of the hazard posed by abandoned waste disposal sites and industrial sites. As part of this work, they have been instructed to check on possible approaches to standardization.

4. RENDERING SAFE AND REHABILITATION

Measures to render safe or rehabilitate sites are always required where the hazard assessment proves or shows the likelihood of a hazard to the environment. Such measures must always aim to dispose of environmental hazards, or at least render them controllable. Environmental hazards can be rendered controllable by means of measures to render them safe, which prevent the contamination from spreading. Techniques to prevent the propagation of pollutants by containment are the main ones used to render abandoned waste disposal sites safe. Subsequent containment can be achieved by sealing the surface and by vertical underground partitions (retaining walls). These techniques have already been used to render abandoned waste disposal sites safe. For particularly troublesome waste disposal sites, special pollutant-resistant sealing materials are currently being developed and tested in R & D projects. Technical development is also being carried out of subsequent horizontal sealing of the base, which is to be used primarily in cases where there is no impermeable bottom layer at an attainable depth on which to base the retaining walls.

In view of the life expectancy of building materials used for containment, it is clear that such measures do not represent permanent solutions, but simply measures for 'gaining time', on the assumption that sealing structures do not in the long term comply with the basic principles of controllability and reparability.

Since containment measures can only have a temporary effect, lasting solutions are required of the problem of actually disposing of environmental hazards, as the only true rehabilitation measure. Technical approaches to rehabilitation may be grouped under the headings of *in-situ* treatment and on-site/off-site treatment, discussion of which follows. The 'on-site' process consists of digging out the contaminated soil—where necessary with suitable groundwater drainage and arrangements for the protection of workers— and treated on the spot in mobile or semi-mobile plants. The cleaned soil is intended for use as backfill. Realization of this concept depends on the groundwater in the surrounding area already being polluted and on the cleaning operation being possible. The 'off-site' process consists of transporting the excavated soil to central treatment plants. In both cases, in addition to the arrangements mentioned above, storage facilities may be provided for the contaminated soil to reduce the load on the treatment plants. Of the many possible treatment processes, the ones which appear to be really feasible for large-scale use at the present state of research and experience for both on-site and off-site operation are the following:

—thermal treatment;
—extractive treatment;
—biological treatment.

In the case of thermal treatment, the pollutants may be eliminated by either direct or indirect heating of the contaminated soil. The resulting polluted gas/air mixture may receive further treatment in the form of high-temperature combustion, catalytic oxidation or gas scrubbing. In the Netherlands, four different thermal treatment processes have been developed to date, and these have already been used for soil rehabilitation. In the USA there are also plants for the thermal treatment of contaminated soil. In the Federal Republic of Germany, plants of a similar design are currently being developed in R & D projects with a view to large-scale application (Züblin AG, Ruhrkohle AG/Deutsche Babcock Anlagen AG).

A very promising technical solution for on-site/off-site treatment of contaminated soil is to use soil washing techniques. These processes consist of the chemical/physical extraction of pollutants by the use of mechanical energy and suitable rinsing fluids. In the USA, the Netherlands and Germany, extractive soil treatment has already been carried out by various processes. Bremer Vulkan AG is working on a mobile pilot plant and another is to enter service as part of an R & D project at the end of the year in Berlin (Harbauer and Co.).

Biological treatment for on-site/off-site rehabilitation is now at the development and trial stage. The process parameters are easier to control for the on-site/off-site process than they are for the biological *in-situ* process. Since 1985 Deutsche Shell AG of Hamburg have been conducting a pilot project on the microbiological off-site rehabilitation of oil-polluted soil at its Harburg refinery. The aim of the project is to develop an economically and ecologically interesting alternative to conventional disposal techniques (combustion and dumping) for oil-contaminated soil, using microbiological degradation. Biological on-site treatment of contaminated soil is also carried out by, among others, the firm Biodetox.

In-situ processes involve the treatment of soil contamination in its natural location. *In-situ* processes can, depending on the hydrogeological relationships in the subsoil and the type of potential contamination, be combined with hydraulic and civil engineering techniques to prevent the contamination spreading. The special problem of *in-situ* processes is that the cleaning and degradation processes are comparatively hard to control. Of the chemical/physical and biological processes which are theoretically possible, appraisals carried out to date find that only the microbiological *in-situ* treatment shows any promise. The particular attraction of this process is that considerable cost savings can be made owing to the fact that excavation is not required and expenditure on equipment is comparatively low. In particular, this process is a clear choice where the ground has already been built on. Its disadvantage is that it can only be used for organic contamination which is susceptible to microbial degradation.

Firms using microbiological *in-situ* rehabilitation include Biodetox and, for the experimental rehabilitation of a petrol station in Berlin, Labor für Umweltanalytik (the Environmental Analysis Laboratory). While it is true to say that only a proportion of the abandoned sites are susceptible to *in-situ* treatment, even using the most promising process to date, namely the microbiological process, new processing developments would be welcomed, at least on economic grounds.

Since on-site or *in-situ* treatment of contaminated soil changes the biological, chemical and physical properties of the soil, criteria and testing procedures need to be developed in a parallel R & D project which will enable decisions to be taken on the possibility of re-using the soil taking account of environmental conditions, the chemical bonding formed by the pollutants and the concentration of pollutants remaining.

An example of combining *in-situ* and on-site processes is the use of hydraulic measures in which contaminated groundwater and percolating water are pumped from wells and released in a clean state after appropriate treatment. These procedures are used either alone or in combination with containment measures. The disadvantage is that such measures can only be used for disposing of pollutants which have already been dissolved and have reached the groundwater and that they must therefore generally be continued for long periods owing to the natural leaching processes.

5.   PROSPECTS

In view of the as yet unknown but already evident extent of the problem of inherited pollution sites (*Altlasten*) the problem of financing is usually raised. Since listing is not yet complete and systematic hazard assessment has not yet been introduced, monetary figures can only be based on estimates. On the basis of the total of about 35 000 listed or estimated suspected sites contained in Table 1, a hypothetical breakdown into 30 000 abandoned waste disposal sites and 5000 abandoned industrial sites was taken as a basis for calculation. Experience to date indicates that it must be assumed that about 10% of the abandoned waste disposal sites and 50% of the abandoned industrial sites are actual *Altlasten*, meaning that more than 5000 *Altlasten* will have to be rendered safe and rehabilitated. Taking account of the unit costs which have been determined for investigation, rendering safe and monitoring alone, there would be a financing requirement over the next 10 years of about DM 17 000 million. Being a conservative estimate, this amount certainly lies on the lower limit. It must be assumed that once accurate listing and hazard assessment data are available as a basis, the *Länder* will draw up definite cost assessments. In individual cases, the polluter will in principle be called upon to provide the finance. However, in the majority of cases the liability of the polluter is not tenable, as the polluter is either no longer in existence or insolvent. This means that in the last analysis rehabilitation measures will become the legal and financial responsibility of government. The Federal and *Länder* governments are at present testing financing models which provide for a reasonable contribution from industry.

Impacts on Soils Related to Industrial Activities: Part II—Incidental and Accidental Soil Pollution

E. E. Finnecy

*Environmental Safety Centre, Harwell Laboratory,
United Kingdom Atomic Energy Authority, Harwell,
Oxfordshire, UK*

SUMMARY

This chapter reviews some of the causes of, and the nature of, incidental and accidental pollution of the soil as a result of industrial activity, and illustrates these by some examples. It is pointed out that accidental soil pollution, as defined, is commonly dealt with more or less promptly, but that incidental pollution often goes unnoticed until some harmful effect is apparent.

There is a brief discussion of the approach to the management of the potential risks from polluted soil, exemplified by the approaches used in The Netherlands, the UK, and parts of the Federal Republic of Germany. Three particular problem areas of significance for the management of polluted soil are identified:

—The difficulty of adequately identifying and quantifying the nature and extent of soil pollution.

—The problems of responding to the identification of a large number of polluted sites in a particular area, and the need to assign priorities for action.

—The lack of effective 'trigger' levels for a wide variety of organic pollutants, the limited range of 'targets' to which existing 'trigger' values refer, and the problems of assigning such values to direct human exposure.

259

1. INTRODUCTION

This chapter is a review of published information and experiences of incidental and accidental soil pollution. This is a vast subject, and in the space available only a few features can be dealt with.

First it is necessary to define the way in which the terms in the title are used.

Incidental pollution arises as a consequence of a deliberate act. Such acts could, for example, include discharges from a chimney that produced a consequential deposition on the soil. The storage of a bulk material, uncontained, directly on the soil will lead to soil contamination. Incidental pollution can also result from a failure to act; thus the failure to repair a leaking valve may, incidentally, result in soil pollution. Incidental pollution may be a totally unconsidered consequence of an action; for example some of the background pollution of the environment with polychlorinated dibenzo-p-dioxins (PCDD) was an incidental, but previously unconsidered, consequence of the widespread use of pentachlorophenol as a wood preservative. Incidental pollution may also result from actions where the consequential contamination of the soil is recognised, but is judged (rightly or wrongly) to be without significance. Given the above interpretation of the term, 'incidental pollution' of the soil may continue, or have lasted, for a considerable time.

Accidental pollution is essentially the consequence of an unplanned event, such as a sudden failure of a storage vessel, a transport accident, etc. Typically accidental pollution will be followed promptly by efforts to clean and restore to normal use the soil polluted by the accident. Though soil clean-ups may not be easy (e.g. Seveso) the existence of a 'problem' requiring remedial action and/or protective measures will usually be apparent fairly rapidly. Accidental pollution often affects only comparatively small volumes of soil, but it may produce very high concentrations of pollutants in the soil.

Soil pollution. In this chapter the word 'pollution' is used to signify the presence of some substances in the soil, as a result of human activity, to such a concentration that there is a significant risk of damage to users of the soil, or a restraint on its free use. The risks can take many forms: impairment of the health of humans, animals or plants using, or growing in, the soil; damage to buildings or structures placed on or in the soil; contamination of groundwaters or surface waters in contact with the soil.

We will consider some of the ways in which incidental and accidental soil pollution occurs, illustrated with examples drawn from experience.

We cannot hope to cover this subject exhaustively in this chapter, and some aspects will not be addressed at all. Thus we will not attempt to make any comment on 'acid rain', contamination by radioactive substances, soil contamination by waste disposal (this is covered by others at this conference), and other aspects will only be mentioned very briefly.

2. WHY DOES SOIL POLLUTION MATTER?

Pollution may restrict our free use of the soil, or require us to engage in expensive and often difficult reclamation methods to restore it to beneficial use. Soil pollution may pose direct threats to our health, and to the health of animals and plants. Soil pollution may cause unacceptable pollution of water resources. These risks are not always obvious, and a considerable time may elapse between the initiating pollution, the deleterious effect, and the identification of the cause of the effect. For example:

Between 26 May and 16 June 1971, four horse arenas in the State of Missouri, USA, were sprayed with waste engine oil as a dust control measure. A horse show was held in one of the arenas 4 days after spraying, and people attending noted an unusual absence of flies. Soon after, an unusual number of dead birds were found in the area, and two horses became ill, followed soon by illness in dogs and cats. Human illness first became apparent in August, at which time the area was still unusually free from flies. One child, who had extensive contact with the soil, was admitted to hospital, suffering from nose-bleeding, diarrhoea and other symptoms. Others, adults and children, suffered similar symptoms, but in a less acute form. All appear to have recovered fully. However of 125 horses exposed at this arena, 58 became ill and 43 died between June and December 1971, there were 26 known abortions to pregnant mares and many foals died at birth or soon after.

It was soon decided that this outbreak was a result of spraying the arenas with the waste oil. Identifying the cause however was not easy. Eventually it was discovered that the 'oil' used had been mixed with waste from the production of 2,4,5-trichlorophenoxyacetic acid (2,4,5-T), and trichlorophenol was found in soil samples from the affected arenas, but the amounts found were not sufficient to explain the severity of the observed effects. It was not until 1974 that 2,3,7,8-tetrachlorodibenzo-p-dioxin (2,3,7,8-TCDD) was isolated in the waste from the trichlorophenol production plant, and soil analysis was able to confirm its presence in samples from the affected horse arenas. Indeed the concentrations found are among the highest yet found in any environmental samples (Reggiani [18]). The consequences of this event have been great. Not only has there been the human illness, and cost of the illness to horses; there has been a considerable level of psychological stress on the exposed populations, and the expenditure of a very large sum of money on repairing the situation. This included buying all the houses in one township and rehousing the inhabitants elsewhere, as well as the costs of the still incomplete rehabilitation of the area.

3.   HOW DOES INCIDENTAL SOIL POLLUTION OCCUR?

It is possible to say that incidental soil 'pollution' is an inevitable consequence of life. Almost all our activities leave traces on, or in, the soil. Without them, our knowledge of our ancestors would be more limited than it is. However, we will confine our discussion to the impact of industrial activity on the soil.

It can be said that all land used for industrial activity will be contaminated by the materials used during the life of the factory. The passage of time after the activity stopped will tend to cause the disappearance of some contaminants by evaporation, migration, or decomposition, while other more stable or immobile materials will tend to spread as the contaminated soil is moved by subsequent redevelopment, or by surface erosion by the action of wind and rain.

This sort of soil contamination can result from:

The delivery of materials. Raw materials and other materials used are usually delivered, whether they arrive by road, rail or water transport, to one place within a factory. Spills can occur at the time the material is transferred from the transport vehicle to the factory store. Over a period of years the soil around this area can be polluted with the materials spilled, to a degree depending on the nature of the surface onto which the spill occurs, the nature of the material spilled, and the vigour with which spills are cleaned up as they occur.

The storage of raw materials. Raw materials, and other materials used in the factory, may be stored loose in bulk, in bulk containers, or in smaller containers. The area in which they are stored may become contaminated in several ways. Thus some bulk materials, e.g. coal, were often stored directly on the earth, and over the years coal can penetrate the earth to a considerable depth. The present author is aware of an old coal carbonisation plant, where the 'earth' in the coal storage area was contaminated with coal to a depth of some 2–3 m below ground level, and where the upper metre of soil contained enough coal to be combustible. Beever [2] reviews the assessment of the hazard of potentially combustible soil. Similarly the ground around oil storage tanks can be contaminated with oil to concentrations of several per cent, as a result of slow leakage over the years, and of losses during transfer. This is probably less of a potential problem at modern factories where such tanks may be placed within a concrete bund—but this practice is not universal for materials such as fuel oils.

Where materials are stored in containers, e.g. 200 litre drums, sacks, boxes, paper bags, bottles, etc., it is likely that over the years, containers will have failed leading to possible contamination of their surroundings. In other cases dispensing the stored material for use may have involved transferring it to another container. This also provides a ready opportunity for spills. The potential consequences of such spills will depend on the material spilled, the surface onto which the spill occurred, and the promptness and efficiency of clean up. In this context it is worth remembering that many old factories did not have specially laid floors—sometimes compacted earth was used as the floor, 'bound' with oil to reduce dust.

Intra-factory movement of materials. Materials will be transported from place to place within the factory by a variety of means, e.g. by handcart, by truck, by pipelines, by rail, by conveyor belt, etc. Each method has a potential to cause soil pollution by spillage or leakage at points along the transport route. Underground pipelines, in particular, can cause considerable levels of soil pollution by leakage from joints. This is not so immediately apparent as when the pipelines are above ground. Underground pipelines are quite common at some factories, and may even be found still full of the material they transported when the factory was working. We have often found high levels of soil pollution immediately surrounding buried pipelines at, for example, old gas works (Parker *et al.* [15]). The soil under conveyor belts used to move solid materials is also likely to be contaminated. Again old gas works provide examples of this where conveyor belts were used to move the iron oxide used to purify the coal gas to and from the purifiers. Residues of the spent oxide are commonly visible in the soil around the conveyor.

Soil pollution as a result of production processes. Many industrial processes themselves contributed to soil pollution to a greater or lesser extent, particularly some that involved bulk handling of large quantities of material. Mineral processing, and ore crushing, for example, contributed to extensive soil pollution in earlier times as a result of what were, by modern standards, inadequate dust control procedures. In addition, soil pollution may have resulted from secondary storage of these materials at the place where they were used. Further, many engineering activities use oils of varying kinds and the places where these processes and materials were used can also be contaminated—often to quite high concentrations.

Storage of products. The areas, warehouses, etc., used for the storage of a factory's products prior to distribution may show soil pollution by mechanisms similar to those outlined for raw material handling.

Waste disposal. While we are not going to discuss further the pollution of soil by waste disposal in this chapter, many factories used part of the land they occupied as a dumping area for at least some of the wastes they generated.

A few general points about the nature of the contamination at old industrial sites can be made that are of importance in assessing the risks the contamination poses, and in planning the remedial measures necessary.

The type of soil contamination we have described so far shows some general properties:

(a)   It is rarely uniformly distributed across the whole site occupied by the factory. Typically the activities that caused the contamination took place at specific places within the factory's land area. While some spreading of contamination may have occurred by diffusion (particularly if the contaminant was a mobile liquid or was water soluble), in general the pattern of contamination will show areas of high concentration of pollutants surrounded by other areas that may be largely free from that pollutant.

(b)   Contamination of the sort we have so far discussed may not spread far from the boundaries of the factory, except when it is extensive, at high concentration, and the contaminant is mobile and of significance at low concentrations, and where the geology and hydrogeology of the site is favourable to migration.

(c)   Where a site is polluted by more than one contaminant they will commonly show different contamination patterns from each other, reflecting the different ways and places of use.

(d) Contamination levels vary in depth as well as on the surface. This may be particularly important when the site has been 'cleared' before a site investigation is conducted. This may have destroyed any links between the surface contamination found and that present at depth.

All the points above are of importance when a survey is to be planned to determine the nature and concentrations of any significant pollutants that may be present.

Industrial land is often contaminated with materials that may, for one reason or another, be omitted from consideration by the team planning an investigation. Two common contaminants come to mind: many sites are contaminated by asbestos and lead. Asbestos was widely used in industry as a structural material for light buildings; in lagging for insulation; and as a filler material. One result of this is that many old industrial sites show extensive surface contamination with asbestos, particularly where buildings were demolished before the introduction of the more recent strict controls on the handling and disposal of fibrous asbestos that have been introduced in several countries. Many industries also made extensive use of lead-based paints. Soil around structures that were painted with these materials often shows elevated lead concentrations, as a result of contamination during painting, and, in particular, caused by stripping old paint when repainting. We have examined sites where lead concentrations up to 5000 mg/kg were found in the annular ring of soil around storage tanks. Other examples of contamination by inorganic paint constituents can arise by similar mechanisms. Thus some paints use cadmium pigments (e.g. cadmium sulphide and cadmium seleno-sulphide), chromium (including hexavalent) compounds, barium, zinc, etc. Some older paints contained mercury fungicides, and others used chlorinated hydrocarbons, including polychlorinated biphenyls, in their formulation. While it is rare, in our experience, for soil contamination by these last two materials to be very extensive, or at high concentration, at sites where the paints were *used*, this may not be the case at old factories where they were *made*.

So far we have concentrated on soil pollution mechanisms that remain, in general, localised to the land occupied by the factory. However, industrial activities can produce soil pollution in areas distant from their originating site by two main methods: the discharge of pollutants to the atmosphere, and discharge to surface waters.

With respect to atmospheric discharges we will concentrate on particulate matter. Factory chimneys may discharge particulate matter as well as gases and vapours. Vapours, such as steam, may condense once they reach cool air. Typically they will condense around solid nuclei, which will then settle onto the land. Even in the absence of condensable vapours, particulate matter will settle on the land eventually at a rate determined by its particle size, density, and on the prevailing meteorological conditions. Rain falling through the plume from a chimney will also wash particles and water soluble gases out of the plume onto the land surface.

A consequence of this is that soil in the vicinity of a chimney will normally show elevated levels of some of the pollutants emitted in the stack gases. Such pollution shows a rather different distribution pattern on the soil from that of the kinds of soil pollution previously considered. In general, ground level concentrations will be low in the soil immediately around the foot of the chimney, will rise to a maximum some distance from the foot of the chimney in the most common wind direction, and then fall to undetectable values with increasing distance. However, it is likely that soil contamination will be found in all directions, following the same general distribution, though the highest concentrations will occur in the prevailing wind direction. The distances from the foot of the chimney to the point of maximum ground level concentration will increase as the chimney height increases. As the chimney height increases the maximum ground level concentration falls, and for very tall chimneys a 'maximum' may be difficult to observe, particularly since variations in meteorological conditions around the discharge point (wind speed and direction, the existence of inversion conditions, etc.) will inevitably result in a general 'smearing' of the simplified picture given above. The ground surface topography, and the existence of other chimneys in the area emitting, perhaps the same pollutants, will also reduce the ease with which the contamination pattern described above is perceived. In heavily industrialised regions this results in a fairly general spread of common pollutants deposited from the atmosphere, such as soot (and its associated polynuclear aromatic hydrocarbons) from combustion sources, metals from metallurgical processes, etc. There have been numerous studies showing local soil contamination as a result of atmospheric emissions from various industrial plants, and others showing generally raised levels of various pollutants in urban soils in comparison with those observed in rural soil.

Discharge of pollutants to surface water can also result in soil pollution where, for example, discharges were made via unlined ditches. For example, we were able to trace the path of an effluent discharge ditch from a tannery that had ceased operating over 100 years ago, and where the ditch itself had been filled with earth over 80 years ago. The tannery had used arsenic trisulphide in one of its processes, and the effluent discharged via the ditch had deposited this material along its length, producing, in 1984, a very sharply defined track of soil contaminated by up to 500 mg/kg of arsenic that extended about 200 m from the site of the tannery to the adjacent river.

Similarly lagoons used to store storm-water run-off from industrial sites may well show quite heavy contamination with materials washed off the surface by heavy rain. Nightingale [14] reports similar findings from lagoons used to store urban storm-water prior to discharge. In particular he records high concentrations of lead and zinc in such lagoons.

Incidental pollution of river sediments as a result of precipitation, settlement, and adsorption of pollutants from effluent discharges is well known, and dredged material from such waters can contain significantly elevated concentrations of heavy metals, hydrocarbons, etc.

The final mechanism for incidental pollution that we wish to address here concerns that caused by or related to land transport. It is well known that soil at the edges of busy roads shows higher than normal concentrations of lead and zinc produced as a result of the combustion of leaded fuel, and from tyre wear (zinc compounds are extensively used in the formulation of tyre rubber). Elevated levels of polynuclear aromatic hydrocarbons (PAH) are also found in these locations. Johnston and Harrison [9] studied the deposition of PAH and lead along a rural stretch of the M6 motorway in NW England. They found that fluoranthene was deposited at a rate of 15 200 ng/m^2/wk at a distance of 3.8 m from the road; that the deposition rate had fallen to 1700 ng/m^2/wk 10 m from the road and that no detectable deposition was measurable at 47 m. They observed that at distances above 30 m from the carriageway deposition from other non-motorway sources contributed significantly to the observed deposition rates. For lead, they found that deposition rates approached the background rate approximately 200 m from the motorway, but even at about 16 m distance from the motorway, the deposition rate for lead (about 0.5 mg/m^2/wk) was less than 10% of that at about 5 m (approximately 7 mg/m^2/wk).

Roadside areas may also be contaminated by materials transported on the roads, particularly if large amounts of a particular material, inadequately contained, are transported along a particular stretch of road. Hemphill *et al.* [8] report elevated lead levels in soils along the edge of a road in Missouri (USA) used for transporting a lead sulphide ore concentrate.

4. ACCIDENTAL SOIL POLLUTION

Accidental soil pollution usually results from a sudden failure of containment, as a result of a collision during transportation, or an explosion or fire. In general it produces a comparatively small volume of highly contaminated soil though this need not always be the case. The accidental soil pollution by TCDD at Seveso involved a large volume of soil. Other accidental releases may also contaminate large volumes of soil. For example, Smith [21] reports the following accidental release in Tennessee (USA) in 1973.

A transformer loaded with a PCB mixture was being moved along a mountain road, 8 km from the town of Kingston, Tennessee, when it started to leak. The truck was stopped by the police and the driver told that he was losing a material that was dissolving the asphalt road surface. The driver dumped the load by the roadside. Heavy rain then helped to spread the by now rapidly leaking PCB. By the time this spill had been dealt with 52 km^2 of soil had been contaminated to varying degrees, 2300 m^3 of contaminated soil contained in 210-litre drums had been moved to Texas for disposal, and over US $1 million has been spent.

Shepherd [20] reports an accident in Maryland, USA, when 3 railcars were derailed spilling 95 m^3 of a hot 90% phenol solution. The hot phenol ran down a hillside towards a tributary of the Potomac river. In the course of the actions taken to clean up this spill, it was estimated that nearly 75 tonnes of phenol were adsorbed in the soil. The contaminated area was surrounded by containment trenches and a reservoir constructed to receive rainwater run-off from the area, as well as any water infiltrating through the contaminated mass. Water from the containment trenches and the reservoir was piped to a specially constructed activated carbon treatment plant. Water entering the activated carbon beds initially contained 100–1000 mg/litre of phenol. Treated effluent contained 0.1–0.5 mg/litre phenol and was discharged to the river. Treatment was continued during the summer and autumn of two consecutive years before the contamination was considered to have been reduced to acceptable levels.

Eisenberg et al. [4] report that groundwater contamination by leakage from underground storage tanks has become a matter of considerable concern in the San Francisco area of California and many municipal authorities are introducing laws requiring that such tanks, including fuel tanks, are monitored. It is estimated that these measures will generate 200–300 new reports of sub-surface leaks in the San Francisco area requiring attention during 1986. Erikson [5] reports one such incident in

Wisconsin, where sub-surface migration of hydrocarbon fuel continued until oil eventually became visible in basements of houses. Investigations showed that large quantities of oil had leaked into the soil from operations at a nearby railway maintenance and refuelling yard. Soil investigations revealed oil concentrations up to 4%, and that about $600\,m^3$ of oil remained in the affected soil volumes. Eventually about $550\,m^3$ of this was recovered over several years by a programme of groundwater pumping. There have been similar European events.

Phimister and Dicken [16] report a leak of about 2·2 tonnes of hexavalent chromium compounds in solution from a large chromium plating installation in Canada. Investigations showed that in the geological systems existing in the area, the chromium remained largely mobile and that little firm binding or adsorption onto minerals was occurring. Again a system of groundwater pumping allowed for the recovery of almost 80% of the chromium from about $900\,m^3$ of affected soil. The residual chromium was fairly strongly bound to soil minerals and was only slowly released.

There have been numerous incidents in which polychlorinated biphenyls (PCBs) have been spilled. We have ourselves taken part in several clean-up operations following such spills. At one about $6500\,m^3$ of soil was contaminated with PCB at concentrations between 30 mg/kg and about 1%, as a result of incidental and accidental events during the life of a factory. Since this soil cannot be disposed of to landfill in the UK (where the limit for PCBs in waste for landfill is 30 mg/kg, and there is no suitable incinerator in the UK that can (or whose operators will) accept this PCB contaminated soil), a special 'tomb' for the PCB contaminated soil is to be constructed to receive the soil.

Where an accident involves a fire or explosion then the immediate damage produced by the fire or explosion will commonly be the major environmental consequence. However, if certain materials are involved in the fire then their release or the release of substances produced by their incomplete combustion or pyrolysis can be of environmental significance. In particular attention has focussed on the distribution of polychlorinated dibenzo-p-dioxins (PCDDs) and polychlorinated dibenzofurans (PCDFs) from fires involving certain chlorinated organic materials, in particular chlorophenols and PCBs.

Both PCDDs and PCDFs include materials that display extreme toxicity to animals, are implicated as possibly carcinogenic, and, when absorbed in soil, are environmentally persistent. Human exposure to PCDDs has produced chloracne, a painful and potentially disfiguring skin disease; effects on the central nervous system and impaired liver function. PCDFs are less well studied but their effects seem likely to be very similar, and they were implicated in the Yusho incident in Japan where many people suffered severely after consuming PCB contaminated oil. The Seveso incident in Northern Italy, where a large area was contaminated by 2,3,7,8-TCDD, following a high pressure release of the contents of a reaction tank in which 2,4,5-trichlorophenol was being made, is well known, and will not be discussed further here. In addition there have been several fires involving PCBs where PCDFs have been distributed over surrounding areas. PCDFs are emitted from municipal waste incinerators, and from almost any circumstance where PCBs are heated in the presence of oxygen to temperatures inadequate to ensure complete combustion. Rappe *et al.* [17] report on accidents involving electrical installations using PCBs giving very brief details on 21 such accidents in Sweden and Finland, and give preliminary information on levels of PCDDs and PCDFs found following an accident at Reims, France, involving a PCB filled transformer. It is almost certainly true that such incidents have occurred in most countries, and have often passed unnoticed and/or unreported. We have been involved in two such incidents in the UK, neither of which has been the subject of published reports. When eventually published, the Proceedings of the 'Dioxin 85' Conference held in Bayreuth, FRG, in September 1985, will contain papers on other incidents.

5. THE RESPONSE TO INCIDENTAL AND ACCIDENTAL SOIL POLLUTION

In both cases the response will be to clean up the contaminated soil or assess the need for clean up. However, the mechanisms by which this response is alerted are often very different. Normally, an accident, or a 'sudden' event, will provoke a response that is more or less immediate, and will commonly involve the emergency services, police, fire service etc. A part of the normal response to accidents of this type will include some measure of 'clean up' of the consequences of the accident. For transport accidents appropriate clean-up procedures are now more likely to be taken as a result of the various labelling, notification, and other requirements introduced under recent Community and National legislation. It is usually the case that several desiderata for an appropriate response will exist at many transport accidents involving many known hazardous substances. Thus:

—the load will be labelled;
—some emergency response information will be carried;
—a contact for advice and assistance will be identified.

Anyone involved in emergency response to transport accidents will know that, valuable though they are, the EEC (and National) regulations are not all-embracing, and that the *use* of the available information is not always easy. However, we cannot go into this subject here. The important point is that an accident is usually followed, quite promptly, by remedial action of some kind.

Incidental pollution is different. Often, by its very nature, its existence is not appreciated by the person or organisation causing it. Even if its existence is known, it may be assessed as of little consequence to that person or organisation. Thus, the incidental pollution may continue for a long time and its extent may not be apparent until the site is redeveloped for some new purpose, or until environmental damage outside the factory fence is apparent. Thus, there are circumstances where the extent of incidental soil pollution, and its significance, may only come to light on redevelopment by an alert and aware developer. It is very likely that redevelopment of polluted land has occurred in the past without the 'clean up' that would, now, seem necessary. For example, Wijnen [22] reports that extensive areas of Rotterdam have been built on land that current Dutch law requires to be decontaminated. The case of Shippam in the UK where parts of the village were on top of old deposits of zinc smelting wastes containing up to 1000 mg/kg cadmium can also be cited. We also know that there are extensive deposits of waste from the Leblanc process for making alkali, used during the first half of the 19th century under the present town of Widnes, and of tannery wastes under Warrington, both in the UK.

For land polluted as a result of an accident therefore, it seems that efforts should be concentrated on improving the speed with which appropriate remedial measures are implemented. Incidental pollution presents more problems, since its existence is not always obvious. This problem is made more difficult by our knowledge that some pollutants may be 'significant' at low concentrations. Thus the US Centers for Disease Control are recommending that the maximum level of 2,3,7,8-TCDD in residential soil should be 1 μg/kg (Kimbrough *et al.* [10]). Many potential pollutants may be significant at concentrations that will not be detectable without extensive chemical analysis, where appearance and odour will not help in identifying their likely presence.

Guidance on the identification of contaminated soil can be given. Finnecy and Pearce [6] for example, summarise procedures they have found useful. In outline, their suggestions start from the idea that *any* industrial land is likely to be contaminated, so that a history of past use of the site should be constructed. Research can then indicate what materials and processes may have been used, and some idea of potential soil pollutants gained. They discuss methods of site investigation, sampling, and some of the problems of chemical analysis that must be faced.

Once the presence of pollutants in the soil is confirmed, with some idea of their concentrations, the next problem applies to all types of polluted soil. It is commonly difficult to answer the questions 'does it matter?', and 'what is a 'safe' concentration?'.

6.   ASSESSMENT OF RISKS FROM POLLUTED SOIL

Once polluted soil is identified and characterised, the decision has to be made whether or not to 'clean' it, and the question 'what do we mean by clean?' must be faced. Polluted soil can give rise to risk to:

humans	—directly by ingestion, skin contact, inhalation; indirectly
animals	by food chain concentration;
plants	—by phytotoxic effects;
water	—by dissolution and transport into waters; by permeation
resources	through pipes carrying water;
construction	—by corrosive attack;
materials	
everything	—by the presence of a sufficient concentration to make the 'soil' combustible;

One of the problems results from the fact that 'soil' is itself very variable in composition, and may contain, even when apparently 'unpolluted', potentially hazardous materials. Nevertheless the assumption is either made, or implied, that unpolluted, 'natural', soils are 'safe'. Even if this assumption is accepted, the problem of natural variation in concentrations is not removed. Thus the Macaulay Institute for Soil Research [12] give the data on the variation in 'natural' concentrations of some potentially hazardous constituents of UK soil shown in Table 1.

The next problem that arises in deciding on 'safe' concentrations is answering the question 'safe for whom or what?' What is 'safe' for a healthy adult, may not be 'safe' for a child, or someone in poor health. What is 'safe' for grass, may not be 'safe' for apple trees or tomatoes. What is 'safe' for concrete may not be 'safe' for aluminium, etc.

There have been several approaches to this problem. The Dutch approach has been to demand that soil retain its 'multifunctionality', that is the soil should be usable for *any* purpose. This approach has been outlined

Table 1

Element	Normal concentration range (mg/kg)
Zinc	10–300
Copper	2–100
Nickel	5–500
Boron	2–100
Chromium	5–500
Cadmium	0.01–1.0
Lead	2–200
Mercury	0.1–0.5
Molybdenum	0.2–5
Arsenic	0.1–40
Fluorine	30–300
Selenium	0.1–2
Cobalt	1–40
Manganese	100–4000

by Moen *et al.* [13]. The assessment is based on three concentrations, individually listed for 7 classes of pollutant (metals, inorganic pollutants, aromatic compounds, polycyclic aromatic compounds, chlorinated organics, pesticides, and 'others'). These three concentration levels are:

A—the reference level—can be accepted as unpolluted;
B—indicative value for further investigation;
C—clean up necessary.

A few examples are given in Table 2.

Table 2

Pollutant	Concentrations (mg/kg)		
	A	B	C
Nickel	50	100	500
Arsenic	20	30	50
Cadmium	1	5	20
Lead	50	150	600
Free cyanide	1	10	100
Benzene	0.01	0.5	5
Phenols	0.02	1	10
Benz(α)pyrene	0.05	1	10
Chlorophenols (total)	0.01	1	10
PCBs (total)	0.05	1	10
Mineral oil	100	1000	5000

Moen *et al.* [13] indicate that while the concentrations listed (particularly the A concentrations) represent desirable targets, circumstances may demand some flexibility in practice. Wijnen [22] draws attention to the difficulty of applying them in an urban area that, historically, has been heavily industrialised.

Moen *et al.* [13] comment that 'A' values should only be considered as 'uncontaminated' when *contaminated* soil is being considered; in other words it is *not* a level to which contamination of clean soil should be .allowed to rise. They also point out that the present 'A' values are not entirely well-founded, and that work continues to establish these values more soundly.

It is the intention of the Dutch Government to establish 'target' values for soil quality based on no-adverse effect levels and maximum tolerable levels for various risks. Thus for the induction of cancer by exposure to chemicals the maximum tolerable level of risk to an individual is set at 10^{-6}, with a no-adverse effect level of 10^{-8}. These targets set the limits for exposure by all routes, not just exposure to contaminated soil, so they imply numerous standards to be met in a multiplicity of circumstances.

Beckett and Simms [1] outline the UK approach. This differs from the Dutch system in that the 'standards' set for contaminated land reclamation are variable according to the immediately intended use to which the land is to be put after reclamation. Thus the Dutch principle of retaining 'multifunctionality' is not applied in the UK. The UK approach seeks to make use of two 'trigger' concentrations:

—a lower, threshold, 'trigger' value: the concentration of a potentially harmful species in soil *below* which the soil can be considered as uncontaminated with respect to *the intended after use*;

—a higher 'trigger' value: the concentration of a potentially harmful species in soil *above* which remedial action is necessary, or the intended use must be changed.

Thus the UK system envisages the concentration of potentially harmful soil contaminants as falling into one of three zones: a zone below the threshold 'trigger' value for the intended after-use; a zone above the higher 'trigger' value for that after-use; and an intermediate zone. The intermediate zone is one where professional judgement, knowledge, and experience is most involved in the decision whether or not remedial measures are necessary in the specific circumstances of the proposed redevelopment, and what form they should take.

Beckett and Simms [1] note that assigning trigger values is difficult, and, so far, only few have been suggested in guidance documents issued by the UK Department of the Environment (Dept. of the Environment [3]). Among these tentative trigger values are those listed in Table 3.

Various cautions are made in relation to the trigger levels below. Thus the values marked * are similar to UK recommendations for the maximum concentrations of these materials in agricultural soils treated with sewage sludge. Thus they should not be assumed to apply in other circumstances or to other potential risk pathways. Similarly, the coal tar trigger value marked † does not refer to the risks that may follow from human skin contact.

Table 3

Contaminant	Intended after-use	Threshold trigger level (mg/kg)
Arsenic	Domestic gardens	10*
	Parks, sports fields	40
Cadmium	Domestic gardens	3*
	Parks, sports fields	15
Lead	Domestic gardens	500*
	Parks, sports fields	2 000
Free cyanide	Domestic gardens, sports fields	50
	Industrial (no landscaping)	500
Coal tar	Domestic gardens	200†
	Public open spaces	500
	Industrial (no landscaping)	5 000

Seng [19] outlines the approach used in the risk assessment of contaminated sites in Baden-Württemburg (FRG). The approach he outlines is based on the notion that contamination levels found in a licensed sanitary landfill (municipal waste only) are 'acceptable'. Four pollutant concentration levels are considered:

A—no risk attached to this level;
B—no remedial or other measures necessary;
C—appropriate control necessary;
D—remedial measures necessary.

This approach is used as a means of ranking sites in order of priority for action as required by the Federal Water Law, and is therefore largely aimed at groundwater protection. Several questions are to be asked in relation to each site, e.g.

—how much pollutant is discharged from the site?
—how much reaches groundwater?
—how sensitive is the groundwater to the pollutant?
—how important is the groundwater?

Hecht [7] describes a somewhat different approach in Hamburg, where the driving force is the same Federal Water Law. In Hamburg it has been decided to adopt two 'trigger' levels: a 'U' value that is the upper level where the soil can be considered to be uncontaminated; and an 'S' value where remedial action is necessary (this is similar to the UK approach). So far only a few 'U' values have been promulgated by the Hamburg Government; these are as shown in Table 4.

Table 4

Pollutant	Guide level above which further investigation is necessary (mg/kg)
Arsenic	50
Lead	300
Cadmium	8
Total Chromium	300
Copper	300
Nickel	300
Mercury	8
Zinc	1 000

It must be noted that these concentrations are presumed to refer to the protection of groundwater, but even then a degree of judgement is allowed with respect to the nearness of a water extraction plant to the contaminated land and the extent of groundwater pollution observed.

Kloke [11] reports on the work to establish soil concentration limits to avoid damage to plants or unacceptable accumulation of heavy metals in the edible portions of plants. He uses three criteria:

—heavy metals should not leach from the soil;
—they should not be distributed by the wind;
—they should not concentrate in the food chain.

Table 5

Pollutant	NOEL (plants) (mg/kg)
Cadmium	3
Chromium	100
Copper	100
Nickel	50
Lead	100
Zinc	300
Mercury	2
Arsenic	20

As a result of experiments he has conducted, Kloke arrived at the concentrations given in Table 5 that constitute no-effect-levels (NOEL) for a variety of plants, with respect to his criteria.

However, Kloke notes (though we should await publication of his paper for full details) that lead concentrations as high as > 5000 mg/kg were apparently acceptable for potatoes and tomatoes, but that 1000 mg/kg seemed to be the limit for edible leaf crops. He also noted that the cadmium content of many Berlin soils was already 7 mg/kg.

7. PROBLEM AREAS

The reclamation of incidentally and accidentally contaminated land raises many problems. Some of these problems are financial and practical in nature, others, however, arise from a lack of adequate relevant scientific knowledge. A few of these problems are:

1. The identification of a contaminated site, and of the contaminants present. In principle this is a practical and financial problem, though the identification of the contaminants has a technical and scientific component. We can assume that *all* industrial land is to some extent contaminated, but this does not remove the difficulty of obtaining an accurate and complete picture of the extent and nature of the contamination. The nature of incidental and accidental soil pollution is such that conventional techniques of chemical analysis of soil samples taken by some statistical sampling processes cannot be relied upon if costs are to be kept within acceptable limits. This can be a particular problem at sites that,

visually, seem largely free from gross contamination. In addition some of the analytical results desired are not easy to obtain, and others are not easy to interpret once they are obtained.

2. There is a political-economic problem if reclamation is demanded of *all* sites, even those in current use, where trigger levels (however set) are exceeded, rather than only applying these standards to sites undergoing redevelopment. Wijnen [22] points out that the demand to restore 'multifunctionality' of sites, including those in use, can produce intractable problems for large urban areas that have been heavily industrialised for a long time. Problems relate (among other things) to the costs of surveying all potentially contaminated sites, and to the complex difficulties of establishing priorities for action to deal with the contamination discovered.

3. Comparatively few 'trigger' levels have been established— particularly for organic contaminants. It is also important to note the 'targets' that these trigger levels will protect, and the routes of exposure that they refer to. Thus few relate to direct human exposure by ingestion, inhalation, or skin contact. Most of the trigger levels (except the Dutch A, B, C values) seem to apply to phytotoxicity and food chain magnification. The Dutch A values refer to some notion of uncontaminated Dutch soil, while the B and C values are, or seem to be, somewhat arbitrary concentrations based on experience, known hazards and risks, and a variety of other factors. No one should underestimate the difficulty of devising trigger values that are not impractically stringent but which will protect the exposed targets adequately. While it is likely that effective trigger values that will protect plants and limit food chain magnification can be devised by sound experiment, this option is not available for direct human exposure. There are real problems in this area. Not only is our database of knowledge of the dose-effect relationships between contaminants and human targets limited, but there are also problems in assessing the amount of exposure, and because of the multiple exposure routes applicable to humans for some contaminants, e.g. food, air, water, occupational exposure, etc., as well as from direct exposure to soil. Lead, for example, is one contaminant that is virtually ubiquitous. The question then is not so much how much exposure should we permit, but how much *additional* exposure is acceptable. Epidemiological studies can be helpful here, but they are difficult, expensive, often controversial, and can lead to ethical dilemmas.

REFERENCES

1. BECKETT, M. J. and SIMMS, D. L. (1985). Assessing contaminated land: UK policy & practice. *Proc. 1st Int. TNO Conf. on Contaminated Soil,* Utrecht, Netherlands, Nov. 1985. Martinus Nijhoff, Dordrecht, pp. 285–93.
2. BEEVER, P. F. (1985). Assessment of fire hazard in contaminated land. In: *Contaminated Soil* (Eds J. W. Assink and W. J. vanden Brink), 1st Int. TNO Conf. on Contaminated Soil, Utrecht, Netherlands, Nov. 1985, Martinus Nijhoff, Dordrecht, pp. 515–22.
3. Dept. of the Environment (1983). Guidance on the Assessment and Redevelopment of Contaminated Land. Interdepartmental Committee on the Redevelopment of Contaminated Land. ICRCL Paper 59/83, Dept of the Environment, London.
4. EISENBERG, D. M., OLIVIERI, A. W. and JOHNSON, P. W. (1985). Investigation and Clean-up of Fuel Tank Leaks in the San Francisco Bay Area—A Regulatory Strategy. *Proc. 1985 Oil Spill Conference,* Los Angeles, USA, Feb. 1985. American Petroleum Institute Publication No. 4385.
5. ERIKSON, B. (1985). North La Crosse Underground Oil Spill. *Proc. 1985 Oil Spill Conference,* Los Angeles, Feb. 1985. American Petroleum Institute Publication No. 4385.
6. FINNECY, E. E. and PEARCE, K. W. (1985). Land contamination and reclamation. In: *Understanding our Environment,* Chap. 4 (Ed. R. E. Hester), Royal Society Chemistry, London.
7. HECHT, R. (1986). Problems in Fixing the Tolerable Concentrations of Hazardous Substances in Groundwater and Soil. *Proc. Int. Workshop on Risk Assessment of Contaminated Soil,* Deventer, Netherlands, June 1986, p. 24.
8. HEMPHILL, D. D., MARIENFELD, C. J., REDDY, R. S. and PIERCE, J. D. (1974). Roadside lead contamination in the Missouri lead belt. *Arch. Environ. Health,* **28** (4), 190–4.
9. JOHNSTON, W. R. and HARRISON, R. M. (1984). *Sci. Total Environ.,* **33,** 119.
10. KIMBROUGH, R., FALK, H., STEHR, P. and FRIES, G. (1984). *J. Toxic. Health,* **14,** 47–93.
11. KLOKE, A. (1986). Soil Contamination by Heavy Metals. *Proc. Int. Workshop on Risk Assessment of Contaminated Soil,* Deventer, Netherlands, June 1986, pp. 42–54.
12. Macaulay Institute for Soil Research (1983). Evidence to House of Lords Select Committee on the European Communities—Sewage Sludge in Agriculture. 1st Report Session 1983-4, HMSO, London.
13. MOEN, J. E. T., CORNET, J. P. and EVERS, C. W. A. (1985). Soil Protection and Remedial Actions: Criteria for Decision Making and Standardisation of Requirements. *Proc. 1st Int. TNO Conf. on Contaminated Soil,* Utrecht, Netherlands, Nov. 1985. Martinus Nijhoff, Dordrecht, pp. 441–8.
14. NIGHTINGALE, H. I. (1975). Lead, zinc and copper in soils of urban storm run-off basins. *J.A.W.W.A.,* **67** (8), 444–6.
15. PARKER, A., PEARCE, K. W., FENNER, E. F. and WRIGHT, M. S. (1984). *J. Haz. Materials,* **9,** 347–54.
16. PHIMISTER, J. P. and DICKEN, R. C. (1985). Containment and Recovery of a

Chromic Acid Loss. *Proc. 2nd Ann. Tech. Seminar on Chemical Spills*, Toronto, Feb. 1985. Environment Canada, Env. Prot. Service, Ottawa, Canada.

17. RAPPE, C., KJELLER, L. O., MARKLAND, S. and NYGREN, M. (1986). Electrical PCB accidents, an update. *Chemosphere*, **15** (9–12), 1291–5.
18. REGGIANI, G. (1980). Localized contamination with TCDD. In: *Halogenated Biphenyls, Terphenyls, Naphthalenes, Dibenzodioxins, and Related Products* (Ed. R. Kimbrough), Elsevier/North Holland Biomedical Press, Amsterdam.
19. SENG, H. H. (1986). Risk Assessment of Contaminated Sites in Baden-Württemburg. *Proc. Int. Workshop on Risk Assessment of Contaminated Soil*, Deventer, Netherlands, June 1986, p. 21.
20. SHEPHERD, A. (1982). Phenol. In: *Hazardous Materials Spills Handbook*, Chap. 12, part 2 (Eds G. F. Bennett, F. S. Feates and I. Wilder), McGraw-Hill, London.
21. SMITH, A. J. (1982). Polychlorinated Biphenyls. In: *Hazardous Materials Spills Handbook*, Chap. 12, part 1 (Eds G. F. Bennett, F. S. Feates and I. Wilder), McGraw-Hill, London.
22. WIJNEN, E. J. E. (1985). Soil pollution in the former industrialised regions of Rotterdam. *Proc. 1st Int. TNO Conf. on Contaminated Soil*, Utrecht, Netherlands, Nov. 1985. Martinus Nijhoff, Dordrecht. pp. 507–13.

Impacts on Soils Related to Industrial Activities: Part III—Effect of Metal Mines on Soil Pollution

L. LAVILLE-TIMSIT

*Bureau de Recherches Géologiques et Minières (BRGM),
Département Gîtes Minéraux, Division Géochimie Appliquée,
BP 6009, 45060 Orléans Cedex, France*

SUMMARY

It is generally accepted that mining activities cause soil pollution, but this assertion should be confronted with the facts. Knowledge acquired through geochemical exploration surveys provides various representative examples which give quantitative data on the distribution of metals in and around metal mines. (1) Even before any mining activities are undertaken, high metal concentrations in relation to mineralizations exist in soil. (2) Pollution can only be defined in relation to its natural environment, and this requires, at the very least, a good knowledge of regional geochemical backgrounds. (3) The dispersion of metals in soil due to the storage of waste rocks and tailings is of limited extent. It is often difficult to distinguish the extent and amount of concentration of this contamination from the disseminated mineralizations occurring around the deposits. (4) These phenomena are therefore of minor importance compared to damage resulting from accidents such as dam breakage and leakage from decanting ponds, which, technically, could be avoided. However, pollution hazards are amplified when a smeltery or transformation industries are established in the vicinity of mining areas.

1. INTRODUCTION

The industrial development of a country calls for a supply of raw materials which, in turn, depends on mining activities.

Although the opening of a mine, which is always a job creator, has major

281

economic effects on a region it also causes various nuisances, which depend in nature and extent on the type of substances mined, the mining techniques and the size of the mine or minefield. The effects of these nuisances must always be seen in connection with the natural environment and economic advantages.

This chapter deals only with the problem of soil pollution by metals in and around metal mines. Examples representing mines of this kind in Western Europe will show how any pollution associated with ore mining and the storage of waste rocks and tailings affect the soil.

2. NATURAL CONCENTRATIONS OF METAL TRACES IN THE SOIL

Mining is only carried out in certain places where the earth's crust contains large concentrations of metals. This mineralization phenomena either appears as outcrops in oxidized form or is concealed by the soil. In this case their detection requires the use of indirect techniques, such as geochemical prospecting, which is based on the identification of metal anomalies generated at the surface by the mineralization phenomena. Systematic geochemical prospecting of regions with mining potential is undertaken in France as part of a National Inventory of Mineral Resources, as shown in Fig. 1. Some of the results involving four of the five subjects presented below (Fig. 2) have come from this.

2.1. Concentration levels in the complete absence of mining activities

At the heart of the Normandy woodland and pasture land, the Beauvain example is a perfect illustration of this phenomenon (Fig. 3). Arsenic levels amounting to several hundred ppm, and even exceeding 1000 ppm at certain points, have been detected in the soil in this sector, which is nonetheless devoid of any mining or industrial activities. Anomalous surfaces stretching over more than 4 km with a width of about 1·5 km form a ring around a granodiorite, where the arsenic levels are less than 100 ppm. This is of granitic and perigranitic differentiation with which mineralization phenomena have often been associated. At Beauvain these molybdenum, tungsten and bismuth mineralization phenomena also produce anomalies at the surface. The natural geological features show through soil by way of a 'geochemical landscape' which represents the 'initial' or 'zero' state.

In this particular case the initial state of arsenic in terms of content is ten

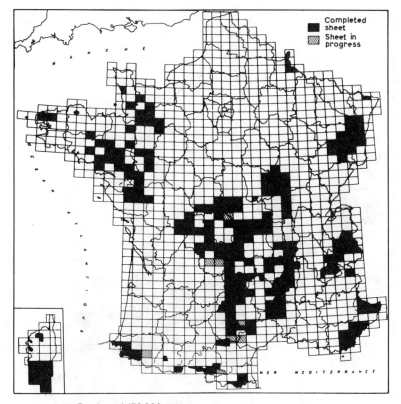

By sheet 1/50 000:
—3 grids/km² (soil, stream sediment);
—22 elements analysed;
—Data processing of all the data
  (geographical, geochemical, type of sampling, etc.);
—Interpretation as work progresses.

End of 1985 ≃ 110 000 km² covered;
  325 000 grids processed.

Fig. 1. Areas covered by the geochemical inventory in France and technical details.

to twenty times greater than the 'threshold value beyond which corrective measures are usually necessary', laid down, for example, for The Netherlands at 50 ppm. (Extract from a document published in August 1983 on the 'implementation of the interim law on the contamination of soil in The Netherlands'.)

This questions the very definition of pollution: if it is defined solely on the

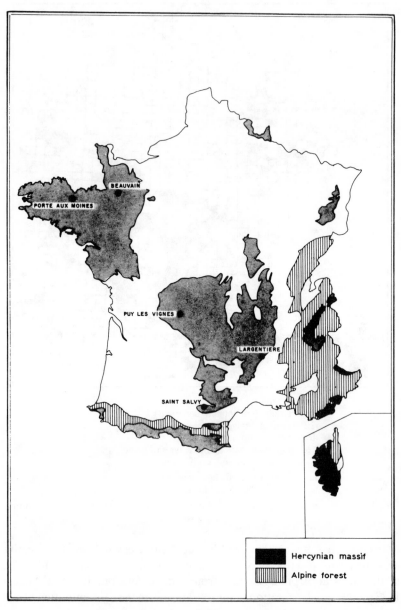

FIG. 2.   Location of reference examples.

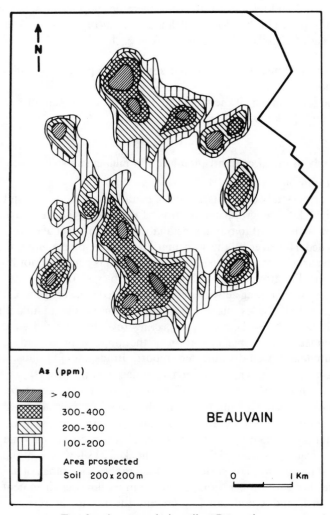

FIG. 3. As anomaly in soil at Beauvain.

basis of total metal concentration levels, numerous sites like Beauvain must be considered as major pollution areas. But this would be out of keeping with the facts since nothing distinguishes it from the general woodland and pasture land and all the forms of life there develop in the same manner.

Pollution should therefore be defined in terms of 'mobilizable elements' and 'risks of dispersion', meaning that in each case:

—it must be measured in comparison with the pre-existing initial state; and
—the forms in which the metals and their ranges of stability are found should be known.

2.2. Study of metal concentration levels in mining site soils

2.2.1. Contribution in the literature

For a correct study of the dispersion of metals in soil in and around a mine, analytical data, geographical location of the samples concerned and their position with regard to mining installations and in particular storage sites are needed. However, all this information does not appear either in mining research articles, where the subject of pollution is generally not broached, or in the literature on environmental problems. In this case articles connected with mining activities are often treated in a very general manner. Fortescue [3], for example, takes an imaginary site to underline the contribution of geochemical prospecting to determine the state of the environment and proper operation of the mine. With the aid of a well-structured and well-documented report, Bradshaw [1] illustrates the progress made by site reclamation techniques over the past 20 years. Sgard [12] presents various methods of restoring iron mine slag-heaps.

Although there are numerous publications on work carried out on special mining sites they are almost always geared to the problems of reclamation. Using comparative tests on a copper mine in Arizona and on sterile terrains, for example, Day and Ludeke [2] show how mining waste can be stabilized by planting legumes, preferably a wide range of ubiquitous varieties for good integration of the site in its floral environment.

Gemmell and Goodman [4] present the results of revegetalization experiments on zinc, copper and sulphur waste at a site in Swansea (former world centre of non-ferrous ore foundry in Wales), which take account of the respective quality of the different types of grassland, fertilization techniques and cutting methods. Hill [5] studies factors limiting plant growth on the sterile dam silt of three mines in Rhodesia excavating respectively for gold, copper and nickel and then proposes the best techniques for establishing vegetation [6].

Peters [11] describes the various stages of reclamation of a vast copper and nickel mine in Canada (Copper Cliff) with, in particular, the gradual conversion of part of the site into a natural reserve, thus creating jobs.

Finally, even though the problem of metal dispersion is tackled, as in the case of Jones *et al.* [8] for silver in derelict Welsh mines, there are no maps showing the distribution of the metal contents levels in the soil.

It is largely this aspect that will be studied below on the basis of results obtained at the BRGM on deposits situated in France (Fig. 2).

2.2.2. Presentation of the individual cases and interpretation

The three cases selected involve mines of different sizes. While two of them, at Puy-les-Vignes and Largentière, now are closed, the third at Saint-Salvy is still in operation.

Puy-les-Vignes (Limousin-Massif Central). This involves wolframite ($FeMnWO_4$) and mispickel ($FeAsS$) mineralization found in vein form in the gneiss and migmatites. The ore, with an average concentration level of around 1% WO_3, was mined from 1863 to 1971. The tailings on this site where storage of waste rocks began in 1938 is now estimated at 150 000 m^3.

This region was covered in 1980 by geochemical prospecting as part of the inventory. An analysis of stream sediment (Fig. 4) shows the existence of beryllium-differentiated granite surrounded by gneiss and migmatites with levels of arsenic varying from 40 to 200–300 ppm. As at Beauvain, this is an arsenic-enriched perigranitic aureola surrounding the Puy-les-Vignes mineralization.

A detailed examination of the mining site (Fig. 5) shows that pollution likely to arise from waste rock dispersion stretches over approximately 1·5 km in two sloping basins draining towards the north-west. However, it will be seen that tungsten concentrations of the same order, varying from 40 to 440 ppm, exist downstream of unworked deposits and that arsenic levels higher than 200 ppm, and reaching 1300 ppm, cover an area of almost 10 km² , far greater than the area of the mine and its waste rocks.

Thus, in the stream sediment of Puy-les-Vignes any pollution due to mining does not really differ from the geochemical landscape, which shows primary mineralization phenomena, largely sub-outcropping.

Let it be said in passing that the analysis of the stream sediment (Fig. 4) shows that the geological map tallies exactly with the geochemical map. In France at least this is very frequent and shows that the geochemical

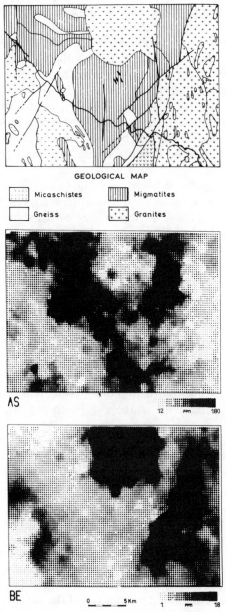

FIG. 4. Geochemical inventory in the region of Puy-les-Vignes: geological chart and map of regional fluctuations in Be and As.

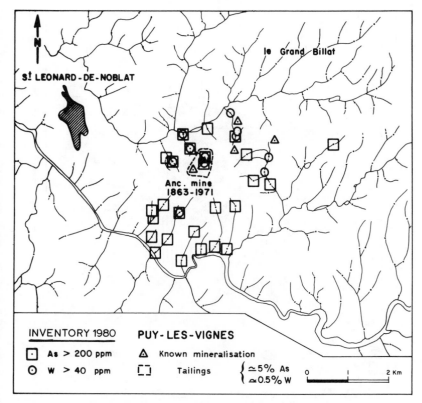

FIG. 5. W–As anomalies around Puy-les-Vignes.

information in the stream sediment is, in certain respects, completely comparable with the soil information.

Saint-Salvy (Albigeois, south of the Massif Central). This is a blende (ZnS) and galena (PbS) vein in Paleozoic schist-sandstone-carbonate formations. Ore mining with 12% Zn and 1% Pb on average began in 1976 on this site, where 2 000 000 m^3 tailings were stored at the end of 1985. The results of geochemical prospecting (Figs 6 and 7) carried out before mining began (1971) and during the period of mining (1978–85), together with the geochemical inventory, indicate:

—lead concentrations of more than 900 ppm over almost 3 km which trace exactly the mark of the vein in the soil; and

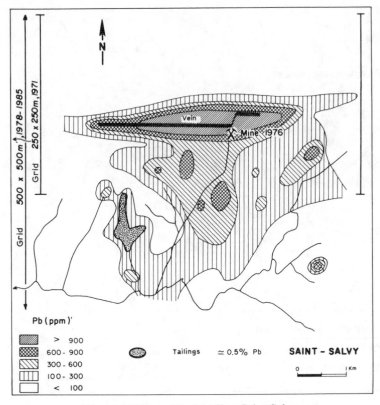

FIG. 6.   Pb anomaly in soil at Saint-Salvy.

—zinc concentrations greater than 1000 ppm at regular intervals over
more than 4 km² downstream of the vein as well as anomalous
aureolas of several hundred ppm towards the east, which are far more
extensive than their lead counterparts.

Various studies have shown that this zinc came from blende dispersed in
the formation close to the vein, thus giving a reference level of regional
spread.

Given that the site of waste rock storage is within this environment, it is
difficult to state the exact degree of pollution, which in any case would only
be to the west and to the south.

The area affected, including all the waste, stretches over a maximum of
2 km north–south and is not more than 1 km east–west. In this area the lead

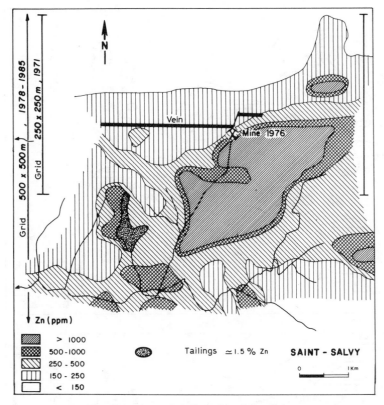

FIG. 7.   Zn anomaly in soil at Saint-Salvy.

and zinc concentrations, amounting to several hundred ppm around the tailings, are less than those engendered in the soil to the north and east just by vein or disseminated mineralizations.

In this case the extent of metal pollution in the soil is actually less than the natural dispersion of the metals resulting merely from geological processes.

Largentière (Cévennes, south-east of the Massif Central). This is a stratiform mineralization with galena (PbS) and blende (ZnS) found at different levels in detritic carbonated sandstone formations of the Triassic base. There are now 5 000 000 m³ of tailings stored on this site, where on average 4% Pb and 0·75% Zn ore was mined between 1964 and 1984.

The results of the geochemical inventory made in 1984, i.e. at the end of the mining period, show that the lead and zinc concentrations in the soil

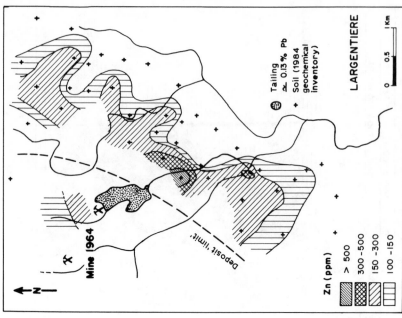

FIG. 9. Zn anomaly around Largentière.

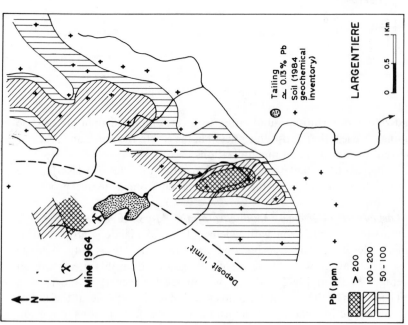

FIG. 8. Pb anomaly around Largentière.

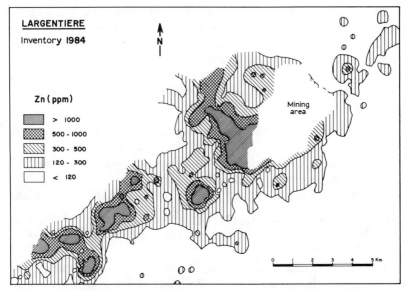

FIG. 10.   Regional distribution of zinc concentrations in the soil in the Largentière area.

downstream of the tailings are greater, respectively, than 200 ppm (Fig. 8) and 500 ppm (Fig. 9), which could result from the mining as such.

This pollution does not, however, stretch for more than 1·8 km from its source. Nonetheless, these same results illustrate that the concentrations form part of an anomalous lead and zinc band, 1–1·5 km wide and more than 5 km long, parallel to the limit of the deposit, i.e. without connection with the mine and the storage site. This is confirmed at regional level (Fig. 10), where the several hundred to several thousand ppm of zinc in the soil is not connected with any mining activities and shows the track of a mineralized formation about 4 km wide and more than 15 km long.

In terms of concentration levels and extent, human-induced pollution is minimal compared with the natural dispersion of metals in the soil.

These examples are not exceptional and reflect phenomena with which geologists are very familiar.

3.   ORIGIN OF NATURAL METAL CONCENTRATIONS IN SOIL

A study of the processes of geochemical anomaly formation and of the behaviour of metals in soil was undertaken at the BRGM as part of mining research programmes. Numerous deposits were examined, following the evolution of the way in which metals manifest themselves from deep primary sulphides to surface materials (soil, sediment, water) through the different horizons of alteration profiles.

With regard to temperate climates, it has been shown (Wilhelm *et al.* [13]) that metals of sulphide origin were found in the soil and sediment mainly in three forms:

—largely associated with gossan fragments, i.e. iron oxyhydroxide resulting from the alteration of sulphide such as pyrite;
—stable supergene minerals such as carbonates, sulphates, phosphates or sometimes even residual primary sulphides; and
—bound to the organic matter which is itself linked to the iron oxides.

Once formed these are stable as long as the physico-chemical balance of the environment does not change, which also ensures the everlastingness of the geochemical anomalies.

However, the natural environment sometimes presents sudden spatial variations, as is the case on the site at Porte aux Moines, where in the space of several hundred metres the soils change from an oxidizing environment to a reducing environment (Laville-Timsit [9]; Laville-Timsit and Wilhelm [10]).

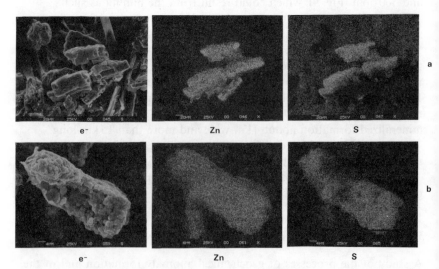

FIG. 11.   Wurtzite neoformation in the soil downstream of Porte-aux-Moines.

The oxidizing environment is characterized by a gossan over a polymetal sulphide mineralization, which gives several hundred ppm of lead in the soil in the form of secondary lead minerals or associated with iron minerals. These forms are stable in this physico-chemical environment, where lead is thus immobilized. Zinc, on the other hand, is leached from the oxidation zone and transported downstream in soluble form over about 600 m, where it precipitates in a morphological hollow occupied by a kind of peat, which constitutes a reducing environment. The zinc concentrations, which reach up to 1%, take the form of low-temperature wurtzite neoformations (Fig. 11). These zinc sulphides are stable in this kind of acidic and reducing environment, which thus acts as a geochemical barrier to this element. The same applies to some of the iron of sulphide origin which, after chemical transfer, settles in the peat (about 10% Fe) on plant blades which it gradually converts into marcasite (iron sulphides, Fig. 12).

All these data, which come from mineralized areas, are nonetheless independent of any mining activity and fully illustrate the value of sound

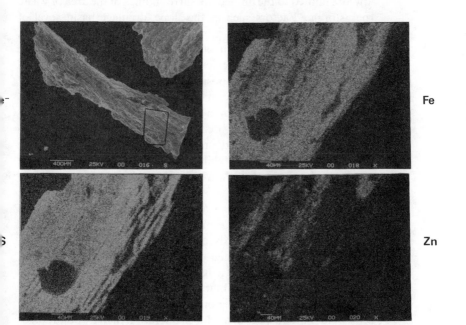

FIG. 12.   Plant blades transformed into marcasite in the soil downstream of Porte-aux-Moines.

knowledge of the various processes taking place on a site in order to quantify the real effect of each of them, including any pollution.

4. CONCLUSIONS

The impact of metal mines on the environment is seen here from the angle of soil contamination by metals. Using specific examples in their geological and geochemical contexts, a certain number of points emerge which must be kept in mind in any study of pollution.

1. In the majority of cases, a mine is established on a site where the soil is naturally enriched in metal because of the very existence of mineral deposits which are to be mined. Knowledge of the initial state of the site is thus indispensable for distinguishing between natural phenomena and any man-induced changes.

2. Metal pollution in the soil brought on by man through mining is low and limited to the immediate surroundings of the area of waste rock storage. Factors such as the slope, drainage basin, level of groundwater, force of the prevailing winds, etc., can, however, change the conditions of metal dispersion in the soil and cause nuisances to varying degrees of severity. Precautions must obviously be taken when choosing conditions and sites for storage. This was not always the case in times gone by when environmental problems did not concern many people, but for several years now preventive techniques have been put forward which allow the risks to be reduced. Johnson and Bradshaw [7] take the view that the best methods consist of stabilizing the waste by vegetation on condition, however, that the vegetation technique best suited to each case is used.

3. The problems of pollution cannot be tackled solely from the concentration level of an element, since this does not give an assessment of the real hazards to which the environment is exposed. The possibility of metal migration must be taken into account, which always requires knowledge of the carrier phases of the metals and their conditions of stability.

4. Experience shows that a region is generally affected more by the economic consequences of a mine in operation than by pollution of it. Finally, although cases of serious soil pollution in a mining context do exist they are the result either of accidents (dam

breakage and the spread of several thousand cubic metres of 'earth' downstream, leakage from decanting ponds, etc.) or from processing activities established on the mining site.

This takes us out of the simple 'soil contamination by mining' and into the far more worrying field of industrial pollution. 'Accidents' in recent years and months bear this out.

REFERENCES

1. BRADSHAW, A. D. (1982). Treatment of mined land. *Coal and Energy Quarterly*, No. 32, 8–18.
2. DAY, A. and LUDEKE, K. L. (1981). The use of legumes for reclaiming copper mine wastes in the southwestern USA. *Minerals and Environment*, **3**, No. 1, 21–3.
3. FORTESCUE, J. A. C. (1981). Environmental geochemistry and exploration geochemistry. *Western Miner*, October, 15–21.
4. GEMMELL, R. P. and GOODMAN, G. T. (1978). Problems of grassland maintenance on metalliferous smelt wastes.
5. HILL, J. R. C. (1977). Factors that limit plant growth on copper, gold and nickel mining wastes in Rhodesia. *Mining Industry*, Sect. A, **86**, 98–109.
6. HILL, J. R. C. (1977). Establishment of vegetation on copper, gold and nickel mining wastes in Rhodesia. *Mining Industry*, Sect. A, **86**, 135–45.
7. JOHNSON, M. S. and BRADSHAW, A. D. (1977). Prevention of heavy metal pollution from mine wastes by vegetative stabilization. *Trans. Inst. Min. Metall.*, **86A**, 47–55.
8. JONES, K. C., PETERSON, P. J. and DAVIES, B. E. (1983). Silver concentrations in Welsh soils and their dispersal from derelict mine sites. *Minerals and the Environment*, **5**, 122–7.
9. LAVILLE-TIMSIT, L. (1986). Néoformation de sulfures de fer et de zinc en milieu réducteur en aval du gisement Pb, Zn, Cu, Ag de Porte-aux-Moines (Côtes du Nord): l'anomalie géochimique Pb–Zn des sols de Kérouaran. *Sci. Géol. Bull.*, **39**, 3, 263–75.
10. LAVILLE-TIMSIT, L. and WILHELM, E. (1979). Comportement supergène des métaux autour du gîte sulfuré de Porte-aux-Moines (Côtes du Nord). Application à la prospection géochimique. *Bull. Bur. Rech. Géol. Min.*, (2), 11, 2–3, 195–228.
11. PETERS, T. H. (1978). INCO Metals Reclamation program. *Environmental Control*, **71**, No. 800, 104–6.
12. SGARD, J. (1979). La mise en végétation des crassiers de mines de fer. *Revue de la Métallurgie*, No. 2, 138–40.
13. WILHELM, E., LAVILLE-TIMSIT, L., LELEU, M., CACHAU-HERREILLAT, F. and CAPDECOMME, H. (1978). Behaviour of base metals around ore deposits: application to geochemical prospecting in temperate climates. *Geoch. Expl.*, *Proc. 7th Inter. Geoch. Expl. Symp.*, Golden, Colorado, pp. 185–99.

Impacts on Soils Related to Industrial Activities: Part IV—Effects of Air Pollutants on the Soil

B. ULRICH

Institut für Bodenkunde und Waldernährung, University of Göttingen, Büsgenweg 2, 3400 Göttingen, Federal Republic of Germany

SUMMARY

In ecological terms air pollutants are acids (and their precursors), potential toxins and nutrients. Their effects on soils are therefore very complex. Owing to the filtering effect of forest canopies, deposition of air pollutants is usually much higher in forests than in areas of low vegetation (pastures and arable land). Soils act either as an inert matrix (e.g. for NaCl) or as sinks (e.g. for heavy metals in neutral soils) or sources (e.g. for acids: equivalent amounts of cations to the number of protons can be leached out) for air pollutants. In non-tillage ecosystems of low cropping intensity (e.g. in forests and heathland) deposited heavy metals accumulate in the surface soil horizon. As a result of acid deposition, forest soils in Central Europe have already lost most of their mobilisable nutrient reserves and are now acid below the rooting zone. Acidification takes place within the seepage layer, endangering the quality of the water from forested areas. The process is not fully understood, and a drastic reduction in emissions of SO_2, NO_x, heavy metals and organic micropollutants is recommended as a regulatory measure.

AIR POLLUTANTS

Air pollutants may be of natural or anthropogenic origin. They can be classified as follows according to their ecochemical function:

Neutral substances: Na^+, Cl^- (sea spray)
Nutrients and essential elements: SO_4^{2-}, I^-, Mg^{2+}, etc. (sea spray); NH_4^+, NO_3^-, Ca^{2+}, trace elements

Acidifiers: SO_2, NO_x
Potential toxins: SO_2, HF, heavy metals, metal oxides, hydrocarbons
Oxidants: NO_x, ozone, hydrocarbons
Radioactive substances

Many anthropogenic air pollutants also occur naturally and are in some cases vital to the development of ecosystems. In Europe, and particularly in Central Europe, anthropogenic emissions do, however, exceed natural emissions by a factor of around 10 in the case of most air pollutants (see Fig. 1).

Different air pollutants have differing effects on the ecosystem depending on their ecochemical function: acids deplete soils of nutrients whereas nutrients encourage plant growth. Nitrogen encourages the action of decomposer organisms in the soil in forest ecosystems with limited nitrogen supply whereas acids and heavy metals have an inhibiting effect. The effects of anthropogenic inputs in the ecosystem which have risen tenfold are hence complex and confusing, as witnessed by the current debate on acid rain.

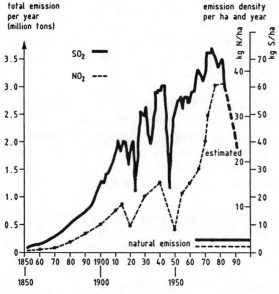

Fig. 1a.   Emission of SO_2 and NO_x in the Federal Republic of Germany since 1850.

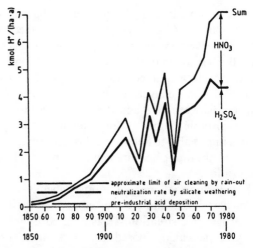

FIG. 1b.   Emission of acidity due to SO_2 and NO_x in the Federal Republic of Germany since 1850.

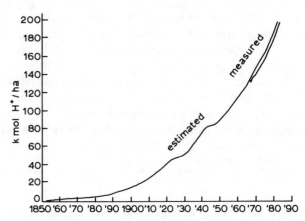

FIG. 1c.   Accumulated acid deposition in stand of fir trees at Solling.

DEPOSITION

Figure 1a shows the trend in emissions of SO_2 and NO_x in the Federal Republic of Germany since the middle of the last century (Ulrich [14]). Figure 1c shows the accumulated acid deposition in a spruce stand in Solling, where deposition has been measured since 1969 (Matzner *et al.* [8]).

Between 1969 and 1985 an average of 3·8 kmol H^+/ha was deposited each year equivalent to a total of 65 kmol H^+/ha. If it is assumed that there was a parallel trend between emissions and deposition the accumulated acid deposition since the beginning of the last century is 200 kmol/ha. The acid deposited would require a base equivalent to 10 000 kg $CaCO_3$/ha as a buffer in the soil and seepage layer. An average of at least 40 kg N/ha was deposited each year between 1969 and 1985 equivalent to a total of 680 kg N/ha. Since leaves are able to directly break down the nitrogen compounds deposited and it is impossible to determine this quantity of nitrogen by the flux balance method, the input measurements are in virtually all cases lower than actual deposition.

The rate of deposition of air pollutants depends not only on the concentration of particulate and gaseous contaminants in the air. Other factors are the orographic situation (areas affected by immissions: outer slopes, ridges and plateaus exposed to wind, areas protected from immissions: sheltered inner slopes and valleys) and the ecosystem's ability to filter air pollutants (high in the case of evergreen softwoods, low in the case of deciduous hardwoods and even lower in the case of agricultural land and grassland). According to available measurements acid deposition in Central Europe is between 0·8 and over 6 kmol H^+ per hectare per year. The acid accumulated in woodland soils since the end of the last century as a result of anthropogenic air pollution is equivalent to a base of at least 2500 kg $CaCO_3$/ha. Accumulated nitrogen deposition is estimated at at least 500–1000 kg N/ha, and is continuing to rise rapidly at an annual rate of 20–80 (average of 30–40) kg N/ha. Owing to its immobility the lead which builds up can be estimated from an analysis of the forest floor. In forests in areas affected by immissions it exceeds 20 kg/ha (Mayer [9]) and in forests close to industrial areas is as high as 60 kg/ha (Gehrmann [2]).

THE SOIL'S FUNCTION AS A SINK/SOURCE

The soil can provide the following functions for substances deposited:

(1) It can act as an inert matrix: the deposited substances leach into the groundwater through the seepage water.
Examples: Na^+Cl^- and other neutral salts ($CaSO_4$).
Acid-soluble heavy metals (Cd, Zn and Mn)
in acid soils and seepage layers.
Nitrates in acid 'nitrogen-saturated' forest ecosystems.

(2) It can act as a sink: the deposited material builds up in the soil.
    Examples: Lead (in top organic soil horizon).
               Heavy metals in neutral soils.
               Nitrogen in 'non-nitrogen-saturated' forest ecosystems.
               Hydrocarbons in acid forest soils where litter break-
               down is inhibited.
               Radioactive isotopes such as caesium and strontium.

(3) It can act as a source: the deposited material accumulates in the soil
    but also causes other substances to be leached with the seepage
    water or discharged into the air.
    Examples: Acids: bases equivalent to the acid input are removed in
               the seepage water or cation acids (e.g. Al ions) are
               released from the soil matrix.
               Cationic acids may accumulate to some degree but will
               migrate in an acidifying soil through the seepage layer to
               the aquifer if acid continues to enter the soil.
               Stable gaseous N and S compounds (e.g. N_2O) are
               released; these may contribute to the breaking down of
               the ozone layer in the stratosphere.

Substances for which the soil acts as an inert matrix and which are leached out in the seepage water will be concentrated in the seepage water as water is lost through the atmosphere by evaporation and transpiration in their journey through the ecosystem.

Depending on the rate of precipitation, concentrations of a factor of 2–10 can be expected. The concentrations in precipitation over stands of spruce trees range between 6 and 27 mg NO_3/litre and 0·6 to over 1 μg Cd/litre. The limit values for drinking water are 50 mg NO_3/litre and 5–6 μg Cd/litre. This would mean that the limit values for drinking water could be reached in water in the case of substances for which the soil does not act as a filter simply if there were emissions under unfavourable conditions.

The rate of accumulation for substances for which the soil acts as a final sink can be estimated from the expected deposition rates. Table 1 gives a range of rates of deposition, accumulation on the forest floor and levels in humus for a number of forest ecosystems in North-west Germany (Schultz *et al.* [12]; Büttner *et al.* [1]).

The immobile (Pb) or less mobile (Cu, Cr) heavy metals occur mainly in organic compounds. This means that in forests they are occurring in increasing concentrations in the soil organic matter. In this form they could impede the breakdown of organic matter. Although there are indications that this is the case considerable research is required in this area.

Since heavy metals become less mobile as the pH rises, areas of higher pH act as sinks for them. In acid soils such areas are roots or even organisms. In the fine root system of spruce concentrations of up to 12 mg Cd, up to 350 mg Pb, 40 mg Cr, 40 mg Cu and 0·55 mg Tl per kg dry matter have been found (Büttner et al. [1]). There are also indications that these heavy metals are partly responsible for root damage but no definite proof has been found.

Owing to their low level of mobility heavy metals will accumulate in neutral soils in the top millimetre to centimetre layer if they are not worked by tilling into a thicker layer of soil.

The fact that, with the exception of grazed grassland and ploughed arable land, the top, even only millimetre to centimetre-thick soil layers act as the sink in which heavy metals usually accumulate will probably have growing adverse ecological effects on soil organisms and seed germination if heavy metals continue to be deposited.

The only scientific data on areas in which radioactive isotopes such as caesium and strontium accumulate concern neutral soils. Like heavy metals, caesium and strontium are immobilised in such soils. No studies have, however, been made to determine how far this isotope accumulates in woodland soils which have become very acid and are seriously depleted of bases.

There is a high level of deposition of polycyclic aromatic hydrocarbons as well as heavy metals in forests. The soil, and in particular the humus layer, act as a sink so none of these substances are washed out. The accumulated material found in Solling amounts to between 15 and 50 times the deposition rate recorded (Matzner et al. [6]; Matzner [7]).

Table 1

Range of rates of deposition, accumulation and levels in humus on forest floor of heavy metals

	Pb	Cd	Cu	Zn	Cr
Deposition on open land (g/ha/yr)	70–140	2–4	10–50	70–300	2–4
Deposition in forests (g/ha/yr)	130–330	3·5–5·5	50–100	300–900	7–22
Accumulation on forest floor (g/ha)	700–60 000 (20 000)	0–200 (50)	200–8 000 (300)	500–40 000 (15 000)	1 000–30 000 (5 000)
Levels in forest floor organic matter (mg/kg)	100–450	0·2–1·4	10–70	50–250	10–300

Most of the pollutants for which the soil acts as a sink also accumulate in the top humus layer where most soil organisms live. In some cases they fall to the ground with leaves or are broken down with the leaf litter into the organic soil matter, i.e. they enter the food chain of soil organisms. A number of pollutants often have synergic effects. Studies which ignore the fact that pollutants are incorporated in the food of soil organisms and that different pollutants which accumulate at the same time will have a synergic effect on the activities of soil organisms cannot give a realistic picture of the possible ecological effects of pollutant accumulation. However, such an approach has not so far been adopted.

Deposited nitrogen is continuing to build up in many forest ecosystems on acid soils in the humus layer. It can be concluded from this fact that the deposited nitrogen is used for plant production but the breakdown of plant litter, i.e. the activity of soil organisms, is impeded. On mull soils where litter decomposition is not or only very slightly impeded the nitrate is washed out. In some cases even more than the amount deposited is washed out (the soil acts as a source and the nitrogen it contains is broken down). Even when litter decomposition is impeded nitrates and perhaps ammonium can be washed out (Büttner *et al.* [1]), especially when the root system and microbial activity would appear to be so severely impaired that the deposited nitrogen can no longer be biologically utilised. The accumulation of deposited nitrogen at an annual rate of 20–50 kg/ha (in extreme cases over 80 kg/ha/yr) would seem to reflect a general balance between trees' ability to take up nitrogen and impeded microbial activity. The accumulation of ever larger quantities of nitrogen in an unstable humus layer raises serious ecological questions. There is a risk that if decomposition is encouraged large quantities of nitrates will reach groundwater or be discharged into the atmosphere in the form of oxides of nitrogen. The fact that at present most of the nitrogen deposited builds up on the forest floor does not mean that this where this emission will stay. This is another reason why continued emissions of NO_x from power stations and traffic and emissions of NH_3 from intensive animal breeding are unacceptable.

A distinction must be made in assessing acid pollution of soils between the self-generation of acid within ecosystems and acid input since both make up the overall acid load. A breakdown of acid pollution and the effects on the soil is shown in Table 2 for non-fertilised forest soils and different acid sources. A distinction can be made between bases (depletion leaching out of nutrients) and acid accumulation (mainly aluminium ions and organic acids) (Ulrich [15]).

Table 2

Acid source	Rate ($kmol\, H^+/ha/yr$)	Effect
Pre-industrial acid deposition	<0·1	—
Leaching of bicarbonates	0–50	Base depletion (lime is dissolved)
	(in calcareous soils)	Water alkalinity
Build-up of new growth	0·3–1·3	Base depletion in root zone
Build-up of forest litter	0–2	Base depletion in root zone
Leaching out of organic anions .(podsol formation)	0–2	Build-up of acids in topsoil
Leaching out of nitrates	0–4	Build-up of acids in root zone
Acid deposition	1–>6	Build-up of acids in soil, subsoil and percolating layer; acidification of watercourses

The only process counteracting base depletion and acid accumulation in the soil is silicate weathering as far as Ca, Mg, K and Na ions are released and protons used up. The rate of silicate weathering is less than 2 kmol ion-equivalent/ha/yr and is usually less than 1·2; for soils derived from loess and sandstone it is ~0·5. A comparison of the proton buffering rate with the proton load rate shows that the buffering rate is more or less used up by new growth but that any additional load, e.g. through acid deposition, inevitably leads to base depletion and acid build-up.

The soil acts as a source vis-à-vis the input of acids: the acid protons will build up in the soil and equivalent quantities of cations bound up in bases or weakish cation acids in the soil will be leached out with the acid anion in the percolating water. The cations bound up in bases are nutrients (calcium, magnesium and potassium). This means that as the soil becomes more acid there is a corresponding loss of nutrients. The degree of base and nutrient depletion or release of toxic cation acids can be determined from the accumulated acid deposition. This is shown in Fig. 1c for the spruce stand at Solling. As already stated, it appears that the accumulated acid input corresponds to a base equivalent of 2500 to over 10 000 kg $CaCO_3$ or 1000 to over 4000 kg Ca/ha depending on tree species and site. In North-west Germany the bases like Ca have been seriously depleted from rooted soil as a result of overutilisation of the ecosystems for centuries (in some cases going back as far as the Bronze Age). The stock of exchangeable bases in the soil was probably already less than 1000 kg exchangeable Ca/ha when industrialisation began. This explains why the exchangeable base supply in forest soils in North-west Germany is today, with very few exceptions, less than 500 kg Ca, in some cases even less than 100 kg Ca/ha and is continuing to fall.

Table 3
Storage of exchangeable K, Ca and Mg in the mineral soil

	Median values		
	K	Ca	Mg
		(kg/ha)	
Hamburg forestry area, 0–60 cm			
Buffer ranges			
Carbonates, 1%	580	> 10 000	> 500
Silicates, 2%	160	8 000	300
Exchangeable substances, 12%	160	1 200	64
Aluminium, 14%	160	270	33
Al/Fe, 71%	120	170	28
North Rhine Westphalia, 0–80 cm			
Soils on limestone	1 300	20 000	550
Soils on clay schist	360	500	100
Loess soils	480	2 200	440
Sandy soils	200	380	40

Table 3 summarises the results of two case studies, one in the Pleistocene sediments in the North German plains (Hamburg forestry authority, Rastin and Ulrich [10]) and the other in North Rhine Westphalia Gehrmann and Büttner [3]). Given are median values. Half of the representative soils studied have stocks well below the values indicated. The soils in the Hamburg forestry area are classed according to buffer ranges (Ulrich [13]). Given is the percentage of soils with the corresponding degree of acidification at the forest area.

In all these soils the acidification front, i.e. the transitional zone from base depletion to the release of toxic cation acids, now lies below the rooted soil in the percolating layer. According to the very little data available the acidification front is now between 1 and 6 m deep (in sandy soils). These soils are obviously virtually totally depleted of bases and nutrients, and a lack of nutrients and acid stress will lead to increasing forest damage. This can be demonstrated by comparing the soil's stocks with the needs of a growing stand of trees which stores a large quantity of nutrients in the growing biomass until the trees reach maturity. In a spruce stand of yield class 9, 400 kg K, 320 kg Ca and 50 kg Mg will be stored in the above ground biomass until age 40, i.e. when the first thinning takes place. Older trees can usually absorb the nutrients they need for growth from deposition even on very poor and acid soils but growing trees need in addition the nutrients from supplies in the soil. Sufficient nutrients cannot now be absorbed from the soil as the majority of forest soils have limited exchangeable nutrient supply. Without regulating measures such as fertilising and liming young stands of trees will be seriously damaged and will even die in the future.

Until recently there was extreme controversy about whether acid rain did make soils acid particularly in forests and even today it is still not generally accepted that it does. This conclusion is based on sound theoretical data (Ulrich [15, 17]) and is backed up by the emission data and long-term input and output measurements in forest ecosystems such as Solling and a comparison of present soil chemistry with earlier data. This theory is of such importance that it needs to be verified in future research programmes.

Soil acidification can be the cause of serious damage to soil organisms and plant organs (roots). The concentrations and interrelationships of individual ions involved in the soil solution are excellent ecophysiological parameters which can be used to assess and analyse such damage. Examples include: H/Ca and Ca/Al ratio as a measure of proton stress and aluminium toxicity (Rost-Siebert [11]; Junga [5]); Mg/Al ratio as a measure of aluminium-induced magnesium deficiency (Jorns and Hecht-Buchholz [4]); and K/Ca ratio as a measure of potassium deficiency or excess calcium (Woodruff [18]). The results of laboratory or greenhouse experiments on the meaning of ion ratios in the solution phase as ecophysiological parameters can be extrapolated to field conditions. Under field conditions the same parameters can be measured as a function of time and space. More research is required here since the data it will provide will make it possible to determine the causal relationships in ecosystems.

Ecosystem elasticity is reduced by soil acidification. Nutrient depletion will show up only in an advanced stage as nutrient deficiency in trees because the nutrient input from anthropogenic air pollutants covers the nutrient needs of existing stands of trees. Damage occurs in particular when growing stands take the last base and nutrient supply from the soil and/or when the external acid load is accompanied by a high acid load generated by the ecosystem itself. This is particularly the case in mull soils with a high humus content where the organic nitrogen supply of the mineral soil is broken down and the nitrogen is leached out in nitrate form.

OUTPUT

The substances which leach out in the percolating water or are discharged in gaseous form into the atmosphere can continue to exert an effect in the receiving medium. The degree to which complex compounds of organic metals can be dissolved or migrate in acid soils and thus lead to heavy metal pollution of groundwater needs to be investigated. Further research is also required into the discharge of stable gaseous sulphur and nitrogen compounds since they can contribute to the breakdown of the ozone layer. Studies are urgently required into the nitrate output and the mechanisms they set in motion in the soil in 'nitrogen-saturated' ecosystems to determine the sources of soil acidification and dangers to groundwater.

SOIL IMPROVEMENT

Productivity is endangered by soil depletion and acidification (Ulrich [17]). Stable ecosystems of high elasticity are required (Ulrich [16]). Such ecosystems require soils with at least an average base supply and a balanced nutrient supply. In theory it is possible to restore soils to this state by liming and fertilising but the soil's original ability to store nutrients can never be restored. Enormous problems are, however, posed in practice. On the one hand forest soils are acid to very great depths, which means that lime (and if necessary fertiliser) has to be dug in to a great depth or—if measures are limited to the surface—it can take decades to reduce the acidity of the mineral soil. On the other hand liming, fertilising and tilling may take substances out of the soil which in turn pollute water (nitrates and heavy metals) and the air (nitrogen oxide). Although much is already known about liming and fertilising forests that is not enough to be able to minimise the negative environmental effects in individual cases.

Heavy metals can in the long term be converted into insoluble compounds by increasing base saturation. This will reduce their ecophysiological effectiveness.

COMPARISON OF THE LOAD ON DIFFERENT TYPES OF ECOSYSTEM

Much fewer air pollutants are deposited on soils on which annual crops are grown or grassland than on forest soils. Where land is worked the accumulating potentially toxic substances are mixed into the soil and the rate of increase in concentration slowed down. The acid input into soils (or produced by the ecosystem itself) can continually be offset by liming. In these soils the load resulting from fertilisation and tilling is much more serious than that arising from deposition.

In woods of all types, even in orchards, a greater percentage of the acid input may in some circumstances reach the soil via the root system: the tree retains the acid in its leaves and draws up the base it requires from the soil. The effect of this is for acid to enter the soil throughout the area around the root system. The resultant acidification in the lower part of the root system cannot be mitigated by applying lime to the surface of the soil. Consequently such deep soil acidification is a specific problem of forest ecosystems. In principle this acidification is reversible. However, it is impossible to offset the harmful effects of acid deposition on tree root systems by liming or fertilisation. The only alternative is to drastically reduce emissions of air pollutants, in particular SO_2, NO_x, heavy metals and hydrocarbons.

REFERENCES

1. BÜTTNER, G., LAMERSDORF, N., SCHULTZ, R. and ULRICH, B. (1986). Deposition und Verteilung chemischer Elemente in küstennahen Waldstandorten— Fallstudie Wingst, Abschlußbericht. Ber. d. Forschungszentrums Waldökosyst./ Waldsterben d. Univ. Göttingen, in press.
2. GEHRMANN, J. (1984). Einfluß von Bodenversauerung und Kalkung auf die Entwicklung von Buchenverjüngungen (*Fagus sylvatica* L.) im Wald. Ber. d. Forschungszentr. Waldökosyst./Waldsterben d. Univ. Göttingen, Vol. 1, pp. 1–213.
3. GEHRMANN, J. and BÜTTNER, G. (1985). Untersuchungen zum Stand der Bodenversauerung wichtiger Waldstandorte im Land Nordrhein-Westfalen. Unveröffentlichter Bericht an die LÖLF, Recklinghausen.
4. JORNS, A. and HECHT-BUCHHOLZ, CH. (1985). Aluminium-induzierter Magnesium- und Calciummangel im Laborversuch bei Fichtensämlingen. *Allgem. Forstzeitschr.*, **41**, 1248–52.
5. JUNGA, U. (1984). Sterilkultur als Modellsystem zur Untersuchung des Mechanismus der Aluminium-Toxizität bei Fichtenkeimlingen (*Picea abies* Karst.). Ber. d. Forschungszentr. Waldökosyst./Waldsterben d. Univ. Göttingen, Vol. 5, pp. 1–173.
6. MATZNER, E., HÜBNER, D. and THOMAS, W. (1981). Content and storage of polycyclic aromatic hydrocarbons in two forested ecosystems in northern Germany. *Z. Pflanzenernähr. Bodenkde*, **144**, 283–8.
7. MATZNER, E. (1984). Annual rates of deposition of cyclic aromatic hydrocarbons in different forest ecosystems. *Water, Air and Soil Pollution*, **21**, 425–34.
8. MATZNER, E., KHANNA, P. K., MEIWES, K. J., CASSENS-SASSE, E., BREDEMEIER, M. and ULRICH, B. (1984). Ergebnisse der Flüssemessungen in Waldökosystemen. Ber. d. Forschungszentr. Waldökosyst./Waldsterben d. Univ. Göttingen, Vol. 2, pp. 29–49.

9. MAYER, R. (1981). Natürliche und anthropogene Komponenten des Schwermetallhaushalts von Waldökosystemen. *Gött. Bodenkdl. Ber.*, **70**, 1–152.

10. RASTIN, N. and ULRICH, B. (1985). Bodenchemische Standortscharakterisierung zur Beurteilung des Stabilitätszustands von Waldökosystemen in Hamburg. Ber. d. Forschungszentr. Waldökosyst./Waldsterben d. Univ. Göttingen, Vol. 10, 1–91.

11. ROST-SIEBERT, K. (1985). Untersuchungen zur H- und Al-Ionen-Toxizität an Keimpflanzen von Fichte (*Picea abies* Karst.) und Buche (*Fagus sylvatica* L.) in Lösungskultur. Ber. d. Forschungszentr. Waldökosyst./Waldsterben d. Univ. Göttingen, Vol. 12, pp. 1–219.

12. SCHULTZ, R., SCHMIDT, M. and MAYER, R. (1986). Heavy metal fluxes in the canopy of a beech and a spruce forest. *Proc. Int. Conf. Environmental Contamination, Amsterdam,* in press.

13. ULRICH, B. (1981). Ökologische Gruppierung von Böden nach ihrem chemischen Bodenzustand. *Z. Pflanzenernähr. Bodenkde.*, **144**, 289–305.

14. ULRICH, B. (1985). Interaction of indirect and direct effects of air pollutants in forests. In: *Air Pollution and Plants* (Ed., C. Troyanowski), VCH-Verlagsgesellsch. Weinheim, pp. 149–81.

15. ULRICH, B. (1985). Natural and anthropogenic components of soil acidification. *Z. Pflanzenernähr. Bodenkde.*, in press.

16. ULRICH, B. (1986). Stability, elasticity, and resilience of terrestrial ecosystems with respect to matter balance. *Ecol. Studies*, **61**, 11–49.

17. ULRICH, B. (1986). Die Rolle der Bodenversauerung beim Waldsterben: Langfristige Konsequenzen und forstliche Möglichkeiten. *Forstwiss. Centralbl.*, **105**, 421–35.

18. WOODRUFF, C. M. (1955). Ionic equilibria between clay and dilute salt solutions. *Proc. Soil Sci. Soc. Amer.*, **19**, 36–40.

Irrigation and Drainage for Improvement of Wet and Saline Soils: Reference to Portuguese Conditions

L. S. Pereira, E. C. Sousa and L. A. Pereira

*High Institute of Agronomy, Technical University of Lisbon,
Tapada da Ajuda, 1399 Lisbon Codex, Portugal*

SUMMARY

This chapter reviews problems and ways of improvement of humid and saline soils, with emphasis on Portuguese conditions. The first part is oriented for a global review of problems and presentation of effects of land reclamation methods for solving or minimizing them according to a Portuguese case study. The second part is oriented for irrigation benefits and needs in Portugal. The analysis is done according to the main type of soil problems, including waterlogging, groundwater and saline situations.

1. INTRODUCTION

Rational utilization of soil and water resources deals with complex problems, the more important as limited the available resources are, either in quantity or in quality. Agricultural development depends more and more on improved production technologies, adapted to existing economic and environmental conditions, with consequent results on increasing production factors. But it is important to point out that technological innovations, for instance the utilization of high yielding varieties, demand a very good management of water and soil.

Portuguese agriculture is facing a very important challenge, its modernization. In fact irrigated areas in Portugal must be increased, but there is also a priority for the rehabilitation of some existing irrigation systems. It cannot be forgotten that water requirements for irrigation

313

include quality which is related to salts and sediment content, that lead to a progressive degradation of soil physical and chemical characteristics, diminishing the yields and damaging the irrigation equipment. This chapter intends to give a brief overview of some of the experiments that have been conducted up to now, in Portugal, where in fact there is a very important gap to be overcome.

2.   PROBLEMS OF WET AND SALINE SOILS

2.1. Plant growth conditions and their improvement

The growth of plants is controlled to a large extent by soil water. Nevertheless, for good root development the soil should have an adequate pore space and within it a balance between air and water. This balance will depend upon the nature of the pore space, mainly the relation macro–microporosity, and the rate of water delivered to and water removed from the soil. In wet soils the pore space occupied by air decreases, therefore reducing the exchange of air between soil and atmosphere and causing oxygen deficiencies. The O_2 deficiency reduces root respiration and root growth and development; increases the resistance to transport of water and nutrients through the root system, as well as the formation of toxic compounds in soil and plants (Scott Russel [42]); and reduces micro-biological activity, thereby decreasing NH_4^+ and NO_3^- availability inducing N deficiency. The low exchange of air affects transport of CO_2 produced in exchange affecting the heat budget so lowering soil temperatures with side-effects on O_2 deficiencies and CO_2 excesses (Wesseling [60]).

In the project of Leziria Grande one of the main influences of drainage which has been evaluated is oxygen availability in the root zone during winter.

Table 1 shows that oxygen availability is greatly influenced by watertable depths and physical characteristics of soils. As oxygen availability has been evaluated by classes, Table 2 presents their definitions, with several interesting aspects: percentage of air in the root zone, waterlogging in the root zone and estimated crop responses to oxygen availability.

As stated before, soil moisture content affects heat budget and soil temperature (Wesseling [60]).

Soil temperature influences germination, seedling emergence and plant growth by its effect on soil moisture relations, availability of nutrient elements in the soil and water uptake by roots. Soil temperature effects are complex but it seems that water uptake is mainly influenced as low

Table 1
Classes of oxygen availability in the root zone according to watertable depths for
different soil types in Leziria Grande
(from Reis *et al.* [38])

Texture of soil surface	Non-capillary porosity	Watertable depth (cm)			
		>80	50–80	30–50	<30
Sandy to sandy-loam	>15%	1	1	2	4
Loamy to silty-clay-loam	8–15%	1	2	3	4
Silty-clay-loam to silty-clay	8–15%	2	2	3	4
Silty-clay-loam to silty-clay	<8%	3	3	3	4

temperatures also affect root development. Plant responses differ widely
but it is known that their reactions have in common a minimum
temperature below which no activity occurs, a maximum above which
activity stops and an optimum temperature at which the highest activity
occurs.

Under anaerobic conditions decomposition of organic matter is
decelerated while part of the available N is immobilized and serious losses
occur resulting from decomposition of NO_3^- by micro-organisms using it
instead of O_2 from air as a hydrogen acceptor, reducing it to nitrite,
nitrogen oxide or nitrogen gas. Therefore N losses are very important after
submergence of the soil. Also, as NO_3^- is quickly lost by flooding, and
mineralization is slow at high moisture following submergence, N
deficiency is partially responsible for the slow recovery of plant growth
after flooding (Wesseling [60]).

Table 2
Definition of oxygen availability classes
(from Reis *et al.* [38])

Classes	Amount of air in soil volume (%)	Number of consecutive days with watertable above 30 cm	Yield decrease for winter cereals (%)	Yield decrease for winter forages (%)
1	>10	<6	—	—
2	>10	6–15	0–5	—
3	6–10	15–30	5–10	0–5
4	<6	>30	>10	>5

2.2. Soil structure and porosity

Soils with a shallow watertable have a deteriorated structure, with a more compact and sticky topsoil. Bad soil structure is also characteristic of saline and alkali soils. Such have been found for Portuguese conditions in the Leziria Grande de Vila França de Xira (Thiadens [51]), in the Mondego Valley (Ramos [36]) and in Loures Valley (Froes *et al.* [13]). In Mondego Valley particular problems of compaction due to successive ploughing in rice fields—plough pan—have been found (Ramos [36]), which characterize soils submitted to long-term rice cultivation (Van de Goor [52]).

The same deteriorated structure occurs in soils submitted to waterlogging by dammed water, having temporary perched watertables (Pinto [35]).

Soil structure affects porosity and soil aeration. Experience demonstrates that drainage improves the structure, increasing the pore space, promoting cracking and aeration of the soil.

Improvement of porosity is in general much higher when soil improvement methods are utilized, together with drainage structures (Eggelsman [12]). Good results are obtained by sub-surface drainage associated with deep ploughing and chemical treatments aiming at control of swelling and shrinkage; in the absence of drainage soil improvement techniques lead to poor results (Caaplar and Feher [7]).

Similar results are expected to be obtained in experiments under development in the ancient rice fields with fine textured saline and sodic soils of Mondego Valley (Kopp [21]) and in soils with a shallow impervious B horizon in Alentejo (Pereira [31]; Pinto [35]).

Drainage influences on soil structure and porosity also affect other physical characteristics of soils related to water movement—hydraulic conductivity and infiltration—and water retention. In fact the betterment of soil structure and the improvement of soil porosity lead to higher hydraulic conductivity and higher infiltration rates and cumulative infiltration. Such conditions are also improved as a consequence of better farming conditions when appropriate cultivation techniques are applied, so soil improvement conditions through drainage is a continuous process of engineering and agricultural management.

2.3. Salinity control

The excess of soluble salts in the soil, known as salinity, has always an adverse effect on plant growth, variable according to salt concentration and plant tolerance to salt water content. So there is a need for improvement of saline and alkali soils by surface or sub-surface drainage for removing salts from soil, together with gypsum application for chemical reaction on the soil complex in case of alkali soils. A good example of Portuguese conditions is given by the experiments in Leziria Grande de Vila Franca de Xira (Maan *et al.* [23]; Vieira *et al.* [59]; Muralha and Pissarra [29]).

Table 3 shows the effects of drainage on removal of salts through the entire profile, with evident improvement of salinity conditions in the root zone. Results are similar to those obtained in other parts of the world, namely in Spain (Martinez Beltran [27]; Van Hoorn [54]). However, data suggest insufficiency of natural leaching during winter, mainly during dry years, as resalinization phenomena are visible. These problems are common in Mediterranean climates (Van Hoorn [54]; Widmoser [61]).

Different results can be obtained when irrigation is added to rainfall for leaching, which leads to higher removal of salts (Sandu and Rauta [41]), namely under rice cultivation (Maianu [25]). Response of saline and alkali soil also varies according to soil texture as influencing other soil characteristics governing salts retention and water movement in soils (Dukhovny [10]; Hoffman and Meyer [16]; Maianu [25]). Given the influences of drainage depth, Dukhovny [11] proposes an increased depth when groundwater salinity is higher.

Influence of gypsum on reclamation of alkali soils is shown in Table 4. Nevertheless, observed differences are smaller than expected. In fact, as stressed by Van Hoorn [54], exchange of sodium by calcium in heavy clay soils is a slow process and so only part of the calcium applied as gypsum has been utilized. To avoid losses gypsum should be applied in small amounts with intervals of several years.

As stated before, data in Table 3 show relatively poor responses to reclamation: the desalinization process is slow, it has slight effects on soil layers below 50 cm and resalinization makes evident a decrease of the leaching efficiency coefficient (Muralha and Pissarra [29]).

Table 5 shows the overall amounts of water which were drained and the salt removed, although the figures for 1979/80, 1980/81 and 1982/83 have been omitted because these were dry years. In fact this was so much so that the drains did not operate largely because of the irregular distribution of the rainfall in the course of those years which even failed to meet the field capacity.

Characterizing the average salinity of drained water, during a certain period, by the ratio of the total amount of leached salt, S, and the total amount of drained water, Dr, (S/Dr), the respective values are those presented in Table 6.

During the first three years, the ratio S/Dr decreased compared to the

Table 3

Soil salinity evolution (ECe-mmhos/cm) in three experimental plots of Leziria Grande
(from Muralha and Pissarra [29])

Field	Layer (cm)	Oct 75	Apr 77	Oct 77	Apr 78	Oct 78	Apr 79	Oct 79	Apr 80	Oct 80	Apr 81	Oct 81	Apr 82	Oct 82	Apr 83	Oct 83	Apr 84
I	0–25	19·0	6·5	13·0	4·1	12·8	6·3	7·1	7·6	8·8	7·5	8·0	3·5	6·5	4·8	5·3	3·3
	25–50	24·0	15·5	14·5	9·9	15·3	14·5	10·3	11·8	10·9	11·0	11·0	8·9	14·5	11·5	10·3	8·3
	50–75	39·0	25·5	19·8	22·3	21·0	21·3	18·5	19·0	17·3	15·5	17·0	16·9	24·3	16·0	13·3	17·8
	75–100	48·3	32·0	33·5	28·3	36·0	27·5	25·3	30·9	21·6	23·3	23·3	21·8	25·5	25·5	22·5	24·3
	100–125	64·5	55·0	41·8	44·5	40·3	36·3	31·0	42·5	25·0	36·3	37·8	37·8	38·0	31·5	43·3	37·0
	125–150	82·5	72·8	55·5	69·3	63·0	59·5	54·3	65·3	46·4	49·0	50·5	52·5	58·0	52·8	69·3	56·0
	150–200	85·3	83·8	70·8	87·5	78·0	81·3	74·5	92·5	62·3	58·5	59·8	67·2	76·3	78·0	80·0	73·8
II	0–25	16·3	12·3	12·0	6·3	10·8	5·9	7·4	8·6	11·1	9·1	13·2	5·5	13·3	8·8	9·8	5·8
	25–50	19·8	24·3	16·5	12·5	12·0	12·9	13·1	15·3	12·0	15·5	15·3	13·7	16·3	11·9	16·5	14·3
	50–75	27·3	33·8	23·7	20·3	20·0	21·8	19·9	20·5	19·5	21·3	22·0	21·3	25·0	14·2	21·5	24·3
	75–100	35·3	40·8	37·5	28·0	24·5	28·0	29·5	30·5	26·5	29·0	30·0	27·0	33·0	20·5	28·8	32·5
	100–125	49·3	49·8	49·5	39·5	31·5	34·8	35·3	42·5	37·5	39·5	40·0	40·0	48·3	27·6	47·3	40·8
	125–150	60·5	63·8	67·7	58·0	41·0	52·8	45·3	55·3	54·0	49·0	53·7	52·3	56·8	40·4	60·3	52·3
	150–200	70·0	74·3	75·5	70·3	52·3	65·8	62·0	64·5	58·8	63·8	60·0	60·3	66·8	55·4	62·5	62·5
III	0–25	8·8	4·3	6·0	4·0	5·7	4·3	4·3	6·5	6·8	4·5	7·9	5·0	8·0	5·9	5·5	5·3
	20–50	13·5	10·0	9·8	6·5	6·4	6·8	7·3	7·7	8·0	7·2	9·5	7·0	8·0	9·0	10·0	7·8
	50–75	18·3	15·3	16·3	11·5	13·0	10·5	9·3	10·8	11·8	10·5	10·8	12·3	12·5	11·3	13·0	11·5
	75–100	24·8	21·8	21·0	16·3	16·0	15·8	15·8	16·8	17·5	16·0	19·6	16·0	37·3	17·6	19·5	16·5
	100–125	34·5	38·3	29·3	30·8	22·8	25·0	24·8	26·3	28·3	26·5	31·2	30·3	28·3	25·7	34·8	24·8
	125–150	48·0	54·5	42·3	48·3	38·5	44·5	40·8	49·3	41·3	41·8	46·0	47·0	42·5	43·5	48·5	39·0
	150–200	61·5	64·0	55·0	62·5	54·3	62·5	58·5	57·5	54·3	52·0	52·4	59·0	50·0	57·7	57·3	53·8

N.B. Gypsum was applied in field-plot I and in half of fields II and III.

Table 4

Influence of gypsum application on the quantity of leached salts (t/ha) in
Leziria Grande de V.F. Xira
(after Muralha and Pissarra [29])

Fields	Gypsum	Winter 76/77	Winter 77/78	Winter 78/79
II	Without	26	26	20
	With	35	30	28
III	Without	24	25	10
	With	34	37	17

preceding year, in all experimental fields, reaching, in the third year, 1/4 of the initial values in fields I and III and approximately 1/2 in field II. This fact is probably due to the soil salinity decrease and/or to a decrease in the efficiency of leaching (Van Hoorn [53]).

As to the period of 1981/82, that relation increased, probably due to the lack of leaching in 1979/80 and 1980/81, with consequent salt accumulation, decreasing once more in 1983/84 (Table 6).

Although these problems are common in fine textured alkali soils difficulties of exchange of sodium by calcium are not sufficient to explain them. Such facts lead to the consideration of management as a part of reclamation of saline and alkali soils.

Two main management questions can be mentioned: (a) The need to consider specific agricultural techniques—crops, cropping techniques, soil improvement measures—during a transition period and an operation and maintenance period after the initial reclamation period as a part of the reclamation itself (Dukhovny [10]; Bellas Rivera [6]), as has been planned for Leziria when considering alternatives and phases of development

Table 5

Amount of removed salt and drained water from November 1976 to April 1984
(after Costa and Pissarra [9])

Fields	Drained water (mm)					Removed salt (t/ha)				
	76/77	77/78	78/79	81/82	83/84	76/77	77/78	78/79	81/82	82/83
I	281	220	306	110	393	116	43	38	26	26
II	273	145	190	101	125	63	28	24	22	24
III	252	266	219	72	40	69	28	14	13	6

Table 6
Ratio S/Dr for the considered periods
(after Costa and Pissarra [9])

Fields	Years				
	76/77	77/78	78/79	81/82	83/84
I	0·41	0·20	0·12	0·24	0·13
I	0·23	0·19	0·13	0·22	0·19
III	0·27	0·11	0·06	0·18	0·15

(Pereira and Bos [33]; Reis *et al.* [39]). (b) The need to consider leaching requirements when irrigation is introduced (Martinez Beltran [27]; Hoffman *et al.* [19]; Jobes *et al.* [20]; Hoffman and Jobes [15]; Hoffman *et al.* [17, 18]).

Drainage effects on soil structure and porosity also affect water movements within a soil by acting on the hydraulic conductivity and by improving infiltration conditions. Related questions for salt-affected soils are also of importance (Martinez Beltran [27]), so they should be included when management is considered.

In the presence of salts hydraulic conductivity decreases as influenced both by the exchangeable sodium percentage (ESP) of the soil and the salt concentration of the percolating solution. ESP affects physical soil properties and water retention conditions, but even for high ESP values hydraulic conductivity is not affected when salt concentration of the water is above a critical (threshold) level, as discussed by Shainberg and Letey [43].

Therefore drainage and leaching success depends on chemical corrections and on ability of soil to release salts to the percolating water in such a way that the threshold level will be respected. Soil responses depend on soil type according to the nature and contents of dispersive and swelling materials.

Regarding movement of water in non-saturated soils, benefits of drainage and desalinization are evident because decrease of salt contents modifies suction–hydraulic conductivity relationships. So infiltration and water redistribution in the soil profile are favoured.

In consequence of improving soil salinity conditions and related physical characteristics of soils, agronomic qualities of soils become improved, namely risks of surface crusting and quality of seed beds. Such aspects have been evaluated in the Leziria Grande project, as shown in Table 7.

Table 7

Risks of surface crusting and quality of seed beds according to salinity conditions (adapted from Beek *et al.* [5])

Texture of soil surface	*ESP < 7%*		*ESP > 7%*	
	OM < 3%	*OM > 3%*	*OM < 3%*	*OM > 3%*
	Risks of surface crusting			
Sandy to sandy-loam	no	no	—	—
Sandy-loam to silty-loam	low	no	—	—
Silty-clay-loam	medium	low	high	—
Silty-clay-loam to silty-clay	high	medium	severe	high
	Quality of seed beds			
Sandy to silty-loam	very good	very good	—	—
Silty-clay-loam	good	very good	medium	—
Silty-clay-loam to silty-clay	medium	good	poor	medium

OM: organic matter.

2.4. Soil workability and trafficability

Soil moisture is a main factor determining timing of farming operations. A wet soil makes the equipment unable to operate for lack of support of traction, or farming operations are done in very bad conditions with poor results and inducing soil compaction and degradation of soil structure. So, as a result of excessive soil moisture, farming operations are delayed in relation to crop requirements, losses occur in production potential also as a consequence of a reduction on time for crop growth, and soil manipulation increases as a result of preceding operations under wet conditions and of consequent degradation of soil conditions (Reeve and Fausey [37]).

Workability is defined as the number of days in a given period suitable for field work. The workability depends on soil moisture in the upper layer of the soil, according to soil type. It is also influenced by microrelief as it may favour ponding. In general, workability should be defined in relation to rainfall conditions, so the probability of a given number of days suitable for field work is related to rainfall probabilities. Also, workability should be defined in relation to crops and to farming operations according to their specific requirements.

Trafficability describes the capability of a soil to permit the movement of a vehicle over the land surface (Reeve and Fausey [37]).

These authors stress the fact that trafficability in agriculture means being able to perform the required farm operations in such a way as to create a

Table 8

Workability classes according to soil surface texture, microrelief and watertable depth in Leziria Grande

(after Reis *et al.* [38])

Texture of soil surface	Microrelief	Watertable depth (cm)			
		>80	50–80	30–50	<30
Sandy to sandy-loam	Levelled	1	1	1	—
	Non-levelled	1	1	1	3
Loam to silty-loam	Levelled	1	1	2	3
	Non-levelled	2	2	3	3
Silty-clay-loam	Levelled	2	3	3	4
	Non-levelled	3	4	4	5
Silty-clay-loam to silty-clay	Levelled	3	4	4	5
	Non-levelled	4	4	5	5

desired soil condition or to get an operation done expediently. So trafficability is more difficult to be quantified and represents a condition of workability.

By improving soil moisture conditions drainage highly increases soil trafficability and workability (Van Wijk and Feddes [56]).

Workability has been studied for the Leziria Grande project (Beek *et al.* [5]; Reis *et al.* [38, 39]) according to soil, microrelief and watertable depth. The results, in Tables 8 and 9, show influences of soil according to texture of surface layers—better conditions in sandy to sandy-loam soils, to microrelief—better in levelled soils, and to watertable depth.

Table 9

Definition of the workability classes

(after Reis *et al.* [38])

Classes	Number of days necessary for reaching the soil moisture suitable for field operations						
	Oct	Nov	Dec	Jan	Feb	Mar	Apr
1	6	9	13	13	13	9	6
2	7	10–11	14–17	14–17	14–17	10–11	7
3	8	12–15	18–23	18–23	18–23	18–23	8
4	9–11	16–18	24–27	24–27	24–27	24–27	9–11
5	>11	>18	>27	>27	>27	>18	>11

In the project of Leziria Grande compaction risks have been evaluated instead of trafficability. It is of interest to both machinery and livestock effects on soil structure by increase of apparent density and decrease of porosity, on infiltration rate, on hydraulic conductivity of soil surface and on availability of water, so inducing problems on growth of roots, O_2 deficiency, risks of waterlogging, worse seed beds and increased workability problems (Beek *et al.* [5]).

Compaction risks have been divided into four classes according to the liquid limit of plasticity and they depend on watertable depth, soil texture and organic matter, as shown in Table 10.

2.5. Coupling irrigation and drainage

Needs for drainage are frequently associated with irrigation. On the one hand, given the nature of irrigated crops, intensive farming and high investments associated with irrigation, the need for avoiding waterlogging problems and related reduction of crop yields, for improving soil conditions and for timing of farm operations are greater in irrigated agriculture. On the other hand, it is necessary to solve specific problems relating to the use of irrigation water by controlling both the effects of irrigation return flows and the consequences of irrigation water quality.

The volume of *irrigation return flow* produced on a farm is the difference between the amount of water delivered from the water supply system and the amount of water held by the soil and consumed by the irrigated crop. It consists of two parts: surface run-off or tailwater produced during irrigation that can be added by stormwater run-off, and sub-surface drainage effluent resulting from deep seepage. The part of the applied water that percolates through the crop's root zone is called the leaching fraction

Table 10
Risks of compaction in Leziria Grande
(Beek *et al.* [5])

Texture of soil surface	*Organic matter*	*Watertable depth*	
		Below 50 cm	*Above 50 cm*
Sandy to silty-loam	>3%	low	low
	<3%	low	medium
Silty-clay-loam	>3%	low	medium
	<3%	medium	high
Silty-clay-loam to silty-clay	>3%	medium	high
	<3%	high	severe

Table 11
Water and salt balance of summer irrigated crops (totals)
(from Costa and Pissarra [9])

Field	Crop	Year	Irrigation water		Drainage water	
			Total (mm)	Salt added (t/ha)	Total (mm)	Salt extracted (t/ha)
I	Melon	1983	180	1	24	5
II	Maize	1984	600	5	144	16
III	Maize	1984	652	4	93	5

and flows into the sub-surface drainage system or adds to groundwater by vertical percolation (Ochs *et al.* [30]).

Drainage requirements for irrigated fields therefore depend on return flow characteristics, quality of irrigation water and soil salinity, all aspects in relation to irrigation methods and water management practices.

Table 11 shows the amount of salt applied through irrigation and that removed by the drain water.

Table 12 shows that, in view of the salinity noted in the soil, before and after irrigation, leaching really did take place in the cross-section down to the level of the drains (layer 0–100 cm) except in the superficial layer (0–25 cm) in fields I and II; that is not usual.

Surface return flow in humid and sub-humid climates is in general of good quality but it also reflects the quality of the applied water. In humid

Table 12
Electrical conductivity (EC 2:1 mmhos/cm) of samples collected in sub-blocks with
irrigated crops
(from Costa and Pissarra [9])

Depth	Field I 1983		Field II 1984		Field III 1984	
0–25	1·5	1·7	1·1	2·1	1·8	1·5
25–50	4·1	3·5	2·9	2·3	3·1	2·5
50–75	5·0	4·9	5·3	3·8	4·5	3·6
75–100	9·7	8·3	8·0	5·7	5·9	5·3
100–125	10·1	15·7	9·5	11·1	8·4	8·3
125–150	20·7	27·8	14·0	16·6	14·3	15·3
150–200	29·4	33·0	18·5	19·3	20·3	19·6

and sub-humid regions sub-surface return flows are generally of good quality and contain a minimum quantity of dissolved solids.

Problems easily occur in arid and semiarid climates. In these regions irrigation efficiency is a prime factor: not only irrigation water is a precious natural resource but risks of salinization through return flows, mainly sub-surface ones, are great. So higher irrigation application efficiency through irrigation scheduling, appropriate irrigation methods and related water management practices should lead to minimum losses and the occurrence of return flows, as well as for avoiding problems of soil salinity or river salinity coming from excess of irrigation water. Good examples of such solutions are given by Law and Skogerboe [22], Skogerboe *et al.* [47] and Van Schilfgaarde [55].

Once return flows are produced, however, they should be drained and reused or disposed of, according to quality. Some quality factors must be considered (Ochs *et al.* [30]).

Salinity. Surface run-off has generally only a slightly higher salinity than irrigation water. In general, it can be reused or disposed of in natural drainage water courses.

The salinity level of sub-surface return flows is considerably higher than that of applied water, which limit its reuse for irrigation and indicates problems of soil (re)salinization when drainage is insufficient.

Suspended sediments. Surface return flows and stormwater run-off may contain large amounts of sediment, which can limit reuse or disposal. Causes are normally related to poor soil management and poor water management practices under irrigation.

Nutrients. Phosphorus normally does not appear in sub-surface run-off as it is readily adsorbed on soil particles, but it can cause problems on disposal of surface return flows. Nitrogen, being very soluble, can be expected in both surface and sub-surface return flows. Nutrients are beneficial when water is reused for irrigation but are detrimental if discharged into a stream, thus promoting undesirable algae and aquatic weed growth.

Pesticides. Significant levels of pesticides can be found in surface run-off but normally degrade rapidly to less toxic compounds. As pesticides are readily adsorbed on the soil particles and so move only short distances, sub-surface return flows usually do not contain pesticides.

Summarizing, surface and sub-surface drainage waters from irrigated fields normally can be reused except when salinity risks occur. But drainage is at all times necessary to avoid risks of waterlogging from surface return flows or watertable rise due to vertical percolation or sub-surface return flows.

Regarding the Portuguese conditions available, results agree with the general aspects mentioned before. Data on quality of surface return flows presented by Alvim and Nunes [4] show that for most of the irrigation schemes no problems exist for reusing run-off waters, but few exceptions exist: waters from Campilhas, Vale do Gaio (Sado) and Roxo are of slightly poor quality (at least when stored volumes are reduced) inducing surface return flows which when used presents risks of soil salinization. Concerning sub-surface drainage return flows, only some data from Leziria and Mondego are available but concern the desalinization process. Nevertheless, data presented by Alvim [2], Alvim and Nunes [4] and Veloso *et al.* [58] show that sub-surface return flows cause salinization risks in soils with poor drainage conditions, mainly those with an impervious B horizon. Drainage and related soil improvement techniques are therefore necessary for some soils in Roxo, Campilhas, Sado, Caia and Loures Valley.

When salinity occurs it is necessary to include a leaching fraction in the irrigation volumes. Drainage is then necessary to carry out the leached salts in the sub-surface return flows. Otherwise leaching would not be accomplished in the entire profile, causing obvious problems.

Leaching requirements depend on crops, on soil characteristics, on salinity of applied water and on saline or sodic conditions of the soil. The present tendency is to improve leaching efficiency, together with improved irrigation methods and irrigation scheduling, so avoiding water losses and related problems (Hoffman *et al.* [19]; Jobes *et al.* [20]; Hoffman and Jobes [15]; Hoffman *et al.* [17, 18]).

Concluding, there is a need for coupling drainage and irrigation, not only when waterlogging problems occur but aiming at the drainage of surface and sub-surface return flows according to irrigation methods and to irrigation management techniques, namely related to leaching and quality management of irrigation waters.

3. IRRIGATION NEEDS IN PORTUGAL

Portuguese agricultural history is marked by our ancestors' skill in overcoming climatic, soil and water limitations, leaving us with wonderful

Table 13

Mean soil water balance. Averages (mm) and variation coefficients of annual rainfall values, of potential climatic evapotranspiration—ET (radiation), of real evapotranspiration, of water deficits and surplus, and of soil water (from Pereira and Paulo [34])

Hydrographic region	Rainfall	Evapo-transpiration		Water deficit		Water surplus		Soil moisture	
		Of reference	Real	Annual	Dry half-year	Annual	Wet half-year	April	June
Minho	1 360	1 106	695	412	408	664	653	69	19
	0·220	0·024	0·115	0·219	0·210	0·405	0·408	0·310	0·726
Douro 1	1 070	1 204	591	613	581	479	471	60	9
	0·229	0·038	0·164	0·217	0·240	0·559	0·560	0·420	0·673
Douro 2	551	1 262	465	800	749	86	86	37	3
	0·293	0·029	0·184	0·149	0·160	1·094	1·094	0·762	0·958
Vouga	1 104	1 176	634	542	508	469	460	62	12
	0·299	0·045	0·176	0·299	0·303	0·531	0·522	0·407	0·751
Mondego	1 108	1 210	638	572	541	458	450	62	11
	0·285	0·039	0·162	0·251	0·262	0·563	0·529	0·391	0·790
Estremadura	765	1 186	531	654	620	234	228	51	5
	0·285	0·028	0·158	0·149	0·164	0·706	0·693	0·460	0·716
Tejo 1	715	1 335	522	813	766	193	193	45	3
	0·252	0·034	0·141	0·137	0·142	0·739	0·739	0·494	0·626
Tejo 2	816	1 439	566	835	787	250	250	50	4
	0·269	0·039	0·141	0·156	0·163	0·701	0·701	0·417	0·828
Tejo 3	711	1 401	535	904	845	176	176	43	2
	0·270	0·044	0·157	0·147	0·150	0·795	0·795	0·573	0·942
Sado	569	1 402	459	942	877	110	110	35	2
	0·277	0·039	0·170	0·130	0·138	0·986	0·986	0·653	1·302
Mira	539	1 317	408	891	833	114	114	36	2
	0·281	0·041	0·161	0·132	0·137	0·985	0·985	0·573	1·556
Guadiana	568	1 521	482	1 040	969	86	86	32	1
	0·284	0·042	0·176	0·134	0·141	1·043	1·043	0·916	1·254
Algarve	504	1 526	443	1 083	999	60	60	22	1
	0·269	0·025	0·158	0·090	0·073	1·178	1·178	0·729	1·150

rural and agricultural landscapes, completely adapted crop systems and a technique able to maximize our available resources.

In recent studies (Henriques [14]; Pereira and Paulo [34]) the sequential monthly water balance was computed for Continental Portugal relative to the period 1951–52 to 1960–61, the results of which are summarized in Table 13, taking as aggregation units the catchment areas represented in Fig. 1.

Table 13 leads to some important conclusions: (a) water deficits and surplus coincide with the dry and wet half-year, respectively; (b) water deficit concerning the dry half-year is very important, with the lesser values registered on north-west (Minho) and on western coastland zone (Douro 1,

FIG. 1.   Hydrographic regions of Continental Portugal.

Vouga, Mondego and Estremadura) and the highest values on inland and south regions (Guadiana and Algarve); and (c) water surplus that represents the potential for run-off presents an inverse spatial distribution, with higher values on western coastland (Minho, Douro 1, Vouga and Mondego) and lower values on south and inland regions, with minimum values on Guadiana and Algarve catchments.

As water is extracted from soil by plant roots and by evaporation, without rainfall and/or irrigation replenishment soil is a limited water reservoir in the root zone. In Portugal the dry half-year is an important cropping season, so owing to the water deficit there is a need to improve water availability in the soil root zone through irrigation. But according to the physical and social characteristics that govern the operation of irrigation systems, irrigation water must be optimally distributed and used, just to meet plant requirements and possible leaching needs, in order to avoid over-irrigation and their results, water waste and salinity risks.

The estimation of irrigated areas (despite including dry farming crops in the rotation) and its relative importance on total arable land (Table 14) show clearly that irrigation was developed where water surplus favoured the existence of shallow groundwater as much as surface water that could be caught without inter-annual regularization works. On the other side, irrigation development was proceeded with where water deficits were less important, that is where it was easier to get a higher irrigation efficacy. In south and inland regions, on the other hand, water requirements are greater and there is only plentiful water when deep groundwater is pumped or when dams are utilized, annually or more frequently.

Table 14
Agricultural and irrigated areas in Continental Portugal

Hydrographic catchment	Agricultural area (1000 ha)	Irrigated area	
		(1000 ha)	(%)
Minho	238·9	132·4	55·4
Douro 1	405·9	147·7	36·4
Douro 2	319·5	23·0	7·2
Douro 3	193·1	26·4	13·7
Vouga	165·0	67·3	40·8
Mondego	276·1	82·2	29·8
Estremadura	174·0	24·2	13·9
Tejo 1	246·3	29·8	12·1
Tejo 2	505·3	67·0	13·3
Tejo 3	750·1	64·4	8·6
Sado	544·1	46·8	8·6
Mira	137·2	8·0	5·8
Guadiana	954·3	41·9	4·4
Algarve	209·8	24·2	11·7
Continent	5 159·7	758·1	14·7

Adapted from Henriques [14].

4.   DRAINAGE NEEDS RELATED TO PORTUGUESE SOILS

The need for drainage in Portugal, especially in the Alentejo region, has been reported by several authors (Silva *et al.* [45]; Teixeira [49, 50]; Vasconcelos [57]). Back in 1945 the late Professor Ruy Mayer [28], in his book on the technique of irrigation, stated that reclamation and drainage in our country were of prime importance because of the climatic pattern of Continental Portugal, the disorderly regime of our surface water courses and the influence that this regime has on underground waters.

The need for both drainage and irrigation is indeed likely depending on the type of soil and the nature of the crop to be grown. It is therefore intended to provide an overview of the soils where drainage needs are more strongly felt and to bring out the special features that make drainage necessary.

For the sake of convenience the soils with drainage problems in Portugal are divided into three groups.

In Group I are put the soils that have a permanent watertable or suffer from temporary waterlogging. They cover about 2900 km^2 and come up to around 3·3% of the total area.

In Group II are placed most of the soils of alluvial origin. They account for about 3500 km^2 and make up nearly 3·9% of the total surface.

In Group III are considered the soils with internal drainage problems due to peculiar characteristics within the pedon. They occupy around 1400 km^2, which means over 1·6% of the total area.

The overall location of these groupings is given in Fig. 2. Although small areas of other soils with eventual problems of excess water or salinity hazard have not been accounted for, it may be said that roughly 7800 km^2, corresponding to about 8·8% of the territory, may be in need of drainage of some sort or other. This may not seem much in absolute figures but it so happens that at least a fifth of the area of Continental Portugal is made up of soils practically unfit for agriculture and a further large percentage can hardly be considered good for farming. On the other hand, the 8·8% under consideration include some of the most promising soils for agricultural purposes. Elimination of their drainage problems may considerably improve agricultural yields in most of them.

4.1. Group I

The soils of this group are mostly in flatlands or slightly concave topography. Their main problem is position in the landscape, which makes it very difficult for surplus waters to find an overland or underground outlet

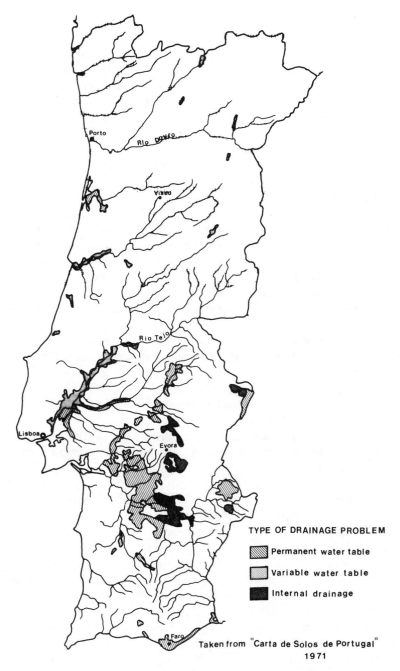

FIG. 2. Soils of Portugal with drainage problems.

so that water accumulates within the pedon during some part of or the whole year. Consequently hydromorphism, whether light or severe, is their most common characteristic. In order of areal importance this group includes Albo-gleyic Luvisols, Gleyic Solonchaks, Gleyic Podzols, Gleyic Luvisols, Eutric Planosols, Histo-humic Gleysols and Gleyic-calcic Luvisols.

Argillic horizons are common to Luvisols and Planosols, while incipient to fairly developed albic horizons appear in Albo-gleyic Luvisols and are well developed in Planosols.

Presence of hydromorphic characteristics within the pedon due to a permanent or temporary watertable imply the necessity for drainage for all the soils included in this group. Many detailed surveys have pointed this out (Teixeira [49]; Marado [26]; Rosa [40]; Constantino et al. [8]; etc.). Removal of excess water will in most cases be enough to provide the necessary balance between water and air space but internal pedon characteristics may sometimes create further difficulties. At least two cases merit special attention: compact argillic horizons and presence of salts in the pedon.

Compact argillic horizons generally show very slow permeability. Downward water flow through them being very slow a short, fairly heavy or protracted low intensity rainfall is enough to give rise to perched watertables on top of these horizons. Where the relief is concave the problem is aggravated by the inflow of water from neighbouring areas. An extreme case is the Planosols.

It is not known whether mechanical disturbance of this horizon will provide a solution since abnormally high amounts of magnesium and sodium have been found in some of them (Madeira and Medina [24]). An example of space analysis following Van de Goor [52], given in Fig. 3, shows that problems may arise below 20 cm and that they are very serious below 50 cm. A different type of profile is shown by the example of Fig. 4. Here the structure of the top horizon seems to have suffered severe damage probably due to cultivation, while another air space deficient horizon is present between about 40 and 60 cm corresponding to the argillic horizon.

When salts are present within the pedon adequate analytical work should always precede drainage. Danger may be envisaged on two fronts: alkalization and cat-clay formation. Those soils that contain carbonates are likely not to require special attention. However, most of the Solonchaks are formed on marine sediments and some of them do have a large proportion of sodium on the exchange complex, while small areas of cat-clays have also been detected. Consequently, where carbonates are absent

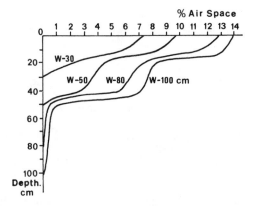

FIG. 3. Influence of groundwater depths (*W*) on soil air space (%). Gleysol, Cova da Beira.

from the soil corrective measures will have to be taken for their reclamation, for instance by the addition of gypsum. Where alkalization has already taken place there may be serious permeability problems due to colloid deflocculation induced by the high sodium content in the exchange complex. Nonetheless, recent studies show that reclamation of most of them, including some that are temporarily under the influence of seawater, is possible (Alvim and Serpa [3]; Alvim [1]). Table 15 shows results obtained in some experimental drainage fields.

However, in some cases the costs involved may be too high whereas in

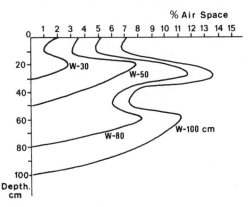

FIG. 4. Influence of groundwater depths (*W*) on soil air space (%). Gleyic Luvisol, Alentejo.

Table 15
Levels of salinity (0·50 m deep layer) in October 1976 and April 1984 in Leziria
Grande
(from Costa and Pissarra [9])

Fields	October 1976		April 1984	
	Ec (mmhos/cm)	Classification	Ec (mmhos/cm)	Classification
I	21·5	Extremely saline	5·8	Moderately saline
II	18·0	Extremely saline	10·0	Highly saline
III	11·2	Highly saline	6·5	Moderately saline

others it is thought that they should not be reclaimed for agricultural
purposes but rather kept as a sanctuary for a specific type of fauna which
otherwise will probably face the risk of extinction.

Air space analysis is given in Fig. 5 for one soil of the Tagus River Valley.
It is a fine textured (silty clay) soil with high salt content (Ec of saturation
extract above 5 mmhos/cm) and sodium in the exchange complex above
20%. As would be expected aeration is deficient throughout the profile but
air space deficiency is further aggravated between about 30 and 45 cm,
probably due to the formation on a plough pan.

4.2. Group II
From the point of view of drainage this is a heterogeneous grouping
ranging from soils practically in no need of drainage to those where

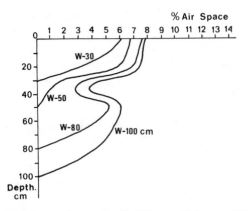

FIG. 5.   Influence of groundwater depths (*W*) on soil air space (%). Solonchak, V.F.
Xira.

drainage is essential. It includes the Fluvisols as well as the Fluvi-eutric Gleysols.

The soils in this group are under the influence of an oscillating watertable and being in more or less flatlands some may suffer from temporary waterlogging even in the surface horizon. They present a varied morphology running a whole range of textural classes, and in the same pedon there may be layers with contrasting textures. They do not have genetically well-marked horizons. Some have carbonates. In addition, sub-surface gleying is typical of Fluvi-eutric Gleysols.

Drainage problems connected with the soils included in this group are mostly related to their oscillating watertable. The lowering of this watertable will normally be enough to push them towards increased agricultural production.

However, in some cases there may be difficulties attached to unfavourable textures and/or unfavourable physical conditions through-out or in some parts of the pedons. Soils without undue amounts of salts or of sodium in the exchange complex, even when limed, may still present highly water-unstable aggregations (Silva [44]). Corrective measures will have to be taken in such cases for drainage to be effective. Figure 6 shows the air space analysis of another such soil from the Tagus Valley. Its surface horizon indicates reduced air space which may have been induced by unfavourable tillage operations. The latter may also be responsible for the further air space reduction at about 20 cm due to the formation of a plough pan.

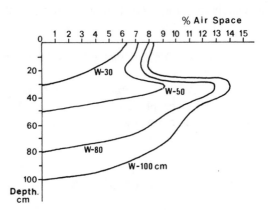

FIG. 6.   Influence of groundwater depths (W) on soil air space (%). Eutric Fluvisol, V.F. Xira.

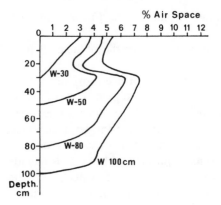

FIG. 7.   Influence of groundwater depths (*W*) on soil air space (%). Vertic Luvisol, Évora.

4.3. Group III
The soils in this grouping have some sort of drainage constraint within the pedon. Foremost among them are Vertic Luvisols, Pellic Vertisols and Calcaro-pellic-Vertisols.

The argillic horizons of Vertic Luvisols, probably due to their admixtures of smectite clays, slow down downward movement of water and so just like some soils of Group I they give rise to perched watertables under short, fairly heavy rainfall or under low intensity rains of long duration. The persistence of such watertables fairly near the surface for a significant period of time is likely to cause severe damage to crops. On the other hand, in years of lower rainfall with long breaks they act as water reservoirs that compensate to some extent the lower water-holding capacity of the surface horizons. In the example of air space analysis presented in Fig. 7 even the generally light-textured surface horizon shows a highly deficient aeration. This is probably due to structure degradation as a result of bad cultural practices, as indicated by the plough pan at about 20 cm. In general, however, it is the argillic horizons that act as a big handicap within the pedon in these soils.

5.   CONCLUSIONS

It may therefore be said that in Continental Portugal, due to its specific type of climate and particular kind of relief, there are appreciable areas that are in need of irrigation and/or drainage. This may be partially responsible for the Portuguese agricultural deficit.

Generally the need for drainage is due to a permanent or temporary watertable whose lowering may sometimes be enough to solve the problem. But there are cases where complementary measures will have to be taken due to special pedon characteristics.

Important among soils with special characteristics are those with compact argillic horizons which give rise to perched watertables. Recent studies suggest that at least in some cases their unfavourable physical conditions may be due to a peculiar chemical status. However, the scarce data available on the genesis of such argillic horizons preclude any definite conclusions as regards actions to be taken towards their amelioration.

Another special case is that of saline and sodic soils and soils in danger of salinization and sodization. Their drainage needs the utmost caution. Besides it may not always be easy due to added unfavourable pedon characteristics.

Vertisols lying in flatlands with difficult external drainage conditions may also present aeration problems for good plant development because of their excessive microporosity induced by significant quantities of montmorillonitic clay.

Finally, there is a need for more experimental work overall, for better understanding of crop growth, development and yield in such different environments.

REFERENCES

1. ALVIM, A. J. S. (1976). Reconhecimento dos Sapais e Salgados de Castro Marim-Vila Real de Santo António. *Pedologia, Oeiras*, **11**(2), 1–99.
2. ALVIM, A. J. S. (1980). Qualidade da água e riscos de salinzação nos perímetros de Campilhas e de Roxo. In: *Planeamento e Aproveitamento dos Regadios* (Ed. L. S. Pereira), DGHEA/Ordem dos Egenheiros, Lisboa.
3. ALVIM, A. J. S. and SERPA, A. M. (1976). Reconhecimento dos Sapais e Salgados do Arade. *Pedologia, Oeiras*, **11**(1), 49–133.
4. ALVIM, A. J. S. and NUNES, J. T. (1984). Qualidade da água e risco de halomorfização do solo nos perímetros de rega do Sul, 1980 a 1982. DGHEA Publ. no. D.78.83, Lisboa.
5. BEEK, K. J., THIADENS, R. A. H., REIS, L. F. C. L., PERDIGÃO, M. V. N. L. and PERDIGÃO, A. M. M. (1980). Projecto da Leziria Grande de Vila Franca de Xira—Avaliação de Terras. DGHEA Publ. R.16.7, Lisboa.
6. BELLAS RIVERA, R. (1984). Underdrainage as means to reclaim saline and alkali soils. Basic information about the low Guadalquivir river marshes. In: *Twelfth Congress on Irrigation and Drainage (Fort Collins, May 1984)*, ICID, Vol. I(B), pp. 571–86.

338   L. S. Pereira, E. C. Sousa and L. A. Pereira

7. CAAPLAR, K. and FEHER, F. (1984). Impacts of surface and subsoil drainage on crop yields and on the physical parameters of cohesive soils. In: *Twelfth Congress on Irrigation and Drainage (Fort Collins, May 1984)*, ICID, Vol. I(B), pp. 729–42.
8. CONSTANTINO, A. T. *et al.* (1977). *Carta dos Solos da Lezíria Grande. Estudo detalhado*, Junta de Hidráulica Agrícola, Lisboa.
9. COSTA, A. P. G. and PISSARRA, A. E. (1985). Drainage experiment data in the Leziria Grande. Interest of their applicability to other alluvial blocks in the Tagus Valley. In: *Seminar on Drainage*, GPIS/DVKW/DGRAH/ISA, Lisbon.
10. DUKHOVNY, V. A. (1978). Saline and alkali soils. Their use, improvement and related problems. In: *State-of-the-Art Irrigation, Drainage and Flood Control*, ICID, New Delhi, no. 1, pp. 391–458.
11. DUKHOVNY, V. A. (1981). Drainage depths effects on the salinization of soil. In: *State-of-the-Art Irrigation, Drainage and Flood Control*, ICID, New Delhi, no. 2, pp. 153–84.
12. EGGELSMAN, R. (1984). Subsurface Drainage Instructions DVWK Bulletin 6, Verlag Paul Parey, Hamburg/Berlin.
13. FROES, J. V. A., PAULO, V. C. and FERNANDO, R. M. C. (1982). Estudo hidropedológico da Várzea de Loures. Rel. Estágio Curso Eng. Agr., ISA, Lisboa.
14. HENRIQUES, A. G. (inv. resp.) (1983). Avaliação dos Recursos Hídricos de Portugal Continental. Contribuição para o Ordenamento do Território. Instituto de Estudos para o Desenvolvimento, colab. Hidrosistemas, Hidroprojecto, Lisboa.
15. HOFFMAN, G. J. and JOBES, J. A. (1983). Leaching requirement for salinity control. III. Barley, cowpea and celery. *Agric. Wat. Manag.*, **6**, 1–14.
16. HOFFMAN, G. J. and MEYER, J. L. (1982). Reclamation of salt affected soils in California. *First Thematic Conference on Remote Sensing of Arid and Semi-Arid Lands* (Cairo, Egypt, Jan. 1982), pp. 147–59.
17. HOFFMAN, G. J., OSTER, J. D., MAAS, E. V., RHOADES, J. D. and VAN SCHILFGAARDE, J. (1984). Minimizing salt in drain water by irrigation management—Arizona field studies with citrus. *Agric. Wat. Manag.*, **9**, 61–78.
18. HOFFMAN, G. J., OSTER, J. D., MAAS, E. V., RHOADES, J. D. and VAN SCHILFGAARDE, J. (1984). Minimizing salt in drain water by irrigation management—leaching studies with alfalfa. *Agric. Wat. Manag.*, **9**, 89–104.
19. HOFFMAN, G. J., RAWLINS, S. L., OSTER, J. D., JOBES, J. A. and MERRIL, S. D. (1979). Leaching requirement for salinity control. I. Wheat, sorghum, and lettuce. *Agric. Wat. Manag.*, **2**, 177–92.
20. JOBES, J. A., HOFFMAN, G. J. and WOOD, J. D. (1981). Leaching requirement for salinity control. II. Oat, tomato, and cauliflower. *Agric. Wat. Manag.*, **4**, 393–407.
21. KOPP, E. (1980). Concepção de ensaios e experimentação para a solução de problemas hidráulicos, pedológicos e de técnica de produção do Vale do Mondego. In: *Desenvolvimento Agrícola do Baixo Mondego* (Ed. L. S. Pereira), DGHEA Publ. M.13.80, Lisboa, pp. 35–45.
22. LAW, J. P. and SKOGERBOE, G. V. (Eds) (1977). Irrigation Return Flow Quality Management. Proceed. Nat. Conference, Colorado State University, Fort Collins, CO, USA.

23. MAAN, M., PISSARRA, A. and VAN HOORN, J. W. (1982). Drainage and desalinization of heavy clay soil in Portugal. *Agric. Wat. Manag.*, **5**, 227–40.
24. MADEIRA, M. A. V. and MEDINA, J. M. B. (1982). Contribuição para o Conhecimento da Influência da Natureza do Complexo de Troca na Permeabilidade de Solos Argiluviados Pouco Insaturados. Pedon, Lisboa, 4, pp. 3–26.
25. MAIANU, A. (1984). Twenty years of research on reclamation of salt-affected soils in Romanian rice fields. *Agric. Wat. Manag.*, **9**, 245–56.
26. MARADO, M. O. B. (1973). Solos e Capacidade de Uso do Solo das Propriedades de Vairão da Estação Agrária do Porto, Boletim de Solos, Lisboa, 14, pp. 92–151.
27. MARTINEZ BELTRAN, J. (1978). Drainage and Reclamation of Salt-Affected Soils. Bardenas Area, Spain. ILRI Publ. 24, Wageningen.
28. MAYER, R. (1945). A Técnica do Regadio. Sá da Costa, Lisboa.
29. MURALHA, J. C. and PISSARRA, A. (1984). A case of drainage of fine textured and saline soils. In: *Twelfth Congress on Irrigation and Drainage (Fort Collins, May 1984)*, ICID, Vol. I(B), pp. 849–65.
30. OCHS, W. J., WILLARDSON, L. S., CAMP, C. R., DONNAN, W. W., WINGER, R. J. and JOHNSTON, W. R. (1980). Drainage requirements and systems. In: *Design and Operation of Farm Irrigation Systems* (Ed. M. E. Jensen), ASAE Monograph no. 3, Am. Soc. Agric. Eng. St. Joseph, MI, pp. 235–77.
31. PEREIRA, L. S. (Ed.) (1980). Drenagem e Conservação do Solo para a Agricultura de Sequeiro Alentejana. DGHEA, Lisboa.
32. PEREIRA, L. S. (1985). The need for drainage with reference to Portuguese conditions. In: *Seminar on Drainage*, GPIS/DVKW/DGRAH/ISA, Lisbon.
33. PEREIRA, L. S. and BOS, M. G. (1982). Land reclamation and agricultural development of the Leziria Grande, Portugal. In: *Polders of the World*, ILRI, Wageningen, pp. 185–206.
34. PEREIRA, L. S. and PAULO, V. C. (1984). Necessidades de Água para a Rega em Portugal Continental. Avaliação da Situação Actual e Previsão da sua Evolução Futura. In: *Simpósio sobre o Desenvolvimento do Regadio*, APRH, Lisboa.
35. PINTO, M. C. (1983). Projecto de Drenagem e Conservação do Solo no Alentejo. Relatório de actividades 1982. DGHEA, Lisboa.
36. RAMOS, F. M. (1980). Contribuição para a caracterização física dos solos do Vale do Mondego. Estudo preliminar. In: *Desenvolvimento Agrícola do Baixo Mondego* (Ed. L. S. Pereira), DGHEA, Publ. M.13.80, Lisboa, pp. 75–126.
37. REEVE, R. C. and FAUSEY, N. R. (1974). Drainage and timeliness of farming operations. In: *Drainage for Agriculture* (Ed. J. Van Schilfgaarde), ASA Agronomy Monograph no. 17. Am. Soc. Agron., Madison, WI, pp. 55–66.
38. REIS, L. C. L., PERDIGÃO, A. M. M. and PERDIGÃO, M. V. L. (1982). Avaliação físico-quantitativa das terras do Projecto da Leziria Grande (Sistema FAO). Parte II— Previsão de evolução com o projecto. Simp. 'A Bacia Hidrográfica Portuguesa do Rio Tejo'. APRH, Lisboa.
39. REIS, L. L., PERDIGÃO, A. M. and PERDIGÃO, M. V. (1982). Land evaluation for agricultural development in Leziria Grande, Portugal. In: *Polders of the World*, ILRI, Wageningen, Vol. 2, pp. 254–66.
40. ROSA, M. P. (1973). Estudo Preliminar da Aptidão para o Regadio dos Terraços

do Rio Sor entre Ponte de Sor e Montargil. Boletim de Solos, Lisboa, 15, pp. 58–69.

41. SANDU, G. and RAUTA, C. (1976). Amélioration des Sols Salins-Alcalins dans la Vallés du Calmatzui. In: *Drainage des Sols Salins*, Bull. FAO Irrig. Drain. no. 16, FAO, Rome, pp. 17–30.

42. SCOTT RUSSEL, R. (1977). *Plant Root Systems: their Function and Interaction with Soil*, McGraw-Hill, London.

43. SHAINBERG, I. and LETEY, J. (1984). Response of soils to sodic and salinity conditions. *Hilgardia*, 52(2), 1–57.

44. SILVA, A. A. (1959). Comportamento Físico de um Aluvião de Alvalade (Sado) e sua Correcção. Estudo Preliminar. *Agronomia Lusitana, Oeiras*, 21(2), 135–54.

45. SILVA, A. A., GARCIA, J. S. and RODRIGUES, J. L. (1965). Alguns Aspectos da Drenagem dos Campos de Évora. Seu Estudo no Posto Experimental de Évora. *Agronomia Lusitana, Oeiras*, 27(2), 155–71.

46. SKAGGS, R. W. (1982). Optimizing drainage system design for corn. In: *Advances in Drainage*, ASAE Publ. no. 12-82, Am. Soc. Agric. Eng., St. Joseph, MI, pp. 50–61.

47. SKOGERBOE, G. V., WALKER, W. R. and EVANS, R. G. (1978). Environmental Planning Manual for Salinity Management in Irrigated Agriculture. US Environmental Protection Agency, Ada, OK.

48. SOUSA, E. C. and PERDIGÃO, A. M. (1985). Soils with drainage problems in Portugal. In: *Seminar on Drainage*, GPIS/DVKW/DGRAH/ISA, Lisbon.

49. TEIXEIRA, A. J. S. (1969). Nota Preliminar sobre Drenagem de Terras Agrícolas. O Caso Particular do Alentejo. *Pedologia, Oeiras*, 4(1), 1–11.

50. TEIXEIRA, A. J. S. (1969). Land Drainage in Portugal. *Pedologia, Oeiras*, 4(2), 163–6.

51. THIADENS, R. A. (1981). Projecto da Leziria Grande de Vila Franca de Xira— Soil Survey and Land Evaluation. DGHEA Publ. R.16.19.81, Lisboa.

52. VAN DE GOOR, G. A. W. (1974). Drainage of rice fields. In: *Drainage Principles and Applications*, ILRI, Publ. 16, Wageningen, Vol. IV, pp. 341–82.

53. VAN HOORN, J. W. (1981). Salt movement, leaching efficiency and leaching requirement. *Agric. Wat. Manag.*, 4, 409–29.

54. VAN HOORN, J. W. (1984). Salt transport in heavy clay soil. In: *Symposium on Water and Solute Transport in Heavy Clay Soils*, ILRI, Wageningen, pp. 92–102.

55. VAN SCHILFGAARDE, J. (1982). The Welton-Mohawk dilemma. *Water Supply and Management*, 6(1/2), 115–27.

56. VAN WIJK, A. L. M. and FEDDES, R. A. (1982). A Model Approach to the Evaluation of Drainage Effects. ICW Technical Bulletin 15, Inst. Land Wat. Manag. Res., Wageningen.

57. VASCONCELOS, F. C. (1969). Carta de Drenagem—Breve Notícia Explicativa. Boletim de Solos, Lisboa, 4, pp. 40–7.

58. VELOSO, M. M., NUNES, J. M. T., GIL, A. S. and MATEUS, M. I. F. (1984). Risco de salinização pela água de rega na Várzea de Loures. *Recursos Hídricos*, 5(3), 3–16.

59. VIEIRA, D. B., PISSARRA, A. E., POMBO, J. G. and PEREIRA, L. S. (1982). Drainage and desalinization studies and design in Leziria Grande, Portugal. In: *Polders of the World*, ILRI, Wageningen, Vol. I, pp. 384–95.

60. WESSELING, J. (1974). Crop growth and wet soils. In: *Drainage for Agriculture* (Ed. J. Van Schilfgaarde), ASA Agronomy Monograph no. 17. Am. Soc. Agron., Madison, WI, pp. 7–37.
61. WIDMOSER, P. (1984). Re-salinization after leaching of coastal soils under Mediterranean climate. In: *Twelfth Congress on Irrigation and Drainage (Fort Collins, May 1984)*, ICID, Vol. I(B), pp. 587–96.

Impact of Recreational and Tourist Activities on the Soil Environment

Costas A. Cassios

Dept. of Surveying and Rural Planning, National Technical University, Athens 15710, Greece

SUMMARY

Recreational and tourist activities cover large areas of different environments ranging from strict nature reserves up to overdeveloped tourist resorts. Soil environment and its characteristics such as erodibility, trafficability, texture, constitute basic environmental variables which must be carefully examined during the planning stage for recreational activities and facilities. The above soil characteristics must be co-examined within recreational and tourist areas with slopes, drainage, flooding conditions, etc. Lack of such analysis could cause severe soil impacts later from use such as: erosion, compaction, loss of organic layers, soil microfauna destruction, soil contamination, etc. More severe impacts occur when slopes are above 25% and the soil is of fine texture. Managerial and engineering measures could be taken in order to prevent and correct soil impacts on recreational and tourist areas. Research on a European level is needed in order to produce soil suitability maps for recreational and tourist activities and facilities.

1. INTRODUCTION

Recreational and tourist activities are human activities which take place on certain naturally attractive areas, during the free leisure time of the people [3, 8].

Leisure time nowadays has become as equally important as other activities, and dominates our lifestyle.

343

With more free time available, higher mobility and more sophisticated recreational equipment the demand for outdoor recreation and tourism from the public is constantly increasing.

In the past, most recreational activities took place on natural areas with outstanding features or on sites with historic and cultural attractions. Today due to the larger numbers of users and the greater demand, more sites are in use for recreation with lesser environmental and natural attraction characteristics.

A result of that, is we face today greater problems of site deterioration and an irreversible impact on the location's soil environments. Soil is a very important variable for the recreational and tourist areas. Its suitability for recreation secures the vegetational longevity and the sustained environmental quality of the recreational and tourist sites.

In order to identify the impacts generated from overuse or bad planning of the soil environment, it is necessary to approach the problem from three different angles:

1. From the type of recreational and tourist areas, i.e. park, outdoor recreational site, nature reserve, tourist complex, etc.
2. From the activities which are taking place in the recreational and tourist regions, by type and intensity.
3. From the time of the year when the areas are mostly in use.

Certain assumptions had to be made in order to avoid the overexpansion of the above topic, such as:

(a) That for recreational areas those are considered which could maintain outdoor recreational activities—mainly forested areas.
(b) That for tourist areas the interest will be concentrated on sensitive environmental areas such as seashores, islands and mountainous regions.

2. SOILS IN RECREATIONAL AND TOURIST AREAS

Soil is an important environmental variable and should be always carefully examined prior to any recreational or tourist development.

It participates in and contributes to the site's environmental quality since it supports vegetation, water regime, sewage, building foundations, etc. Certain soil types are suitable for recreation while other types require proper management and sometimes engineering treatment. The properties which should be examined for soil suitability for recreation according to Lynch and Hack [9], Rutledge [14], Cassios [1], US Forest Service [15], and several other researchers especially from the US and Canada are the following:

1. Soil erodibility.
2. Soil texture.
3. Soil drainage.

Variables directly related to the soils for recreational and tourist areas are:

1. Slope and exposure conditions of the recreational areas.
2. Climatic conditions for the region.
3. Flooding conditions.
4. Type and degree of density of the vegetation.

According to the soil classification for recreational uses by L. Partain [12] there are four types of soils suitability:

(i) Soils without problems—suitable—no limitations.
(ii) Soils with moderate problems. Slight soil limitations.
(iii) Soils with problems—less severe—soil limitations.
(iv) Soils with serious problems—severe—soil limitations.

The degree of soil suitability depends upon the degree of the combination of the above mentioned variables, their values and severity. As an example soils in recreational areas with slopes more than 15% and in a rainy region with clay soils—bad drainage—could be classified as 'severe' and thus are unsuitable for recreational uses.

The above soil categories vary according to the recreational activities, i.e. camping, picnicking, horse riding, etc., regarding their suitability and the degree of deterioration, something which is related to the intensity of site use by recreationists.

As a general remark it could be stated, that even with severe soil deterioration several months during the year without use, for outdoor recreational areas would contribute to a natural recovery. Close monitoring of the soil physical condition in the recreational areas, helps the managers to decide when to stop the site from being used and directs them on what technical and engineering actions should be taken in order to assist the soil environment to recover.

Soil types suitable for recreational uses are those which are: well drained and excessively drained, coarse textured, not subjected to flooding and up to a maximum slope of 25%. Poorly drained soils with fine texture or very stony and subject to flooding and more than 25% slopes are not suitable.

3. RECREATIONAL AND TOURIST AREAS

Recreational and tourist areas are those areas which in one way or another attract people for short visits to spend their leisure or vacation time and provide recreational and accommodation facilities.

Earlier it was pointed out, that forested or natural areas are those which can be used for recreation and this environment suffers of course at the hands of the users.

Tourist outdoor recreational areas could be categorised according to their degree of development and uses as following:

I. Recreational areas:
  1. Nature reserves.
  2. National parks.
  3. Natural parks.
  4. Regional parks.
  5. Aesthetic forests.
  6. Recreational parks.
  7. Wilderness areas.
  8. City parks.
  9. Neighbourhood parks.
II. Tourist areas:
  10. Ski resorts.
  11. Summer house resorts.
  12. Historic sites.
  13. Therapeutic or spa sites.
  14. Sea water oriented areas.

Each of the above recreational areas according to the accepted criteria of the International Union of Nature Conservation (IUCN) must have certain minimum environmental chracteristics and certain levels of maintenance.

The Council of Europe accepting the above criteria has since 1962 established the European Diploma—an award granted to the most outstanding European natural wonders—and recognises four natural recreational areas as Category A, B, C and D.

Among the basic established criteria is the soil environment, its properties and characteristics which must be maintained in order that the site can belong to one or the other recreational category.

The different recreational areas according to their natural status and their special location can withstand certain recreational activities and a certain degree of use without deterioration of their natural character.

However for sites such as city parks and neighbourhood parks the environmental variables could be modified or be engineered, because of their role which sometimes is not resource oriented but social or economic [7].

For tourist areas where more intensive land uses and environmental alterations take place, because of their needs for permanent facilities, soil environment is examined for its suitability for sewage, foundations and landscape development.

Usually the tourist areas are developed with economical orientation. The result of such site selection is often to have environmental problems.

Overuse of the sites creates erosion problems, losses of soil fertility, run-off and quite often environmental deterioration.

Areas such as those in Greek and Mediterranean islands or on mountainous areas for ski resorts have sometimes suffered severe soil environment destruction. The reason for that is the insufficient soil examination prior to the tourist site development and since economics dominate tourist installations, maintenance has been kept minimal.

In general tourism develops besides economic welfare, a series of other environmental and social impacts which in the long run, if proper planning and management is not carried out, could work against the region.

The example of the Greek islands or South Italy is useful in order to draw some conclusions about the danger of tourism on the environment.

In most of those islands, because of tourist needs an extensive change in land uses took place. Farmlands which were sometimes extremely fertile changed character and were used for building tourist facilities. The result was that agricultural products once produced abundantly locally now, because of lack of agricultural land, have to be imported from the mainland. Since tourism requires seasonal working and pays very well, traditional farmers gave up farming and now work in tourist enterprises. As a result farmlands were abandoned and thus became idle maqui and guarique vegetation lands suitable only for grazing and thus losing their fertility. Soil erosion from run-offs and winds, now occurs very often and an irreplaceable valuable resource, the soil, is almost lost. The water regime in those dry regions, because of the high requirements of tourism for water became scarce.

Finally the demand for summer houses in those tourist islands and because of the existing legislation that makes it more easy to build on farmlands than forested ones resulted in the elimination of the agricultural soils.

And while the problems for the seashore regions because of tourism are continuously increasing, the problem is no less serious for the mountainous areas used for winter sport development.

The need for access in high mountain regions requires the development of roads, highways, ski routes, telephonic installations, etc., and it is difficult to estimate the damage to soil environment. Erosion problems are often a phenomenon in those areas.

Summarising the known experiences of impacts on the soil environment by the above mentioned recreational categories it could be stated that:

1. Nature reserves and wilderness areas; because of the controlled management condition and the selective users (mainly researchers) soil impacts are only those from usual natural processes (erosion, run-off, flooding, wild fires).

2. National and Natural Parks: these constitute large natural areas and attract large numbers of visitors. Impacts on the Park's soil could be found on locations within the park limits, which are assigned to visitors such as: trails, walkways, concession areas, campgrounds and information centres, etc. Since National Park managers have a close interest in their environment, deterioration of the soil is mostly preventable through managerial actions and engineering treatments. In general for National Parks the impacts are mainly on small areas in relation to the parks' sizes. However lands adjacent to National Parks—sometimes called peripheral or Buffer Zones—do suffer most from the visitors, since all the intensive recreational uses are distributed outside of the parks and into these zones. Soil problems in these zones appear from erosion, roadside banks, housing development, campground development, hotels, motels, parkways and mainly from the change of agricultural and forest fertile lands into tourist and recreational areas. Strict zoning regulations based upon soil potentials and slope condition are required in order to prevent environmental deteriorations and to avoid soil destruction.

3. Regional, Aesthetic and Recreational Parks: these are areas with less attractive environments, but because of their location close to large cities or main traffic routes could attract large number of visitors and maintain a diversity of recreational activities with longer timespan for use during the year.

Impacts on soils such as: erosion, compaction, loss of soil's organic layers, poor aeration, soil microfauna destruction, poor sewage, flooding, are among the common problems, which if allowed to accumulate through the years cause serious and irreversible damages to the environment. It is often observed that infrastructure development into these areas such as roads, parkways, utility lines, structures, etc., damage the landscape and create permanent alterations to it.

Quite often and in the warmer regions of Europe, Spain, South France, South Italy and Greece the extensive use of those areas cause forest fires. These forest fires have a tremendous impact on soil environment and cause severe erosion problems [1, 11]. It has been observed that it is in those areas that most of soil environment problems occur.

4. City Parks and Neighbourhood Parks: these categories of recreational areas are under constant pressure by users and are mainly socially oriented [7, 14]. Their environmental variables including soil environment are closely monitored and very often receive technical and engineering treatment to be maintained. However for certain sports in these areas the soil is permanently damaged from excessive use such as in playing fields, concession areas, bike trails, etc.

5. Tourist Areas: in the tourist areas because of the need for accommodation and recreational facilities for large numbers of visitors in a comparatively small area the soil environment is altered and often irreversibly damaged.

Tourist areas designed without careful consideration of the environmental factors such as slopes, climate, soil characteristics, could cause severe environmental damage when developments take place on mountainous areas (ski or winter facilities, cottages, intensive campsites, etc.). There is required for tourist areas, an environmental suitability map to be developed in order to identify the environmentally sensitive areas which can thus be excluded or used for suitable developments, thus avoiding future problems.

In certain Greek islands where tourist development took place without careful environmental consideration extensive landscape impacts and thus environmental destruction have occurred.

Archaeological and historic sites often suffer irreparable damage to their soil since large number of visitors have to be guided through the same walkways and routes all year round. A good example is the trail going through the historic parts in the Delphi region where the calcareous soil on slopes running up to 60% has been washed out and bedrock has appeared.

In such cases managerial schemes are required with routes which could be alternately closed and open in order to let the soil recover and to use material for soil support (mulching, fertilising, etc.).

4. RECREATIONAL AND TOURIST ACTIVITIES

The recreational and tourist activities which take place in the recreational areas are mainly those which cause most of the soil damage.

Distribution of the activities and design of their required facilities in the above mentioned recreational areas, without a careful examination of the site's natural chracteristics and environmental requirements cause environmental damage and nature destruction.

Usually in recreational planning, recreational suitability maps are required, using the normal technique of land use overlays or the more sophisticated computerised graphics and 3-D presentations in order to achieve the optimum land use and activity requirements [1, 2, 3, 14].

Soil maps, aerial B/W and I/R photography, satellite photographs and ground photographs are useful to understand the environmental variables of the recreational area.

Soil characteristics which should be examined besides the ones mentioned earlier are: bedrock depth, drainage, erosion, flooding or ponding, permeability rockiness or stony, texture, water table, inherent fertility, drought, blowing, slope [10, 14, 17].

This information—quite often not required for all recreational area categories or all recreational activities—is evaluated and gives a picture of the existing soil condition, its carrying capacity and problems. Recreational activity allocation in the recreational area is a result of such an investigation.

The outdoor recreational activities are divided into two major categories according to the energy which the users consume in their participation. One category is the so called 'passive recreational activities' while the other is 'active recreational activities' [3]. As 'passive' are considered activities such as photography, bird watching, sightseeing, nature study, picnicking. 'Active' recreational activities are those such as hiking, climbing, horseback riding, camping, playing group games, hunting, fishing, back packing. The above activities in order to be exercised need special environmental conditions, as well as certain facilities to be developed, to satisfy the users, which constitutes the recreational output. The passive activities do not require many technical infrastructures and their impacts are almost minimal to the environment. On the other hand, the 'active' activities generate single or accumulated impacts on the environment when uses exceed the site's carrying capacity. Common observed soil impacts from such activities are:

1. Hiking: soil compaction, erosion, removal of organic material, bad soil aeration, water depth, soil moisutre.
2. Camping: compaction, top soil removal.
3. Group games—playingfields: compaction, top soil removal, reduction of soil fauna.
4. Roads, parking, buildings, sewage: soil removal, erosion, soil contamination [1].

All the above soil impacts are directly related to the previously mentioned site soil characteristics along with slopes and climate, where it should be examined beforehand in order that the recreational and tourist areas maintain environmental quality. In general the most severe soil impacts occur from all activities which take place on soils with bad drainage, slopes (above 25%) and of fine texture [12].

5. RECREATIONAL TOURIST AREAS AND TIME

The carrying capacity for a recreational area could be considered, which is the number of users which at a given area and for a given period of time it could maintain, without losing its environmental quality. It has been found [5] that the length of use of a recreational area seasonwise as well as the number of users for that period are linearly related to impacts on the site's environment.

City and neighbourhood parks which receive all year round large numbers of users suffer the most impacts. The soil in recreational areas suffers most from such long periods of use and quite often is damaged irreversibly. Hiking trails, campsites, horse riding trails, ski routes, open playingfields, concession areas, archaeological sites, natural monuments are among the victims of such overuse through time. Soil compaction, erosion, bad drainage, loss of moisture, soil fauna destruction are among the usual damage which the soil environment receives.

Through the years recreational and tourist area managers have developed methods and techniques along with the use of chemical or mechanical treatments in order to maintain the soil's fertility and properties.

In order to avoid large scale soil damage in the recreational areas soil monitoring plots should be established in order to provide early warnings for managers to prevent soil damage.

6. SOIL IMPACT PREVENTIVE MEASURES

As was earlier pointed out, a good knowledge of the recreational and tourist areas soil environments constitute the basic elements for a sound planning and development.

Identification of soil limits could prevent future severe damage by the users.

There are two kinds of actions which must be taken in order to prevent or correct soil damage. One is of managerial type while the other is of soil treatment and engineering.

1. The managerial actions could be:
   1.1 Establishment within the area soil quality monitoring plots, which could give an early warning about soil condition.
   1.2 Design auxiliary alternative facilities (trails, campsites, picnic sites, etc.) within the area in order to avoid overuse or to be used as overflows.
   1.3 Regulate seasonal site uses according to the soil's seasonal sensitivity.
   1.4 Use of efficient signs informing users of soil problems and risks.
   1.5 Fences and controls on areas with severe soil problems in order that there is a sufficient period of time to assist recovery.
2. Technical and engineering actions could be:
   2.1 Soil mechanical aeration when compaction had exceeded certain levels for the soil type.
   2.2 Soil fertilisation, soil ameliorators and soil cultivation.
   2.3 Use of organic mulches (bark mulch, sawdust, etc.) for sites and facilities with excess use such as hiking trails, horse riding, routes, parking lots, etc.
   2.4 Development of sufficient drainage system on flooding locations within the recreational area.
   2.5 Establishment of proper vegetational types to withhold soil on erodible areas.
   2.6 Terracing the area if slope grades exceed facility requirements.
   2.7 Application of hydroseeding techniques and special plastic nets for steep banks to withhold the soil, to avoid erosion.

2.8 Use of safe sewage systems to avoid soil contamination when soils appear to have poor drainage.
2.9 For soil improvement and during the off-use season seeding and planting of the recreational areas.

7. CONCLUSION

Recreational and tourist activities because of their attraction of large number of users generate multiple impacts on the soil environment on the areas which are used.

Careful soil examination and evaluation is needed during the planning stage for such areas. Neglecting to do so could cause irreversible soil damage which results in the site's destruction and thus environmental deterioration.

Soil mapping along with clear recreational development zones are needed for all countries, but mainly for those which rely upon tourism such as south France, Greece, Spain, Italy, Portugal. Erosion, compaction, bad aeration, loss of organic material, soil fauna loss, soil contamination are among a few of direct damages to the soil which have been found on recreational and tourist areas.

Implicit soil damage could also be produced because of recreation and tourism, when productive farmlands are changed to recreational lands or when farmers are abandoning farming to turn to tourist employment. On a European level more research is needed in order to identify the scale of damage which so far recreation and tourism have caused to soil resources. Soil suitability maps for tourism and recreation should be produced for the southern European countries with heavy tourist impacts, along with the establishment of a soil environment monitoring network.

An Environmental Impact Study to produce clear specifications for any major tourist development will help to prevent serious and irreparable soil environmental damage.

REFERENCES

1. Cassios, C. (1972). Mt Ossa-Greece, recreation site selection and development. M.Sc.-thesis. University of Wisconsin, Madison Library (unpublished).
2. Dechiara, J. and Koppelman, L. (1978). *Site Planning Standards*, McGraw-Hill, New York.
3. Douglass, R. (1975). *Forest Recreation*, 2nd edn, Pergamon Press, New York.

4. Environmental Impact Assessment (1983). PADC-Env. Series No. 14. NATO ASI series.

5. ESPESETH, R. (1984). Site planning of park areas. ORPR-36. University of Illinois-Urbana.

6. GARTHWAITE, P. F. (1971). Forest management for conservation, landscaping, access, sport. Res. paper No. 81 by Forestry Commission.

7. GOLD, S. M. (1980). *Recreation Planning and Design*, McGraw-Hill, New York.

8. IUCN (1968). Landscape planning. Papers of Conference in Argentina. Morgres, Switzerland.

9. LYNCH, K. and HACK, G. (1984). *Site Planning*, 3rd edn, MIT Press, Cambridge, Massachusetts, USA.

10. MCHARG, I. L. (1971). *Design with Nature*, The American Museum of Natural History, Doubleday, New York.

11. METZ, L. and DINDAL, D. (1980). Effects of fire in soil in North America. In: *Soil Biology as Related to Land Use Practices*, USEPA, Washington, D.C. EPA-560.

12. PARTAIN, L. (1969). Use of soils knowledge in recreational planning. Symposium on management of recreation, Washington, D.C.

13. Proceedings (1971). National symposium on trails. Dep. of Interior, U.S. Printing Office, Washington, D.C.

14. RUTLEDGE, A. J. (1971). *Anatomy of a Park, the Essentials of Recreation Area Planning and Design*, McGraw-Hill, New York.

15. US FOREST SERVICE (1963). *Recreation Management, Manual and Handbook*. Title 2300, USDA, Washington, D.C.

16. WAGAR, A. J. (1964). *The Carrying Capacity of Wild Lands for Recreation*, Forest Science Monograph, p. 24.

17. YASSOGLOU, N. J., NOBELI C. *et al.* (1969). A study of biosequences and lithosequences in zone of brown forest soil in N. Greece. Morphological physical and chemical properties. *Soil Sci. Ann. Proc.*, Vol. 33.

Soil Losses from Erosion in Relation to Urbanism and Infrastructure

D. GABRIELS

Faculty of Agricultural Sciences, University of Ghent,
Coupure Links 653, 9000 Ghent, Belgium

SUMMARY

Intensive practices such as road construction, residential development, dams, drainage canals and other structures cause severe erosion which endangers the structures themselves or causes serious damage in the watershed.

Empirically derived regression models and physical process models can be used to estimate soil losses from specific and localised problem areas where erosion rates are high. Surface run-off and erosion can be controlled or reduced on the bare slope by producing a persistent vegetative cover quickly.

The hydroseeding technique by which a mixture of water, grass seed, fertiliser, an organic filler and a soil stabiliser is sprayed on the soil surface is finding increasing favour when vegetation needs to be established.

1. INTRODUCTION

The enterprising man is active in developing new highways, canals, dams, harbours, buildings and other structures where most of the natural vegetation must be destroyed. The soil mantle loses its strength when roots deteriorate after denudation. Bare soil and slopes are hazards in soil erosion. Intensive land management practices such as road building and residential development cause an increase in soil mass movement and decrease the slope stability by mechanically oversteepening or overloading slopes. Frequently soils are used that stimulate erosion; the use of sandy soils with a low organic-matter content will promote wind erosion and the

355

use of heavier soils with a low organic matter content will promote water erosion.

The intensive practices generally accelerate erosion to a greater extent than do extensive practices within a given area. It has been found that in comparing soil loss rates between a steep skid road and a steep agricultural field, the skid road can produce up to 50 times more sediment.

Roadways in forest and rural land disturb an otherwise unerosive landscape with bare soil cut slopes, fill slopes and road surfaces. Disturbances to soil, vegetation, slope inclination and drainage patterns create further potential for accelerated erosion and increased sediment yield. In some cases increased sediment yields are absorbed or buffered by the undisturbed land, in other cases the roadway is situated so that eroded materials easily reach streams and lakes (Ward [16]). Non-vegetated roadsides and riverbanks are recognised as potentially significant sources of watershed sediment. Accumulation of sediment in rivers and harbours hinders navigation, chokes aquatic plants and animals and necessitates costly dredging operations. Roadside erosion presents a serious problem in the maintenance of a network of roads. If erosion is not controlled, the roadway itself will erode to a point that requires significant repair. Gullies developing on the roadsides endanger traffic and, in some cases, make sections of a road completely untrafficable.

A bare soil surface has to be protected from the denuding action of elements as wind and water.

2. SOIL PROTECTION, EROSION CONTROL AND SLOPE STABILISATION

Steeply sloping earth banks are liable to landslides. A bank which is stable when dry can become unstable when wet because it has then a greater weight and a reduced resistance to deformation. Therefore drainage of the sloped bank is necessary (Hudson [7]).

Surface run-off and erosion can be controlled or reduced during and after slope grading by creating a persistent vegetative cover quickly so as to break the energy of falling raindrops, to break the velocity of the run-off water and to break the velocity of the wind at the surface. For a satisfactory germination of species there are three bio-technical requirements, and these are ample supplies of water, of nutrients and of seed.

Conventional practices of 'finish' or smooth and hard grading slopes inhibit water infiltration and enhance water run-off erosion and provide the poorest soil surface environment for revegetation of disturbed areas. The application of topsoil, although a costly procedure, has often improved the standard of vegetation especially on shallow slopes and infertile and droughty soils. Vegetation can be established in plantpits with good soil and fertilisers.

One of the problems in the establishment of vegetation is the continual movement of the soil. This is particularly the case in low rainfall areas with sandy soils which have been subjected to wind erosion. Young trees, shrubs or grasses, planted on a moving soil surface often die through root exposure or suffocation under deposits of eroded soil. If seeding is attempted, the seeds are windblown or washed away and germination prevented from taking place. Organic mulches such as straw and hay have been used successfully in reducing soil movement provided they were evenly spread and when necessary fastened on the slope. A technique which is finding increasing favour is the hydroseeding method. A mixture of water, grass seed, fertiliser, an organic filler and a soil conditioner or stabiliser, organic or synthetic, is sprayed as a film on the soil surface.

Control of streambank erosion with vegetation is seldom sufficient as protection against the scour of the stream, so mechanical protection is usually necessary. The unstable bank itself may be protected from rain erosion and surface run-off through the establishment of a vegetation.

3. RESEARCH

Most erosion monitoring has focussed on agricultural lands and related problems. There has also been a growing effort to address the problem of erosion to specific and localised problem areas where erosion rates are high. Two general approaches are currently in vogue: empirically derived regression models and physical process models. The universal soil loss equation (USLE), an empirical relationship between soil loss and site characteristics and originally developed for agricultural lands, has been proposed for other applications, including roads in general (Farmer and Fletcher [4]).

When applying a model based on physical processes for estimating sediment yields, the soil detachment parameters, one for rainfall and one for overland flow are difficult to estimate. On large plots the contribution from raindrop splash is hidden by the larger overland flow component, on small plots the raindrop splash component dominates the overland flow (Ward [16]).

Also differences in erodibility between surface soils and sub-soils should be considered. Exposed sub-soils have a tendency to form crusts more rapidly with the detachability of soil particles reduced (Meyer *et al.* [12]). If soil losses need to be estimated the rain-erosivity factor should be based on storm events rather than on a period of several years.

For the analyses of processes, specific erosion and sedimentation mechanisms, laboratory research can be as important as field experiments. For rainwash, splash, sealing and infiltration, the use of rain simulators is central. Simulated rainfall offers the advantage (Meyer [11]) that it yields more rapid results, it is more controlled and specific parameters such as the influence of soil type or the degree of slope can be tested in a system in which all factors but one are kept constant.

Temporary soil stability and reduction of soil movement can be achieved by applying a stable film of a soil stabiliser on the soil surface. Although soil stabilisers have been used for many years on newly seeded areas, research into their efficiency has not produced a base for modelling their effects. Work has concentrated instead on investigating, for the numerous conditioners now commercially available, appropriate application rates, the length of time for which stability is provided, effects on plant emergence and the cost of application per unit area (Kirkby and Morgan [8]).

Wind tunnel experiments were carried out to test the effectiveness of several spray-on materials and organic compost for wind erosion control (Lyles *et al.* [10]; Armbrust and Dickerson [2]; Knottnerus, [9]; Armbrust [1]). Tests on the use of soil stabilisers, synthetic and organic, for water erosion control were done with a laboratory rainfall simulator (Morin and Agassi [13]; Gabriels *et al.* [5]; De la Pena and Gabriels [3]) or on small or larger field plots (Morin and Agassi [13]; Yli-Halla *et al.* [17]; Gorke and Stoye [6]; Stalljann [15]; Nille [14]). The soil conditioners, in solution or emulsified form, applied as mulches or mixed with the upper soil surface layer can preserve the stability of the soil until the development of the vegetation cover.

4.   CONCLUSIONS

Planning, designing and the development of urban areas and infrastructure must include practices for run-off and erosion control. These will not only protect the site but also prevent undesirable off-site effects like channel pollution and an increase of the overall watershed sediment. It is seldom possible to avoid all erosion while an area is being developed. In too many instances the first construction operation is to clear the entire area, exposing large expanses of bare soil to the erosive power of the rain and wind. Sometimes the construction may not be completed on parts of the area for several years. The construction plan should keep the areas of exposed bare soil to a minimum for the shortest time. Therefore the areas subject to erosion should be stabilised with vegetation as soon as possible. The hydroseeding method has proven to be a useful technique for this.

REFERENCES

1. ARMBRUST, D. V. (1977). A review of mulches to control wind erosion. *Trans. Am. Soc. Agric. Engrs*, **20**, 904–5, 910.
2. ARMBRUST, D. V. and DICKERSON, J. D. (1971). Temporary wind erosion control: cost and effectiveness of 34 commercial materials. *J. Soil Water Conserv.*, **26**, 154–7.
3. DE LA PENA, E. and GABRIELS, D. (1976). Evaluation of soil conditioners for protection of steep slopes from water erosion. *Proc. 3rd. Int. Symp. on Soil Conditioning, Ghent, Belgium, 1975. Meded. Fakulteit Landbouwwet. R.U. Gent*, **41**, 327–34.
4. FARMER, E. E. and FLETCHER, J. E. (1977). Highway erosion control systems: and evaluation based on the universal soil loss equation. In: *Soil Erosion: Prediction and Control*, Soil Cons. Soc. Am., Ankeny, Iowa, pp. 12–21.
5. GABRIELS, D., MOLDENHAUER, W. C. and KIRKHAM, D. (1973). Infiltration, hydraulic conductivity and resistance to waterdrop impact of clod beds as affected by chemical treatment. *Soil Sci. Soc. Amer. Proc.*, **37**/4, 634–7.
6. GORKE, K. and STOYE, D. (1986). Soil stabilizing and grassing—a new hydroseeding process. *Proceedings of the Symposium on Assessment of Soil Surface Sealing and Crusting, Ghent, Belgium, 1985*, pp. 302–10.
7. HUDSON, N. (1971). *Soil Conservation*, Batsford Ltd, London.
8. KIRKBY, M. J. and MORGAN, R. (1980). *Soil Erosion*, John Wiley, Chichester.
9. KNOTTNERUS, D. J. C. (1976). Stabilization of a wind erodible surface by hydroseeding (town refuse compost). *Proc. 3rd. Int. Symp. on Soil Conditioning, Ghent, Belgium, 1975. Meded. Fakulteit Landbouwwet. R.U. Gent*, **41**, 69–72.
10. LYLES, L., ARMBRUST, D. V., DICKERSON, J. D. and WOODRUFF, N. P. (1969). Spray-on adhesives for temporary wind erosion control. *J. Soil Water Conserv.*, **24**, 190–3.
11. MEYER, L. D. (1965). Symposium on simulation of rainfall for soil erosion control. *Trans. Am. Soc. Agric. Engrs*, **8**, 63–5.
12. MEYER, G. J., SCHOENEBERGER, P. J. and HUDDLESTON, J. H. (1975). Sediment yields from roadsides: An application of the universal soil loss equation. *J. Soil Water Conserv.*, **30**, 289–91.
13. MORIN, J. and AGASSI, M. (1976). A method to evaluate the efficiency of soil conditioners to fight water erosion. *Proc. 3rd. Int. Symp. on Soil Conditioning, Ghent, Belgium, 1975. Meded. Fakulteit Landbouwwet. R.U. Gent*, **41**, 301–16.

14. NILLE, B. (1986). Cultivation of extremely adverse sites. *Proc. Symp. Assessment of Soil Surface Sealing and Crusting, Ghent*, Belgium, 1985, pp. 343–50.
15. STALLJANN, E. (1986). Recultivation by the hydroseeding method. *Proc. Symp. Assessment of Soil Surface Sealing and Crusting, Ghent*, Belgium, 1985, pp. 311–19.
16. WARD, T. J. (1985). Sediment yield modeling of roadways. *Proc. Int. Conf. Soil Erosion and Conservation*, Jan. 16–22, 1983, *Honolulu*, Hawaii.
17. YLI-HALLA, M., ERJALA, M. and KANSANEN, P. (1986). Evaluation of various chemicals for soil conditioning in Finland. *Proc. Symp. Assessment of Soil Surface Sealing and Crusting, Ghent*, Belgium, 1985, pp. 294–301.

SESSION III

The State of European Soils—
Recovery, Protection and Strategies for Prevention

Treatment and Reutilization of Contaminated Sites

RENÉ GOUBIER

*National Agency for the Recovery and Elimination of Waste
(ANRED), 2, Square la Fayette, BP 406, 49004 Angers Cedex, France*

SUMMARY

In recent years contaminated land has become of major concern in industrialized countries. Contamination arises from past or present use of the land for industry or for the disposal of hazardous waste and from wind or water dispersion of contaminants. High level soil contamination occurs usually in such circumstances and is mostly connected with surface and/or groundwater pollution and other negative environmental effects. Therefore, this paper deals with operations in which soil reclamation is part of a wider problem of site decontamination.

Remedial action for hazardous and contaminated sites usually includes three main steps:

(1) Identification of potential problems—contaminated sites inventories.
(2) Evaluation of site contamination—risk assessment.
(3) Treatment of contaminated sites—land recovery.

This third step presents the main cleanup processes available at the present time and classified as follows:

—cleanup by hazardous and contaminated material extraction and off-site treatment;
—cleanup in which hazardous and contaminated materials remain on the site after or without treatment.

The possibility of reuse of contaminated land will obviously be dependent of the site restoration which has been carried out. Cleanup involving on-site containment of hazardous material could appear at first less expensive, but would leave long-term safety problems and severely limit site reutilization.

1. INTRODUCTION

In recent years there have been numerous cases of contamination of natural and urbanized sites in industrialized countries. These situations are mainly due to the polluting action of industrial activities which have led to:

—accidental or voluntary discharges of pollutants on the soils and in the waters; and
—uncontrolled dumping of dangerous and toxic products and waste.

More often than not the contamination affects the entire environment in question: soil, surface and groundwater, air, vegetation; and, sometimes, the risks may also concern people and goods when the problem is due to the controlled storage of explosive or flammable products. Accordingly, soil pollution forms part of the wider problem of contaminated sites.

Initially in industrialized countries, the first contaminated sites were the result of accidents and pollution of the environment. This led to critical situations for the population, which caught the attention of the general public and political leaders, who then realized the need to take voluntary national measures in order to come to terms with these problems. National policies vary in extent in the various industrialized countries, but in all cases the approach remains the same, with three essential stages:

—identification of contaminated sites—inventories;
—diagnosis—evaluation of contaminated sites;
—treatment of contaminated sites.

2. STAGES OF CONTAMINATED SITE TREATMENT

2.1. Identification of contaminated sites—inventories
Identification of contaminated sites may be based on two principal methods: surveys and detection techniques.

2.1.1. Surveys
(a) *Surveys among producers, transport operators, disposers of dangerous waste and administrative authorities responsible for supervision.* Logically,

attention should be focussed on the latter for obtaining information on the existence of risk sites. However, although often quite considerable, the effectiveness of this measure is generally limited since in the past the follow-up of industrial activity in respect of environmental protection has not usually been examined for:

—technical reasons: lack of knowledge of the phenomena in question;
—legal reasons: non-existence or failure to adapt regulations;
—financial reasons: lack of supervisory means in terms of staff and equipment;
—political reasons: little interest shown by the government (in the majority of countries the creation of ministries of the environment has been relatively recent). Furthermore, even though transport supervision and the disposal of dangerous waste has been improved considerably in recent years it has still not gone far enough.

Surveys among polluter companies and transport and disposal operators of waste can obviously be a source of valuable information, but again their effectiveness can be limited for the following reasons:

—disappearance of the firm which used the site;
—no recollection of the operation within the firm: there have been discharges in the past, without written trace, by members of staff who have since left the company;
—reticence on the part of the firm which may prefer to reveal nothing of a practice which calls into question its responsibility and, once discovered, could oblige it to undertake costly decontamination of the site.

(*b*) *Surveys among bodies and persons with special knowledge of the territory and protection of the environment.* These sources of information may vary from one country to another, but by way of a guide the following non-exhaustive list is given:

—municipalities or groups of municipalities responsible for the management of the territory;
—public or private bodies responsible for the monitoring of surface and groundwater and for the production and distribution of drinking water;
—environmental protection associations;
—user associations, e.g. fishing associations.

(*c*) *Utilization of data from water quality observation networks.* The existence of polluted sites has sometimes been revealed by contaminated groundwater. It might therefore be of use to examine the results of observations from groundwater monitoring networks in order to identify possible instances of pollution that can be attributed to the presence of polluted sites.

(*d*) *Studies on the disposal in the past of certain groups of dangerous waste and/or waste from certain types of industry.* This would be a study into the background of how certain groups of pollutant waste are produced in order to determine the quantities produced, the places of production and the methods of disposal (and hence possible dumping). This type of study must be assigned to a consultancy bureau with in-depth knowledge of the industrial activities concerned.

2.1.2. Detection techniques

(*a*) *Visual observations.* In view of its unwieldiness, this elementary method cannot be used systematically for identification. It could be used for a limited study of the territory where there is evidence of site contamination, but it would be more of an additional tool to the surveys mentioned above.

(*b*) *Interpretation of aerial photographs.* Once again, this procedure cannot be put to systematic use because of the poor cost price/effectiveness ratio. In contrast, an analysis of existing aerial photographs in the archives and the mounting of a specific campaign over a polluting industrial zone could be extremely useful in revealing unknown waste deposits.

2.1.3. Conclusions on inventory methods

In short, there is no single efficient method of identifying all the contaminated sites on a territory, but an inventory of this kind can be reliably set up by coordinating surveys, visual observations and aerial photographs.

These inventory methods can also be applied by type of dangerous waste by means of surveys of companies which produce it and have produced it in the past.

Given the vast quantity of information to be collected, arranged and processed, these inventories are major operations requiring considerable resources in terms of experienced or specially trained staff.

2.2. Diagnosis—evaluation of contaminated sites

Where a suspect site is found incidentally or by inventory the following questions should be asked:

Is the site the cause of nuisances or risks of nuisances?
If so, which?
How can this be remedied?

The diagnosis—evaluation of a site and its impact—aims at answering the first two questions. The findings will come from a comparison between the site (potential danger) and its environment (vulnerability) and will thus be based on data acquired on the suspect site and its environment.

2.2.1. Data on a suspect site

(*a*) *General points.* Let it first be said that dangerous sites can be found in different circumstances and from different aspects, these being in the main:

—abandoned industrial terrain (industrial wasteland) where polluting industries were established which have discharged or dumped dangerous and toxic substances;
—industrial terrain in comparable situations to the ones above, but with polluting industries which are still totally or partly active;
—uncontrolled discharge or dumping of pollutants, abandoned or still active.

The difference between the first two categories is mainly a legal one, and in the case of the second it will be easier to illustrate responsibilities and bring action to bear on the industrial companies themselves. The third case could be found in the natural or rural environment whereas the first two are typical of industrialized areas.

In terms of defining contaminated sites, these are often industrial structures (chemical reactors, reservoirs, piping, etc.) left abandoned. However, they may also be lagoons or pools of liquid or pasty products, waste drums, demijohns, bags, assorted containers or even conventional discharges of urban waste in which toxic products have been discharged.

In practice, site evaluation comprises a description, background, quantities, physical state, conditioning and nature of the products the site contains. It will be largely on the basis of these data that the existence and extent of the potential danger of suspect site can be determined: possible presence, nature and characteristics of unstable, explosive and pollutant products.

From the point of view of methodology, experience has shown that data on a suspect site can be acquired in two successive and complementary phases.

(*b*) *Collecting and processing of the information available.* This information can be collected from the sources named above for the inventories:

—persons responsible for the use of the site and supervision of these persons;
—external bodies or persons with knowledge of the pollution developed on the site.

As for the inventories, the value and scope of the information collected can vary appreciably from one case to another. It will always be useful to collect the maximum number of written documents: commercial and administrative archives, photographs (especially aerial photographs). Witnesses by word of mouth may be valuable but must be treated with caution since some might be lying (to hide or, on the contrary, exaggerate the gravity of a problem) or, more simply, might no longer have any real recollection of what actually happened.

Experience shows that, as with inventories, a rational comparison of these different sources of information provides the best data on a suspect site. At the end of this survey, which will be completed by a visual examination of the site, a first description with its background will be established. This evaluation on the basis of the data available can sometimes highlight the possible dangers of a site, but more often than not it is not sufficient to draw conclusions, and special on-site investigations are then necessary.

(*c*) *Special investigations to determine a suspect site.* These investigations, which complete the information acquired and the visual observations of the preceding phase, may be direct or indirect.

Direct investigations. The nature and characteristics of dumped products may be determined by samples and analyses which will depend in type and extent on the information collected from examining the data available and on the complexity of the problem. It will not be possible in certain, particularly dissimilar, cases, to collect exhaustive data on the site, and there will be specific sampling in order to determine the essential characteristics. Analyses should highlight the potential danger of the dumped products: stability, content of toxic substances and their ability to spread in the environment. In this respect, leaching tests could be conducted in order to evaluate the possibility of water contamination (predictive behaviour test).

With a view to the subsequent decontamination of the site in question, it would also be useful to undertake analyses of suspect products, not only to determine their potential danger but also the possibility of treatment (e.g. incineration).

Given the inherent danger of some products, site inspections and especially sampling must be under good safety conditions, with persons working on the site suitably equipped.

Indirect investigations. Quite often dangerous waste is concealed, invisible and not readily accessible. It may then be worth using indirect investigation procedures to determine the geometry of sites and discharges and to identify some of their components (in particular, those contained in metallic recipients—drums). The main methods are:

—interpretation of aerial photographs in natural light and infra-red and, more generally, multispectral airborne remote sensing;
—magnetic and electrical prospecting and, more generally, ground-level detection techniques.

The use of these methods, which might usefully be coordinated and require specialists, often provides valuable information which then helps to optimize direct investigations.

2.2.2. Data on the environment of a suspect site

The aim is to define the environment of a dangerous site by determining its situation vis-à-vis people, property and the medium (soil, air, water) which may be affected by the nuisances it produces or can produce.

In this context there is a need to determine:

—the general characteristics of the environment;
—human activities, buildings, roads, utilization of soils and waters;
—the climatology (temperature, wind, precipitation), geology, pedology, surface and underground hydrology.

The main characteristics of the environment of a dangerous site can nearly always be determined fairly accurately for the preliminary evaluation of the risks involved by collecting the data available: geographic and hydrographic, geological and hydrogeological maps, climatological charts and inspections.

This approach should also indicate the factors representing the particular vulnerability of the environment, such as houses close by, cultivated land and meadows, wells for collecting water to be used in food downstream of the site.

2.2.3. Diagnosis—evaluation of a site and its impact
In some cases the discovery itself of a site will have been the establishment of accidents to, or pollution of, the environment; in others, it will be the investigations described above that reveal the existence of nuisances attributable to a contaminated site.

In general, the diagnosis of a dangerous site is based essentially on the comparison of its potential danger, as evaluated in the 'site data' study, with the characteristics of its environment, as determined by the 'environment data' study. This diagnosis could come to very varied conclusions: from discharges where the potential danger is negligible the inconvenience is largely aesthetic to the discharge of highly toxic and water-soluble waste immediately upstream of drinking water collection or irresponsible storage of flammable and explosive products near houses. Between the two there can be a whole range of situations, worrying to different degrees, which warrant definition both for a better evaluation of the impact of a dangerous site and, accordingly, for better determination of measures to be taken to re-establish safety and clean up the site in question.

When this evaluation forms part of a general operation of decontaminating risk sites (at country or regional level) it should highlight urgent needs and priority action.

This can be provided:

—by a study of any nuisances identified;
—by precise study of the impact of the deposit on its environment, mainly by sampling the air, water, soils and plant life likely to be contaminated. This will often call for probes and piezometers to evaluate and monitor the impact on groundwater.

Once again direct investigation methods could be usefully backed by indirect prospecting techniques: geophysical and electrical prospecting of the sub-soil affected, multispectral airborne remote sensing.

2.3. Treatment of contaminated sites
Where studies indicate nuisances or significant risks of nuisances in respect of a site, work must be carried out on that site in order to restore the safety of the environment.

The nature and extent of this work will vary appreciably from one case to another. However, despite this diversity the restoration techniques can be classified into two large categories according to whether the dangerous products are removed or left on the site.

2.3.1. *Restoring the safety of the site by removing the dangerous products*

In this case the various pollutants must be identified and a system of treatment selected for each of them. Selecting a system of this kind generally comes down to the traditional problem of eliminating toxic waste for which there are technical possibilities on the market (incineration, physico-chemical treatment and discharging), bearing in mind the possibility of reutilization.

However, this work will generally be more difficult than for untreated waste. Unfortunately, the sites to be treated are often irregular, disorganized, with the state of packaging varying (drums). It will often be necessary to set up a real identification work station with sorting, repackaging and dispatch of pollutants and contaminated earth. The certain or presumed presence of dangerous products will call for suitable safety measures to protect site personnel and, possibly, the surrounding neighbourhood (risks of explosion, fire, air pollution, etc.).

This treatment by removal and then elimination of the waste on the site is particularly suitable for surface sites (stores, storage areas, etc.) and sometimes underground sites if the products can be easily located, identified and extracted. The existence of dangerous waste disposal installations at short distances from the site to be treated can have a considerable influence on whether or not this type of work is undertaken.

2.3.2. *Restoration of safety with pollutants left on the site*

(a) *Improvement of the site—confinement.* Fairly often, pollutants have been buried, mixed with earth, urban waste or inert materials under conditions that make it very difficult (for technical and economic reasons) to envisage extraction. In addition, the soils and waters are also contaminated most of the time.

Under these conditions the proposal may be to confine the dangerous products, i.e. to upgrade the site in order to prevent noxious substances from settling in the waters or in air. The search for a solution could be aided by certain characteristics of the site. For example, a shallow impermeable geological formation beneath the site would be favourable to confinement in respect of the groundwater.

In general, confinement consists at least of an impervious cover (layer of compacted clay or synthetic membrane). Sometimes the system is completed by a side wall (injected). Complete artificial confinement (top + wall + base) is also a possibility.

However, work of this kind must always be completed by the establishment and monitoring of a surveillance network to show that the confinement system remains efficient. In the majority of cases (risk of groundwater contamination) this is a piezometric surveillance network. Similarly, the use of reclaimed land must be strictly controlled and hence very restricted in order to prevent disturbances to the confinement system. It should be remembered that treatment of this kind neutralizes the risks of nuisances but does not make them disappear altogether.

Another possibility, likewise applied exclusively or as a complement to a confinement procedure where this does not make pollution disappear completely, consists of an in-depth study of the site and its environment followed by the establishment of a network that can collect all the contaminants which are then treated by suitable techniques. The main drawback of this procedure is the need to continue recovering contaminants up to complete leaching, which can be an extremely long process.

Finally, the use of a confinement system on a contaminated site may be seen not as a final solution to restoring safety but as a transitional solution prior to final treatment.

(b) *Treatment of the waste on the site: Physico-chemical stabilization.* Processes currently exist which are used mainly to treat pasty or liquid waste (lagoons) of an assorted nature (sewage sludge, hydrocarbon waste, acid tar, etc.). They consist of *in situ* treatment with suitable reagents (neutralizing properties, chemical fixation, solidification) of the waste in question with the aim of mechanical and chemical stabilization. Laboratory studies conducted by the 'Institut de Recherche Hydrologique de Nancy' (Nancy Institute of Hydrological Research) have shown that the commercial treatment processes available on the French market generally give good mechanical consistency and are effective in trapping mineral contaminants, although generally less good for organic compounds.

Correct application of these processes, which are currently being examined with a view to improving their performance and determining the limits on their use, will always call for specific adjustment to the product being treated (addition of reagents specially developed by laboratory trials).

Treatment is generally as follows:

—extraction of the products to be treated;
—mixture with reagents added in advance;
—replacing of the products treated on the site.

There have been far fewer incidences of injecting reagents into the mass of waste to be treated.

Treatment of this kind would normally be completed by a system of monitoring the treated site (observational well) and by surveillance of the quality of the groundwater concerned (piezometric surveillance and samples for analysis).

Subsequent use of the land should also be controlled.

Other types of treatment. Dangerous waste and contaminated products can also be treated *in situ* with the help of mobile installations.

Accordingly, mobile incinerators have been designed, in particular in the United States, some with outstanding technical performance levels (capacity to treat PCBs and other organochlorides). Mobile water or soil purification stations have also been used, and, in other cases, especially when borne out by the quantities to be treated, a special purification installation has been set up on a given site.

Finally, *in situ* decontamination by microbiological processes (bacteria) has sometimes been employed (case of waste contaminated by phenols in France), but the view is that these processes still need to be developed further.

2.3.3. Contaminated soils and waters

In a good many cases dumping or discharging waste and pollutants has led to contamination of the soil and waters close to, and downstream of, the site. Polluted soils and waters must therefore be treated as part of the restoration of site safety.

Polluted waters, by their nature mobile, must be recovered, generally by pumping, and then treated in purification facilities on the site or elsewhere. Polluted land could be treated together with the pollutant waste by incineration, physico-chemical stabilization or confinement. In some cases specific treatment could be applied after removal and elimination of the pollutants from the site. This specific type of treatment carried out *in situ* or in fixed installations may require thermal, physico-chemical (extraction and elimination of contaminants) or microbiological (technique being developed) procedures.

3. CONCLUSIONS

The various techniques of treating contaminated sites have been presented separately for the convenience of the report. It is clear that in some cases,

Table 1

Conditions for applying methods—advantages and disadvantages

Decontamination technique	Initial favourable factors	Advantages	Disadvantages
Extraction and elimination of pollutant products	Pollutants easy to identify and extract. Environment highly vulnerable because of: —the use of the site and nearby human activities; —the presence of water resources that cannot be protected; Existence of a treatment plant close by	Definitive decontamination of the site: re-use of the land without problems	Often difficult (site safety) and costly operation
Physico-chemical stabilization of pollutants	Pollutants relatively similar in physical and chemical terms. Environment averagely vulnerable	Considerable limitation of the risks of pollution. Possible re-use of the site for certain applications. Medium cost	Not applicable to certain toxic and particularly mobile components. Surveillance of the treated site necessary
Confinement of pollutants	Highly diversified pollutants difficult to identify, extract and treat. Pollutants fairly immobile. Environment not very vulnerable. Suitability of the site for confinement	Comparatively cheap	Extremely limited re-use of site. Need for surveillance and maintenance of the treated site

particularly complex sites with diversified contamination, these techniques could be used successively or simultaneously in accordance with the nature, physical state and packaging of the products to be treated.

In some cases several different techniques will be applicable and a choice will have to be made between them. Economic criteria may then prevail, but if this is the case it should be remembered that:

—Restoration of a site to safety by extraction and elimination of the pollutant *in situ* or in external installations will generally be a costly but definitive solution permitting subsequent use without reservation of the reclaimed land.

—Decontamination by physico-chemical stabilization of the pollutant components may be an effective solution since it reduces the potential pollution to a considerable degree and restores the mechanical qualities of the treated site, but it must be monitored and subsequent use of the site must be adapted.

—Decontamination by confinement of dangerous products, generally the least costly in terms of initial investment, may restore the safety of the site in question, but the potential danger remains and, especially if the confinement is regarded as final, safety cannot be maintained except by medium and long-term surveillance and maintenance of the site for which utilization will be severely limited.

Table 1 provides a summary of the conditions for applying these methods of treatment along with their advantages and disadvantages.

BIBLIOGRAPHY

HENIN, S. Réaction des sols aux pollutions—Symposium. Protection des Sols et Devenir des Déchets, La Rochelle, France, Novembre 1983.

GODIN, P. Les sources de pollution des sols: essai de quantification des risques des eaux éléments traces—Symposium. Protection des Sols et Devenir des Déchets, La Rochelle, France, Novembre 1983.

COLIN, F. Evaluation des performances des procédés de fixation de boues utilisées en France—Symposium. Protection des Sols et Devenir des Déchets, La Rochelle, France, Novembre 1983.

VÍRAPIN, F., LAFFITE, P. and GOUBIER, R. (1985). La résorption des sites de déchets polluants—Agence Nationale pour la Récupération et l'Elimination des Déchets (ANRED), Angers, France.

RULKENS, W. H., ASSINK, J. W. and VANGEMENT, W. J. (1985). On site processing of contaminated soils. NATO, *Contaminated Land* Vol. 8.

SANNING, D. E. (1985). In situ treatment, NATO, *Contaminated Land* Vol. 8.

CHILDS, K. A. (1985). In ground barriers and hydraulic measures. NATO, *Contaminated Land* Vol. 8.

HOOGENDOORN, D. and RULKENS, WIM (1985). Selecting the appropriate remedial alternative. Conference contaminated soils, Utrecht (NL).

HATAYAMA, H. K. (1985). Site Investigation: a review of current methods and techniques. Conference contaminated soils, Utrecht (NL).

VAN DEELEN, C. L. (1985). Assessing the risk of soil contamination in the case of industrial activities. Conference contaminated soils, Utrecht (NL).

Land Suitability Evaluation in Major Agroecological Zones and its Application in Land Use Planning and Nature Protection

W. H. VERHEYE

Geological Institute, University of Ghent, 281 Krijgslaan, 9000 Ghent, Belgium

SUMMARY

The suitability of a land depends on a combination of physical factors (mainly soil, climate and landform) and socio-economic options. The former may hereby be considered as rather stable components while the latter are highly variable in time and space.

Land suitability evaluation in terms of physical potential deals with a comparative classification of the land for a given range of utilization types, which may cover as well agricultural uses such as forestry, recreational or nature preservation options. The procedure hereby used is based on a matching exercise between the specific growth requirements and the environmental conditions. Those conditions can be derived from the soil and topographic maps and from climatic station data. However, because similar soils may have different potentials or because similar slope gradients may present different erosion hazards under variable climates their interpretation has to be made in a fully integrated soil-climatic context. The concept of agro-ecological zones meets this requirement as it combines the characteristics of a climatic zone with a soil and landform map unit.

When conceived as such the land suitability evaluation per agro-ecological zone and per land utilization type may provide a series of scenarios which can be used by decision makers for future land use implementation in its broadest context.

1. INTRODUCTION

The potential of a land to grow plants and to produce crops is determined by its inherent physical suitability and by the willingness or skill of the farmers to use the given opportunities. Both aspects are closely linked and are of almost equal importance, in the sense that potentially good agricultural land which is not cultivated is as useless as unsuitable fields on which high technology and know-how is wasted.

Land suitability has different meanings, and its interpretation varies often with the users' scientific or professional background. The concept of suitability for the economist is always linked to a cost/production factor, while for the agronomist suitability relates mainly to production and yields, expressed in quantity and quality. Unfortunately, the relation between both appreciations is not always evident because high yielding crops which have to be sold at low prices are not necessarily an example of a successful economic operation. Even for one and the same land use type, i.e. vegetable growing, suitability may be largely time-dependent: early vegetables, even produced in small amounts, yield indeed often a better income for the farmer than the much higher mid-season productions which have to be sold at a moment when the consumers' market is fully saturated.

Land suitability evaluation is, however, not the exclusive domain of agronomists or economists, but refers also to those who, in one way or another, are involved in land use planning: town and country designers, landscape architects, road and sewage engineers, environmentalists....

In Europe, with its high population density and intensive agriculture and industry, the growing pressure on the land over the past decennia and the dramatic increase in the use of agro-chemicals, especially fertilizers and pesticides, has led to a serious concern about the undesirable side-effects of those actions and on their overall impact on the environment. The movement in favour of a more organized nature protection has often resulted in a conflicting situation between environmentalists and farmers' associations. The concept of suitability used by the environmentalists has hereby completely shifted from the economic or productivity sphere towards more moral and affective grounds. High value biotopes, in terms of biological or environmental appreciations, may hereby cover a full range of otherwise economically interesting agricultural or forestry land, including also poorly drained marshes, abandoned dune ridges or heath zones as well as top-class agricultural areas.

In this context it is obvious that land suitability evaluation may have a wide scope and that areas which are suitable for one land use type are not necessarily so for another one. An initial definition of the procedure and purpose for which the suitability evaluation is made is therefore of primary importance for a better understanding and implementation of the results.

2. DEFINITIONS AND BASIC CONCEPTS

Land suitability evaluation hereby considered deals with a comparative suitability rating of the land for a given range of utilization types, using an objective and scientifically sound evaluation scale which covers the major growth requirements. This procedure leads to a classification of the actual suitability potential, emphasizes the constraints of the land and/or comments on the possibilities for an improved future suitability.

Although it is understood that the suitability of an area depends both on physical and on socio-economic factors, and those are closely linked, it is recommended that as a first step the physical suitability procedure should be separated from the socio-economic evaluation, mainly because of the different nature and impact of the factors involved in both approaches. Indeed, the physical parameters of the land, dealing mainly with soil, climate and landform properties, are relatively stable components and, once inventorized, do not quickly change. The socio-economic context, however, dealing with topics such as land tenure, technical know-how and availability of manpower, distance to consumers, market conditions, etc., is highly variable both geographically and in time; this context may even be abruptly modified by political or economic decisions, both at national or international level.

In order to be compatible with the concepts outlined above, the physical land suitability evaluation should result in a comparative rating of a series of alternatives in terms of crops or land utilization types. Those alternatives should then be evaluated against the background of the momental socio-economic or political situation, so as to provide a series of scenarios out of which decisions can be taken. It is hereby emphasized that the term 'land utilization type' has to be interpreted in its broadest sense (cf. Ref. 3), covering not only agricultural uses (crops and cropping systems at a given management level) but dealing also with forestry or recreational purposes, nature preservation or environmental options. Hence, when conceived as such the physical land evaluation may provide a practical tool for an overall land use planning which covers all aspects of the occupation of the land.

3. PROCEDURES FOR PHYSICAL SUITABILITY EVALUATION

Physical suitability evaluation deals with the factors of the natural environment, i.e. climatic, soil and landform conditions which have a direct impact on the growth and the production of the plant.

For optimal germination and vegetative development the plant needs a minimal temperature, a good moisture supply and an equilibrated nutrient availability in the root zone; additionally, a reasonable energy regime is needed for optimal photosynthesis and biomass production.

Plant or crop growth requirements in terms of moisture, energy and nutrient supply may vary from crop to crop or, in a broader context, as a function of the land utilization type. Those requirements may also vary at the different growth stages and need therefore to be linked to the crop calendar. Hence, at a first stage in the evaluation procedure, land utilization types have to be defined and the growth requirements listed for the plants or crops related to them. For each parameter hereby considered optimal and marginal conditions have to be estimated, covering a gradual limitation scale.

In the case of agricultural land uses the crop requirements are relatively well known. They refer to specific climatic requirements, such as temperature, air humidity, sunshine or radiation data, as well as moisture needs at different stages of the growing season. They deal also with landform properties (related to the use of machinery and to erosion hazards), with soil requirements in terms of profile depth and nutrient status, and with a series of physical or chemical soil conditions which affect root development and biomass production. Requirements in the broad agricultural sense include also the specific climatic conditions at the time of field preparation and harvesting.

For non-agricultural uses growth requirements are often less well known or are less extensive in number. Specific conditions needed for forestry production may be quite similar in type and number of criteria involved for agriculture, but this is no more the case of forest recreation reserves where minimal requirements of temperature, sunshine and moisture supply (in relation to species development) are often, although not always, of minor importance as compared to site locations and social pressure. The situation changes again in cases of land use types concerned with specific nature protection goals, whereby sometimes a small number of site-specific parameters are at the origin of the occurrence of valuable biotypes. Examples of those refer to specific marshland or tidal swamp species which could disappear when land drainage and/or empoldering is operated. At a larger scale one can also refer to the heath-vegetation which in Belgium, Holland and Germany is almost exclusively linked to very poor sandy soils with low pH values (4·0–4·5). Calluna, Carex, Molinia and other species which populate these heath zones are very pH-dependent and will not survive in an environment where the soil nutrient status and pH increases and, hence, the invasion of other plants (grasses mainly) is promoted. Protection measures for such an environment must not only be based on the knowledge of the plant growth parameters, but should also consider the nature and evolution of the surrounding land and the eventual risks they provide for the pollution of the protected zones and for the buffering capacity of the heath area itself. This brings us to the next step in the procedure.

Once the growth needs have been established they are matched with the environmental conditions. The procedure hereby used has been initiated by the FAO Framework for Land Evaluation and the subsequent Agro-ecological Zones Projects [3, 4], and was further developed by several other authors [1, 5, 7, 9, 12]. The matching exercise enables definition for each given plot of land the compatibility of the individual environmental parameters with the plant growth requirements and, as such, to make an overall suitability classification on the basis of the nature and degree of the constraints.

The suitability classification hereby obtained refers to the actual situation. The procedure can, however, also be extended towards the potential effects of environmental changes and towards their eventual impact on the quality of the surrounding zones. Drainage of the land in one area will indeed often influence the level of the watertable in neighbouring fields. Sandy soils occurring in a high rainfall region—and extensive areas in Belgium, Holland and Germany are examples—are potential sources for groundwater pollution in neighbouring regions due to the infiltration of excessive fertilizer applications to those soils. In both cases the predictable scenarios can be evaluated and their eventual impact calculated or anticipated.

4. THE CONCEPT OF AGRO-ECOLOGICAL ZONES

A key factor in the methodology described above refers to the possibility of matching the plant requirements and the environmental conditions. The latter data have therefore to be inventorized and regrouped into a form which permits effective and proper matching.

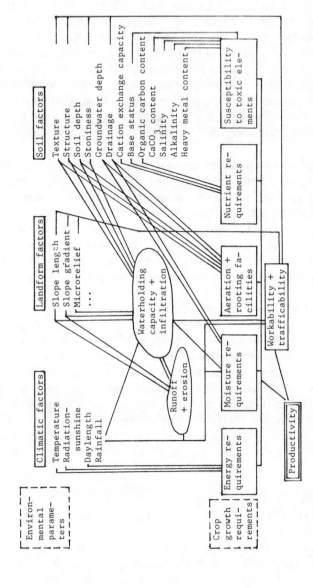

Fig. 1. Relationship between observed and/or measured environmental parameters and crop growth/production requirements.

Environmental parameters deal with climatic, soil and landform factors. Some of those, like temperature, air humidity and insolation, correspond directly with the crop needs (Fig. 1). Others do not, however, like moisture availability, for example, which depends on rainfall (amount, distribution and intensity), soil depth, infiltration rate and moisture retention capacity of the profile, depth of the groundwater table (and related capillary rise), and even on landform characteristics such as slope and run-off intensity. In other words, one and the same soil may have a completely different growth potential depending on the rainfall regime to which it is subjected. In the same sense the critical slope level for erosion will vary depending on the rainfall distribution and rain intensities of the region. And the same reasoning holds true for other parameters.

A definition of the environment in separate terms of soil, climate and landform is thus not of great use, and with these considerations in mind FAO [4] has introduced the concept of the agro-ecological zone, which combines the characteristics of a climatic zone with those of a soil map unit (which often includes already a landform parameter).

Linked to the introduction of those agro-ecological zones was the initiation of the concept of the growing period [4], which corresponds to the time in the year that the crop or plant has no limitations on temperature or moisture for its full development and maturation. This period is conventionally calculated as the time in the year (expressed in days) that rainfall exceeds half the potential evapotranspiration (calculated by the modified Penman method [6]), extended by the time that an arbitrarily chosen amount of 100 mm moisture stored in the soil is consumed by the plant. This period is subdivided into a pre-humid ($PET > P > 0.5PET$), a humid ($P > PET$) and a post-humid time ($0.5PET < P < PET$), which corresponds largely to the different plant development stages. Interruptions in this moisture-based growing period may occur due to minimal temperatures below which the plant moves into a dormancy period.

In the intertropical belt the growing period will mainly be defined on the basis of the moisture criteria, while in mid-latitudes the temperature factor will gradually become more important towards the cold temperate belt.

Each agro-ecological zone may thus be defined by a proper start and length of the growing period, based on easily available precipitation (calculated), potential evapotranspiration and temperature parameters. This exclusively climatic definition can at a later stage be completed when the different soil units within the zone (or agro-ecological zones in the proper sense) are taken into consideration and the soil-specific moisture retention is more accurately evaluated.

A proper definition and interpretation of agro-ecological zones needs thus a climatic and soil data base, whether available in maps or in any other form.

5. AGRO-ECOLOGICAL ZONES IN EUROPE

Information on soils, climate and landform characteristics is largely available in Europe. The nature and form wherein these data are stored, the density of observations and the methods by which maps and map legends have been designed is, however, extremely variable between one country and another.

In terms of soil data collection the new 1/1 000 000 EC soil map [2] constitutes a unique information tool, describing on a standardized basis the major properties and geographic distribution of the soils in the 12 member states. The map contains 312 units (and 7 phases) which, due to the complexity of the soil pattern and as a result of the imperatives of the scale, correspond to soil associations with variable composition. Each map unit is therefore defined in terms of dominant and associated soils and/or its eventual inclusions, with their relative importance indicated. The major soil properties of each unit are outlined or can be derived from the classification and the type profile characteristics published in the explanatory textbook. They refer to texture, soil depth, gravel or stone content, slope, drainage, salinity–alkalinity content and nutrient level; they include the necessary background information for an evaluation of the moisture storage capacity, fertility status, workability–trafficability conditions, erosion sensitivity or any other matter which may be of importance in evaluation work at continental broad-scale level.

Climatic data are available from station records and from maps. Raw synoptic information is obtained without major problems from the National Meteorological Offices, but due to the relatively recent introduction of the agro-ecological zone concept, using the FAO-adapted Penman calculation for PET, this parameter needs always to be calculated. Hence, an agro-climatic data base and related map for the EC as a whole is not yet available. In this context a joint effort is now being undertaken by the EC Agricultural and Environmental Research Divisions (DG VI and XI) to establish in the first place an overall climatic data base and an agro-climatic map, meeting the goals for agro-ecological zoning.

The proposed agro-climatic map will include approximately 25 zones, defined on the basis of the start and length of the growing period as a whole, the importance of the humid period and a cumulative degree-day-factor, which is an expression of the energy status of the environment. Each zone is hereby identified by a reference station with a synoptical sequence which is representative for the area concerned.

Preliminary results obtained from this agro-climatic research have shown that in France, for example, eight zones can be distinguished, many of which occur also in the Mediterranean (Spain and Italy), Oceanic (Ireland and the UK) or more continental parts of Europe (Belgium and Germany).

The Mediterranean zone including the Provence, Languedoc and coastal sector between the Italian and Spanish borders is mainly characterized by a winter precipitation and by mild temperatures, so that the growing period is almost completely determined by the rainfall pattern. This period starts in mid-September and lasts for approximately 200 days until the end of April, with a short growth stop between the end of December and mid-February due to mean temperatures below 65°C. The humid period is concentrated between early October and February and permits not only maximal water storage in the soils during that time but implies also a risk for sheet erosion and mass movements in sloping areas. The radiation regime shows a high cumulative degree-day figure (3316°C above 5°C in Marseille–Marignane and 3256°C in Montpellier), 30 to 45% of which is registered during the growing period.

These overall climatic conditions already restrict the range of species which can be successfully grown in this region, especially when the length of the growing period and the moisture availability are considered. Depending on the soil-specific moisture storage potential the growing period can more or less be extended in each of the different soils or agro-ecological zones situated within this climatic area. As such, one can conclude that perennial plants like fruit crops or forestry species with all-year-round water needs have to be excluded from freely drained upland soils, and have to be restricted to those agro-ecological zones which have a groundwater table within the reach of the root pattern. Crops or plants with a vegetative cycle of less than 200 days can more or less successfully be grown, but those with a growing period surpassing the purely climatic waterbalance can only develop in soils or agro-ecological zones which, at the end of the calculated growing period, can obtain a supplementary moisture supply from the soil storage reserve or through capillary rise from shallow groundwaters. Such conditions are generally met in the deep alluvial soils of the Rhône Valley (units 6 and 10 of the EC soil map), as most upland areas (units 29, 69 and 87) have limited water retention capacities as a result of textural and/or profile depth constraints. Irrigation is hereby needed to overcome the natural drought-stress conditions.

The northwestern part of France, Brittany and Normandy in particular, has a completely different climate, in the sense that precipitation is moderate but continuous over the year and that temperatures are mild. Similar climatic conditions may be found in southern Ireland and in parts of Cornwall, Devon and Wales in the UK.

The growing period in that part of Europe is generally not restricted by a rainfall deficit, neither by a winter dormancy period. The radiation regime is, however, definitely lower than in the Mediterranean regions and hardly exceeds 2000°C degree-days (above 5°C). Winds are however strong.

Under these conditions there is no proper drought period and thus no real start to the growing season. In other words, it is evident that there are few limits for an almost continuous vegetative growth all through the year. For economically interesting crops best production results will, however, mainly depend on a crop calendar which meets best the photoperiod requirements, but perennials like forest trees and grass will not be affected by those.

Variations in the land use potential will, however, be observed in the different agro-ecological zones, mainly as a function of inherent soil properties like texture, nutrient status and aeration of the root zone. On soils with a limited depth (cf. the lithic phase on the soil map) tree growth could be hampered due to combined constraints of poor root anchorage facilities and strong wind action. It may finally be emphasized that under continuous rainfall conditions coarse-textured soils represent a serious danger for infiltration and pollution through excess fertilizers applied on those fields.

The climate of the Paris basin differs from both other areas by the annual occurrence of a relatively weak but obvious rainfall deficit in summer time and of a cold winter period during which plant growth is interrupted. Hence, in this climatic area the growing period starts in the first week of August and its length ranges between 225 and 250 days, partly due to the summer moisture deficit but mainly as a result of the low winter temperatures between mid-November and mid-March. The cumulative degree-day figure above 5°C ranges between 2000 and 2250°C.

Autumn-sown crops will obviously be little affected by this situation, but for perennials like grass and tree crops a moisture deficit is to be expected, requiring additional irrigation for optimal production. Large forest species with an extensive root system exploring a large volume of soil can to some degree overcome this situation; spring and summer crops, however, which extend their vegetative cycle beyond mid-June, are by preference cultivated in agro-ecological zones with high water-holding capacities in the profile as in mapping units 1, 3 and 51 (Orthic and Gleyic Luvisols).

In other words, the different soils within the area, or the agro-ecological zones, will depict a completely different crop potential not only in terms of crop types but also in yields. Because the latter are in many cases a direct result of a temporary water-stress farmers in the area are considering the economic impact of the installation of sprinkler systems.

6. CONCLUSIONS

The procedure outlined above is based on a matching exercise between quantified environmental conditions and crop-specific growth requirements, using a scientifically solid framework which leaves little space for subjective interpretations. The integrated approach linked to the concept of the agro-ecological zone overcomes, moreover, the shortcomings related to a too narrow single soil survey interpretation or a too exclusively agro-climatic conclusion.

The system provides also a tool to compare the suitability of the land for different crops or land utilization types in general and, hence, provides alternatives for different scenarios. Such alternatives may also cover non-agricultural purposes. In the European context, and particularly in the light of the future agricultural policy, the results and alternatives proposed through this approach could be helpful in conflicting situations between farmers and environmentalists and as such provide a basis for discussion on different scenarios and on their impact on the evolutions or protection of the lands involved.

REFERENCES

1. DUMANSKI, J. and STEWART, R. B. (1983). Crop production potentials for land evaluation in Canada. Land Resources Research Institute, Agriculture-Canada, Ottawa, 80 p.
2. EC (1986). Soil map of the European Communities at scale 1/1 000 000. Directorate General for Agriculture, Coordination of Agricultural Research, Brussels, 7 sheets + explanatory text, 124 p.
3. FAO (1976). A framework for land evaluation. *FAO Soils Bull.*, **32**, Rome, 79 p.
4. FAO (1978). Report on the agro-ecological zones project. Vol. I: Methodology and results for Africa. World Soil Resources Report 48, Rome, 158 p.

388

5. FAO (1983). Guidelines land evaluation for rainfed agriculture. *FAO Soils Bull.*, **52**, Rome, 237 p.
6. FRERE, M. and POPOV, G. F. (1979). Agrometeorological crop monitoring and forecasting. FAO Plant Production and Protection Paper 17, Rome, 64 p.
7. HIGGINGS, G. M. and KASSAM, A. H. (1981). The FAO agro-ecological zone approach to determination of land potential. *Pedologie*, **31**, 147–68.
8. MAGALDI, D. and RONCHETTI, G. (1984). Report on the developing project for land evaluation in Italy on a 1:1 million scale. In: *Progress in Land Evaluation* (Eds Haans, Steur and Heide), A. A. Balkema, Rotterdam–Boston, pp. 57–63.
9. SYS, C. (1978). Evaluation of land limitations in the humid tropics. *Pedologie*, **28**, 307–35.
10. VERHEYE, W. (1984). Land evaluation methodology and interpretation of the EC soil map of Europe. In: *Progress in Land Evaluation* (Eds Haans, Steur and Heide), A. A. Balkema, Rotterdam–Boston, pp. 67–75.
11. VERHEYE, W. (1986). Quantified land evaluation as a basis for alternative land use planning in Europe. *Proc. ISSS Workshop on Quantified Land Evaluation, Washington, DC*, May 1986 (in press).
12. VERHEYE, W. (1986). *Principles of Land Appraisal and Land Use Planning within the European Community. Soil Use and Management*, Oxford (in press).

Optimizing the Use of Soils: New Agricultural and Water Management Aspects

J. H. A. M. STEENVOORDEN

Institute for Land and Water Management Research (ICW),
PO Box 35, 6700 AA Wageningen, The Netherlands

and

J. BOUMA*

Netherlands Soil Survey Institute (Stiboka),
PO Box 98, 6700 AJ Wageningen, The Netherlands

SUMMARY

Recovery, protection and strategies for prevention of soil pollution or degradation are to be based on quantitative analyses of soil inputs and outputs as a function of soil and water management. Groundwater and surface water pollution with nitrogen and phosphates can not only be restricted by reducing fertilization rates, but also by an intensification of agricultural practices including application of (sprinkling) irrigation, drainage and manipulation of water-table levels: a vigorous crop will adsorb more nutrients from the soil. Soil structure degradation may be reversed by deep loosening of compacted layers followed by new cropping systems leaving organic residues. Minimum or zero tillage may result in a more favourable soil porosity for plant growth, and may reduce surface erosion owing to higher infiltration rates. Optimization of soil use can be based on a quantitative analysis of new management aspects by computer simulation of soil water and solute transport in association with crop growth. Soil types represented on soil maps contain soil horizons that can be used to focus basic physical and chemical sampling for simulation. Thus, vulnerability maps can be derived from soil maps and predictions can be made for the effects of different scenarios. Major

* Present address: Bodenkunde en Geologie, Landbouw Hogeschool, PO Box 37, 6700 AA Wageningen, The Netherlands.

differences may exist within the same soil unit, however, because of the occurrence of different soil structures as a function of different past and present land-uses.

1. INTRODUCTION

Agricultural production within the countries of the European Community has strongly increased during the last decades as a result of, amongst other factors, increased use of fertilizers and pesticides and widespread mechanization of farming operations. This increase has been beneficial to our community in many ways. However, there is increasing evidence of adverse effects of modern agricultural practices on soil and groundwater quality. This, in turn, has in some cases unfavourable implications for public interests such as water supply, food quality, recreation, forestry and nature conservation. Public awareness of possible adverse effects of intensive agriculture has provided an impulse towards developing modified or new production systems. To do so, fundamental knowledge about basic physical, chemical, hydrological and biological processes in soil has to be available. To allow communication among researchers within and among countries, it is necessary to define unified and quantitative research procedures. This will stimulate the development of optimized new agricultural, soil and water management practices at different scales. In this chapter some examples are reviewed of such practices and of the use of quantitative modelling and simulation techniques. The usefulness of soil survey data for obtaining data for models is discussed, because soil surveys are available for many countries within the Community.

2. MANAGEMENT OF SOIL CHEMICAL ASPECTS

2.1. Introduction
Management of soil chemical aspects can be defined as the application of available knowledge to reduce or to prevent pollution of soil and water. Soil and water can be polluted by many compounds used to improve or to protect agricultural production like heavy metals, crop nutrients and pesticides [19]. The consequences of the use of these compounds for soil and water quality depend on the added amount and processes taking place in the soil. Processes may be rather complex sometimes but more and more

fundamental knowledge becomes available to quantify environmental consequences. One of the present challenges in agriculture is to develop agricultural management systems which take care of the desired quality of the environment. These qualities may be different from region to region. Possibilities of introducing new agricultural and water management techniques for optimizing the use of soils will be illustrated for nitrates and phosphates as examples.

2.2. Nitrate pollution

For the indication of vulnerability for nitrate leaching, soils are normally divided into groups based on soil type and soil use. In this approach a certain nitrogen input on arable land on sandy soil results in a fixed nitrate leaching loss. However, crop uptake appears to play an essential role in the fate of the applied fertilizers. Sandy soils differing in crop production potential will cause different nitrate loads on groundwater at the same fertilization level (Fig. 1). In the same year higher nitrate concentrations are

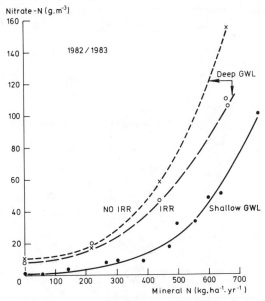

FIG. 1.   Influence of mineral N-supply (kg/ha/yr as N) on nitrate leaching (g/m^3 as N) in the period of October 1982 till April 1983 below grassland fields with shallow groundwater (roughly 1 m below soil surface) and deep groundwater (more than 2·5 m) with and without sprinkler irrigation.

found below grassland with a deep groundwater level than with a shallow groundwater level. The higher nitrate concentration at the deep groundwater level can be explained by lower N uptake in the grass due to a lower capillary rise. Moreover, indications are found that higher denitrification at shallow groundwater levels is partly responsible for lowering of the nitrate leaching loss [29, 15]. An additional supply of water in dry periods on the grassland field with deep groundwater level improves crop uptake and reduces leaching. The crucial role of crop uptake can be demonstrated by relating nitrate leaching to the difference between nitrogen input (I) and crop uptake (C) instead of input alone. The amount of nitrate leached (L) can be described as a function of this difference: $L = F(I - C)$, where the function F will depend on climate, soil type and water management. In this case the function F appears to be a linear one (Fig. 2). Comparable results have been found for arable land [29, 16]. The given formulation for leaching demonstrates the possibilities of manipulating nitrate leaching.

Crop growing conditions can be improved by changing soil physical conditions (see Section 4) or by means of irrigation in dry periods. An extra crop uptake can be realized by growing a cover crop on arable land after

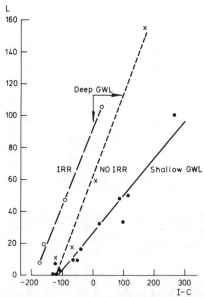

FIG. 2.   Relation between mineral N-supply (I) minus crop uptake (C) (kg/ha/yr as N) on nitrate leaching (L, in g/cm^3 as N). For explanation see text for Fig. 1.

harvest of the main crop [31, 2]. Fertilization advice could be made more site-specific by taking into account differences in soil physical properties. Advice is also based on field experiments carried out during a restricted amount of years. Climatic differences among years are responsible for great variation in the results. Fertilizing advice could have a more firm base by using hydrological and nitrogen models for simulation of crop growth over a number of years. Surface application of slurries gives rise to large uncertainties with respect to nitrogen recovery in crops due to large variations in ammonia volatilization [4]. Injection of slurry or direct ploughing after surface spreading leads to a better utilization of slurry N and a replacement of mineral fertilizer N. Moreover ammonia deposition on surrounding forests and nature areas is strongly reduced. It has already been discussed that manipulation of water level influences nitrate leaching. The effect of water-table depth on leaching can be translated in a reduction coefficient by which nitrate leaching from deeply drained soils must be multiplied (Table 1).

In regions where the groundwater level can be manipulated a systematic approach should be developed to improve planning, design and operation of surface water management systems [3].

2.3. Phosphate pollution of ground and surface waters
The given formulation for nitrate leaching applies in principle also to the phosphate leaching problem. Accumulation in the soil of excess P, however, is responsible for a time lag between overdosing and leaching [19]. Phosphate pollution of surface waters from fertilizing activities can take place via leaching and surface run-off. Surface run-off plays a role in periods with high precipitation rates on soils with a reduced permeability for water due to soil physical characteristics or a low storage capacity resulting from a high groundwater table. During transport over the soil surface nutrients are dissolved from the soil or from fertilizers on the soil surface resulting sometimes in high loads on surface waters [26, 28]. The best method to reduce surface run-off pollution is by putting fertilizers in the soil. Injection

Table 1
The reduction coefficients by which the nitrate leaching from deeply drained soils must be multiplied as a function of the mean depth of the groundwater table in winter [25]

Mean water-table depth in m—surface	0·0	0·2	0·3	0·4	0·6	0·9	>1·0
Reduction factor	0·04	0·10	0·15	0·22	0·41	0·73	1·0

of slurry on grassland is already practised on sandy soils in the
Netherlands. Groundwater transport partly may take place via shallow soil
layers where phosphate accumulation has already occurred. In this
situation we deal with soils which have a sub-optimal groundwater level
from the agricultural point of view. Depending on groundwater level class
the water table may occur for many months above 20 cm below the soil
surface [27]. Shallow groundwater transport may be responsible for an
important extra P load on surface waters. Lowering of the groundwater
table will reduce this risk of surface run-off and shallow groundwater
transport [30].

3.   MANAGEMENT: MECHANICAL AND PHYSICAL ASPECTS

3.1. Introduction

Mechanical and physical soil degradation phenomena are well known and
have been researched at many research centres. Modern high-technology
agriculture relies on mechanization and restricted crop rotations. This may
result in soil compaction and puddling as well as in a decrease of soil
stability due to decreasing organic matter contents. Lower stability may
enhance soil erosion and slaking, which, in turn, reduces infiltration rates
and soil aeration.

Various new soil and water management procedures are being tested in
an attempt to characterize and reverse physical soil degradation. Reference
is made to several excellent review papers [1, 13, 21, 22, 36]. Some Dutch
examples will be discussed as case studies.

3.2. Reversing soil compaction

When discussing recovery and strategies for prevention of soil degradation,
it is crucial to realize that the soil processes involved are highly correlated
with each other and that these processes are quite different in different areas
of the EC. Restricting attention to the northern countries, a tendency can
be perceived to increasingly emphasize a biological component in the
overall management scheme. This may mean a more systematic inclusion of
green manure in the cropping cycle and uses of various soil covers and/or
minimum or zero tillage. Increased soil stability due to higher organic
matter contents also reduces the erosion hazard. Formation of macropores
by biological activity increases infiltration rates, and creates a heterogeneous pore system that is more stable than the one in which large pores
between clods are formed by soil tillage. The case study to be discussed here

centred on the effect of sub-surface compaction by agricultural machinery, which had resulted in a plough pan below the Ap horizon in a sandy loam, Dutch marine clay soil. Deep ploughing to disrupt the plough pan was applied locally, with poor results. Several soil-structure types could be distinguished during the soil survey of this particular soil type. These types, which corresponded with certain tillage practices, could be observed reproducibly by different surveyors. Differences were expressed as soil-structure phases for a particular soil type during soil survey (e.g. [20]). The hydraulic conductivity of the primary plough pan was an average low 8 cm/day, but the secondary plough pan had a significantly lower average value of 3 cm/day. Meadowland of the same soil type, had a relatively high biological activity and a much higher conductivity of 150 cm/day owing to macropores made by roots and earthworms (Fig. 3). Comparable structures

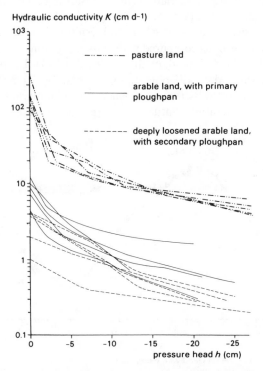

Fig. 3. Hydraulic conductivity curves derived from five samples of each soil: pasture land, arable land with a primary plough pan and deeply loosened arable land with a secondary plough pan formed in three years. Depths of all samples: 25–45 cm (from [20]).

can be obtained when using zero or minimal tillage and when incorporating crop residues in the soil. The question as to the effect of these various structures on particular land qualities, such as workability, moisture deficit and aeration status can be studied by computer simulation of the water regime using physical characteristics for the different soil-structure types. Simulation requires availability of basic physical data such as moisture retention and hydraulic conductivity for major horizons, as well as rooting patterns (Fig. 3). The only alternative for simulation is long-term monitoring of the water regime in field experiments with different management schemes which is, unfortunately, not feasible because of obvious practical and budgetary reasons. Measured physical data were used in a water simulation model applied to potato growth [23]. Results indicated that deep ploughing of a plough pan was not effective because a new secondary plough pan had formed within 3 years when common planting and harvesting practices were not changed. A simulation run was made using physical data for grassland and results indicated significant and favourable differences as compared with soils containing plough pans as is illustrated in Fig. 4 for the land quality workability [23]. The probability of having a workable day will potentially increase by 20–30% in crucial periods as a result of soil-structure regeneration. Comparable results were obtained for the aeration status. These results encourage attempts to

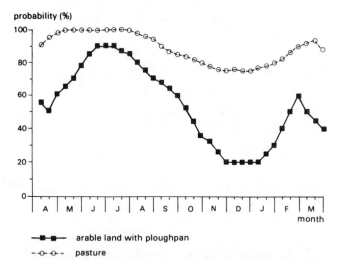

FIG. 4.   Probability of occurrence of workable days with adequate trafficability on a sandy loam soil with a biologically active soil structure (pasture) and with a plough pan (from [23]).

modify soil management by incorporating cropping schedules and soil tillage practices that allow a higher degree of biological activity than is associated with conventional tillage. Simulation, in fact, allows the prediction of soil physical conditions as a function of management as expressed by soil structure.

3.3. Effects of drainage or water extraction

A detailed study on the effect of soil drainage on arable crop yield was recently reported by Van Wijk and Feddes [35]. They used different simulation models to evaluate the effects of different drain depths on potato yields in two soils (Table 2). Drainage of wet soil allows earlier tillage and planting and emergence of the crop. Relative yields in spring are therefore higher. Relatively deep water tables in summer may, however, result in lower yields owing to reduced unsaturated flow from the water table to the root zone (Table 1). The overall effect indicates optimal water-table levels at drain depths of 120 cm in both soil types. Data indicate average values for a 30-year period, that are composed of very different years. These yearly values are of course obtained by simulation. This example illustrates use of simulation techniques for defining optimal water management schemes. The procedure can also be used to analyse irrigation practices, including a

Table 2

30-year averages of relative yields of potatoes due to effects of drainage in spring (Q_{pot}/Q_{max}), in summer (Q_{act}/Q_{pot}) and the total combined effect (Q_{act}/Q_{max}) at five different drain depths for a sandy loam and a 40 cm loam on sand

$D\,(cm)$	$\dfrac{Q_{pot}}{Q_{max}}$	$\dfrac{Q_{act}}{Q_{pot}}$	$\dfrac{Q_{act}}{Q_{max}}$
Sandy loam			
60	0·83	0·97	0·81
90	0·92	0·95	0·87
120	0·98	0·94	0·92
150	1·00	0·90	0·90
180	1·00	0·87	0·87
40 cm loam on sand			
60	0·93	0·92	0·86
90	0·96	0·90	0·86
120	0·99	0·88	0·87
150	1·00	0·85	0·85
180	1·00	0·81	0·81

characterization of fluxes down to the water-table which are important for groundwater pollution.

A second example, relating to the effects of water-table levels on soil use, is derived from groundwater extraction practices for industrial and municipal uses [8]. Extraction of groundwater implies lowering of the water table which, in turn, may result in lower yields due to reduced upward, unsaturated flow from the water table to the root zone. Some soils are more vulnerable than others (see Fig. 5). Soils with deep water tables (type 4) are not affected. Soils with a thick surface horizon having a high water holding capacity (type 1) are less affected than soils with a shallow surface soil and a coarse sub-soil that allows little upward, unsaturated flow (type 3, and to a lesser extent type 2). The more loamy texture of the soil of type 2 allows more upward unsaturated flow, while surface soil has a higher water-holding capacity. The graphs in Fig. 5 illustrate the relative vulnerability of the different soils for water extraction in terms of reduced yields. Availability of such data for major soils in an area allows optimization of the extraction procedure. It also allows calculation of cost, in terms of reduced yields, that results from the extraction procedure.

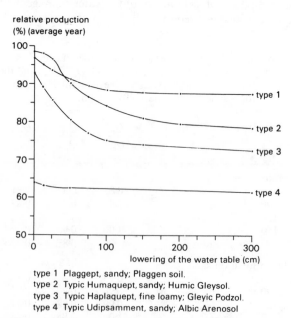

type 1 Plaggept, sandy; Plaggen soil.
type 2 Typic Humaquept, sandy; Humic Gleysol.
type 3 Typic Haplaquept, fine loamy; Gleyic Podzol.
type 4 Typic Udipsamment, sandy; Albic Arenosol

FIG. 5. The effect on relative production of lowering of the water table on four soil types [8].

3.4. Need for an integrated analysis

The rather arbitrarily chosen examples presented above, demonstrate the close interaction between physical, chemical and biological processes in soil. Development of a scientific basis for soil protection within the European Community is in our opinion only feasible when the effects of these processes are expressed in an integrated manner. Simulation and modelling are eminently suitable for obtaining an integrated approach, particularly because they are based on basic scientific processes that are valid everywhere, even in all EC countries. Difficulties that arise when attempts are made to combine various more qualitative national systems of soil characterization should not be underestimated. The examples presented also illustrate another important aspect which is the definition of effects of various soil and water management schemes for varying weather conditions. Thus, optimal management schemes can be selected from a wide range of possibilities. These possibilities are, however, a direct function of the soil type being studied. Each of these has a limited, characteristic range of properties to be realized by management. The major challenge for soil survey in the future is to define these characteristic ranges for different soil types.

4. MODELLING AND SIMULATION

4.1. Introduction

Soil protection strategies should be based on application of available knowledge about mechanical, physical, chemical and biochemical processes in soil. Complexity and interdependency of processes have stimulated the construction of models for various purposes. Some have been developed for a better understanding of processes while others can give results useful for management purposes on field or regional scale. Nitrogen can be used as an example to illustrate the complexity and interdependency of processes and to present some recent developments in modelling.

4.2. Modelling of nitrogen processes

Nitrogen in soil can be present in different forms: dissolved (NH_4^+, NO_2^-, NO_3^-, organic), solid (organic) and gas (NH_3, N_2, N_2O). Conversion of nitrogen compounds takes place via physical, chemical and biochemical reactions, which may be interrupted by adsorption and desorption. Nearly all reactions are influenced by pH, temperature, moisture and organic carbon. Moisture content, in turn, depends on climatic factors, soil physical characteristics and the regional hydrological situation. Soil compaction by mechanization might lead to changes in soil physical characteristics. It can be concluded that crop uptake and nitrate leaching are influenced by an overwhelming amount of processes and variables and that modelling is the only remedy to gain maximum profit from available knowledge.

Three main fields for application of models may be distinguished: study of processes, advisory work and forecast of environmental consequences. The understanding of the processes in which nitrogen compounds are involved has initially been an important justification for modelling [17]. Recently comparison and evaluation has taken place of available models for the assessment of the nitrogen fertilizer requirement of crops [24]. More accurate fertilizer advice prevents excessive mineral fertilizer application and reduces leaching losses. Some of the models made for fertilizing advisory work also produce information on leaching losses. The results can be applied on a field level. Regional hydrological models are the basis for the calculation of the regional environmental impact of activities in the field of water management and agricultural nutrient inputs. One example is the study into long-term effects of different nitrogen fertilization scenarios on nitrate concentrations in groundwater for 152 public water supply stations [32]. In this study an analysis has been made of the consequences of the fertilization level in the protection zone and in the region outside the protection zone. Quantification has taken place of the number of supply stations where the maximum acceptable nitrate level for

Table 3

Number of catchments in which the mean concentration of the infiltrating water in the aquifer exceeds the maximum accepted concentration of $11 \cdot 3 \, g/m^3 \, NO_3\text{-}N$ for municipal water supplies at different restrictions in fertilizer application [25]

Protected area with residence time for groundwater $\leqslant 100$ years	Total region outside protected area				
	No restrictions. Slurry production of 1982	No dumping	Spring application	Sub-optimal fertilization 75%	50%
No restrictions	75	—	—	—	—
No dumping	72	66	—	—	—
Spring application	61	58	41	—	—
75% N-requirement	52	48	34	25	—
50% N-requirement	42	35	22	13	10

drinking water was exceeded (Table 3), of the nitrate removal costs and of the loss of income in agriculture at lower fertilization levels. A second example is the study carried out into the effects of water management on the often competing interests of agriculture, nature and public water supply. For this study amongst others a set of models has been developed to support decision making on a regional level [14]. One of the models, the agricultural nitrogen model, ANIMO, deals with simulation of nitrogen processes including water transport, aeration and organic carbon processes [5]. Information on regional environmental effects can be calculated by coupling ANIMO to a regional hydrological model. It can be concluded that a lot of effort is put into modelling of nitrogen processes and that the results look very promising with respect to development of strategies for soil protection and taking into account the different interests of soil users.

5.   APPLICATION OF SOIL SURVEY

5.1. Introduction

Optimizing the use of soil requires first a definition of actual soil and water conditions as well as potential conditions to be achieved by various agricultural and water management practices. Optimal conditions usually require a delicate balance between demands by production-agriculture on the one hand and ecological requirements on the other. Optimal conditions are characteristically different in different soils, not only because of different physical and chemical soil properties, but also because of different climates. Soil surveys record soil properties which can be used to define agricultural and environmental land qualities, as has been discussed in this symposium by Tavenier, Lee and Verheijen. In addition, modern modelling and simulation techniques can be quite effective to predict the effects of different soil and water management schemes in quantitative terms. Such predictions can be used to define optimal conditions and associated management practices (e.g. [7, 35]). Some examples of modelling were presented in the previous section. The purpose here is to review contributions to be made by soil survey to modelling, which is seen as an attractive scientific procedure for deriving various management scenarios for soil protection.

Soil maps are based on legends that consider differences among soils on the basis of soil formation (pedology), such as, for example, leaching of clay, in-situ weathering or accumulation of organic matter, which has resulted in the formation of a sequence of specific soil horizons. Soil formation is a slow process that may have taken thousands of years. When using soil maps for modern interpretations, one should be aware of the fact that some reported features have no direct bearing on interpretations in a modern context emphasizing soil protection. Not all pedological differences are also functionally different. However, data being gathered during soil survey, such as soil texture, structure and organic matter content, are important by themselves and can be used to predict other more complex soil characteristics that are more difficult to measure. We can think, for example, of hydraulic conductivity and moisture retention characteristics that are basic data for simulation of the soil-moisture regime, but also of soil-chemical characteristics such as Phosphate Sorption Capacity, Denitrification potential, etc.

5.2. Soil survey data for modelling

Soil survey reports are a source of basic soil data that are presently inadequately being used. Some examples are:

1. Relatively simple soil data that can easily be obtained in the field (e.g. texture, organic matter, but also identification of soil horizons) can be related to more complex properties, such as hydraulic conductivity or moisture retention. Functions obtained are called Transfer Functions [9]. These functions are of crucial importance for increasing the use of soil survey information in the future. Various applications require different transfer functions. An illustration is provided in Fig. 6 showing a Dutch plaggen soil with eight pedological horizons. Different soil characteristics, as collected during soil survey by field and laboratory measurement, are used to estimate the Cation Exchange Capacity (using clay and organic matter content), the travel time (using measured permeabilities in well defined soil horizons) and the Phosphate Sorption Capacity (using Fe and Al contents) [10].

2. Theories and procedures to simulate water and solute movement and uptake by plant roots, are not always suitable to represent field conditions in terms of occurrence of peds, macropores and contrasting soil horizons (e.g. [33]). Soil structure data can be used to define optimal sample volumes for physical or chemical measurements, placement of monitoring equipment in situ, choice of methods and for defining flow patterns in well structured soils with macropores (e.g. [6, 11]). We have found that the principle 'Look and measure' is useful for avoiding or at least diminishing unreliable measurements.

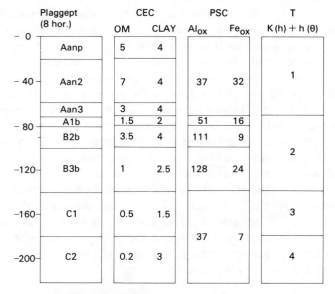

FIG. 6.   Illustration of the procedure to derive functional criteria from soil survey data, using soil horizons as 'carriers' of information. Different applications result in different groupings of soil functions. Different transfer functions are defined (see also [9]).

3.   Soil horizons can be characterized by physical or chemical methods facilitating areal interpolation of data because of the three-dimensional character of soil horizons in a landscape. Sometimes, different pedological horizons have identical physical or chemical properties. Combining them reduces the number of distinctions to be made and, therefore, calculation time (e.g. [34]). The example in Fig. 6 illustrates the principle. All pedological horizons were not functionally different. Eight horizons were still distinguished for the CEC, but only five for the PSC and four for the travel times.

5.3. The problem of scale
Soil protection and strategies for prevention of soil pollution and degradation are being considered at different levels of scale. For the EC, a

map scale of 1:1 000 000 is appropriate. Larger scales are used in individual countries. In the Netherlands, for example, a national soil survey at scale 1:50 000 is being carried out that can be used at regional level. On farm level a more detailed scale of 1:10 000 or larger is necessary to define management support systems. Definition of 'representative profiles' for mapping units, allows land evaluation for areas of land. However, a problem may be the unknown internal variability. In principle, mapping units that are relatively variable allow a less detailed interpretation as compared with those that are relatively homogeneous. Generally, mapping units of small scale maps are likely to have a higher internal variability than units of larger scale maps which cover a smaller area of land. Interpretations should therefore be less detailed. Detailed definitions of 'representative profiles' for small-scale maps are therefore often quite misleading as they suggest the possibility of making rather specific interpretations for extensive areas of land. Determination of realistic levels of generalization when interpreting soil maps is still a major, so far rather undeveloped, area of study within soil survey.

5.4. Showing effects of different management scenarios

Soil maps are excellent geographical information systems as such. The use of modern computer facilities (e.g. [18, 12]) allows almost instant production of interpretative maps and use of all gathered soil survey data if these data are stored in a data base. This rapid availability is of crucial importance for modern applications in the context of land use planning, where alternative scenarios have to be made at short notice. Geographical information systems have specific functions at different scales. New 1:250 000 maps of the United Kingdom and the Netherlands have been used effectively to define environmental land qualities on a national level. In the Netherlands, this related to, for example, phosphate sorption capacity, trafficability for manuring in winter time and early spring and potential nitrate pollution of groundwater. In addition to soil data (see Section 5.2), other environmental data are needed, such as weather data, land-use data and a record of fertilization practices. Remote sensing techniques can be quite useful to derive land-use data. Application of simulation models and soil information systems offers great opportunities for applications of soil survey data in the future. However, the link with the field should never be lost. Sometimes it appears that feeding simulation models is more difficult than their conception and development. To obtain truly representative physical, chemical and biological basic data for soils in the field is a continuing challenge for soil survey in the future.

6. CONCLUSIONS

1. Soil protection strategies should be based on an integrated consideration of chemical, physical, hydrological and biological aspects of soil behaviour. Simulation and modelling are useful tools for this approach which can yield new insights. For example, chemical fertilization in combination with an optimized water regime using drainage and irrigation techniques, may result in more effective crop uptake of nutrients and less environmental pollution.

2. Recovery, protection and strategies for prevention of soil pollution and degradation require information about actual soil conditions as well as a range of potential conditions to be achieved by various soil and water management scenarios. Such ranges are characteristically different for different soils and climates.

3. Simulation and modelling are necessary quantitative tools for defining soil conditions as a function of different soil and water management scenarios because they use basic scientific principles that are universally valid. In contrast qualitative or semi-quantitative approaches are very difficult to unify.

4. Soil protection against pollution from agricultural sources can partly be realized at farm level by educating farmers and supplying them with the necessary information for management decisions. This information may concern: climate, water supply of crops, fertilization, soil tillage, pest control, etc. Computer software should be developed to support management decisions of farmers.

5. Decisions have to be made as to the most desirable scale at which research on soil protection has to be executed within the EC. Small scale investigations covering large areas of land (e.g. the 1:1 000 000 soil map of the EC) allow only highly generalized conclusions. Are these to be used in EC policy? Larger-scale maps at national, regional or local level are associated with evermore detailed data needs and yield increasingly specific conclusions. Emphasis should perhaps be placed on the development of unified procedures at different scales, to be applied in different countries.

6. Soil surveys are a widely available source of basic soil data. However, many data that are needed for a dynamic characterization of soil, as required for soil protection strategies, cannot directly be derived from soil survey. Transfer functions are being defined that relate soil survey data to other more complex

characteristics (e.g. hydraulic conductivity, phosphate sorption capacity, denitrification potential). This activity needs more emphasis in soil survey research.

REFERENCES

1. ALLMARAS, R. R. and DOWDY, R. H. (1985). Conservation tillage systems and their adoption in the United States. *Soil and Tillage Research*, **5**, 197–222.
2. ASLYNG, H. C. (1986). Nitrogen balance in crop production and groundwater quality. In: *Agricultural Water Management. Proc. EEC. symposium* (Eds A. L. M. van Wijk and J. Wesseling), A. A. Balkema, Rotterdam, pp. 291–303.
3. BAKEL, P. J. T. VAN (1986). Planning, design and operation of surface water management systems. Thesis, Agricultural University, Wageningen.
4. BEAUCHAMP, E. G., KIDD, G. E. and THURTELL, G. (1982). Ammonia volatilization from liquid dairy cattle manure in the field. *Can. J. Soil Sci.*, **62**, 11–19.
5. BERGHUYS-VAN DIJK, J. T. (1985). ANIMO. Agricultural Nitrogen Model. Nota 1671. Institute for Land and Water Management Research (ICW), Wageningen.
6. BOUMA, J. (1983). Use of soil survey data to select measurement techniques for hydraulic conductivity. *Agric. Water Managmt*, **6**, 2/3, 177–90.
7. BOUMA, J. (1984). Estimating moisture related land qualities for land evaluation. In: *Land Use Planning Techniques and Policies*, SSSA Special Publication 12. American Society of Agronomy, Madison, Wisc., pp. 61–76.
8. BOUMA, J., DE LAAT, P. J. M., VAN HOLST, A. F. and VAN DE NES, TH. J. (1980). Predicting the effects of changing water-table levels and associated soil moisture regimes for soil survey interpretations. *Soil Sci. Soc. Amer. J.*, **44**(4), 797–802.
9. BOUMA, J. and VAN LANEN, H. A. J. (1986). Transfer functions and threshold values: from soil characteristics to land qualities. In: *Quantified Land Evaluation*. Proc. ISSS/SSSA workshop, Washington, DC (in press).
10. BREEUWSMA, A., WÖSTEN, J. H. M., VLEESHOUWER, J. J., VAN SLOBBE, A. M. and BOUMA, J. (1985). Derivation of land qualities to assess environmental problems from soil surveys. *Soil Sci. Soc. Amer. J.*, **504**(1), 186–90.
11. BURKE, W., GABRIELS, D. and BOUMA, J. (Eds) (1986). *Soil Structure Assessment*, A. A. Balkema, Rotterdam–Boston, 91 pp.
12. BURROUGH, P. (1986). *Principles of Geographical Information System for Land Resources Assessment*, Monographs on Soil and Resources Survey 12, Oxford University Press, Oxford.
13. CANNELL, R. G. (1985). Reduced tillage in Northwest Europe—a review. *Soil and Tillage Research*, **5**, 129–79.
14. DRENT, J. and VAN WALSUM, P. E. V. (1986). Decision support system for water management in regions with conflicting interests. Institute for Land and Water Management Research (ICW), Wageningen (in preparation).
15. DUYNISVELD, W. A. M. and STREBEL, O. (1981). Tiefenverlagerung und

Auswaschungsgefahr von Nitrat bei wasserungesättigten Böden in Abhängigkeit von Boden, Klima und Grundwasserflurabstand. Landwirtsch. Forsch. Sonderheft 4.

16. DIJK, T. A. VAN (1985). De uitspoeling van mineralen op bouwland waaraan jaarlijks drijfmest wordt toegediend. Rapport 2–85. Instituut voor Bodemvruchtbaarheid, Haren (English summary).

17. FRISSEL, M. J. and VAN VEEN, J. A. (1981). Simulation of nitrogen behaviour of soil-plant systems. Centre for Agricultural Publishing and Documentation (PUDOC), Wageningen, 277 pp.

18. GILTRAP, D. J. (1984). MIDGE, a microcomputer soil information system. Proc. Soil information systems technology. 6th meeting ISSS working group on Soil information systems (Eds P. A. Burrough and S. W. Bie), PUDOC, Wageningen, pp. 112–19.

19. HAAN, F. A. M. DE (1986). Chemical degradation of soil as the result of use of mineral fertilizers and pesticides. Aspects of soil quality evaluation. Symposium on Scientific basis for soil protection in the European Community, Berlin, October 1986.

20. KOOISTRA, M. J., BOUMA, J., BOERSMA, O. H. and JAGER, A. (1985). Soil structure variation and associated physical properties of some Dutch typic Haplaquents with sandy loam texture. *Geoderma*, **36**, 215–29.

21. LAL, R. (1985). A soil suitability guide for different tillage systems in the tropics. *Soil and Tillage Research*, **5**, 179–97.

22. LAL, R. (1986). Impact of farming systems on soil erosion in the tropics. Transactions plenary papers XIII ISSS congress. Vol. 1, German Soil Science Society, Hamburg, pp. 97–111.

23. LANEN, H. A. J. VAN, BANNINK, M. H. and BOUMA, J. (1986). Use of simulation to assess the effects of different tillage practices on land qualities of a sandy loam soil. Soil and Tillage Research (submitted).

24. NEETESON, J. J. and DILZ, K. (1985). Assessment of nitrogen fertilizer requirement. *Proc. 2nd meeting N.W. European Study group for the assessment of nitrogen fertilizer requirement,* Institute for Soil Fertility, Haren, 168 pp.

25. RIJTEMA, P. E. (1986). Nitrate load and water management of agricultural land. In: *Agricultural Water Management. Proc. EEC symposium* (Eds A. L. M. van Wijk and J. Wesseling), A. A. Balkema, Rotterdam, pp. 303–13.

26. SHERWOOD, M. and FANNING, A. (1981). Nutrient content of surface run-off water, from land treated with animal wastes. In: *Nitrogen Losses and Surface Runoff from Landspreading of Manures. Developments in Plant and Soil Sciences*, Vol. 2 (Ed. J. C. Brogan), Martinus Nijhoff/Dr. W. Junk Publ., The Hague, pp. 5–18.

27. SLUIJS, P. VAN DER and GRUIJTER, J. J. DE (1985). Water table classes: a method to describe seasonal fluctuation and duration of water tables on Dutch soil maps. *Agric. Water Managmt*, **10**, 109–25.

28. STEENVOORDEN, J. H. A. M. (1981). Landspreading of animal manure and runoff. In: *Nitrogen Losses and Surface Runoff from Landspreading of Manures. Developments in Plant and Soil Sciences*, Vol. 2 (Ed. J. C. Brogan), Martinus Nijhoff/Dr. W. Junk Publ., The Hague, pp. 26–34.

29. STEENVOORDEN, J. H. A. M. (1985). Nutrient leaching losses following application of farm slurry and water quality considerations in the Netherlands.

In: *Efficient Land Use of Sludge and Manure* (Eds Dam Kofoed, Williams and L'Hermite), Elsevier Applied Science, London, pp. 168–77.

30. STEENVOORDEN, J. H. A. M. and BUITENDIJK, J. (1980). Oppervlakte-afvoer. In: *Waterkwaliteit in grondwaterstromingsstelsels*. Rapporten en Nota's 5. Comm. Hydr. Onderz. TNO, 's-Gravenhage, pp. 87–92.

31. STEFFENS, G. and VETTER, H. (1983). Stickstoffverlagerung nach Gülledüngung mit und ohne Zwischenfruchtanbau. *Landwirtsch. Forsch. Sonderheft*, **40**, 354–62.

32. Werkgroep Nitraatuitspoeling in Waterwingebieden (1985). Nitraatproblematiek bij grondwaterwinning in Nederland. Onderzoek naar alternatieve maatregelen. Rapport ns 12, Institute for Land and Water Management Research (ICW), Wageningen, 49 pp.

33. WHITE R. E. (1986). The influence of macropores on the transport of dissolved and suspended matter through soil. In: *Advances in Soil Science*, Vol. 3 (Ed. B. A. Stewart), Springer Verlag, Berlin, pp. 95–121.

34. WÖSTEN, J. H. M., BOUMA, J. and STOFFELSEN, G. H. (1985). The use of soil survey data for regional soil water simulation models. *Soil Sci. Soc. Amer. J.*, **49**(5), 1238–45.

35. WIJK, A. L. M. VAN and FEDDES, R. A. (1986). Simulating effects of soil type and drainage on arable crop yield. In: *Agricultural Water Management. Proc. EEC symposium*. (Eds A. L. M. van Wijk and J. Wesseling). A. A. Balkema, Rotterdam, pp. 97–113.

36. WIJK, A. L. M. VAN and WESSELING, J. (1986). *Agricultural Water Management. Proc. EEC symposium*, Arnhem, Netherlands 1985, A. A. Balkema, Rotterdam, 325 pp.

Soil Protection—Changes in Soil Characteristics and Species Dynamics: Landscape and Land Use Planning Requirements

G. KAULE

Institut für Landschaftsplanung, University of Stuttgart,
Keppler Strasse 11, 7000 Stuttgart 1, Federal Republic of Germany

SUMMARY

The various characteristics of soils are important determinants, not only for the cultivation of plants but also for the distribution of almost all species of plants and animals. Next to the primary land use (forest, agriculture, settlement) and structure of the landscape, the characteristics of the soil, especially the nutrient regime and soil moisture, influence species dynamics.

Along with the destruction of a variety of ecosystems through changes in land use, the nation-wide equalisation of soils into uniform, nutrient-rich, medium moist habitats has led to a population reduction in ca. 80% of all species. For the most part, these are species that were widely distributed as little as 30 years ago. These species are most certainly of quantitatively greater importance in ecosystems than those that are naturally rare.

The ecological effects of species loss cannot be properly evaluated in the short term, and cannot be predicted in the long term. The process of nutrient enrichment and the equalisation of the soil water budget has been going on for the last 150 years, but has been especially accelerated in the last couple of decades. It must be stopped, and in wide stretches of the country reversed. Addition of more nutrients must be curtailed, and greater changes in the water balance must be tolerated in order to increase the habitat variety available to plants and animals in this country.

This, of course, is not the only measure required to ensure the survival of our soils and plant and animal species; it should be regarded as an integral part of a wider programme including nature reserves, pollution control, and more controlled urbanisation and road construction.

INTRODUCTION

In the Federal Republic of Germany there are some 2700 vascular plant species and between 45 000 and 50 000 species of animals, excluding soil micro-organisms. Ever since the early days of environmental research it has been generally recognised that, for all species, soil plays a central role in ecosystems. Among other things, it is responsible for recycling.

Due to taxonomic difficulties and the complicated observation methods required, we know less about life in the soil, strictly speaking, than about the larger species that live on the surface, or only partly, underground. We generally consider it as a kind of 'black box'.

This is often necessary even in detailed studies, let alone in surveys of the large areas of land with which land use and landscape planning is concerned.

Consequently I can identify the effects of changes in soil characteristics on larger organisms but not on soil micro-organisms themselves. It might, however, be assumed that conditions do not generally differ to any great extent.

In the course of their evolution species have adapted to their habitats but habitats have also changed. In other words, there has been a co-evolution of species and soils.

The Federal Republic's red lists show that some 50% of all species across the whole spectrum of taxonomic groups are endangered to a greater or lesser extent. The situation is very similar in the other Member States of the European Community.

In the case of animals interesting to hunters, or identifiable as their competitors (wolves and bears), direct 'persecution' is a proven cause of species reduction. For endangered plant species (red list species), the reasons can be found in the eradication of specific habitats through drainage and land use changes (see Table 1).

In theory, it is easy to do something for endangered rare species (i.e. conservation of their habitats, designation and preservation of protected areas). This is not, however, the case for species which have large or very large territories.

A survey of biotopes which need protection has been undertaken in the Federal Republic of Germany. A system of protected areas is being planned. All that is now required is the political will to apply this system (3% to 10% of land area depending on natural features and use, in exceptional cases such as many Alpine areas or tidal flats up to 50% of land area; see Kaule [8]).

The decline and extinction of rare species has been documented but we are also increasingly noting a reduction in the population of species which are not yet threatened (see Meisel [12, 13]).

Twenty-five years ago the average number of species per $100\,m^2$ on meadowland used for farming purposes (at the time still a common practice) was 30. Today we often find only 10–15 species, in extreme cases only five.

The reduction is of the same order of magnitude on cultivated land. Before the period of more intensive farming (1950–55) there were 30–50 species compared with only 5–10 now. In the process, problem weeds have remained and harmless ones killed off (Kaule [8]).

A detailed study has been made of ditches in the Vier und Marschland areas (marsh land) by Martens *et al.* [11]. They differentiate between 11 types of ditch according to age (since last ditch clearing) and location in fen or clay soil, which provide a habitat for up to 73 characteristic species. With increasingly intensive use of land, and in particular eutrophication, the ditches contain only dominant common Reed and Reed Canary Grass and nettles; the number of species is reduced to between 10–20.

The same phenomenon has been observed in hedges. Instead of regional differences in the composition of species in hedgerows there is now a standard central European type (Reif [14]; Weber [18]; Kaule [8]).

Average losses are in all cases between 75% and 85% (see Kohler [10]). The red lists of endangered plant communities—such as that drawn up for Schleswig-Holstein—show that communities which were not even discussed 20 years ago are now threatened (Dierssen [3]).

It would be useful here to look at the factors which determine species distribution. Local environmental factors (water and nutrient supply, soil chemistry, heat balance throughout the year (continental or maritime climate) and light (sunlight)) are the dominant natural factors which, coupled with the features of the area (accessibility, migration levels and distribution mechanisms), determine actual population levels. These local conditions created a distribution pattern of communities in the original natural landscape which, as a result of human use and adoption, has been turned into a cultural landscape.

(1)   This pattern of use (forest, agriculture and settlements) has created a framework of different types of woodland, open land and synanthropic land, i.e. land used directly for human purposes.

(2)   Within this framework intermediate areas ('border structures') form a more detailed structure, usually known as the micro-structure (Söhngen [15]; Auweck [1]), either in the form of a

Table 1

Nature conservation and landscape planning: causes of decline in species
(Sukopp et al. [17])

Plant information

Causes

- Number of endangered species
- Percentage of endangered species

- Coastal vegetation and inland salt marshes
- Non-Alpine rock vegetation
- Alpine vegetation
- Areas of hygrophilous therophytes
- Areas of field weeds and short-lived ruderals
- Areas of long-lived ruderals and herbaceous plants and felled areas
- Areas of ground-cover plants
- Areas of couch grass/dry areas
- Dry/semi-dry grassland
- Vegetation of eutrophic waters
- Vegetation near springs
- Oligotrophic moorland, wooded moorland and moorland lakes
- Low shrubland and mat grass
- Meadows and pasture
- Wet meadows
- Sub-Alpine vegetation
- Wet woodland areas
- Xerothermic woody vegetation
- Deciduous and coniferous woods on acid soils
- Mesophilic deciduous woods including fir woods

No.	Influence	n	%	Data
3.	Mechanical influences	99	17·0	6 12 1 39 8 29 5 5 5 7 1
4.	Use of defoliants, land clearing, burning	81	13·9	3 4 18 1 10 5 5 3 2 29 14 10
5.	Changes in use	123	21·2	2 4 6 42 1 9 24 4 54 7 2 3
6.	Discontinuation of certain uses	172	29·6	1 6 25 1 5 50 1 1 20 26 7 60 7 2
7.	Introduction of exotic species	10	1·7	9 1 1
8.	Abandonment of certain crops	5	0·9	5
9.	Discontinuation of activities which damage the soil	42	7·2	20 3 6 7 3 6 1 1 1
10.	Drainage	173	29·8	1 2 12 2 1 3 27 5 97 5 1 34 10 10 2
11.	Eutrophication of soil	17	2·9	1 2 9 1 2 1
12.	Eutrophication of surface water	56	9·6	5 52 1 1
13.	Air and soil pollution	8	1·4	8
14.	Water pollution	31	5·3	26 10
15.	Destruction of ecotones	210	36·1	7 2 2 12 28 11 17 6 86 1 1 5 9 5 41 3 23 3 1
16.	Development of watercourses and lakes	69	11·9	1 3 5 2 4 1 36 4 19 2 3 1
17.	Creation of artificial watercourses and lakes	7	1·2	1 2 1 2 2
18.	Demolition and levelling	112	19·3	1 2 1 1 65 3 21 1 24
19.	Filling in and building	155	26·7	5 1 16 5 2 7 1 68 15 23 1 2 27
20.	Urbanisation of villages	20	3·4	7 11 3

'variegated' landscape or large uniform areas. Uniform landscapes have a different collection of species than variegated landscapes. This breaking up of the landscape also creates transition areas (ecotones) and many central European species are these ecotone species (see Sukopp *et al.* [17]).

(3) The type and frequency of use (e.g. grazing intensity and frequency), the number of mowings, the number of times land is worked and the amount of fertiliser and pesticides used determine the pattern of plant cover, which is also referred to below as the 'microstructure'.

In the last 150 years very little woodland area has been turned into farming land but the area under settlement has virtually doubled and now accounts for 12% of land. The percentage of forest area made up of coniferous trees increased; the size of fields in agricultural areas was enlarged at the cost of microstructure.

These changes in land use cannot, however, totally account for the disappearance of many species and the threat to virtually half the remaining flora and fauna (see red lists, data and bibliography in Blab *et al.* [2]).

In addition to quantitative changes in the amount of land used for different purposes and the disappearance of microstructures, there are at least two major qualitative changes that have taken place almost everywhere: trophic levels (i.e. the nutrient balance) and the water balance. For many plant and animal species, however, light and temperature are the decisive habitat determinants rather than nutrient levels (Ellenberg [4]). The majority of plant species now endangered require lots of sunlight but can tolerate low nutrient levels (Ellenberg [5]). They retreat to nutrient-poor sites, the only locations where they are not displaced by faster growing plants which overshadow them. Moisture and nutrient parameters determine—through the mediation of the site-adapted plant cover—to a large degree the solar radiation balance, and therefore heat balance of a site. This in turn significantly influences the living conditions for animal species which prefer comparatively warm conditions, but in the past few decades have had to 'settle' in areas with only partial or minimum ground cover.

The soil moisture and nutrient needs of vascular plant species are well known (Ellenberg [6, 7]). Over 80% of the indigenous species of central Europe (excluding Alpine species) need dry or damp to wet habitats with low or at most average nutrient levels. Over half of these species have adapted to extreme conditions.

It is, however, precisely these soil characteristics which have over the last 150 years, and in particular the last 30 years, changed virtually everywhere—in moving in the direction of nutrient-rich and medium moist habitats. To take the example of higher plants again, there are only 15% of potential species which can live in such habitats.

Over the 4000 years in which land has been cultivated in central Europe man has exposed the soil and herbaceous ground cover to light by cultivating meadows, pasture land and farm land. In many areas nutrients that were removed from the soil through the harvest of crops were only partly replaced by leaf litter, coppicing and grassing over. Only on certain small areas (gardens, land on which cattle are kept, wasteland near habitations and some fields) did nutrient enrichment take place.

Soil characteristics have tended to become uniform in our modern industrial and agricultural society as a result of the following:

1. The improvement of agricultural land (*inter alia* to turn grassland into arable land) and development of watercourses to ensure flood water is carried away faster.
2. Large-scale lowering of the watertable due to the abstraction of drinking and process water.
3. Increased fertilising of farming land (an average increase from 25 kg N/ha (1950) to over 100 kg N/ha). Peak levels of 350 kg N/ha have been recorded on arable land. There is also an annual input of 25–45 kg N/ha of additional nutrients throughout the country as a result of general air pollution (Ellenberg [5]; Weller [19]).

DEVELOPMENTS OVER THE LAST FEW DECADES USING THE EXAMPLE OF THE SAARLAND

In a joint study Kaule and Ellenberg [9] investigated and made a quantitative analysis of these factors for soils in the Saarland and three types of organisms (vascular plants, birds and butterflies). An attempt was made to take an integrated look at the following sets of data: local and state-wide land-use statistics, forestry statistics, agricultural surveys, geological, hydrogeological and soil maps, floristic and faunistic surveys; red lists of extinct and endangered species, statistics on the use of fertilisers and biocides, measurements and calculations of area-wide air pollution, indicator values for vascular plants, and habitat requirements for animal species and groups of species.

Table 2 shows that some 350 of the 1012 vascular species of plant studied

Table 2

Changes in areas in the Saarland with soils with different trophic and moisture characteristics and respective number of plant species

Soil characteristics	Very low to moderately low nutrient level			High to very high nutrient level		
	Number of species (red list species)	Surface area (km²) 1950	Surface area (km²) 1980	Number of species	Surface area (km²) 1950	Surface area (km²) 1980
Dry/very dry	300	90	65 125ᵃ	35	< 1	20
Moist	197	1 000	500	146	580	1 100
Wet/very wet	223	220	110	111	600	450

ᵃ Slag heaps with little plant cover.

(only grasses and herbaceous plants) are concentrated in areas with soil characteristics (trophic levels and moisture content) which account for some 60% of the total surface area. It should also be noted that the majority of these species of plant require a high level of exposure to light at a height of 0–50 cm. We have relatively few forest species (Table 3).

With the large amount of fertiliser used modern varieties of cereals grow quickly and soon shade the ground, thus displacing small field weeds which require light. There are virtually no more low-growing species in wet meadows as a result of high-nutrient flood water. In fallow fields (15–20% of land in Saarland) there are almost only tall species which cast shadows.

A comparison of these trophic-moisture and trophic-light needs of plant species in the Saarland with the available land shows the following:

250 of the species can grow on 90% of the total surface area;
766 of the species can grow on 5% of the total surface area;
5% of the total surface area is sealed (roads, buildings, etc.);
30 years ago 766 of the species could grow on 50% of the total surface area.

Figure 1 plots the habitat requirements of species of bird indigenous to the Saarland against moisture and trophic gradients. A double peak can be seen in the curve for extinct or endangered species in wet gradients with a slump in the moist range where species which are not endangered dominate. In the trophic gradients the red list species peak in the low-nutrient range.

Table 3
Changes in areas with defined levels of light exposure in the Saarland
and respective number of plants

	Number of species	*Surface area (km²)*	
		1950	*1980*
Very sunny to sunny (0–50 cm)	663	1 600	250
Semi-shade	246	125	650
Dark to very dark	115	700	1 400
Roadside, courtyards, roof terraces	—	80	200

Species which are not endangered 'prefer' or tolerate high-nutrient
conditions. In relation to the land available this means that in 1950
1310 km² and in 1980 677 km² could support endangered species in the
Saarland.

An important biotope requirement for butterflies is for light to be able to
penetrate the entire herbaceous layer so that as many flowers as possible are
produced and the vegetation structure can be formed. 'Modern' meadows
with dense grasses and few large herbaceous plants will be visited as little as
fields of crops with densely packed high haulms. Most 'edge' biotopes are

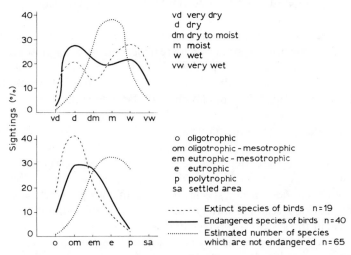

FIG. 1. Habitat requirements of species of birds in the Saarland in trophic and
moisture gradients calculated from number of sightings (Kaule [8]).

Table 4
Change in amount of land in the Saarland which can support
endangered species of butterfly

Light reaching lower herbaceous layer and soil	Number of endangered species	Surface area (km²)	
		1950	1980
Strong sunlight	79	1 600	250
Semi-shade	33	125	100 (650)[a]
Dark to very dark	0	700	1 400
Sealed surface	—	80	200

[a] Additional area which cannot be used since it is made up of
intensively farmed arable fields.

now also overshadowed by a few tall species. There are very few species of
butterfly with forest habitats (Table 4).

In the old cultural landscape approximately 70% of the land served as
biotopes for butterflies compared with only 10% now. In other words, there
has been an 85% reduction over 30 years in land which can be used by
butterflies as habitats or partial habitats.

CONSEQUENCES AND ACTION REQUIRED

In a system of protected areas where the direct input of nutrients is
prevented, some of the species which are dependent on a very low nutrient
level may survive. A second prerequisite is for measures to be taken to
reduce air pollution and to drastically reduce diffuse nutrient input. This is
not, however, enough for populations with large territories.

In no way could a system of protected areas which would cover 3–10% of
land fulfil the ecological functions of the soil which we would expect it to
have on a large scale.

Species can only be of quantitative importance to the natural balance in a
particular landscape where they are distributed over a wide area. The trend
identified encourages a few species throughout Europe which obviously
cannot provide a regional variety even if they are still found in 'left-over'
areas, i.e. in the system of protected areas.

The only measures which will have any real impact are to reduce the
nutrient input and to allow and encourage a variety of habitats from dry to
wet in all parts of the landscape.

The improvement and fertilising of soils to increase yields to supply the basic needs of the inhabitants of a country is quite certainly justified. It is another question, though, when, in order to satisfy non-essential luxury needs, habitat variety is reduced and species are eradicated or their populations radically reduced on a regional level.

But it is completely incomprehensible and immoral to allow and even officially subsidise over-production, with all the administrative costs this involves, which still has to be disposed of at the other end of the chain in the form of waste at the expense of future generations.

Without needing to, we have not only reduced the populations of other living beings but also the usefulness of natural resources for future generations. What we are doing is living on the land's principle and not its interest (Spaemann [16]: the chances of future generations are being reduced).

Obviously the landscape factors discussed here are the developments which man has initiated by improving and fertilising soils are only one portion of the range of factors which determine species distribution, affect the physical landscape balance and reduce the usefulness of natural products in the long term.

Other factors include the use of land for urban and residential development and traffic, the carving-up of the landscape, land-use changes, waste disposal and input of dangerous substances into the soil and water, and pesticides.

Clearly any measures in these areas to halt the decline in species which might be suggested in other chapters will only be effective if the groundwater and nutrient balance are also taken into account and developed in a manner which encourages species distribution.

Protected areas and specific technical measures such as sewage treatment plants will make only a limited, although important, contribution to solving environmental problems. Many individual measures such as retaining basins for floodwater will even be harmful in toto because although they will bring about improvements in specific areas they will have a negative effect over a large area. This reduces the pressure to develop more comprehensive solutions.

Another of these aspects which needs rethinking is the much discussed land-use planning objectives of large-scale zoning instead of greater mixed land use. The idea of large-scale differentiation into intensive farming areas and semi-natural landscapes is tempting to the European Community or large countries since it means that land in disadvantaged areas could be taken out of production and put into some kind of management programme, while at the same time industrialised agricultural production could be pursued in areas appearing suitable.

The ecological and social consequences would be enormous: farmers would be divided into producers and 'landscape gardeners' and there would be even greater separation of populated and recreation areas (at least in the Federal Republic of Germany where large cities are surrounded almost entirely by existing or potential areas of intensive farming). A decline in the usefulness of natural resources, for example groundwater, would result. There are large groundwater reserves under soils which could be used for intensive farming. When they can no longer be used there will be increased pressure for water to be imported from distant areas and 'dry out' these areas with all the ecological consequences this entails.

Soil erosion, being a physical phenomenon, first initiated the limits on intensive farming. As in the case of air pollution, sensitive species such as lichen, followed by firs, show reactions, followed by further species of tree until finally whole forest biocenoses were seen to be affected. It was not until this happened that the public became aware that important ecological functions had been lost.

Similarly, the effects of the trend towards uniformity of habitats were first noticeable in the water regime and the effects of excess fertilisation were only seen initially in individual species. Now we see them in many biocenoses. It will not be until later that the effects on the ecosystem's interrelationships, including soils and the functions they are expected to perform, will be seen.

It is also for these reasons that we cannot afford to abandon, in ecological terms, large areas of land. We must demand and carry through the development of carefully managed areas of ecological priority. Yes, this is the same old story about the need for large conservation areas that we have heard in the past.

However, this must not lead to intensive farming. In addition to a network of small protected areas and a general pattern of small structures such as hedges, balks, river banks, wet grassland, etc., throughout all farming areas the use of fertilisers and pesticides must be contained and reduced.

REFERENCES

1. AUWECK, F. (1978). Kartierung von Kleinstrukturen in der Kulturlandschaft. *Natur und Landschaft*, **53**(3), 84–9.
2. BLAB, J., NOWAK, E., TRAUTMANN, W. and SUKOPP, H. (1984). Rote Liste der

gefährdeten Tiere und Pflanzen in der BRD, erweiterte Neubearbeitung. Kilda, Greven (Reihe Naturschutz aktuell 1), 4. Aufl.

3. DIERSSEN, K. (1983). Rote Liste der Pflanzengesellschaften Schleswig-Holsteins. Schriftenreihe des Landesamtes für Naturschutz und Landschaftspflege Schleswig-Holstein 6, Kiel.
4. ELLENBERG, H. Jr (1983). Gefährdung wildlebender Pflanzenarten in der BRD. Versuch einer ökologischen Betrachtung. *Forstarchiv*, **54**(4), 127–33.
5. ELLENBERG, H. Jr (1985). Veränderung der Flora Mitteleuropas unter dem Einfluß von Düngung und Immissionen. *Schweizerische Zeitschrift für Forstwesen, Zürich*, **136**(1), 19–39.
6. ELLENBERG, H. (1979). Zeigerwerte der Gefäßpflanzen Mitteleuropas. *Scripta Geobotanica IX* (1. Aufl. 1974), Göttingen.
7. ELLENBERG, H. (1982). Vegetation Mitteleuropas mit den Alpen in ökologischer Sicht. Ulmer.
8. KAULE, G. (1986). *Arten- und Biotopschutz*, UTB, Ulmer, Stuttgart.
9. KAULE, G. and ELLENBERG, H. Jr (1985). Ursachen und Folgen der Artendynamik in Mitteleuropa, dargestellt und diskutiert am Beispiel des Saarlandes. *Zoologisch-Botanische Gesellschaft in Österreich*, **123**, 175–98, Festschrift, Wendelberger.
10. KOHLER, A. (1986). *Landnutzung und Artenschutz. Schriftenreihe Agrar- und Umweltforschung in Baden-Württ.*, Ulmer, Stuttgart.
11. MARTENS, J., KURZ, H. and GILLANDT, L. (1985). Biotopschutzkonzept in den Vier- und Marschlanden als Grundlage für das Artenschutzprogramm Hamburg. Ökologie Forum 7.6.84, Hamburg, Unweltbehörde, im Druck.
12. MEISEL, K. (1983). Veränderungen der Ackerunkraut- und Grünlandvegetation in landwirtschaftlichen Intensivgebieten. Landespflege und Landwirtschaft. *Dtsch. Rat für Landespflege*, **42**, 168–74.
13. MEISEL, K. (1984). Landwirtschaft und 'Rote-Liste'-Pflanzenarten. *Natur und Landschaft*, **59**(7/8), 301–7.
14. REIF, A. (1982). Vegetationskundliche Gliederung und standörtliche Kennzeichnung nordbayerischer Heckengesellschaften. Laufener Seminarbeiträge, ANL Laufen 5/82.
15. SÖHNGEN, H. H. (1975). Die Bewertung von Landschaftsbestandteilen für die landschaftspflegerische Begleitplanung in der Flurbereinigung. *Natur und Landschaft*, **50**(10), 274–5.
16. SPAEMANN, R. (1983). Technische Eingriffe in die Natur als Problem der politischen Ethik. *Ökologie und Ethik*. Reclam, Stuttgart, pp. 180–206.
17. SUKOPP, H., TRAUTMANN, W. and KORNECK, D. (1978). Auswertung der Roten Liste der gefährdeten Farn- und Blütenpflanzen in der Bundesrepublik Deutschland für den Arten- und Biotopschutz. *Schriftenreihe für Vegetationskunde*, **12**.
18. WEBER, H. E. (1982). Vegetationskundliche und standortkundliche Charakterisierung der Hecken in Schleswig-Holstein. Laufener Seminarbeiträge 5, 9–14. ANL Laufen.
19. WELLER, F. (1983). Stickstoffumsatz in einigen obstbaulich genutzten Böden Südwestdeutschlands. *Z. Pflanzenernährung Bodenkunde*, **146**, 261–70.

The Soil Protection Plan in the Federal Republic of Germany

BERND DELMHORST*

*Federal Ministry of the Environment,
Nature Conservation and Nuclear Safety,
Graurheindorfer Str. 198, 5300 Bonn 1, Federal Republic of Germany*

SUMMARY

Soil protection plays a central role in environmental policy in the Federal Republic of Germany today. On 6 February 1985, the Federal Government therefore passed a comprehensive Soil Protection Plan (Official Journal of the Bundestag 10/2977) which sets the framework for striking a balance between claims on using the soil on the one hand and the prevention of damage, hazards and long-term risks on the other. At the same time it states that natural resources are worth saving for their own sake. The Soil Protection Plan identifies two areas in which action is primarily to be taken: (i) the reduction of inputs of substances which pose a problem either because of their quality or because of the quantities involved, and (ii) changing land-use patterns including the exploitation of raw materials and changes to the natural state of habitats. Implementation of the Soil Protection Plan will entail integrating soil protection into all the Federal Government's existing legislation and regulations in preparation and working out a package of soil protection measures between the Federal Government and the Länder. This package of soil protection measures covers four areas: legislation, administrative implementation, research and information.

*Paper presented by Mr R. Schier.

Soil protection is a relatively new addition to the Federal Republic of Germany's environmental policy but it is steadily growing in importance for three main reasons:

1. The soil is overwhelmingly the start and finish of all human activity and as such has long been the centre of economic concern.
2. Germany's population density and intensive economic activity mark it as one of the industrialized nations which places by far the greatest demands on its environment and natural resources.
3. Despite the undeniable successes in reducing emissions of pollutants at source, in controlling sewage and waste disposal, and in setting up nature reserves to safeguard water supplies and biotopes, not enough has been done to prevent soil pollution.

After extensive preliminary studies to ascertain and assess all the chief factors affecting the soil, on 6 February 1985 the Federal Government therefore adopted a comprehensive Soil Protection Plan setting out the fundamental policy objectives plus a large number of lines of action covering legislation, planning, administrative implementation and scientific work. It provides a framework for striking a balance between the demands made on the soil by the various uses and the prevention of damage, hazards and long-term risks. At the same time it clearly states that, irrespective of their use to mankind, natural resources are worth saving for their own sake.

The idea behind the Soil Protection Plan is to forge the environmental protection schemes introduced in separate sectors into an interdisciplinary programme covering many different areas, i.e. to establish soil protection as the central component of an environmental policy which puts ecological considerations first.

Two central lines of action emerge on consideration of the areas of policy with an impact on the soil.

1. Minimization of inputs of substances from industry, commerce, transport, agriculture or the domestic sector which pose a problem either because of their quality or because of the quantities involved

This implies constantly striving to strike a balance, at the lowest possible level, between inputs of pollutants on the one hand and the natural regenerative functions of the soil on the other, by taking measures to limit emissions of pollutants at all sources and using land in a way designed to inflict minimum damage on the environment. In the long run, therefore, waste management and control of the cycle must, as far as possible, put an end to discharges of harmful substances into the soil, whether directly or indirectly via the air and water. Above all, rules on prevention and recycling are needed.

2. *Changing the pattern of land use*

More must be done to use land for purposes to which natural conditions at the site are best suited. This includes agricultural applications. Raw materials must be used sparingly and efficiently, for the sake of the economy and the environment as a whole. As a general rule, any remaining undisturbed areas, or areas still in almost their natural state, should be safeguarded. Before any more land is built on and developed, schemes should be promoted to maintain and regenerate disused land in urban areas, to make more careful use of land when building and to upgrade existing transport links where possible. More careful use of land would reduce the demands made on the soil and soil pollution. To achieve this, greater importance must be attached to ecological considerations throughout all the planning procedures.

These basic ideas have already been given firmer shape, in particular in the form of:

(i) protection of foodstuffs from hazardous substances, and in particular from substances which accumulate in the soil, are directly harmful or which enter the food chain;

(ii) protection against further acidification of the soil by persistent widespread emissions of sulphur dioxide and oxides of nitrogen at levels in excess of the soil's buffer and breakdown capacity;

(iii) protection of groundwater resources against any further damage by nitrogenous compounds in fertilizers or by plant protection agents;

(iv) cleaning up past pollution, particularly disused dumps and industrial sites;

(v) examination of all building schemes requiring a large amount of land;

(vi) more careful cultivation and working of farmland;

(vii) careful exploitation of raw materials deposits.

The Federal Government has decided on three measures in connection with soil protection legislation:

1. A general review of tangible measures to tailor legislation to protecting the soil and of ways of making fuller use of the existing legislation.

The individual protection measures will be directed towards foreseeable hazards. In the process, soil protection must take priority wherever public health or natural resources essential to life are in danger. Soil protection must be the decisive restraint wherever human activity could lead to pollution of the food chain or groundwater resources or to lasting disturbance of other functions dependent on the soil. This implies, in particular, setting limit values for persistent pollutants from all sources. Accordingly, future environmental protection instruments must take account not only of the primary objectives but also, explicitly and directly, of the side-effects on the soil. Priority must be given to taking precautions against avoidable damage wherever there are good reasons for fearing it, even if there is as yet no scientific proof of any cause-and-effect relationship between the factors which might have an adverse effect on the soil.

2. A special review of, *inter alia*, the Federal Immission Control Law, the Federal Clean Water Law, the Federal Nature Conservation Law, the Federal Law on Mining and the planning legislation.

3. An inventory of the categories of pollutants which, according to the information available, must be dealt with first in the soil protection measures, including heavy metals, substances which form acids, persistent organic matter, plant protection agents and solvents.

Implementation of the Soil Protection Plan will entail action at two levels.

1. Integration of soil protection in all planned and existing Federal Government measures

In this context, the measures particularly concerned include the laws on clean air (the 1985 Technical Instructions for Maintaining Air Purity and the Order limiting Emissions of Highly Volatile Hydrocarbons), on water (the Fifth Law amending the Federal Water Act), on waste (the Fourth Law amending the Waste Disposal Act), on nature conservation (the amendment to the Federal Nature Conservation Act, concerning the protection of species), on the protection of crops (the Plant Protection Act), and on building and planning permission (draft Building Statute Book, the Land Use Act, the Law on Trunk Roads and the 1985 Federal Transport Plan). Beyond that, if the soil is to be protected effectively, reliable information is needed, particularly for risk assessments and drafting decisions to protect the soil, whether by means of clean-up or preventive measures or by using land for applications best suited to the site, causing no pollution and saving land. To achieve this, fundamental knowledge is needed of a large number of different areas. But priority must be given to country-wide information on emissions, breakdown and residues of problem substances in the soil and groundwater. Any current backlog in assessing and collating the available data, including those on soil properties and land uses, must be eliminated. Accordingly, the research will concentrate on:

(i) emissions into the soil;
(ii) past soil pollution and clean-up measures;
(iii) farming and nature conservation;
(iv) land use;
(v) the practicability, efficiency and acceptability of soil protection measures;
(vi) harmonization of soil analysis procedures.

A special joint working party, combining representatives of the Federal and Länder (states) authorities, has been asked to gather the basic data needed specifically for soil protection measures, to harmonize data collection, processing and evaluation methods, and to standardize data storage.

2. Cooperation with the Länder

The Federal authorities and the Länder have agreed to prepare a comprehensive, coordinated package of soil protection measures covering four areas: legislation, research, information and administrative implementation. They have already produced a rough draft setting out measures, in order of priority, on all the central issues, including farming, nature protection, emissions into the soil and groundwater, past pollution, land use, protection of soil resources and basic information.

On a wider front, the chief objective of these talks between the Federal and Land authorities is to firmly establish soil protection as one of the all-pervading tasks of environment policy, from the technical, legal and organizational points of view.

Soil Protection Programmes and Strategies in Other Community Member States: Examples from The Netherlands

J. E. T. MOEN and W. J. K. BRUGMAN

Ministry of Housing, Physical Planning and Environment,
Soil and Groundwater Division,
PO Box 450, 2260 MB Leidschendam, The Netherlands

SUMMARY

In The Netherlands the introduction of the Soil Protection Act in 1987 marks the transition from a period of developing legislation and strategies for a soil protection policy to a period of implementation of this policy. Soil policy is embedded in general environmental policy, characterized by an integrated approach. A short review of the available instruments is given, with a description of the policy objectives, along the two tracks of effect-oriented and source-oriented policy. On the first track emphasis is laid on defining parameters for a good soil quality in quantitative terms. Source-oriented policy distinguishes between local sources and diffuse sources. Local sources have to meet the criteria of isolation, monitoring and control; for diffuse sources a link is made between the application of substances and the maintenance of a good soil quality. As an example, a short description is given of the forthcoming regulation on manuring.

1. INTRODUCTION

In The Netherlands in January 1987 a Soil Protection Act will come into force that will provide the legal basis for an integral protection of soil quality, soil in this respect being defined as including the groundwater, gases and soil organisms. This event marks the transition from a period of developing legislation and strategies for a soil protection policy to a period of implementation of the now available instruments. A major effort will be put into activities to prevent (further) deterioration of soil quality. The confrontation with a large number of contaminated sites in recent years has led to a large operation to investigate and to clean up these polluted sites, with a yearly budget of about Dfl 200 million [1]. It is expected that until the year 2000 some 5 million kg of polluted soil has to be cleaned, in situ or after excavation, at an estimated total cost of Dfl 3 million. The costs of this cleaning-up operation have stressed the importance of preventive measures for preserving the desired soil quality, as shown also by the consequences of the intensification of agriculture during the last decades for the present status of soil and groundwater. Fertilizers, manure and pesticides cause a substantial deterioration of soil and groundwater quality. At present special attention is paid to the nitrate problem. The EC permissible level for drinking water of 11·3 mg N/litre has been reached in 1985 in two groundwater pumping stations. In several private wells concentrations have been observed of more than 20 mg N/litre. To prevent further growth of the manure surpluses, at the end of 1984 the Interim Act Restriction Swine and Poultry Farms was adopted (see Section 4).

Soil policy in The Netherlands is embedded in general environmental policy, characterized by an integrated approach: measures taken to maintain the desired quality of one part of the environment, air, water or soil, should not lead to problem-shifting to another compartment. An example of such problem-shifting forms the problems of sewage sludge and the pollution of underwater soils which are now becoming recognized. Solutions to these problems must be found in a multi-sector approach both on a national and international level.

Environmental policy in The Netherlands is laid down in the Environmental Programs, a co-production of three departments, published yearly by the Dutch Government (also available in English) [2].

In this chapter a short description is given of the main axioms and objectives of environmental policy in general and soil policy in particular, with an indication of the measures being taken or foreseen.

2. INSTRUMENTS

An exhausting and detailed description of legal and other instruments falls beyond the scope of this chapter. It is therefore only a short overview.

Legal instruments. Aspects of soil protection are covered by a number of legal instruments such as the Chemical Waste Act, the Waste Substances

Act, the Mines Act, the Surface Waters Pollution Act, the Pesticides Act, the forthcoming Fertilizer and Manure Act, and instruments on a provincial and municipal level. Soil clean-up operations are performed and financed mainly on the basis of the Soil Clean-up (Interim) Act that will be incorporated into the Soil Protection Act before 1989. The Soil Protection Act will form an integrating framework for soil policy. It will provide the basis for general administrative orders on the regulation of activities that may lead to pollution or impairment of the soil, and on soil quality standards, applying to The Netherlands as a whole or to areas of certain categories (general protection level). According to this Act the provinces have an important role in soil protection, and in indicating areas that require a specific protection level, the so-called soil protection areas and groundwater protection areas. Soil protection areas are areas of particular value where the chemical, physical and biological properties of the soil are not or only to a slight extent adversely affected by human activities. Specific restrictions may be imposed on harmful activities. The provinces are obliged to initiate an inventory on the presence of such areas in their territory. The national government may provide guidelines to define such areas. Provincial authorities are obliged to draw up a plan for protection of the quality of the groundwater in their province, with a view to water supply, indicating areas that require specific protection to this end. They shall provide regulations to realize this specific protection level. Co-operation between provinces and national government has recently led to uniform models for both plan and regulations.

Other instruments. An important instrument in general environmental and soil policy in The Netherlands is planning. For a description the reader is referred to Ref. 2. Another instrument is research and monitoring of environmental data. Recently a national groundwater quality monitoring network was completed, in addition to the monitoring of the groundwater pumping stations. At the request of the Minister of Housing, Physical Planning and Environment this network is managed by the National Institute of Public Health and Environmental Hygiene. It consists of about 400 observation wells, located in relation to geohydrological situation and to land use.

Extended sampling in the coming years will make it possible to carry out trend analysis. In addition to this network, a monitoring program for soil quality is in preparation. In The Netherlands soil protection research has become a priority area of Dutch science policy. In addition to existing research programs, in May 1986 a coordinated soil research program concerning the Ministers of Science, Environment, Agriculture and Water Management came into being. It covers a total budget of Dfl 56 million for a period of four years. Major topics are basic research, mainly on soil ecology, effects of substances on soil functions and technology development both for prevention and clean-up operations. The monitoring and research programs will provide data for regulations concerning the soil quality to be maintained and potentially harmful activities.

3. POLICY OBJECTIVES

A central concept in Dutch environmental policy is the concept of being a 'guest in our own environment'. We have the right to arrange our living surroundings as we wish and to enjoy them, but in addition we also have the duty of ensuring that they are handed over in good condition to future generations. To this end a two-track policy is being developed, an effect-oriented and a source-oriented policy.

The effect-oriented policy must make clear which objectives are being pursued with respect to the quality of the environment in The Netherlands as a whole or of certain parts, and the tasks for target groups (e.g. industry and agriculture) implied by those objectives. In addition, a source-oriented policy is necessary that makes clear the way and the tempo in which the behaviour of target groups will be 'corrected', with respect to the environmental quality objectives and tasks formulated in the framework of the effect-oriented policy. The priorities defined in following those tracks are formulated in the yearly Environmental Program.

3.1. Effect-oriented policy
Generally described as an environmental quality, it must be maintained such that the health and well-being of people and the preservation of animals, plants, goods and forms of use are ensured. In soil policy this goal has been translated into the task of preserving the properties of the soil which are of importance for various possible functions, the so-called multifunctionality concept. The Netherlands is a densely populated country where the soil has many functions, which exist close to each other or even above each other. Land use is frequently changing, human activities often lead to irreversible changes in soil structure and chemical composition. The strategy of preserving the multifunctionality of the soil is aimed at 'keeping all options open' for future generations. A change in the use of the soil after some time must remain possible (general protection

level). A soil of good quality may then be defined as follows: it poses no harm to any use by human beings, plants and animals, it can function without restriction in natural cycles (e.g. the hydrological cycle) and it does not contaminate other parts of the environment.

From a purely scientific point of view 'multifunctionality' is a complex, abstract concept. A complete and unambiguous description of a multifunctional soil in terms of measurable chemical, physical and biological characteristics may be a long and arduous task. But the development of a soil protection policy does not fully depend on the availability of such a complete description. It may primarily be focused on those human activities with irreversible effects on the structure and composition of the soil that might endanger any use of the soil by human beings, plants and animals. Examples are pollution of the soil with heavy metals, several organic chemicals and nitrates in groundwater.

An important step in defining good soil quality in quantitative terms is the formulation of reference standards. Recently the Dutch Government has published a discussion note on provisional reference values for certain parameters [3] that take into account the heterogeneity of the soil environment, an improvement to previously published reference values solely intended for use in clean-up operations [4]. The note is aimed at stimulating discussion in scientific and in a second step political circles on soil quality. At the request of the Minister of Housing, Physical Planning and Environment, this scientific discussion is organized by the (provisional) Technical Soil Protection Committee, an advisory body foreseen in the Soil Protection Act. To this end a symposium on soil quality will be held in December 1986.

For some major pollutants, such as nitrate and phosphate, provisional values have been published in the Environmental Program 1986–90, indicating the desired quality of the upper levels of groundwater, based upon the EC directive on the quality of drinking water.

3.2. Source-oriented policy
Ideally the desired soil quality sets goals, for example emission reduction from sources of soil contamination. In Dutch environmental policy in general (potential) sources of soil pollution are divided into two main categories, local or point sources and diffuse or non-point sources. Examples of local sources are solid waste disposal sites, storage tanks and industrial premises. Diffuse sources are, for instance, airborne deposition, the contamination of underwater soil or some agricultural activities.

For local sources only a remote link with soil quality is made as the approach for those sources is 'isolation': they have to meet the so-called ICM criteria (isolation, monitoring, control). Priority is laid on applying those criteria in the construction of new storage or dumping sites. To be able to standardize ICM provisions for different materials and different soils, general and source-specific standards for the acceptable risk levels of a possible failure of such provisions are being developed. A research program to obtain the necessary methodology is in progress, mainly concentrating on storage facilities for chemical fluids and mineral oil, and on risk analysis on a factory level. The intention is to provide in several months a check list for a risk analysis of possible soil contamination on industrial premises.

For the discharge of liquid and solid wastes, underground storage tanks for liquid fuels and the use of waste materials for construction purposes a more detailed specification of the ICM criteria in general administrative orders based on the Soil Protection Act is now in preparation.

Diffuse sources have to meet other criteria. One objective is to define a link, mathematical or otherwise, between the application of substances and the maintenance of a good soil quality. For long-term preservation of soil quality an equilibrium between input and output of substances must be reached, which may be only possible with non-accumulating or non-persistent substances, degraded in a natural way, or substances removed by crops. Most heavy metals and many organic pollutants which do not meet these requirements are placed on a 'black list' of substances that should not be allowed to enter the soil. It is based on the EC Guideline on Groundwater Protection. This list forms one of the bases for the selection of so-called priority substances. A systematic selection of priority substances taking into account the risks involved is now in preparation. For these substances basis documents will be drawn up to provide the scientific basis for (integral) measures with regard to soil, air and water emissions. These basis documents will appear in English translation [2].

An example of the approach to diffuse sources, the manuring of agricultural land, is given in the next section.

4. REGULATION OF MANURING

The Netherlands has about 5 million cows grazing on grassland, 13 million pigs and 85 million poultry, fed mainly on imported feed. This has led to manure surpluses measured in millions of tons that have to be disposed of. These surpluses lead to leaching of nitrate and phosphate to groundwater and other water reservoirs. The evaporation of ammonia from manure is considered to contribute about 30% to acid rain and its effects. At the end of 1984 the Interim Act Restriction Swine and Poultry Farms was adopted. This law forbids the operation of new swine and poultry farms and, in the districts with the most intensive animal husbandry, forbids even the extension of existing farms for a two-year period. In other parts of the country only limited extensions are permitted. For the regulations of manuring, as for other aspects of soil protection relevant to agriculture, a close co-operation exists between the Minister of Environment and the Minister of Agriculture. In short, further regulation is planned along the following lines: the quantity of manure to be used in relation to type and composition and to use of the soil, the manner of application and the circumstances and location for application. A draft order on the basis of the Soil Protection Act has been prepared; the definite order should come into force in April 1987. The approach to the manure problem is divided into three phases of five years each. For practical reasons the regulation is largely based on the phosphate content of the manure; ultimately no more phosphate in the form of animal manure should be given than is necessary for the desired crop yield. When the phosphate level in the soil is sufficient a dose equal to the amount withdrawn by crops can suffice.

Table 1 shows the reduction in manure application foreseen in the forthcoming regulation.

In the year 2000 the final standard should be reached. The exact value of this final standard remains to be defined (about 125 kg P_2O_5/ha). Of course a parallel reduction in the nitrate load has to take place. A levy on surpluses will provide the means for provisions on handling, marketing and in the future possibly export of the manure products. With close co-operation between government and agricultural sector research is going on to establish the best way to handle the large mass of surpluses on an environmentally and economically sound base.

Table 1
Regulation of manure application

	Maximum of manure to be applied per hectare (kg P_2O_5)			
	1987–91	*1991–95*	*1995–2000*	*2000–*
Crops	125	125	125	Final standard
Maize	350	250	175	
Grassland	250	200	175	

The incentive for these source-oriented measures is the intended quality for phosphate and nitrogen compounds in groundwater. Provisional values for these compounds have been formulated in the Environmental Program 1986–90.

Purification of groundwater is seen as only a temporary measure; the emphasis must lie on prevention. Environmental monitoring will show the effectiveness of the measures foreseen.

REFERENCES

1. EIKELBOOM, R. T. and VON MEIJENFELDT, H. (1985). The soil clean-up operation in the Netherlands; further developments after five years of experience. In: *Contaminated Soil. Proc. First Int. TNO Conference on Contaminated Soil*, Martinus Nijhof Publishers, Dordrecht, pp. 255–68.
2. MINISTRY OF HOUSING, PHYSICAL PLANNING AND ENVIRONMENT, MINISTRY OF AGRICULTURE AND FISHERIES, MINISTRY OF TRANSPORT AND WATER MANAGE-MENT (1985). *Environmental Program of the Netherlands 1986–1990*, Staatsuit-geverij, The Hague.
3. MINISTRY OF HOUSING, PHYSICAL PLANNING AND ENVIRONMENT (1986). Discussienotitie bodemkwaliteit (only available in Dutch). The Hague.
4. MOEN, J. E. T., CORNET, J. P. and EVERS, C. W. A. (1985). Soil protection and remedial actions: criteria for decision making and standardization of requirements. In: *Contaminated Soil. Proc. First Int. TNO Conference on Contaminated Soil*, Martinus Nijhoff Publishers, Dordrecht, pp. 441–8.

Soil Protection Programmes and Strategies in Other Community Member States: Setting up of an Observation Network for Soil Quality in France

A. Gomez and C. Juste

*Station d'Agronomie, INRA, Centre de Recherches de Bordeaux,
Domaine de la Grande Ferrande, 33140 Pont-de-la-Maye, France*

SUMMARY

*The quality of agricultural land can deteriorate, for which there may be a
number of causes:*

—*unsuitable farming practices or use of products containing toxic
compounds (heavy metals, pesticides);*
—*industrial or urban activities contributing to the accumulation of toxic
compounds (atmospheric fallout, use of sewage sludge or domestic waste
composts).*

*In order to detect and diagnose the causes of soil quality deterioration at a
sufficiently early stage and to take action in good time, the French Ministry of
the Environment decided to set up an 'Observation network for soil quality'.
Accordingly, a feasibility study was assigned to the 'National Institute of
Agricultural Research' (INRA).*

*The summary drafted on completion of this work stresses the following
points:*

—*the term soil quality must be taken in its broadest sense, since it covers
both the phenomena of physical and chemical degradation and problems
of toxicity (heavy metals, pesticides). For this reason, the aims of the
observation network are to study heavy metals, other physico-chemical
and chemical properties, plant health products, biological properties of
soils, erosion and physical degradation other than erosion;*
—*the task of this network must not be limited to simply reporting the*

present situation, it must also monitor the quality and, in time, predict developments;

—the observation network must be based on a permanent structure with specific financing and the help of scientific institutions.

Finally, the report insists on the gradual establishment of this network with a preparatory training phase for staff and the development of specific sampling techniques.

Priority has been given to 'heavy metals' because of the seriousness of the problem presented by their accumulation in soils and the mastery of techniques for determining these elements. In this respect a soil sampling study was conducted jointly by the INRA and ACTA (Association for Agricultural Technical Coordination) with the aid of the Ministry of the Environment.

The observation network currently covers the following regions: Aquitaine and Lorraine (1985), Brittany and Nord-Pas de Calais (1986). The fields covered in these regions are as follows: heavy metals, physico-chemical properties of the soil and plant health products. This makes in all for 37 sites.

1.   INTRODUCTION

The quality of agricultural land depends on a number of physico-chemical, chemical and biological factors. These can change as a result of various causes:

—unsuitable farming practices; cases in point include premature deforestation leading to erosion, bringing to the surface by work techniques of the soil of underlying horizons naturally richer in pollutants such as toxic metals, and dilution of the organic matter from the surface horizon by excessively deep ploughing, etc.;

—spreading of toxic metals or other micro-pollutants; to the surface of agricultural land due both to industrial or urban activities (atmospheric fallout) and to the use of certain fertilizers (phosphate fertilizers rich in cadmium, sewage sludge, household waste compost) or products to treat crops (pesticides) which can sometimes contain pollutants.

This alteration in soil quality can lower the crop yield and also have an effect on the quality of the food chain by contaminating crops or groundwater. These risks become even greater when they are in areas with a high degree of industrialization.

With a view to assessing the present situation and then evaluating any deterioration in soil quality so as to take measures in good time, the French Ministry of the Environment decided to set up an 'Observation network for soil quality'. To this end a feasibility study was assigned to the 'National Institute for Agricultural Research' (INRA).

2. FEASIBILITY STUDY OF AN OBSERVATION NETWORK FOR SOIL QUALITY

This study was conducted in 1982–83 by a team of research workers from the INRA. It gave rise to a report consisting of seven documents, a general summary and six reports on the following topics: heavy metals, chemical and physico-chemical developments (other than in relation to heavy metals), plant health products, soil biology, erosion, and physical soil degradation (other than erosion).

The conclusions of this report are as follows:

— the term 'soil quality', must be seen in its broadest sense, covering, as it does, both the phenomena of physical, chemical and biological degradation and problems of toxicity, heavy metals and pesticides;
— this network is not limited to simply observing the present situation, it is designed to monitor this quality, and, at least in time, predict developments;
— it should be based on a permanent administrative structure which has both specific financing and help from scientific institutions.

Since it does not cover the entire country, but only a small number of representative sites, this type of network puts a limit on financial undertakings.

In the current state of the art this system could become operational at different rates:

— in the short term, for heavy metals, physico-chemical properties and certain fragmentary elements of physical degradation, including erosion;
— in the medium term, for the whole of physical degradation and initial interpretation of the data relating to biological aspects and pesticide residue;
— in the long term, for predictions of biological soil behaviour and pesticide residue developments.

The proposed methodology should allow for both rapid availability of data that can be interpreted in terms of soil quality for some of its components and preparation of the future for the others.

As regards plant health products, this type of observation network will contribute towards the necessary improvement, in time, of the laws relating to the approval of toxic molecules.

There should be harmonization of the different types of 'network' already set up or being studied, whether they concern soil, water or air and regardless of their exact designation and current head. These will often be the same 'experts' who participate in the various study groups. By tackling the problems on a broader scale they can help to improve the effectiveness of the resources generally available today.

Although extensive, financing could nonetheless be staggered, especially as it is technically necessary to provide for a gradual establishment of this network, with a preparatory training phase for staff and the development of specific sampling techniques.

The need for a permanent structure with specific financing as part of a multi-discipline scientific framework calls for a formula which has not yet been fully defined.

This kind of 'structure' should bring together representatives of the many bodies involved, assume the responsibility and management of the observation network and conclude multi-annual agreements with persons and bodies making a special contribution.

The outcome of this study was a decision to set up the observation network for soil quality on a gradual basis while observing the report's recommendations. In view of their immediate feasibility, physico-chemical properties and heavy metals will be the first subjects tackled. Given what will probably be a fairly moderate amount of heavy metals in the majority of soils cultivated in France, the group of experts recommended that the actual establishment of observation sites be preceded by a study designed to determine the best methods of soil sampling. This work, which was assigned to the 'Agronomics Department' of the INRA in Bordeaux, is summarized below.

3. SOIL SAMPLING TO MONITOR TRENDS IN HEAVY METAL CONTENT

In view of the low concentrations of these trace elements likely to be found in the soils and the probable slowness of their development, the problem of representative sampling takes on considerable importance. The methods used for conventional estimates of soil fertility (AFNOR X31100 standard) cannot be used in this instance.

The aim of the study assigned to the 'Agronomics Department' of the INRA in Bordeaux was to define the minimum number of samples to be taken, how far apart they should be and what they should consist of, so that the contents measured are as close as possible to the actual average values of that particular observation site.

To do this, a large number of elementary samples (100) have been taken on two lots (measuring 1 ha each) over a regular grid covering the whole of each lot, both being analysed separately. The results of these analyses and the average values obtained by calculation were compared with those provided by different, lighter sampling systems.

The lots in question represent two characteristic situations, these being respectively:

—high, diffuse pollution dominated by atmospheric fallout (Lème, départment 64);
—high pollution due to the proximity of an industrial site (Evin-Malmaison, départment 62).

The data were processed by conventional statistical analysis and by the method of regionalized variables developed by the statistical department of the ACTA in an attempt to lower the number of samples by introducing the possible structure of spatial dependence for a given parameter.

This study resulted in a lot sampling system, given the acceptable results regarding the validity of a composite sample and the fairly good accuracy of the estimated mean from the 25 samples taken from a uniform configuration of a 1 ha lot and, of course, in view of the desire to limit analysis costs. The following procedure was adjusted to monitor the average content in a given metal:

—25 soil samples (0–20 cm);
—25 sub-soil samples (20–40 cm);
—5 composite samples for soil and sub-soil from the 25 samples dried beforehand in air and sifted to 2 mm;
—each composite sample (500 g) is made up of the most uniform possible mixture of 20 g of earth taken from each initial sample;
—separate analysis of the five composite samples in order to ensure low analytical error and low make-up error of the composite sample (the present study showed that uniformity deviation is low and that the

analytical error is likewise negligible if all the analysis operations are taken over a short period of time);

—where an apparent average density of soil and sub-soil can be assessed with sufficient accuracy the analysis results can be expressed in g metal/ha;

—conservation of all the soil and sub-soil samples (total of 50) taken separately and of the 10 composite samples.

In the second observation of the lot (set at time $t + 5$ years) the new composite samples will be analysed at the same time as those taken at time t so as to eliminate any analytical spacing drift over a period of time, which appeared quite clearly in the sampling study. The introduction into the batches analysed of reference samples (such as those available at the Community Bureau of References) is obviously one way of surmounting this type of snag.

The aim of keeping all the samples taken is to confirm any trend which might appear just after the analysis of the composite samples, and even, if necessary, to illustrate it by means of a map of the observation site, using geostatistical methods.

The main question that remains to be answered is the time lapse between two observations. The arbitrary period chosen in this instance of 5 years is possibly not long enough, in the case of metal pollution which is diffuse and hence of modest intensity, for the amplitude of the effective variation in concentration between the two periods of observation not to be masked by uncertainty of evaluation due to sampling and analysis.

4. PROGRESS ON THE ESTABLISHMENT OF THE OBSERVATION NETWORK FOR SOIL QUALITY IN 1986

To take account of the financial requirements and the recommendations of the feasibility study, the network will be set up gradually by regions. Accordingly, financing was found to allow Aquitaine and Lorraine to commence work on the project in 1985. Respectively, eight lots were selected in Aquitaine and 13 in Lorraine for the study of physico-chemical and chemical properties and heavy metals. Further regions, Brittany and Nord-Pas de Calais, were provided with financing in 1986 for the same topics, with, respectively, 6 and 10 lots, i.e. 37 in all.

At this stage of development a distinction cannot be made between a situation where a variety of subjects are studied and one where one to two subject areas are investigated. At the instigation of the 'Centre of

Agricultural Mechanization of Agricultural Engineering of Waterways and Forestry' (CEMAGREF) the study into what becomes of plant health products began in 1986 on six of the eight sites in Aquitaine.

The same procedures will be applied throughout the observation network for sampling and treatment of samples, and the analyses will be performed by the soil analysis laboratory of the INRA in Arras. Accordingly, it will be possible to compare the results and to evaluate the situation in the various regions.

At this point the results obtained will give an assessment of the current situation, trends not emerging until the next sampling campaign in 5 years time.

There are plans to extend the network to other regions, but this will depend in large measure on the results obtained in the first regions studied by the observation network.

Finally, there is the problem of using, conserving and disseminating the information obtained, which in any case will be kept in the data file of the STIPA (System for the Transfer of Pedological and Agricultural Information, which in turn is attached to the International Network for the Processing of Data on Soils).

BIBLIOGRAPHY

GOMEZ, A., JUSTE, C., DESENFANTS, C., BRUN, T. and LOPEZ, C. (1986). Echantillonnage des sols en vue du suivi de l'évolution des teneurs en métaux lairds. Rapport de synthèse Min. Environnement/INRA-ACTA, 10/4/86, 52 pp.

INRA (1983). Etude de faisabilité d'un observatoire de la qualité des sols. Secr. Etat Environnement et Qualité de la Vie, Sept., 110 pp.

Soil Protection Programmes and Strategies in Other Community Member States: The UK Approach to Soil and Landscape Protection

P. J. W. SAUNDERS

Chief Scientist Group, Department of the Environment,
Romney House, 43 Marsham Street, London SW1P 3PY, UK

SUMMARY

The United Kingdom is a densely populated island with a long tradition of careful husbandry of its land, that husbandry being responsible for the shaping of the landscape as we know it today. The soils, landscape and land uses of the UK are reviewed briefly. The structure of land-use protection is then described in the context of land-use planning with reference to safeguarding agricultural land, conserving areas of landscape value, urban expansion and minerals extraction. More targeted approaches to soil protection exist in government policies for management of the countryside, the conservation of wildlife and habitats, and measures to combat pollution. There are also guidelines and regulations covering such matters as special problems of soil erosion, soil management and the contamination of soils with toxic substances. The underlying UK approach is one of flexibility so that appropriate responses can be made to changing circumstances.

1. INTRODUCTION

As a densely populated island with a long tradition of the careful husbandry of its land, the UK greatly welcomes the initiative of the CEC in holding this Symposium. Soil protection stands for the protection of the soil from pollution and from degradation by other means. Soil for what? For food and forest production; for water storage; for the support of natural flora and fauna; and therefore, incidentally, of our landscape heritage.

Sometimes these requirements of the soil are in conflict. We must therefore adopt an open-minded approach so as to ensure that appropriate solutions can be found for the problems we encounter.

Soil has been studied traditionally as an agricultural resource—as a medium for the culture of crops and for forage. The main objectives of management have been the enhancement of soil quality to maximise productivity. In the past the soil quality has been measured according to coarse but effective criteria on the assumption that management can achieve the main objectives in most situations. A similar approach is adopted in forestry. More recently problems of soil loss, structural damage, nitrate contamination of groundwaters, acidification of soil and surface waters, etc., have become recognised as important considerations in formulating environmental protection policies and strategies.

We recognise that differences in physical geography can lead to differences in national approach even where the final objectives are the same, but we believe that there are valuable lessons to be learnt by contrasting different approaches. This chapter describes the UK approach as it has evolved to date, the principles employed, some of the remaining problems and the prospect for their solution.

2.   UNITED KINGDOM—SOILS AND LANDSCAPE

It is appropriate to begin with a brief description of the UK's physical geography. The underlying geological strata run approximately north-east to south-west, with the older rocks to the north and west so that crossing Britain from west to east traverses major changes in geology in a relatively short distance; from the granites of the Scottish Highlands and Cornwall, through the carboniferous limestones of central England, to the chalk, clays and sands of the south and south-east. The uplands lie mainly in Scotland and Wales and along the Pennines, which form a central backbone to England. Over 50 major river systems flow from these watersheds to the sea. No part of Britain is more than 150 km from the sea, and even the longest river, the Severn, is only some 350 km long.

The wide range of underlying geology combined with variations in climate that tend to be emphasised by the UK's topography have produced a wide range of soil types. We have soil profile and chemical data from 5500 rural and 1000 urban sampling points plus a further 270 000 points to identify the limits of particular soil types. Sixteen main categories of agricultural land use associated with about 200 main soil types are identified. These data can be augmented by additional sampling when required, together with other geographical information (e.g. Ordnance Survey data) to identify 32 classes of land based on soil type, geology, topography, climate, vegetation cover and use. This compares with the 4–8 general land classes used for most agricultural and forestry purposes. The pattern of these soils is complex and detailed mapping is normally required at a local level. It has therefore been found that for some purposes it is most practical to think of soil quality indirectly, not only in terms of its detailed physico-chemical structure but of the quality and diversity of the vegetation and life that it can support.

Some 80% of the UK land area is devoted to agriculture. For the reasons outlined above, the land to the west and north tends to be devoted to the rearing of livestock and to dairy farming, whereas to the east and south is mainly arable cultivation, particularly the production of cereals. Technological developments have tended to decrease the extent of mixed farming.

A further 8% of the land area—mainly in the northern and western uplands—is devoted to forestry, principally pine and spruce. The broadleaf forest, the climax vegetation for much of central and southern Britain, has been displaced by centuries of agricultural, industrial and urban development.

The remaining 12% of land area is developed to non-agricultural uses. Urban expansion and other physical development has caused a major loss of quality soils in the UK, whereas other mechanisms such as acidification, erosion or pollution have resulted in isolated losses of soil quality. It is not therefore surprising that in the UK probably the most important single element of protection of soil and landscape quality is the system of land-use planning.

3. STRUCTURE OF LAND-USE PROTECTION

The UK's system of land-use planning, to be described below, provides the strategic framework within which many different approaches to land-use protection are integrated. These approaches, which range from voluntary agreements to prescriptive regulation, are chosen to be the most appropriate for each particular use of the land.

Separate planning legislation applies to England and Wales, to Scotland and to Northern Ireland, based on similar principles and procedures and administered by separate government departments. The Department of the

Environment (DOE), headed by the Secretary of State for the Environment, is responsible at central government level, among many other things, for the administration in England of the Town and Country Planning Act 1947, which introduced a comprehensive land-use planning system for the United Kingdom.

The primary responsibility for the formulation and implementation of land-use policies for urban and rural areas rests with the local planning authorities (LPAs). They carry out this task within the framework of the legislation and government policy for which the Department of the Environment is responsible in England. The department takes account of representations received from a wide range of bodies, some set up by statute, including the Countryside Commission, the Nature Conservancy Council and the Development Commission, all of which have significant executive as well as policy advisory functions in their own right.

Positive strategies for managing conflicting demands on land and other resources in the countryside are set out in county structure plans, which must be approved by the Secretary of State for the Environment. Policies may be set out in more detail in local plans prepared by districts. The structure and local plans themselves provide a basis upon which the local authorities can decide upon applications for individual planning permission for the development of land, for housing and industry. They may attach conditions to a permission or refuse permission altogether, although all applications must be considered on their merits even if they do not accord with the development plan. 'Planning permission' (which is quite separate from building permission) is required for individual development projects anywhere in the country. 'Development', precisely defined in the Town and Country Planning Acts, means, in general terms, any building, engineering or other works, or material change of use.

4. SAFEGUARDING BETTER QUALITY AGRICULTURAL LAND

It is government policy for the countryside to ensure that, as far as possible, land of higher agricultural quality is not taken for development where land of a lower quality is available, and that the amount of land taken is no greater than is reasonably required for carrying out the development in accordance with proper standards.

Responsibility for planning control lies initially with the Local Planning Authority (LPA) and, ultimately (on appeal or at his discretion), with the Secretary of State for the Environment. The Ministry of Agriculture,

Fisheries and Food (MAFF) does, however, have a most important role to play in ensuring that proper account is taken of the government's policy for the protection of agricultural land. At the strategic level, MAFF regional officials are consulted by the LPAs when they are drawing up structure and local plans. This enables MAFF to advise LPAs and the Department of the Environment of the agricultural implications of these development plans and to ensure that these are fully taken into account in the final development plans. When it comes to individual planning applications, the LPAs are required to consult MAFF on any proposals which would lead to the loss of 4 ha of agricultural land (or on proposals where, although the loss will be of less than 4 ha, the LPA considers that the development is likely to lead to further agricultural land loss in the future) and which do not conform to an agreed development plan. LPAs sometimes also consult MAFF on proposals for agricultural development (farm cottages, large farm buildings, etc.) to enable them to come to a decision as to the needs for a particular proposal.

In advising LPAs on proposals for development of agricultural land, MAFF relies heavily on the Agricultural Land Classification system (ALC). Under this system all agricultural land in England and Wales is classified into five grades by identifying the inherent qualities of the land, irrespective of its present use, by means of readily identifiable long-term physical characteristics of soil, topography and climate which are considered to impose limitations on the agricultural use to which the land might be put. When consulted on a planning application, MAFF officials make a survey of the site and then advise the LPA on the quality of the land according to the ALC system and highlight any other facts that might have a bearing on the agricultural implications of the development. More detailed studies may be required where the soils of a site exhibit unusual physical and chemical characteristics, especially in relation to drainage. MAFF also has a system for classifying hill and upland areas according to their potential for improvement; the Forestry Commission has a similar system for assessing forest potential. This information is used by LPAs, National Parks and others involved in land-use planning and conservation issues in these areas.

5. CONSERVING AREAS OF HIGH LANDSCAPE VALUE

The responsibility for the designation of National Parks and Areas of Outstanding Natural Beauty (AONBs) rests with the Countryside

P. J. W. Saunders

Commission, subject to the confirmation of the Secretary of State for the Environment. Proposals for development in these areas are subject to particularly stringent planning control policies which aim to preserve their special character and attractiveness. Policies of similar stringency may be applied in Heritage Coast areas although no statutory designation is involved.

However, planning controls serve mainly to restrict 'urban'-type development in the countryside. The use of land for agriculture or forestry is specifically excluded from the definition of development in the Town and Country Planning legislation. Thus planning controls do not extend to changes of use from one type of farmland to another or from farmland to forestry or vice versa, nor, with one exception, to the cultivation of previously uncultivated land. Buildings on agricultural land are defined as development under Class VI of the Town and Country Planning (General Development) Order 1977. However, a wide range of agricultural developments, including most farm buildings, are normally exempt from specific development control unless they exceed the area threshold of $465\,\text{m}^2$. There are, in addition, other conditions which relate to the proximity of the buildings to roads, aerodromes and existing farm buildings. This means that planning applications are not normally required for most farm buildings except where the local planning authority brings the development within the scope of control by a specific direction. In those areas to which the Landscape Areas Special Development Order 1950 applies, which include some of the most attractive parts of our National Parks, the Local Planning Authority can require details of the design and appearance of a proposed farm or forestry building of less than $465\,\text{m}^2$ area to be submitted to it for approval before the development proceeds; the government proposes to extend this order to cover all National Parks in England and Wales.

6. URBAN EXPANSION

Although urban expansion and physical development in rural areas cannot realistically be prevented, it is government policy that full and effective use should be made of land within existing urban areas. The expansion of a town into the surrounding countryside is objectionable if it creates ribbons or isolated pockets of development, or reverses accepted policies for separating villages from towns and from one another, or if it conflicts with national policies for the protection of the environment such as those concerning Green Belts, National Parks, Areas of Outstanding Natural Beauty, good farming land or nature conservation. Green Belts have now been established around nearly all of Britain's major cities and cover an area of more than 18 000 km², playing an important role in checking urban sprawl. The implementation of this policy is through a variety of different means, including Town and Country Planning legislation.

7. MINERALS EXTRACTION

Mineral working has in the past been a major cause of dereliction of land and damage to soil. It is important both that existing dereliction is cleared and that current and future mineral working does not lead to dereliction in years to come. Provided great care is taken with the stripping and storage of the soils which overlay the mineral, with the replacement of these soils after the mineral has been extracted, and with the management of the land after it has been initially restored, it is often possible to return the site of mineral workings to useful agricultural production. Even where it is not possible to return the land to agriculture, its use for some other purpose, be it housing, industrial or amenity or forestry use, is of indirect benefit to high quality soils in that it relieves pressure for such developments on other agricultural land in the area.

In seeking to reconcile mineral working with other—often conflicting—claims on land, the main objectives of minerals planning policies are:

(a) to ensure that the needs of society for minerals are satisfied with due regard to the protection of the environment;

(b) to ensure that any environmental damage or loss of amenity caused by mineral operations and ancillary activities is kept to an acceptable level;

(c) to ensure that land taken for mineral operations is restored at the earliest opportunity so that wherever practicable the land is capable of an acceptable use after working has come to an end;

(d) to restore land as near to its original quality as possible if being returned to agriculture; and

(e) to prevent the unnecessary sterilisation of mineral resources and ensure their optimum use.

The development controls available through the Town and Country Planning Act 1971, as modified by various statutes and regulations, provide the principal means of achieving these objectives. Provision is made for conditions to be attached to permissions for mineral working requiring restoration of sites when activities have ceased. The powers available to prevent dereliction arising from mineral working were extended by the introduction of the Town and Country Planning (Minerals) Act 1981. Section 5 of the 1981 Act gives to Mineral Planning Authorities the power to impose 'after-care conditions' where the intended after-use of the site is for agriculture, forestry or amenity. The basic premise is that not only should the mineral operator replace the topsoil when the mineral working has ceased; he should also look after the land for up to five years afterwards.

Before imposing an agricultural condition, the Mineral Planning Authority must consult the relevant minister responsible for agriculture. MAFF undertakes an assessment of the site to identify physical characteristics prior to working. Where forestry after-use and after-care are involved the Forestry Commission must be consulted. There are certain other sections of the 1981 Act which have recently been brought into force and are likely to have particular relevance for the restoration and after-care of mineral workings. Section 3 requires Mineral Planning Authorities to undertake periodic reviews of all mineral working sites which are currently active or have been so in the previous five years, and to take action if appropriate.

It is government policy that land of a higher agricultural quality should not be worked for minerals where land of a lower quality is available and that, where mineral deposits are overlain by high-quality agricultural land, the feasibility of restoration to agriculture should be taken fully into account before such land is allocated for mineral extraction. In the National Parks, Areas of Outstanding Natural Beauty and other areas given protection for environmental conservation, any application to work minerals is subject to the most rigorous examination.

A closely related policy area is that relating to the landfill disposal of controlled waste materials. Over 80% of controlled wastes in England are disposed of to landfill, and mineral excavations form the most common type of disposal site. Such disposal is controlled under two main pieces of legislation—the Town and Country Planning Act 1971 and the Control of Pollution Act 1974.

8.   TARGETED APPROACHES TO SOIL PROTECTION

The planning system outlined above constitutes the major instrument for controlling the use of the land, taking full account of the likely damage to the soil of any proposed use, and provides for legally binding requirements set at the time of granting planning permission. Other measures are available to ensure that the designated use is exercised with proper regard to soil condition, and more specific legislation on control of pollution is used to control some activities and uses of land which have potential to damage soils. These measures range from management agreements with the owners to voluntary agreements. A number of important examples can illustrate these principles.

9. MANAGEMENT OF THE NATURAL LANDSCAPE

The government takes the view that the most effective way to secure landscape and nature conservation in the countryside is through the negotiation of voluntary management agreements between the relevant authorities and the landowners or occupants of the land. The legislative framework for the voluntary approach is provided in Part II of the Wildlife and Countryside Act 1981. Section 39 of the Act is concerned specifically with landscape conservation. It permits a County Planning Authority to enter into a management agreement with any person having an interest in an area of land, for the purpose of conserving or enhancing its natural beauty or amenity. The purpose of such an agreement might, for example, be to encourage a farm to retain land in grazing rather than converting it for arable use. Where management agreements are made under Section 39 (or Section 41 of the Act—see below), grant aid towards the cost of maintaining them is available at a rate of 75% in the National Parks and 50% elsewhere.

The Countryside Commission will continue to play an important role in encouraging the use of management agreements to safeguard areas of prime landscape quality. A number of demonstration farm projects designed to show how the interests of farming and landscape conservation can be reconciled will be supplemented by encouraging local farming and wildlife groups to set up 'link farms' also exhibiting good practice in the integration of farming and conservation, but on a less formal basis. Conservation advisory work will be assisted by grant aiding the costs of employing countryside advisers. Direct conservation action on the farm will be encouraged by the provision of grant aid for planting of trees and shrubs, planting and managing small woodlands for landscape and amenity purposes, and conserving other small-scale features. The

Countryside Commission already makes a substantial financial contribution to the costs of amenity tree planting and woodland management— more than £1·4 million in the financial year 1983/84. Government departments and agencies are also funding research to develop improved methods for monitoring changes in land use and landscape.

On 1 April 1985 an experimental scheme was introduced to protect a unique area of grazing marshes in the Norfolk Broads for a 3-year period until 31 March 1988 with an overall budget of £1·7 million. The scheme was set up by the Countryside Commission under Section 40 of the Wildlife and Countryside Act 1981 and run in partnership with the Ministry of Agriculture, Fisheries and Food. The budget is borne equally between the Commission and MAFF. Payments of about £120/ha/annum are offered to those landowners and full agricultural tenants who agree to their land being managed under the scheme's grazing guidelines which are designed to encourage a continuation of traditional grassland practices. Some 90% of the eligible area was covered in the first year of the scheme, which was extended to two new areas of grazing marshes on 1 April 1986.

10. PROTECTING AND CONSERVING IMPORTANT WILDLIFE HABITATS

In seeking to protect wildlife and its habitats the government's main objective is to foster awareness of the importance of nature conservation and to ensure that the nation's natural heritage is taken into account fully in all decisions affecting rural land use and development. In developing these policies the Secretary of State for the Environment is advised by the Nature Conservancy Council (NCC), an independent government-funded body whose statutory duties include the establishment of nature reserves, notification of Sites of Special Scientific Interest (SSSIs), research, and education of the public about nature conservation. The NCC also has powers to make agreements for the management of nature reserves and SSSIs and to give grants for any nature conservation purpose. The NCC's functions, powers and duties derive principally from the National Parks and Access to the Countryside Act 1949, the Countryside Act 1968, the Nature Conservancy Council Act 1973, and the Wildlife and Countryside Act 1981.

The Wildlife and Countryside Act 1981 seeks to build on the previous legislation by providing more effective mechanisms for protecting wildlife habitats within a framework of goodwill and voluntary co-operation. Part

II of the Act includes a requirement (under Section 28) for the NCC to give notice of SSSI proposals to owners and occupiers and to consider any representation or objections before formalising the notification. Once sites are notified, owners and occupiers are required to give notice to the NCC of any potentially damaging operations they may wish to carry out—these are specified in the notification. This enables the NCC to consider any effects on the wildlife interest and, where appropriate, to use their existing powers to negotiate management agreements, with compensation for loss of profits, or other forms of safeguard. The Secretary of State for the Environment is also given reserve powers under Section 29, to impose a restraining order for a period of up to 12 months to enable a means of protection to be negotiated. The Act also places a duty on agriculture ministers to further conservation in considering applications for farm capital grant under a national grant scheme (as distinct from a scheme which implements European Community policy) and on Water Authorities (WAs) and Internal Drainage Boards (IDBs) in considering proposals relating to the discharge of their functions. In this respect, in 1983 MAFF, DOE and the Welsh Office jointly issued conservation guidelines to WAs, IDBs, NCC and the Countryside Commission, and these are being updated at present. The guidelines advise WAs and IDBs on fulfilling their objectives under Section 22 of the Water Act 1973 as amended by Section 48 of the 1981 Act. They also explain the role of the NCC and Countryside Commission and aim to improve co-operation between the bodies concerned.

Since October 1980 it has been a condition of the farm capital grant schemes administered by MAFF that a farmer in a National Park or SSSI shall notify the relevant authority before carrying out any work on which he intends to claim grant. This enables the authority to comment or suggest modifications for conservation purposes or, if agreement cannot be reached, to object to the payment of grant if a claim is made. If, in consequence of such an objection, a grant under a national grant scheme is refused, then the Wildlife and Countryside Act (Section 32 for SSSIs or Section 41 for National Parks) requires the relevant authority to offer a management agreement to the owners or occupants of the land under which restrictions on the work are agreed in return for compensation on a loss-of-profit basis.

An Agriculture Bill currently before Parliament provides the necessary power for the implementation of Article 19 of the EC Agriculture Structures Regulation permitting the designation of areas of environmental sensitivity. This was a provision for which UK agriculture ministers fought particularly hard during negotiation of the regulation. The measures will be administered by agriculture departments and it is envisaged that, once areas are designated, farmers within them will be encouraged to enter into agreements under which they will receive incentive payments in return for respecting certain conditions in the management of their land. These might include the maintenance of stone walls and hedges and restrictions on the use of fertiliser, silage cuts, etc. It is hoped that the first areas will be designated by 1 January 1987. The Bill also includes amendments to the Wildlife and Countryside Act which would extend to Community capital grant schemes the duty on agriculture ministers to further conservation when considering applications for a grant, and the obligation to offer a management agreement where a grant is refused following an objection by the relevant authority in a National Park or SSSI.

11.   MEASURES TO COMBAT POLLUTION

The Department of the Environment (DOE) has the major responsibility for policies protecting the environment from pollution arising from industrial and other non-agricultural sources. The United Kingdom has adopted what is called the principle of 'the Best Practicable Environmental Option', which is to find the optimum method of disposal so as to limit damage to the environment to the greatest extent achievable for a reasonable cost. Landfill disposal of waste requires a licence which is withheld if there is a threat of water pollution or if public health is endangered. Site licences can be made subject to conditions such as the kinds and quantities of waste received, the methods of dealing with them and recording information. Planning consent for the landfill operation can include conditions as to the after-care and restoration of the site when operations have ceased.

In the countryside, agriculture itself can be a major source of pollution. This relationship was explored by the Royal Commission on Environmental Pollution in their 7th Report, 'Agriculture and Pollution'. The Royal Commission came to the conclusion that, although the general responsibility for safeguarding the environment rests with the DOE, the initiative for reducing the pollution caused by agriculture should lie with the Ministry of Agriculture and that the Ministry should be responsible for promoting research on the polluting effects of agricultural practices, whether or not such work appears likely to bring benefits in terms of farming economies. The government has accepted both these recommendations. On farms, the government sees the initiative as laying mainly in the encouragement of good agricultural practice so that farmers adopt high standards of environmental protection which are consistent with an efficient and competitive industry and secure a satisfactory balance between agriculture and the environment. This is achieved through farm visits, on-farm events, general publicity and a wide range of advisory literature. There is, in addition, a range of legislation which makes it an offence to cause pollution of land and water.

Under Part II of the Control of Pollution Act 1974, the main provisions of which came into force on 25 July 1984, the Minister of Agriculture and Secretary of State for Wales, after consulting the Secretary of State for the Environment, issued a Code of Good Agricultural Practice which helps farmers to identify and avoid types of activity which are likely to cause water pollution. In addition, arrangements have been made to ensure that agricultural development which might have a water pollution impact is assessed by the relevant water authority before MAFF makes any payment of a grant under the capital grant schemes. There has been a steady increase in research and development on the agriculture/environment interface and there is close consultation with all the agencies involved. Many millions of pounds are currently being spent annually.

As regards air pollution by agricultural activity, amendments have been proposed to Class VI of the Town and Country Planning General Development Order 1977 which will include measures to secure an appropriate degree of control over the siting of intensive livestock units, particularly in proximity to residential and other non-agricultural property. In addition, MAFF has issued guidelines on housed livestock units which, amongst other things, includes advice on the steps necessary to avoid causing pollution.

The government's responsibility for pesticides has operated for many years and involved non-agricultural as well as agricultural ministers in a scheme which sought to submit to the most detailed scrutiny all pesticide formulations which manufacturers wish to market in the UK. The scheme was recently put on a statutory footing in the Food and Environment Protection Act 1985. The properties of the different products are examined in relation to toxicity to animals, to wildlife and to man; and in relation to persistence in food and water, soil and air. It assesses the conditions under which the product may safely be marketed, and grants commercial clearance only in relation to specific packaging, labelling, application techniques and crops. Biological control agents are included in this scheme with special protocols for their scrutiny.

12.   SOIL EROSION

Erosion by both wind and water causes soil loss and damage to growing crops, and in some cases danger or nuisance to adjoining urban areas. The risk of erosion is governed by soil type, topography and climate. The soils most susceptible to wind erosion are peat and sands in the drier eastern part of the country. The Government's Agricultural Development and Advisory Service (ADAS) gives advice on methods of control which include shelter belts, protective cropping, straw planting, sub-soil mixing, synthetic stabilisers and cultivation techniques. The Forestry Commission (FC) also provides advice on similar issues to the forest industries.

Erosion by water occurs more widely, and there is evidence that some recent changes in agricultural practice have increased the risk of erosion by water. These factors include the cultivation of steeper slopes, increase in field size, surface compaction by heavy machinery and the use of permanent wheelings in arable crops. In addition to the direct effects of loss of soil and damage to crops, erosion can cause blocking of ditches and streams, and debris on roads. Soils with a high silt content are particularly susceptible to erosion by water, but are also very productive under arable cropping, so that protection by perennial cropping is economically unattractive. ADAS advises farmers on the best practical means of minimising the risk of erosion by water, principally by improved soil management, and cropping systems which provide crop cover during susceptible periods. Research is also in progress with the Soil Survey for England and Wales, and with universities.

13.   SOIL MANAGEMENT

The introduction of heavier machinery has increased the risk of soil compaction; at the same time, the need for cultivation to control weeds has decreased. New cultivation techniques have been introduced both to minimise the time and energy input required for seed bed preparation, and also to maintain and improve the structure of both topsoil and sub-soil. These developments have been most marked in the cultivation of upland soils for forestry and of lowland clay soils for winter wheat production. More recently, experimental and advisory effort has focused on adaptation of the techniques to allow the incorporation of chopped straw in arable land.

The roots of many annual crops penetrate to a depth of 1 m or more, and farmers are increasingly aware of the need for good soil management to enable the crop to fully exploit this depth of soil. Associated with this is the need for the installation and effective maintenance (including secondary treatments) of drainage systems on the extensive areas of arable soils of low hydraulic conductivity.

14.   TOXIC ELEMENTS IN SOIL

Most metals are strongly retained by soils. It is therefore possible for potentially toxic elements to accumulate in soils from small additions over a long period of time, to a point where damage to plant growth occurs. There are also possible risks to consumers in the accumulation of toxic elements in food plants. This type of soil degradation is essentially irreversible, since there is no known method of removing the toxic elements from soils.

There are small inputs of toxic elements to soils from the atmosphere and from fertilisers, but the greatest input is from sewage sludge and other waste products. For this reason, there are guidelines agreed between the Ministry of Agriculture, Fisheries and Food and the Department of the Environment on the maximum amounts of the most common elements which can be added to soils over a long period. Monitoring of toxic element concentrations in sewage sludge and in the soils to which it is applied is carried out to ensure that these maxima are not exceeded.

A large number of industries use or produce toxic substances. The regulations controlling the handling of these substances and the disposal of toxic wastes are strict and are enforced. However, accidental spillages on site do occur and the resulting pollution must be contained. Considerable effort is devoted by the Department of the Environment to research on the rehabilitation of such sites—derelict and active.

15.   SOIL PROTECTION—CONTINUING EVOLUTION

A feature of UK soil protection policy has been its evolution with the changing problems of the countryside. It remains essential that a flexible approach is adopted so that appropriate responses can be made to changing circumstances. The close proximity of land and water in the UK especially emphasises the need to consider simultaneously the impacts of possible changes in land use (such as increased forestry) and in pollution inputs (such as acid deposition) upon water catchment yield and quality, and the alternative routes for disposing of the wastes produced by society. The need for a multi-media approach (implemented in the UK through the medium of the Best Practicable Environmental Option principle) is essential to ensure that soil quality is not degraded unnecessarily, thereby maintaining the national soil resource.

Finally, the variation in conditions within the UK has left us firmly committed to a local approach to soil protection. However, this does not mean that valuable lessons about the basic principles and practices of soil protection cannot be learnt by the sharing of experience. There is a particularly important role for those developed nations with the good fortune of northerly and temperate climates to transfer their experience of soil science to those who have to combat the additional problems of severe erosion and of desertification.

Soil Protection in Switzerland

RUDOLF HÄBERLI

National Soil Research Programme,
Schweizerischer Nationalfond zur Förderung der wissenschaftlichen
Forschung, Bundesrain 20, 3003 Berne, Switzerland

SUMMARY

Current soil protection measures in Switzerland are described. These include:

Measures to limit settlement areas and give particular protection to farming (land).
Measures to limit pollutant immissions from non-agricultural sources and from agriculture itself.
Application rules and advisory services for farmers.
A properly organized waste management system.
A surveillance system for monitoring the level of pollution in the soil.

This statutory national system is to all intents and purposes complete. It remains to be seen whether it will be sufficient to guarantee the desired degree of protection; for in Switzerland, too, numerous growth and intensification processes are continuing unabated which present a potentially serious threat to the soil.

As a result, the Swiss Bundesrat has announced a national research programme on 'Land utilization in Switzerland'. The object of this programme is to draw up proposals for using land sparingly. The integrated programme is to last 5 years and examine land in its function as natural environment, as building land and settlement area and as a commercial and legal commodity. Proposals for specific measures are to be based on the studies. They will be aimed at a wide audience, for effective soil protection is possible only if supported by the general public. Public relations work is therefore an essential part of a soil protection strategy.

461

1. INTRODUCTION

Switzerland has a surface area of about 41 000 km². Two-thirds of this are mountains, rocks and glaciers, alpine pastures, lakes and rivers or forest. Only one-third of the surface area, about 13 500 km², is available for agriculture and habitation, roads, railways and airports. Approximately 6·4 million people live in these 'usable' parts of the country, which is equivalent to a population density of about 500 inhabitants per km². There is therefore a scarcity of land suitable for intensive use in Switzerland, and potential environmental damage is high, in line with the high level of production and consumption.

Soil protection entails five tasks:

1. Space conservation: the extension of settlements is to be restricted and the remaining agricultural land and largely undisturbed natural environments protected.
2. Emissions from settlements, industrial and commercial activity and transport must be limited.
3. Pollution from agriculture must be limited.
4. Damaging forms of cultivation must be avoided.
5. The level of pollution in the soil (immissions) must be monitored; where limit values are exceeded, stricter measures must be taken if necessary.

A soil protection concept formulated and adopted by the Government does not exist. However, a number of measures have been adopted with statutory effect nationally which can be regarded as part of such a concept.

2. NATIONAL MEASURES IN FORCE

2.1. Protection of arable land

The Federal Land-use Planning Act [1] has been in force since 1980. It requires the Federal authorities, the Cantons and the local authorities to ensure that land is used 'economically', i.e. in a sparing and non-destructive manner. The amended Land-use Planning Order of 1 May 1986 [2] has improved the level of protection for good agricultural land. The Cantons are required to draw up a land register for every community, charting and giving figures for the remaining suitable arable land, or so-called crop rotation areas. The Bundesrat, the Swiss Federal Government, stipulates the minimum area of arable land to be secured and the distribution by

Canton. The Cantons take the necessary land-use planning measures to ensure that their allotted share of arable land is preserved permanently. If a Canton does not fulfil its obligations, the Bundesrat has the power to implement the requisite measures right down to community and landowner level. The aim is to reserve at least 450 000 ha of the best arable land for permanent agricultural use. This is no less than one-third of the 13 500 km² of the land area mentioned above as suitable for settlement in the strict sense.

2.2. Limitation of emissions

The Federal Environment Protection Act [3] came into force on 1 January 1985. Its purpose is to protect man, fauna and flora, their communities and habitats from disturbance and harm and to preserve the fertility of the soil.

All environmental protection measures are based on the combined principles of preventive action and maximum pollution tolerance (Fig. 1).

Irrespective of the actual pollution in an area, the principle of prevention is to be applied and all measures taken that are possible and economic in the state of the art.

If pollution levels exceed the stipulated limit in an area, additional measures must be taken to prevent emissions at source.

A number of Orders implementing the Environment Protection Act had been adopted by 1 September 1986, specifying the measures in detail. (Fig. 2).

The Substances Order [4] contains *inter alia* provisions on plant protection products, fertilizers and plant protection in forestry. Certain applications of plant protection products have been restricted or

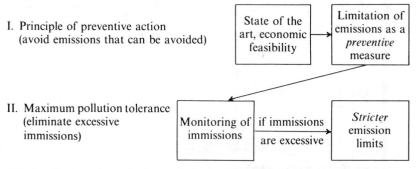

FIG. 1. Strategy for reducing pollution according to the Federal Environment Protection Act.

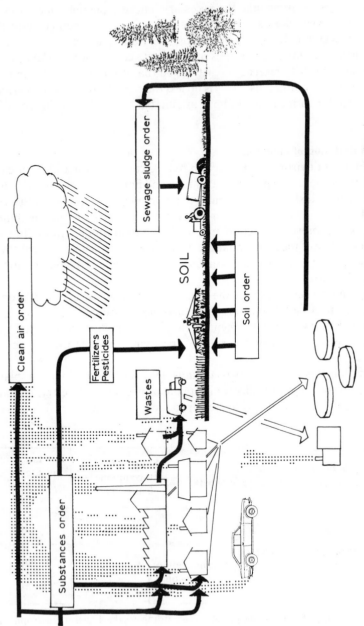

FIG. 2.　Soil protection on the basis of the Federal Environment Protection Act.

prohibited, and substances such as arsenic, some halogenated organic compounds and mercury wholly or partly prohibited. In the case of fertilizers, application rules have been adopted and maximum limits stipulated for chromium, vanadium and cadmium. The use of herbicides and fertilizers in forests is banned completely.

The Clean Air Order [5] contains a whole list of emission limits for stationary and mobile sources. The thresholds at which emissions into the atmosphere are excessive and consequently further measures must be taken at source are stipulated in the form of limit values for sulphur dioxide, nitrogen dioxide, carbon monoxide, ozone, suspended dust and dust deposition, and heavy metals in the dust.

The Sewage Sludge Order [6] stipulates maximum limits for heavy metals and contains provisions on disinfection, storage, transport and application of sewage sludge on agricultural land.

The objective of the Substances Order, Clean Air Order and Sewage Sludge Order is to prevent, as far as possible, environmentally toxic substances from being brought into circulation, disseminated through the atmosphere or applied to the soil by farmers.

2.3. Further measures

These emission control regulations are supplemented by several other laws and measures designed to promote soil conservation. Examples are the Federal laws on nature conservation and forestry, health and protection of the aquatic environment. The Protection of the Aquatic Environment Act also contains provisions on waste management.

In addition, there are six national agricultural research institutes and corresponding agricultural advisory services in the Cantons. One of their tasks is to develop environmentally compatible, 'integrated' cultivation methods and to educate farmers in their use. Advice to farmers is considered crucial to the successful implementation of the laws and regulations quoted.

2.4. Monitoring and controls

Even if all these regulations are observed, it cannot be excluded that immissions in individual areas might accumulate to such an extent that excessive overall pollution levels occur locally.

A national soil monitoring network is to be set up under the Soil Order [7] to deal with this risk. Soil samples are to be taken every 5 years at 100 carefully selected points throughout the country and analysed for pollutants. In high risk areas the national monitoring network is to be supplemented by cantonal measuring stations.

The Soil Order contains guide values for maximum acceptable levels of pollutants in the soil, notably heavy metals and fluorine. If the guide levels are reached or exceeded at any point, the authorities must identify the source of pollution and take more stringent measures, primarily to deal with the emissions side of the problem.

2.5. Balance of measures in force

The soil protection system already in place is basically complete. It comprises:

—measures to restrict settlement areas and give special protection to agricultural land;
—measures to limit pollutant immissions from non-agricultural sources and from agriculture itself;
—application rules and advisory schemes for farmers;
—a properly organized waste management system;
—a surveillance system to monitor pollution levels in the soil.

Does this mean that we have solved our soil protection problems before they really materialized? Do we have 'a grip on' developments in Switzerland?

This conclusion would be something of an oversimplification and certainly premature. Consumption of land for housing, employment, leisure and transport is increasing in Switzerland, too, although the population is no longer growing. Every day more agricultural land is sacrificed for building purposes, up to 3000 ha every year. Furthermore, production and consumption of manufactured goods, which generates pollutants and wastes, is increasing. There is also a trend in agriculture towards applying growing quantities of auxiliary materials to increasingly uniform crop successions, and heavy machinery is thoughtlessly used to work the land even in unfavourable weather conditions. Unfortunately, and farmers are no exception, the most elementary application rules and stipulations are often ignored.

3.   THE NATIONAL SOIL RESEARCH PROGRAMME

In the light of this unmistakable threat to the soil and in awareness of the mechanisms influencing development, the Swiss Bundesrat launched a national research programme on 'land utilization in Switzerland' on 27 February 1985 [8]. The object of the programme is to draw up proposals for economical use of land. The 5-year programme has three specific overriding aims:

—soil fertility must be preserved in the long term;
—losses of undisturbed soil must be reduced;
—land utilization must be more evenly balanced.

The resource 'land' is to be studied in an integrated overall programme to illuminate its function as part of nature, as development and settlement land and as a commercial and legal commodity (Fig. 3).

The section of the programme on 'soil as part of the natural environment' concerns pollutant levels and pollutant balances, intensive stock-rearing and use of farmyard manure, soil compaction, soil erosion and soil organisms. Soil erosion in particular, until recently a negligible factor in our farming areas, has shown a marked increase. In addition, meadows with a high species count, hedges and other largely undisturbed natural environments are to be examined and their value as compensatory and regenerative areas in intensively cultivated farming country demonstrated. The feasibility of using agricultural policy measures to slow down the use of auxiliary materials and other intensification aids is also to be examined.

The section on 'land as building and settlement land' is intended to collect and develop ideas on how space can be saved in the construction of residential accommodation and industrial installations, in the services and public sector and in the layout of leisure facilities. Proposals for better utilization of existing buildings are also to be drawn up and the practicability of these proposals examined in towns, suburban communities and villages. Established development planning principles such as the zoning system and building regulations will be critically reviewed to ascertain whether they impede rather than promote space-saving building.

The part of the programme devoted to 'land as a commerical and legal commodity' is concerned with the mechanisms of the land market, pricing and the motives of those involved, and will appraise important regulatory areas such as taxation, agricultural policy and land-use planning to establish whether they can contribute to using land more sparingly. This research work is now starting to get under way. Initially 39 research projects are concerned [9]. The results of these basic projects will be collated in a second phase when particular attention will be focussed on reciprocal effects and repercussions. Finally, proposals are to be drawn up for specific measures.

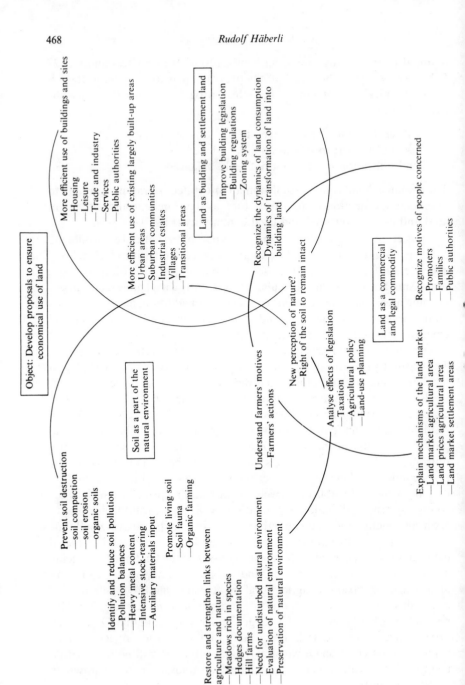

These recommendations will be aimed at various target groups:

—farmers and gardeners working directly with the soil;
—architects, engineers and planners engaged in the preparatory work for, or construction of, buildings and other structures;
—investors, banks and insurance companies who control the use of land through their investments and loans;
—authorities at all levels who set the scene for land use through laws and their implementation;
—the general public, who in the final analysis determined land use through their actions and normal democratic processes.

We are convinced that the soil protection debate must be carried beyond the circle of experts into politics, public life and schools. Effective soil protection is only possible if our attitudes to the environment and nature change. This cannot be done by regulatory means, but the collective consciousness of this issue must gradually develop. Knowledge and legislation are not enough on their own, they must be reflected in action. Effective public relations work is vital to the success of soil protection, which is a long-term challenge.

REFERENCES

1. Federal Land-use Planning Act, 22 June 1979, SR 700.
2. Land-use Planning Order, 26 March 1986, SR 700.1.
3. Federal Environment Protection Act 7 October 1983, SR 814.1.
4. Order on Environmentally Toxic Substances, 9 June 1986, SR 814.013.
5. Clean Air Order, 16 December 1985, SR 814.318.142.1.
6. Sewage Sludge Order, 8 April 1981, SR 814.225.23.
7. Order on Pollutants in the Soil, 9 June 1986, SR 814.12.
8. Execution plan for National Research Programme 22: Land use in Switzerland, Bern, February 1985, 20 p. Swiss National Fund for the Promotion of Scientific Research, Post Box 2338, CH-3001 Bern.
9. Bulletin 'Boden/Sol/Suolo' of the National Soil Research Programme. The bulletin, which is published 2–4 times annually (in German and in French) can be obtained free of charge from the Directorate, NFP Boden, Eigerstrasse 65, CH-3003 Bern.

Strategies for Soil Protection under Intensive Irrigation in Israel

Amos Feigin

*Institute of Soils and Water, ARO, The Volcani Center,
Bet Dagan, Israel*

SUMMARY

*In Israel, where water supplies are limited, brackish water and sewage effluent
are increasingly used for irrigation. Such low quality water may cause
problems of salinity and sodicity in arid and semiarid soils, which require
proper management of cropping, irrigation and nutrition. Suitable methods
have been used to eliminate or mitigate salinity. In addition, mulching and the
use of appropriate chemicals have been shown to be effective in protecting soil
against crust formation. Improved tillage operations reduced deterioration of
soil physical properties in irrigated soils. Intensive agriculture involves the use
of large quantities of fertilizers and pesticides, with concomitant potential
accumulation in soil. The leaching of salt and agrochemicals from the root
zone of crops by rainfall and irrigation water may result in pollution of both
ground and surface water. Well-controlled fertilizer and pesticide application
strategies can reduce these problems. Such strategies, derived from a
knowledge of the behaviour and response of both the soil and the plant, are
embodied in appropriate cropping–irrigation–fertilization management
systems. The use of the minimum leaching concept is promising as a means for
increasing irrigation efficiency and for reducing pollutant transport into the
groundwater. Successful use of these techniques depends on adequate
information on the relevant soil–water processes.*

1. INTRODUCTION

Due to the sub-humid to arid climate in Israel, irrigation has become a dominant factor in its agriculture. The water used includes marginal water, with a consequent wide range of quality. This, coupled with a high evapotranspiration potential that greatly exceeds rainfall, is responsible for the development of salinity and sodicity conditions. Such conditions also occur in non-irrigated soils. On the other hand, the same climatic conditions are suitable for very intensive cropping systems, provided soil moisture and fertility levels are adequate. Heavy application of fertilizers and pesticides is common, and increases the hazards of impaired fertility conditions and groundwater pollution. In addition, the use of heavy equipment for intensive cultivation of irrigated soils is a threat to the soil structure.

As the transport of salt and chemicals in soils depends on rain and irrigation water, soil and groundwater quality are clearly related. New techniques based on recent scientific and technological innovations (e.g. drip irrigation, fertigation, double cropping, moisture maintenance in part of the field) need to be developed, in addition to standard soil conservation operations (e.g. water ponding, drainage, pitting and grassed waterway for run-off water).

Protection and conservation of soil and groundwater under conditions of intensive irrigation present a difficult challenge under semiarid conditions, where water is both limited and expensive. Some examples will be presented that illustrate the strategies adopted and the importance of basic and applied research in the development of efficient solutions.

2. CLIMATE, SOIL PROPERTIES AND QUALITY OF IRRIGATION WATER

The cultivated land of Israel is distributed between the humid and sub-humid Mediterranean climatic zones (having a mean annual rainfall of 500–1000 mm) in the north to semiarid (300–500 mm) and arid (less than 300 mm rainfall/yr) in the south. Many soils are calcareous, with a predomination of montmorillonite clay, and the pH is usually higher than 7. Naturally saline soils occur in the dry parts of the country and the proportion of adsorbed Na (ESP) is also greater in soils of the arid regions.

The rainy season (between November and March) is followed by a dry, hot summer. Irrigation is consequently essential for the production of perennial and intensively cultivated crops. High quality water is fully used for irrigation, and the sources of additional water are sewage effluent and low quality brackish and flood waters.

The relationships among climate, soil properties and salinization were widely studied in Israel [7, 8]. The combination of low quality water and high evapotranspiration potential is conducive to the development of saline conditions. Such situations may also occur in some areas where good quality irrigation water is added to closed valleys with poor drainage.

The proportion of Na in arid and semiarid soils in Israel is often considerable and the presence of Na in low quality water aggravates the problem. Shainberg [29] has shown that crust formation takes place even at low exchangeable sodium percentage (ESP), and reduced infiltration and increased run-off occur at ESP levels as low as 1 in the absence of electrolytes. Thus, raindrop impact on such soils can cause crust formation.

3.   STRATEGIES SUGGESTED FOR THE PROTECTION OF SOIL AND WATER

3.1. Management of irrigation with brackish water

Much attention has been paid to salinity problems in Israel and the subject was recently reviewed by Meiri [25], Meiri and Plaut [26] and Shalhevet [30]. The strategies suggested for minimizing yield reduction under saline conditions are: (i) control of root zone salinity; (ii) reduction of damage to crop growth; and (iii) reduction of damage to plant density. The control of root zone salinity is obtained by irrigation and leaching.

Evaporation increases soil salinity and maintenance of low soil salinity levels (usually measured as the electrical conductivity of soil-saturated paste extract, EC_e) is obtained by leaching of the extra salt below the root zone. The minimum leaching fraction (the relative volume of water that carries salt out of the root zone, LF) that keeps soil salinity below the required level is the leaching requirement (LR). The response of crops to salinity is usually calculated by means of the following equation [24]:

$$Y/Y_{max} = 1 - b(\overline{EC_e} - a)$$

where Y = yield and Y_{max} = yield under non-saline conditions, a = threshold (salinity level at which $Y = Y_{max}$) and b = slope indicating the decline in yield as a result of salinity. Maas and Hoffman [24] and others presented relevant data showing the response of different crops to salinity.

Suitable salinity management [26, 30, 31] involves careful timing of

Table 1
The effect of irrigation method on the response curves of potato plants
(extracted from Meiri and Plaut [26])

Irrigation method	Threshold (a) (dS/m)	Slope (b) (% yield/dS/m)	Zero-yield salinity (dS/m)
Sprinkler	1·1	11·7	10·1
Drip	2·6	6·3	17·1

leaching, cultivation practices, irrigation intervals and water supply. Other aspects involved in crop management under saline conditions are plant stand and row spacing, climatic conditions, irrigation method and fertilization; the last aspect was recently reviewed by Feigin [11] and Kafkafi [22]. The influence of management factors on crop yield under saline conditions is shown in Table 1; the avoidance of direct salt damage to the leaves resulted in salinity alleviation. Similar effects were found in many other cases. Meiri and Plaut [26], Shalhevet [30] and Shalhevet *et al.* [31] have evaluated the efficiency of different techniques suggested for a successful use of brackish water for irrigation. Shalhevet [30] concluded that the recommended method for applying brackish water is drip/trickle irrigation. This irrigation method has attracted much attention recently [5]. The flexibility of this system is shown in Fig. 1. The wetting front and salt distribution can be controlled. The data presented in the figure are the result of model prediction, but they match well with field measurements.

Successful use of brackish water for irrigation requires experimental evidence, as in the case of a choice between the following options of brackish water [30]: (a) using high quality water for sensitive crops and low quality water for tolerant ones; (b) blending waters of different quality; (c) applying good and poor water alternately; and (d) applying high quality water during part of the growing season and low quality water during the rest of the season.

3.2. The use of chemicals to prevent crust formation
Crust formation due to the presence of exchangeable Na is widespread in the arid and semiarid soils of Israel. Such impermeable crusts increase run-off and soil fertility deterioration, and at the same time reduce water infiltration into the root zone of the plants. Shainberg [29] concluded that crust formation during rainfall is the consequence of two complementary mechanisms: (i) mechanical breakdown of soil aggregates due to the

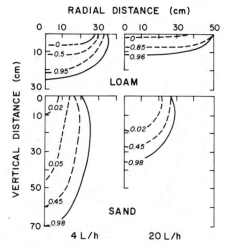

FIG. 1. Calculated wetting front (solid lines) and relative concentration (broken lines) in sandy and loam soil drip-irrigated at two different intensities: 4 and 20 litre/h (after Bresler *et al.* [6]).

raindrop impact, followed by compaction of a thin soil layer; and (ii) chemical dispersion of the clay particles. Dispersion depends on the ESP of the soil, and even 3–5% exchangeable Na can cause soil dispersion, crust formation and very low infiltration rates. The low electrolyte level in rainwater is a major factor responsible for the formation of crust. At relatively high ESP values (>4–5), clay removal occurs and the porosity of the soil is reduced.

Experiments carried out to prevent crust formation and to improve infiltration have shown that straw mulch is an effective solution, but not practical due to sanitation problems. Some chemicals, like industrial phosphogypsum (PG), were also successfully used. Since PG is an effective and enduring source of suitable electrolytes, its application exerts a positive influence on soil structure. Agassi *et al.* [1] have found that application of 5 or 10 mg phosphogypsum/ha improved infiltration to a large extent (Fig. 2) and water run-off and erosion were greatly reduced (Table 2). The yields of crops which depended on natural rainfall were considerably improved due to the higher infiltration and water use efficiency.

3.3. Improved tillage to preserve soil physical properties
The intensive cropping that is possible under the local climatic conditions requires intensive cultivations. This, plus the use of heavy agricultural

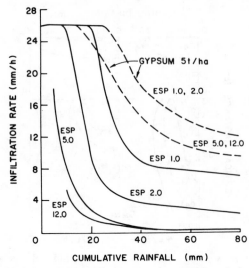

FIG. 2. Effect of exchangeable sodium percentage (ESP) and phosphogypsum on the infiltration rate of a sandy soil (after Shainberg [29]). Aggregate size of soil = 0–4 mm and rain intensity = 26 mm/h.

machinery, has had detrimental effects on soil physical condition, as shown in Table 3 [19]. Soil specific resistance, which reflects the energy demand for tillage operations, has recently increased greatly. At the same time, a clear tendency toward reduced infiltration, soil crust formation and increased run-off was related to unsuitable tillage operations. As a consequence, crop growth and yields have been reduced. Figure 3 [20] shows the clear effect of

Table 2

Rainfall, level of phosphogypsum (PG), erosion and wheat yields at some experimental sites in southern Israel
(extracted from Agassi *et al.* [1])

Location	Rainfall (mm)	PG level (mg/ha)	Run-off (mm)	Soil erosion (mg/ha)	Yield (mg/ha)
Devir	229	0	39	1·4	2·55
		5	11	0·3	3·14
		10	6	0·1	3·14
Alumim	217	0	23	—	0·98
		5	12	—	1·22
		10	13	—	1·28
Qedma	362	0	51	11·2	1·65
		10	23	3·4	2·20

Table 3
The relationship between tractor size and some physical properties of
the soil
(from Hadas [19])

Soil property	*Tractor size*	
	20–30 kW	*90–130 kW*
Specific resistance (kg/cm^2)	0·6–0·7	1·2–1·5
Infiltration rate (mm)	10–20	2–5
Soil crusting	Nil	A growing problem
Run-off	Low	Increasing

soil bulk density on cotton plant stand and a consequent increase in seed cotton yield.

Improved tillage operations are essential for the preservation of soil structure and the maintenance of soil fertility. Strategies that require fewer tillage operations and which employ suitable equipment (smaller tractors) and permanent traffic lanes have been suggested. Table 4 summarizes recommended cultivation methods, ways for their rapid application and topics where further research is needed.

3.4. Efficient application of agrochemicals

The increase in water use efficiency in Israel during the last decades [33] is presented in Fig. 4. The irrigated area has remained stable in recent years

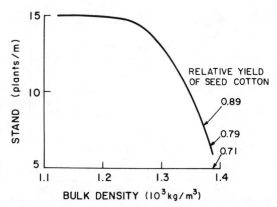

FIG. 3. The relationship between soil bulk density, stand of cotton plants and the relative yield of seed cotton (redrawn after Hadas *et al.* [20]). The relative yield of seed cotton was calculated from experimental response curves of the crop.

Table 4
Strategy for preservation of physical properties in cultivated soils
(after Hadas [19])

Implementation of available information on the effect of tillage operation on soil:
Reduced traffic and tillage practices result in less compaction, and in savings of
labour and energy

Rapid application of research results by inclusion of skilled, educated farmers on
research teams, close liaison with extension officers; inclusion of demonstration
fields at the farm level

Further research, the main objectives of which are to improve tillage practices in
order to reduce run-off, decrease resistance to tillage and improve root penetration,
lessen the number of tillage operations and consequent soil compaction, optimize
energy expenditure and tractor weight and size

but the quantity of water used per unit area has decreased greatly. Statistics
show that crop yield has not been reduced and in many cases has even risen.
The tremendous increase in water use efficiency was *inter alia* a result of
relevant research work. The improvement in other farming operations also
played a role; efficient use of agrochemicals was one of these factors.

Improved fertilization efficiency leads to smaller losses of fertilizers and
fewer residues available for leaching and water pollution. Much attention
has been directed to NO_3^- pollution. However, recent information has
shown that small quantities of P reaching surface water storages may be
responsible for eutrophication [2]. Other ions derived from fertilizers may

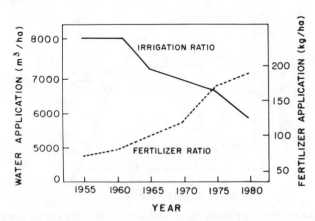

FIG. 4. Trends in irrigation and fertilizer use in Israel (redrawn from Waldman
and Shevah [33]).

also reach groundwater. Excessive application of fertilizers can also be detrimental to soils and crops. Consequently, efficient use of fertilizers is important for both economic and environmental reasons.

The rate of fertilizer application has recently risen considerably (Fig. 4). This is the result of increasing demand by the plants. Fertilizer use efficiency in Israel has been improved by applying up-to-date information on crop demand obtained from both empirical data and models [21]; and the Bet Dagan and Gilat long-term field experiments [12] were important sources of data. Combined application of fertilizers and water (fertigation) resulted in improved control over nutrient distribution in soil. Nitrate distribution was well controlled even under sand dune conditions [3] (Table 5). Drip irrigation was also used successfully for irrigation with sewage water, provided clogging was prevented [4]. Nitrate residues were less under fertigation in comparison with solid application [14]. Small sprinklers were also successfully used to obtain well-controlled N distribution in citrus groves [9, 10]. A combination of double cropping (e.g. wheat and cotton in a single year) and limited wetting of the root zone (Carmi and Plaut, personal communication) results in more efficient fertilizer and water uptake and in lower residual fertilizer levels in the soil. Letey *et al.* [23] concluded that the most positive approach to regulating nitrate-nitrogen in irrigated agriculture is through the control of leaching. A well-controlled irrigation–fertilization management system based on adequate research information is an efficient means of increasing fertilizer efficiency. Such systems will prevent accumulation of chemicals in the soil and at the same time reduce the threat of water pollution. Pratt and Jury [28] gave the following requirements for a management system to maximize N use by crop and to minimize NO_3^- leaching: (a) efficient use of root system for

Table 5
The effect of different water and N application rates on the distribution of mineral N in two sandy soil layers, and at different distances from the emitters ($0 = a$; 13 cm $= b$), 76 days after tomato seeding
(extracted from Bar-Yosef [3])

Soil depth (cm)	Water (litre/emitter/day) and N (mg N/emitter/day)					
	3·2	320	6·0	640	12	1280
	a	b	a	b	a	b
0–20	4	3	6	4	12	11
20–40	4	4	4	4	9	6

NO_3^- uptake; (b) efficient irrigation–fertilization methods in which there is no increase in NO_3^- leaching even though adequate quantities of N fertilizer are added to obtain maximum yields; (c) information on the N requirements of crops that will ensure maximum yield or maximum economic return so that N excess can be avoided; and (d) adjusting water movement through the soil when large amounts of NO_3^- are available for leaching. Their statement that 'This ideal system can serve as a goal and also as a model against which research and practice can be compared' is to be commended.

Recent research on pesticide use in Israel has shown that well-controlled application, taking into account soil properties (such as clay and organic matter content) and using improved irrigation technology (e.g. drip irrigation), can increase pesticide efficiency while reducing its availability for water pollution [16, 17, 18, 27, 34].

3.5. Use of sewage effluents for irrigation

As partially treated sewage effluent contains micro-organisms, organic matter, nutrients and different ions (e.g. NA^+, CL^-, HCO_3^-), its use should be based on reliable information concerning health, environmental and economic aspects. Effects of sewage effluent on soil, plant and groundwater are directly relevant to agriculture.

The shortage of high quality irrigation water has led to increased interest in sewage water for irrigation. Many field experiments on sewage water were conducted in Israel in the last decade, mainly with grass, fodder, grain and industrial crops (mostly cotton). The results have shown that efficient irrigation–fertilization management is possible for successful use of domestic sewage effluents. Pertinent information on the materials introduced into the soil is necessary to reduce water pollution. Data from a recent field study (A. Feigin and J. Dag, unpublished data) show that NO_3^- losses from soil can be greatly reduced while maintaining high yields of seed cotton. Figure 5 shows that the yield of seed cotton grown continuously during the years 1977–81 was maintained without any additional N fertilizer in the effluent-irrigated treatment (having a concentration of 50 mg N/litre), while the yield in the freshwater-irrigated plots declined when fertilizer was not added. Figure 6 represents N uptake by the crop during the 1981 season. The N uptake by the effluent-irrigated plants was not affected by the addition of fertilizer, while the other plants were dependent on the addition. An N balance was estimated (Fig. 7) for five irrigation treatments: freshwater standard irrigation (RI), and effluent standard irrigation (R1), 1·2 times the RI treatments (HV); and effluent at

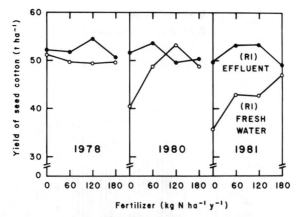

FIG. 5. The effect of secondary sewage effluent versus fresh water, each combined with four levels of N fertilizer, on the yield of cotton grown in a clay soil during three different seasons (A. Feigin and J. Dag, unpublished data).

shorter (HF) or longer (LF) irrigation intervals. Each irrigation treatment was combined with four fertilizer levels (0, 0·5 × standard, standard, 1·5 × standard, arranged in this order in Fig. 7). Figure 7 shows that while almost no N was lost from the non-fertilized freshwater-irrigated plots, increasing N losses were detected where fertilizer was added; a combination of effluents and fertilizer resulted in the greater losses, due to the total large N

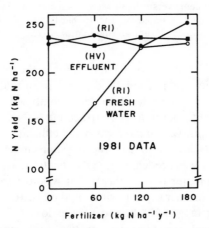

FIG. 6. Uptake of nitrogen by cotton plants irrigated with secondary sewage effluents versus fresh water, each combined with four levels of N fertilizer, during the 1981 season (A. Feigin and J. Dag, unpublished data).

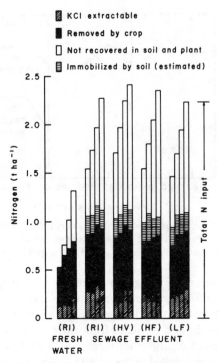

FIG. 7. Estimated nitrogen budget for a cotton plant–soil system irrigated with secondary sewage effluents versus fresh water, each combined with four levels of N fertilizer. The N budget is based on the 1978–81 data (A. Feigin and J. Dag, unpublished data).

quantities added. Yet the N not accounted for in the unfertilized effluent-irrigated treatments was not greatly different from that measured in the standard or $1\cdot5 \times$ standard fertilizer level combined with fresh water. Figure 8 shows that the level of KCl-extractable N $(NH_4^+ + NO_3^-)$ in a soil profile of $1\cdot2$ m depth was greater when the effluent was combined with fertilizer. At the N level found in the relevant effluent (50 mg/litre) no additional fertilizer was needed. Figure 8 indicates that leaching probably accounted for the major part of the N that was lost from the soil–plant system. However, another experiment carried out with the same soil in the greenhouse, using tagged-N, showed greater losses (7% of the total N added) in the effluent-irrigated versus nutrient solution-irrigated soil. The level of P in the soil was increased in the effluent-irrigated soil, but did not exceed the recommended levels.

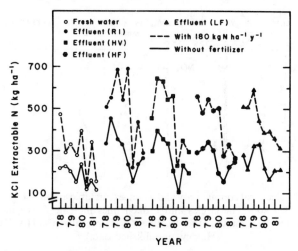

FIG. 8. The effect of irrigation with secondary sewage effluents versus irrigation with fresh water, each combined with two nitrogen fertilizer levels, on the level of KCl-extractable N in soil. For explanation of the effluent treatments, see text (A. Feigin and J. Dag, unpublished data).

Waldham and Shevah [33] concluded that the pollution hazard in the most densely populated area in Israel (the Coastal Plain) and other areas underlain by aquifers supplying fresh water is considerable. However, they believe that proper planning and management can ensure that pollution does not exceed acceptable levels. Other experiments with sewage effluents confirmed the idea that proper irrigation–fertilization management can greatly reduce pollution hazard [13, 15, 32].

3.6. The minimum leaching approach

Available data indicate that efficient irrigation with minimum leaching results in water, energy and labour savings, and in smaller chances of pollutant and salt transport into groundwater [6]; the field data of Letey *et al.* [23] agree with this approach. The potential benefits to be obtained from the minimum leaching concept have stimulated research work in Israel (conducted by A. Feigin and J. Halevy, ARO, The Volcani Center). Long-term field experiments in which (a) a wide range of N, P, K fertilizers (the Bet Dagan experiment) and (b) N, manure and city refuse treatments (the Gilat experiment) have been tested for 25 years provide a sound basis for the research. Large differences in yield and consequently in transpiration are probably associated with considerable differences in

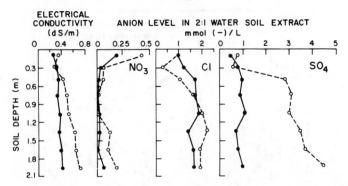

FIG. 9. The electrical conductivity and water-extractable anion (NO_3^-, Cl^-, SO_4^{2-}) levels in a clay soil. The circles and broken lines indicate heavy nitrogen and phosphorus fertilization; the other treatment was unfertilized soil (A. Feigin and J. Halevy, unpublished data).

leaching. The study by Feigin and Halevy will include measurement of long-term effects on the distribution of ions within a deep soil profile (up to 2·1 m) and their relation to yield, and the possibility of predicting ion distribution using available models. The research should provide quantitative data on the relation between yield and ion flux below the root zone of the crop. Preliminary results (Fig. 9) show large differences in the anion distribution within the soil profile between no fertilizer addition and heavy application of N and P. It is too early to draw any conclusions, but Fig. 9 suggests that there are large fertilizer effects on the ion distribution in soil. In addition to differences in the level of fertilizer added, differences in precipitation (SO_4^{2-}), uptake, leaching and denitrification (NO_3^-) appear to be influential. It is assumed that this study will provide information on optimal fertilizer treatments that combine high crop yield with minimum residues in the soil and minimum leaching to the groundwater.

4. CONCLUSIONS

Intensive cropping with low quality water and high loads of agrochemicals can have adverse effects on soil physical, chemical and biological properties, and on groundwater quality. Inappropriate tillage operations impair soil physical properties.

The protection of soil is closely associated with the control of surface and groundwater quality.

Effective strategies for soil protection under intensive cropping conditions must be based on a high level of control and monitoring of the agricultural systems, as well as on appropriate research effort to deal with the problems as they arise.

REFERENCES

1. AGASSI, M., SHAINBERG, I. and MORIN, J. (1985). Infiltration and runoff in wheat fields in the semi-arid region of Israel. *Geoderma*, **36**, 263–76.
2. AVNIMELECH, Y. (1984). Behavior of phosphate in the unsaturated zone. In: *Pollutants in Porous Media* (Eds B. Yaron, G. Dagan and J. Goldschmid), Springer-Verlag, Berlin.
3. BAR-YOSEF, B. (1977). Trickle irrigation and fertilization of tomatoes in sand dunes: water, N, and P distribution in the soil and uptake by plants. *Agron. J.*, **48**, 486–91.
4. BIELORAI, H., VAISMAN, I. and FEIGIN, A. (1984). Drip irrigation of cotton with treated municipal effluents. I. Yield response. *J. Environ. Qual.*, **13**, 231–4.
5. BRESLER, E. (1977). Trickle-drip irrigation: principles and application in soil–water management. *Adv. Agron.*, **29**, 343–93.
6. BRESLER, E., MCNEAL, B. L. and CARTER, D. L. (1982). *Saline and Sodic Soils: Principles–Dynamics–Modelling*, Springer-Verlag, Berlin, 236 pp.
7. DAN, J. and KOYUMDJISKY, HANNA (1986). Formation and distribution of salinity in the soils of Israel. *Israel J. Earth Sci.* (in press).
8. DAN, J. and YAALON, D. H. (1982). Automorphic saline soils in Israel. *Catena Suppl.*, **1**, 103–15.
9. DASBERG, S., BIELORAI, H. and ERNER, Y. (1981). Partial wetting of the root zone and nitrogen effects on growth, yield and quality of Shamouti oranges. *Acta Hort.*, **119**, 103–7.
10. DASBERG, S., ERNER, Y. and BIELORAI, H. (1983). Nitrogen balance in a citrus orchard. *J. Environ. Qual.*, **13**, 353–6.
11. FEIGIN, A. (1985). Fertilization management of crops irrigated with saline water. *Pl. Soil*, **89**, 285–99.
12. FEIGIN, A., HALEVY, J. and KAFKAFI, U. (1974). Cumulative effects of manures applied with and without fertilizers, on nutrient availability in the soil and on yields of crops under irrigation. *Hassadeh*, **65**, 1351–84 (Hebrew, with English summary, and table and figure legends).
13. FEIGIN, A. and KIPNIS, T. (1976). Improving N uptake by Rhodes grass (*Chloris gayana*) from treated municipal effluents to reduce fertilizer requirements and prevent pollution. In: *Agrochemicals in Soils* (Eds A. Banin and U. Kafkafi), Pergamon Press, Oxford, pp. 387–94.
14. FEIGIN, A., LETEY, J. and JARRELL, W. M. (1982). Nitrogen utilization efficiency by drip irrigated celery receiving preplant or water applied N fertilizer. *Agron. J.*, **74**, 978–83.
15. FEIGIN, A., VAISMAN, I. and BIELORAI, H. (1984). Drip irrigation of cotton with treated municipal effluents. II. Nutrient availability in soil. *J. Environ. Qual.*, **13**, 234–8.

16. GERSTL, Z. (1984). Adsorption, decomposition and movement of oxamyl in soil. *Pestic. Sci.*, **15**, 9–17.
17. GERSTL, Z. and ALBASEL, N. (1984). Field distribution of pesticides applied via a drip-irrigation system. *Irrig. Sci.*, **5**, 181–93.
18. GERSTL, Z. and YARON, B. (1983). Behavior of bromacil and napabromide in soils: adsorption and degradation. *Soil Sci. Soc. Am. J.*, **47**, 474–8.
19. HADAS, A. (1983). Tillage research in Israel—a case of rapidly changing tillage practices. *Soil Tillage Res.*, **3**, 317–20.
20. HADAS, A., WOLF, D. and RAWITZ, E. (1985). Soil compaction effect on cotton stand and yields. *Trans. Am. Soc. Agric. Engrs*, **28**, 691–6.
21. HADAS, AVIVA, BAR-YOSEF, B., KAFKAFI, U. and YARIV, J. (1979). A model to determine N, P, K fertilization recommendations under field conditions. Pamph. Agric. Res. Orgn, Bet Dagan 217, 59 pp.
22. KAFKAFI, U. (1984). Plant nutrition under saline conditions. In: *Soil Salinity Under Irrigation—Processes and Management* (Eds I. Shainberg and J. Shalhevet), Springer-Verlag, Berlin, pp. 319–38.
23. LETEY, J., BLAIR, J. W., DEWITT, D., LUND, L. J. and NASH, P. (1977). Nitrate-nitrogen in effluent from agricultural tile drains in California. *Hilgardia*, **45**, 289–319.
24. MAAS, E. V. and HOFFMAN, G. J. (1977). Crop salt tolerance—current assessment. *J. Irrig. Drainage, Div. ASCE*, **103**(IR2), 115–34.
25. MEIRI, A. (1984). Plant response to salinity: application of experimental results. In: *Soil Salinity Under Irrigation—Processes and Management* (Eds I. Shainberg and J. Shalhevet), Springer-Verlag, Berlin.
26. MEIRI, A. and PLAUT, Z. (1985). Crop production and management under saline conditions. *Pl. Soil*, 253–71.
27. MINGELGRIN, U. and GERSTL, Z. (1983). Reevaluation of partitioning as a mechanism of non-ionic chemical adsorption in soils. *J. Environ. Qual.*, **12**, 1–11.
28. PRATT, P. F. and JURY, W. A. (1984). Pollution of the saturated zone with nitrate. In: *Behavior of Pollutants in the Unsaturated Zones* (Eds B. Yaron, G. Dagan and J. Goldshmid), Springer-Verlag, Berlin, pp. 53–67.
29. SHAINBERG, I. (1985). The effect of exchangeable sodium and electrolyte concentration on crust formation. *Adv. Soil Sci.*, **1**, 101–22.
30. SHALHEVET, J. (1985). Management of irrigation with brackish water. In: *Soil Salinity Under Irrigation—Processes and Management* (Eds I. Shainberg and J. Shalhevet), Springer-Verlag, Berlin, pp. 298–318.
31. SHALHEVET, J., HEUER, B. and MEIRI, A. (1984). Irrigation interval as a factor in the salt tolerance of eggplants. *Irrig. Sci.*, **4**, 83–93.
32. VAISMAN, I., SHALHEVET, J., KIPNIS, T. and FEIGIN, A. (1982). Water regime and N fertilization for Rhodes grass irrigated with municipal waste water on sand dune soil. *J. Environ. Qual.*, **11**, 230–2.
33. WALDMAN, M. and SHEVAH, Y. (1984). Prevention of groundwater pollution on a national scale: Israel as a case study. In: *Pollutants in Porous Media* (Eds B. Yaron, G. Dagan and J. Goldshmid), Springer-Verlag, Berlin, pp. 237–46.
34. YARON, B., GERSTL, Z. and SPENCER, W. F. (1985). Behavior of herbicides in irrigated soils. *Adv. Soil Sci.*, **3**, 121–211.

Chemical Fate Modelling in Soil Systems: A State-of-the-Art Review

MARC BONAZOUNTAS

Department of Civil Engineering, National Technical University,
5 Iroon Polytechnicon Str., 157 53 Zograton, Athens, Greece

SUMMARY

This chapter can be read as a separate document; it describes principles of the state of knowledge of the mathematical terrestrial (land, soil, groundwater) modelling for use in environmental quality studies and decision making. Chemical fate modelling in soil systems has been traditionally performed for three distinct environments: the land surface (or watershed), the unsaturated soil (soil) zone and the saturated (groundwater) zone of a region. Three types of model categories and models are available: miscible (or dissolved pollutant) models, immiscible fluid (or non-aqueous phase liquid, NAPL) models and chemical equilibrium (or speciation) models. A fourth category (ranking models) deals with the environmental rating of waste sites, in cases where detailed modelling is not desirable or possible. Detailed information on the above issues is presented in the following sections.

Unsaturated soil zone models can simulate moisture flow and quality conditions of a soil zone profile extending between the ground surface and the groundwater table. The traditional modelling for dissolved pollutants employs either the time-dependent diffusive convective mass transport differential equation in homogeneous isotropic soils or simplified analytic expressions. Modelling for immiscible fluids employs the two-, three- or four-phase flow (air, moisture, NAPL, dissolved pollutant) equations in porous media. Compartmental mathematical modelling for dissolved and NAPL compounds is an area of current research. Equations are principally solved numerically or analytically when seeking solutions for simplified environments.

Groundwater models describe the fate of contaminants in aquifers.

Ironically enough, although the number of model types is large, only a few basic processes are modelled, mainly via the convective, dispersive, adsorptive, reactive, pollutant transport equation for a dissolved pollutant in a saturated porous medium. This equation, or the equation systems, is solved via analytical, numerical or statistical techniques, and for one pollutant at a time. Modelling of NAPL in groundwater follows the principles of soil modelling.

An evaluation of the fate of trace metals in surface and sub-surface waters requires detailed consideration of complexation, adsorption, coagulation, oxidation–reduction and biological interactions. These processes can affect metals solubility, toxicity, availability, physical transport and corrosion potential. As a result of a need to describe the complex interactions involved in these situations, various models have been developed to address a number of specific situations. These are called equilibrium or speciation models, because the user is provided (model output) with the distribution of various species.

There are two basic approaches to the solution of the species distribution problem: (1) the equilibrium constant approach and (2) the Gibbs free energy approach. Most models use the former approach, which utilizes measured equilibrium constants for all mass action expressions of the systems. The latter approach uses free energy values. In both cases, the most stable condition is sought, and a solution to a set of non-linear equations is required. Aquatic equilibrium models are at a developmental stage. Current versions are steady-state models and are formulated for one or multiple compartments.

This chapter presents information on pollutant sources and emissions, fate of miscible and immiscible pollutants in the soil environment, environmental factors and chemistry, mathematical modelling, dissolved pollutant modelling, immiscible contaminant modelling, speciation modelling, parameters affecting contaminant migration through soils, and model applications.

1. INTRODUCTION AND BACKGROUND

1.1. Introduction

This chapter can be read as a separate document; it describes concepts, uses and limitations of state-of-the-art mathematical environmental pollutant fate modelling for use in both environmental studies and analyses of environmental quality or human exposure.

The purpose of this chapter is to help readers understand modelling complexities, identify specific modelling packages and select a documented mathematical model.

This chapter describes the principles of the state-of-the-art mathematical terrestrial (land, soil, groundwater) modelling for use primarily in environmental quality studies and secondarily in environmental decision making. Terrestrial modelling of inorganic species is complex and must deal with many uncertainties; therefore, it has not been extensively employed in decision making.

Subsequent sections provide a short background on terrestrial modelling, describe the principal processes considered by models, summarize the differences between organic and inorganic pollutant computerized FORTRAN codes, list speciation codes and illustrate applications of certain models.

Each section of this chapter is fairly self-contained; it includes an introduction, a discussion of key physical and chemical issues related to each modelling category, a discussion of important environmental interactions of inorganic pollutants with the specific environment, an outline of model requirements for input data and model applications, and examples of model applications. References for each modelling category are presented at the end of the sub-section for that category.

1.2. Regulatory background

1.2.1. Legal/regulatory issues

Models are used in a variety of ways to assist in decision making and will be used to a greater extent in the future; specific statutes or regulations require the use of models in certain situations. Additionally, provisions of the National/Environmental Policy Act of 1969 (EPA), as well as judicial decisions constituting NEPA and other environmental statutes, should facilitate the increased use of mathematical models.

There are a number of reasons for the increasingly widespread use of mathematical and computer models in environmental decisions. For example, the US Congress mandated the use of computed models in the 1977 amendments to the Clean Air Act. Under the 1977 amendments, models must be used in connection with the Prevention of Significant Deterioration (PSD) for air quality and for designation of non-attainment areas. Under the PSD regulations one must obtain a permit prior to commencing construction; this must be preceded by an analysis for air quality impacts projected for the area as a result of growth associated with such facility. The analysis must be based on air quality models specified by regulations promulgated by the EPA. The 1977 amendments also require conferences on air quality modelling every 3 years, to ensure that the air quality model used in the PSD program reflects the current state-of-the-art in modelling.

The increasing use of quantitative models, particularly computer models, has placed a new burden upon the courts in their review of environmental decisions based on those models. This burden is a part of the 'new era' in environmental decision making and reflects the increasing involvement of scientific and technical issues in legal decisions. Problems in judicial review arising from the use of computer models and other quantitative methodologies in environmental decision making are described by Case (1983), whose publication is the source of a considerable amount of the information presented here.

The problems arising from the use of an environmental model in regulatory use are twofold; first, its use may actually increase the likelihood that a substantially incorrect decision will be reached. This greater probability of error generally can be traced to the inability of environmental decision-makers to deal with certain aspects of the use of models in making such decisions. Second, the use of a model increases the danger that wrong environmental decisions may not be detected and corrected by the reviewing courts. Additionally, certain aspects of the institutions of the environmental agencies, the courts and their relationship further contribute to the difficulties of judicial access to technical resources to assist in analysing the issues involved, the limits on the court's ability to supplement or go outside the record, and the traditional deference which the courts give to administrative decisions.

Nevertheless, model development and application will increase in the future. Beyond any legal mandates, increasingly complicated and intractable environmental problems will compel the greater use of quantitative models by environmental decision-makers. Many experts believe in environmental fate, exposure and risk modelling, since models contribute to scientific understanding of the environmental quality.

1.2.2. Purpose of modelling

Modelling mobilization and fate of chemicals in soil systems is conducted for three purposes: (1) assessment of environmental quality; (2) assessment of human exposure; and (3) decision-making, including control strategies for environmental and human protection.

There are four major exposure pathways for contaminants from uncontrolled hazardous waste disposal sites; for example (Ehrenfeld and Bass [24]): groundwater/leachate, surface water, contaminated soils, and residual waste and air. The environmental setting for an uncontrolled disposal site located above the water table is shown in Fig. 1. The potential pathways to human and ecological receptors are depicted in Fig. 2. A

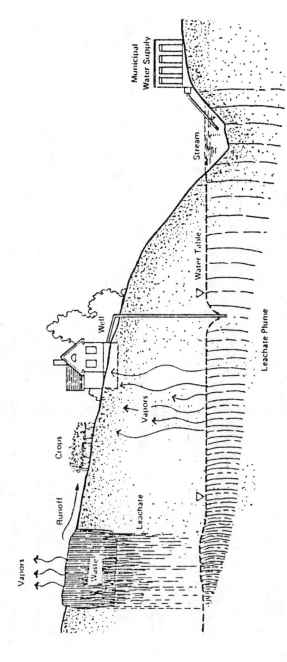

FIG. 1. Environmental pathways from a generalized hazardous waste site (Ehrenfeld and Bass [24]).

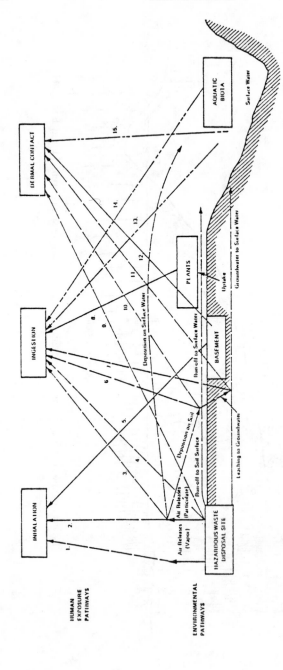

FIG. 2. Exposure pathway from hazardous waste site (Arthur D. Little, Inc.). Description of exposure pathways: 1, inhalation of vapours; 2, inhalation of particulates; 3, swallowing of larger particulates; 4, ingestion of soil on site (particularly children); 5, inhalation of vapours from seepage in basements; 6, ingestion of soil off-site (particularly children); 7, ingestion of groundwater; 8, ingestion of plants (home gardens); 9, dermal contact with soil on-site; 10, dermal contact with soil off-site; 11, dermal contact from household use of groundwater (bathing, etc.); 12, dermal contact w/seepage in basements; 13, ingestion of surface water; 14, dermal contact with surface water (swimming, etc.); 15, ingestion of aquatic biota.

variation of the above would be a site where the waste was buried below the water table; in this case, the leachate plume and groundwater are coincident. The exposure pathways are essentially the same in both cases.

Remedial actions are designed to reduce exposure to humans and the environment to acceptable levels either by containing pollutants originating from the waste site in place or removing the hazardous substances from the immediate environment.

Modelling can play an integral role in waste clean-up, agricultural development and other environmental protection studies. A model is a decision tool which, if applied properly, can greatly assist decision-makers in effectively dealing with complex issues at uncontrolled waste sites. Today, five basic categories are used: (1) emission models, to quantify release (quantity) or pollutant emissions in the environment (e.g. air emissions or leaching from a waste site); (2) fate models, to estimate concentrations of pollutants in the environmental media (e.g. fate of pollutants in the soil and groundwater); (3) exposure models, intended to convert environmental concentrations to absorbed doses by humans (e.g. in inhalation); (4) risk models, known also as dose-response models, for the extrapolation of animal carcinogenicity data to humans and the estimation of probable human risks to cancer; and (5) cost/effectiveness models or analyses (e.g. reduction of human risk) when imposing alternative actions or strategies (e.g. remedial actions) at waste sites.

1.2.3. Modelling concepts

Terrestrial chemicals fate modelling has traditionally been performed for the three distinct sub-compartments: (1) the land surface (or watershed); (2) the unsaturated soil (or 'soil') zone; and (3) the saturated (or 'groundwater') zone of the region (Fig. 3). In general, the mathematical simulation is structured around two major processes, the hydrologic cycle and the pollutant cycle, each of which is associated with a number of physico-chemical processes. Land surface models also account for a third cycle, sedimentation.

Land surface models describe pollutant fate on land (known as watershed), the unsaturated soil zone of the region and the pollutant contribution to the water body of the area. Unsaturated soil zone models simulate both (1) soil moisture movement and (2) soil moisture and soil solid quality conditions of a soil zone profile extending between the ground surface and the groundwater table. Groundwater models describe the fate of pollutants in aquifers.

When used properly and with an understanding of their limitations,

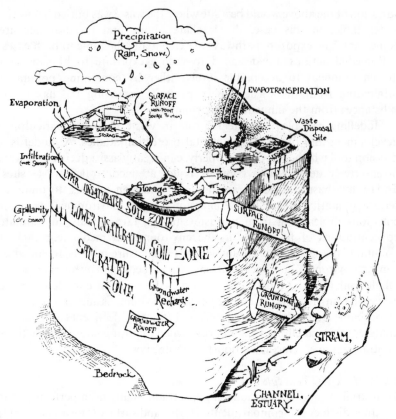

FIG. 3. Schematic presentation of the soil compartment (Bonazountas and Wagner [16]).

mathematical models can greatly assist decision-makers in determining the importance of pollutant pathways in the terrestrial environment. For this reason, the use of models has grown dramatically over the past decade. Although the number of terrestrial model types is very large, there are only a few fundamental modelling concepts.

Soil zone modelling is highly complex, because the physical and chemical dynamics of a soil sub-compartment—in contrast to those of a water or air sub-compartment—are governed by external ('out-compartmental') forces such as precipitation, air temperature and solar radiation. Water and air modelling are generally simpler, because the dynamics of these compartments are governed by 'in-compartmental' forces.

Chemical modelling in soil systems provides information in the distribution of elements (e.g. metal species) within a soil matrix consisting of soil-solids, soil-moisture (in the soil zone) or soil-water (in groundwater) and soil-air (in the soil zone) (see Fig. 6) for the soil zone of a region. The objective is to determine the amount of pollutants in the solid phase, aqueous phase and/or gaseous phase at a given point and time.

1.2.4. Principal processes—general overview

An evaluation of the fate of inorganic compounds in soil and groundwater requires a detailed consideration of the physical, chemical and biological processes and reactions involved, such as complexation, absorption, precipitation, oxidation–reduction, chemical speciation and biological reactions to determine the free metal concentration in soil solutions. These processes can affect such characteristics as species solubility, availability for biological uptake, physical transport and corrosion potential.

To describe the complex interactions involved, various kinds of models have been developed. These are, for example, 'adsorption' models (which utilize mathematics involved in specific adsorption, surface complex formation and ion exchange), surface complexation models, constant-capacitance models, cation-exchange models, and overall fate-modelling packages that take into account the effects of one or more geochemical processes. One category ('equilibrium' or 'speciation' models) is designed to determine the distribution of inorganic species in the soil water. Chemically-based computer models of soluble trace metal speciation are being employed increasingly in decision making, such as in studies related to sewage and effluents applied to agricultural land.

The transport of particular 'species' in terrestrial systems is of interest to a variety of scientists, since measurement or reporting of 'total' concentration of a particular inorganic compound in the soil may be misleading in many environmental management situations. Toxic effects of trace metals, for example, may be affected more by their chemical form than by their total concentration. Therefore, mathematical computer models capable of simulating the distribution of inorganic pollutant species in soil and groundwater systems are valuable tools for analysing contaminant pathways.

1.2.5. Models—general overview

Mathematical fate modelling and models are available for (1) a dissolved (miscible) organic or inorganic pollutant, (2) an immiscible contaminant (with one or more pollutants), and (3) a metal. At this stage of intensive research on terrestrial chemical modelling, we can group the prevailing concepts into two major categories: (1) geochemical or, more appropriately, pollutant fate models, and (2) speciation equilibria models. This terminology is not standard but is employed here for convenience.

Geochemical models may be applied for a simplified overall inorganic pollutant fate estimation via an adsorption process (routine) or for many chemical and biological processes that govern the fate of the total pollutant mass from dissolved or partially dissolved (immiscible) fluids. Speciation models estimate the distribution of metal in various forms, but certain 'speciation' codes can simulate the fate of individual dissolved pollutant species and the total mass of dissolved pollutants, since they combine fate and speciation codes in one package.

The variety of geochemical (pollutant fate) models has dramatically increased during the last decade. These models can be employed for both organic and inorganic pollutants, since the chemical processes are simulated via their overall equilibrium coefficients (e.g. adsorption coefficients). Information on such models is provided in Section 8.

The status of geochemical speciation equilibria models for inorganic pollutants in soils are described in Section 8.2. Speciation computer codes account for additional processes—redox reaction, adsorption, complexation and others (e.g. James and Rubin [38]). Speciation equilibria models are based on chemical thermodynamic principles. Some of the computer codes available for speciation calculations in soil and groundwater aqueous systems are described in Section 8.3.

Calculations and procedures are described in Section 8.4. Steps in speciation calculations are not of concern in the following section, which focuses on integrated computerized modelling codes.

Excellent state-of-knowledge reviews of chemical equilibria codes (models) of inorganic pollutants in soils are presented by numerous investigators. The following sections have drawn on these reviews. Readers interested in details should refer to the original publication.

2. SOURCES AND EMISSIONS

Through numerous human activities, pollutants are released to the soil compartment. The particular particle significantly influences the fate of pollutants in the soil and groundwater zones. Releases include both point source and area loadings. They may be intentional, such as landfills and spray irrigation of sewage; unintentional, such as spills and leaks; or indirect, through pesticide drift or surface run-offs. The point of release may be at the soil surface, or from a source buried deep in the soil. Substances released are in liquid, semi-liquid, solid or particulate form. In some cases a waste material will be pretreated or deactivated prior to disposal to limit its mobility in soil. The rate of release may be continuous, such as at a municipal landfill; intermittent; or on a 'batch' basis such as practised by some industries, or as a one-off episode such as the uncontrolled disposal of barrels of waste or a spill.

Soil and groundwater contamination are commonly encountered problems at uncontrolled hazardous waste sites; they result from the migration of leachates originating from a wide variety of waste management facilities.

Table 1 indicates primary pollutant sources and waste models, and Table 2 indicates the primary and secondary sources and associated pollutants. The primary sources of soil contamination include land disposal of solid waste, sludge and wastewater, industrial activities, and leakages and spills, primarily of petroleum products. The solid waste disposal sites include dumps, landfills, sanitary landfills and secured landfills.

Land disposal sites result in soil contamination through leachate migration. The composition of the substances produced depends principally on the type of wastes present and the decomposition in the landfill (aerobic or anaerobic). The adjacent soil can be contaminated by direct horizontal leaching of surface run-off, vertical leaching, and transfer of gases from decomposition by diffusion and convection. The disposal of domestic and municipal wastewaters on land takes place through septic tanks and cesspools; sewage sludge from primary and secondary treatment plants often spread on agricultural and forested land (land treatment); liquid sewage, either untreated or partially treated is applied to the land surface, by spray irrigation, disposal over sloping land, or disposal through lagooning of sewage sludge.

Landfills are principally disposal sites for municipal refuse and some industrial wastes. Municipal refuse is generally composed of 40–50% (by weight) of organic matter, with the remaining consisting of moisture and inorganic matter such as glass, cans, plastic and pottery. Under aerobic decomposition, carbonic acid that is formed reacts with any metals present and calcareous materials in the rocks and soil, thus increasing the hardness and metal content of the leachate. Decomposition of the organic matter also produces gases, including CO_2, CH_4, H_2S, H_2, NH_3 and N_2, of which CO_2 and CH_4 are the most significant soil contaminants.

Agricultural practices affect soil quality in many ways, principally by increasing soil nutrients and by rooting chemicals to the groundwater or to the adjacent water bodies via the washload (sediment).

Table 1

Sources and wastes contaminating soils
(Bonazountas [13])

Pollutant source	Waste-water impound-ments	Solid waste disposal sites	Waste-water spray irrigation	Land applica-tion	Injection or disposal wells	Septic tanks and cesspools	Pits	Infiltration/surface run-off	Leaching from storage sites
Industrial									
Wastewater	×		×						
Sludge					×				×
Solid waste		×		×					
Municipal									
Wastewater			×						
Sludge		×		×					
Solid waste		×							×
Household wastewater						×			
Agricultural feedlot		×						×	
Mining	×				×		×	×	
Petroleum exploration					×		×		
Cooling water					×				
Buried tanks and pipelines									×
Agricultural activities								×	×

Table 2

Primary sources of soil contamination and associated pollutants
(Bonazountas [13])

Source	Type of pollutants
Industrial sources	
Chemical manufacturers	Organic solvents
Petroleum refineries	Chlorinated hydrocarbons
Metal smelters and refineries	Heavy metals
Electroplaters	Cyanide, other toxics
Paint, battery manufacturers	Conventional pollutants
Pharmaceutical manufacturers,	Acids, alkalis, other corrosives;
paper and related industries	many are highly mobile in soil
Land disposal sites	
Landfills that received sewage sludge,	BOD, inorganic salts, heavy metals,
garbage, street refuse, construction	pathogens, refractory organic compounds,
and demolition wastes	plastics; nitrate; metals including iron,
	copper, manganese suspended solids
Uncontrolled dumping of industrial	
wastes, hazardous wastes	
Mining wastes	Acidity, dissolved solids, metals, radioactive
	materials, colour, turbidity
Agricultural activities	BOD, nutrients, faecal coliforms, chloride,
	some heavy metals
Agricultural feedlots	
Treatment of crops and/or soil with	Herbicides, insecticides, fungicides, nitrates,
pesticides and fertilizers; run-off or	phosphates, potassium, BOD, nutrients,
direct vertical leaching to septic tanks	heavy metals, inorganic salts, pathogens,
and cesspools	surfactants; organic solvents used in cleaning
Leaks and spills	
Sources include oil and gas wells;	Petroleum and derivative compounds; any
buried pipelines and storage tanks;	transported chemicals
transport vehicles	
Atmospheric deposition	Particulates; heavy metals, volatile organic
	compounds; pesticides; radioactive particles
Highway maintenance activities	
Storage areas and direct application	Primarily salts
Radioactive waste disposal	
Eleven major shallow burial sites exist	Primarily ^{132}Cs, ^{90}Sr and ^{60}Co
in US; three known to be leaking	
Land disposal of sewage and wastewater	
Spray irrigation of primary and	BOD, heavy metals, inorganic salts,
secondary effluents	pathogens, nitrates, phosphates, recalcitrant
	organics
Land application of sewage sludge	
Leakage from sewage oxidation ponds	

3. FATE OF CONTAMINANTS IN SOIL SYSTEMS

3.1. General

The mechanisms of contamination from various activities and the transport of organic contaminants in groundwater are described by Hall and Quam [36], Mackay and Roberts [57], Mercer [65], Faust [28], Abriola and Pinder [1], Fried *et al.* [34] and many other researchers. Information presented below is mainly obtained from the work of Mackay and Roberts and Hall and Quam.

Organic contaminants can reach the groundwater zone either dissolved in water or as organic liquid phases that may be immiscible in water. These contaminants travel with the soil moisture and are retarded in their migration by various factors.

The sub-surface transport of immiscible (NAPL) organic liquids is governed by a set of factors different from those for dissolved contaminants. However, some components of organic liquids can dissolve into the groundwater; therefore, the process of dissolved pollutants has been of primary importance in the past. Metals are subject to different processes.

3.2. Nature of contamination

Hall and Quam [36] report on the nature of contamination from petroleum products, for example, that one of the most important factors in contamination of groundwater is the extremely low concentration of the product which can give rise to objectional tastes and odours. The specific aspects of contamination may be broadly classified into

(a)   the formation of surface films and emulsions; and
(b)   the solubility in water of certain petroleum products.

The problems associated with surface films are minimized due to the ability of aquifers to adsorb or absorb much of the product. However, this phenomenon magnifies the problems associated with the soluble components of the product since hydrocarbons held in this manner are subject to leaching as water passes over them.

Surface films may affect the aesthetics and interfere with treatment or industrial processes. They may also be toxic to animal or plant life if they emerge into surface waters.

The water-soluble components of petroleum products which give rise to taste and odour problems are the aromatic and aliphatic hydrocarbons. Phenols and cresols are examples of these compounds, and are known to generate taste and odours at concentrations as low as 0·01 mg/litre. When chlorine is added to drinking water, as in most municipal water supplies, it reacts with the phenols to form chlorophenols which give rise to objectional tastes and odours at concentrations as low as 0·001 mg/litre. It is thus apparent that very small quantities of hydrocarbons can give rise to widespread contamination of water resources.

3.3. Transport and distribution of contaminants

According to Mackay and Roberts [57], major transport and transformation processes for organic dissolved contaminants are advection, dispersion, sorption and retardation, and chemical and biological transformation. The migration of an immiscible organic liquid phase is governed largely by its density, viscosity and surface-wetting properties.

Density differences of about 1% are known to influence fluid movement significantly. With few exceptions, the densities of organic liquids differ from that of water by more than 1%. In most cases the difference is more than 10%. The specific gravities of hydrocarbons may be as low as 0·7%, and halogenated hydrocarbons are almost without exception significantly more dense than water.

Mackay and Roberts [57] designate organic liquids less dense than water as 'floaters', which spread across the water table, and organic liquids more dense than water as 'sinkers', which may plummet through sand and gravel aquifers to the underlying aquitard (relatively impermeable layer) where present.

There is extensive evidence from field studies that low-density organic liquids float on the water table. It is also important to recognize that migration of dense organic liquids is largely uncoupled from the hydraulic gradient that drives advective transport, and that the movement may have a dominant vertical component even in horizontally flowing aquifers. Such a liquid will sink through the saturated zone as an immiscible phase, displacing the groundwater as it descends (Fig. 4). If the contaminant is slightly soluble in water, a plume develops by dissolution of the contaminant liquid retained in the aquifer pores as well as by dissolution of the pool of contaminant liquid residing on the bottom of the aquifer.

In order to explain the contamination of wells and springs, it has been often found necessary to explain the principles of groundwater flow and the relationship between groundwater and surface waters. In this way, the greater sensitivity of recharge areas can be explained and the need for early detection of leaks can be stressed. However, for reasons outlined above, this correlation has not always proved to be successful.

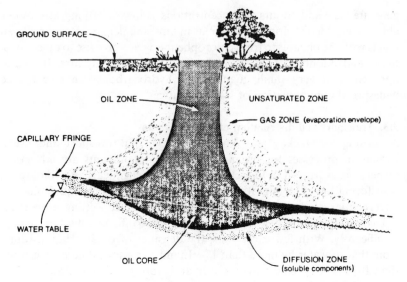

GROUND SURFACE

OIL ZONE

CAPILLARY FRINGE

WATER TABLE

UNSATURATED ZONE

GAS ZONE (evaporation envelope)

OIL CORE

DIFFUSION ZONE
(soluble components)

FIG. 4.   Cross-section schematic of petroleum spill (Abriola and Pinder [1]).

4.   ENVIRONMENTAL FACTORS AND CHEMISTRY

4.1. General

The chemical, physical and biological properties of a substance, in conjunction with the environmental characteristics of an area, result in physical, chemical and biological processes associated with the transport and transformation of the substance in soil and groundwater. These processes are shortly described in the following sections, along with some representative mathematical methods or models employed in the literature. Information is mainly obtained from Bonazountas and Fiksel [18].

The rates of each of the environmentally important chemical processes are influenced by numerous parameters, but most processes are described mathematically by only one or two variables. For example, the rate of biodegradation varies for each chemical with time, microbial population characteristics, temperature, pH and other reactants. In modelling efforts, however, this rate can be approximated by a first-order rate constant (in units of time).

Soil models tend to be based on first-order kinetics; thus, they employ only first-order rate constants with no ability to correct these constants for environmental conditions in the simulated environment which differ from the experimental conditions. This limitation is both for reasons of expediency and due to a lack of the data required for alternative approaches. In evaluating and choosing appropriate soil and groundwater models, the type, flexibility and suitability of methods used to specify necessary parameters should be considered.

The physical, chemical and biological behaviour of a chemical determines how the chemical partitions among the various environmental media and has a large effect on the environmental fate of a substance. For example, the release into the soil of two different acids (with similar chemical behaviour) may result in one chemical mainly volatilizing into the air and the other chemical becoming mainly sorbed to the organic fraction of the soil. The physical behaviour of a substance therefore can have a large effect on the environmental fate of that substance.

The processes and corresponding physical parameters that are important in determining the behaviour and fate of a chemical are different in the analysis of dissolved trace-level contaminants to analyses of contaminants from large-scale releases (e.g. spills).

Models are available for organic dissolved compounds, immiscible fluids and inorganic compounds. Physical, chemical and biological processes important to the above modelling categories are given below.

4.2. Miscible/organic compounds
Physical processes affecting dissolved organic compound fate in soil systems are mainly

(1) advection, dispersion, volatilization;
(2) sorption, ion–cation exchange.

Chemical processes affecting dissolved contaminant migration are mainly

(3) ionization;
(4) hydrolysis;
(5) oxidation/reduction;
(6) complexation.

Biological processes affecting dissolved contaminant migration are mainly

(7) bioaccumulation;
(8) biodegradation.

Information on the above processes and algorithms related to mathematical modelling are presented overleaf.

4.2.1. Advection, dispersion, volatilization

In sand and gravel aquifers, the dominant factor in the migration of a dissolved contaminant is advection, the process by which solutes are transported by the bulk motion of flowing groundwater. Groundwater velocities range between 1 and 1000 m/yr.

In general, the processes of advection, dispersion, volatilization and sorption are most important to both trace-level analyses and large-scale release analyses. Bulk properties (e.g. viscosity, solubility) are usually only important in simulations involving large amounts of contaminants.

Dissolved contaminants spread as they move with the groundwater. Dispersion results from two basic processes, molecular diffusion in solution and mechanical mixing. The process results in an overall net flux of solutes from a zone of high concentration to a zone of lower concentration.

Diffusion in solution is the process whereby ionic or molecular constituents move under the influence of their kinetic activity in the direction of their concentration gradient. The process of diffusion is often known as self-diffusion, molecular diffusion or ionic diffusion. The mass of diffusing substance passing through a given cross-section per unit of time is proportional to the concentration gradient (Fick's first law).

Volatilization refers to the process of pollutant transfer from soil to air and is a form of diffusion, the movement of molecules or ions from a region of high concentration to a region of low concentration. Volatilization is an extremely important pathway for many organic chemicals, and rates for volatilization from soil vary over a large range. This process is less important for inorganic than for organic chemicals; most ionic substances are usually considered to be non-volatile.

Many models are available in the literature, and some of these models can be applied only to specific environmental situations and only for chemicals for which they were developed. Obviously all models do not provide the same numerical results when employed to provide answers to a particular problem, so care must be taken in choosing an appropriate unsaturated zone model, or when specifying a volatilization rate. For modelling algorithms, and numerical examples, the reader is referred to the work of Lyman *et al.* [55], Bonazountas and Wagner [15] and others listed in the references section.

In many cases dispersion may be of minor importance. If dispersion is important, and in the absence of detailed studies to determine the dispersion characteristics of a given field situation, longitudinal and transverse dispersivities must be estimated based on prior field work in similar hydrogeological systems (Mackay and Roberts [57]).

4.2.2. *Sorption/ion–cation exchange*

Adsorption is the adhesion of pollutant ions or molecules to the surface or soil solids, causing an increase in the pollutant concentrations on the soil surface over the concentration present in the soil moisture. Adsorption occurs as a result of a variety of processes with a variety of mechanisms and some processes may cause an increase of pollutant concentration within the soil solids—not merely on the soil surface.

Adsorption and desorption can drastically retard the migration of pollutants in soils; therefore, knowledge of this process is of importance when dealing with contaminant transport in soil and groundwater. The type of pollutant will determine to what kinds of material the pollutant will sorb. For organic compounds it appears that partitioning between water and the organic carbon content of soil is the most important sorption mechanism (Mackay 1982).

Sorption and desorption are usually modelled as one fully reversible process, although hysteresis is sometimes observed. Four type of equations are commonly used to describe sorption–desorption processes: Langmuir, Freundlich, overall and ion or cation exchange. The Langmuir isotherm model was developed for single-layer adsorption and is based on the assumption that maximum adsorption corresponds to a saturated monolayer of solute molecules on the adsorbent surface, that the energy of adsorption is constant, and that there is no transmigration of adsorbate on the surface phase.

The Langmuir model is described by

$$ds/dt = K_{sw} \times (s_s - s) \qquad s_s = Q^0 \times bc/(1 + c)$$

The Freundlich sorptive isotherm is an empirical model expressed by

$$s = x/m = K \times c^{1/n}$$

where ds/dt = temporal variation of adsorbed concentration of compound on soil particles; s = adsorbed concentration of compound on soil particles; K_{sw} = Langmuir equilibrium soil-water adsorption kinetic coefficient; s_s = maximum soil adsorption capacity; Q^0 = number of moles (or mass) of solute adsorbed per unit weight of adsorbent (soil) during maximum saturation of soil; b = adsorption partition coefficient; t = time; c = concentration of pollutant in soil moisture; x = adsorbed pollutant mass on soil; m = mass of soil; K = adsorption (partitioning) coefficient; c = dissolved concentration of pollutant in soil moisture; and n = Freundlich equation parameter. At trace levels, many substances (particularly organics) are simply proportional to concentration, so the Freundlich isotherm is frequently used with $1/n = 1$. For organics, K_{oo} (the adsorption coefficient on organic carbon) is often used instead of K. These coefficients are related by $K = K_{oo}$ (% organic carbon in the solid)/100.

Ion exchange (an important sorption mechanism for inorganics) is viewed as an exchange with some other ion that initially occupies the adsorption site on the solid. For example, for metals (M^{2+}) in clay the exchanged ion is often calcium:

$$M^{2+} + [clay] \times Ca \leftrightarrow Ca^{2+} + [clay] \times M$$

Cation exchange can be quite sensitive to other ions present in the environment. The calculation of pollutant mass immobilized by cation exchange is given by

$$S = EC\,MWT/VAL$$

where S = maximum mass associated with solid (mass pollutant/mass of soil); EC = cation exchange capacity (mass equivalent/mass of dry soil); MWT = molecular (or atomic) weight of pollutant (mass/mole); and VAL = valence of ion ($-$). For additional details see Bonazountas and Wagner [15].

4.2.3. Ionization

Ionization is the process of separation or dissociation of a molecule into particles of opposite electrical charge (ions). The presence and extent of ionization has a large effect on the chemical behaviour of a substance. An acid or base that is extensively ionized may have markedly different solubility, sorption, toxicity and biological characteristics than the corresponding neutral compound. Inorganic and organic acids, bases and salts may be ionized under environmental conditions. A weak acid HA will ionize to some extent in water according to the reaction

$$HA + H_2O \leftrightarrow H_3O^+ + A$$

The acid dissociation constant, K_a, is defined as the equilibrium constant for this reaction:

$$K_a = [H_3O^+][A^-]/[HA][H_2O]$$

Note that a compound is 50% dissociated when the pH of the water equals the pK_a ($pK_a = -\log K_a$).

4.2.4. Hydrolysis

Hydrolysis is one of a family of reactions which leads to the transformation of pollutants. Under environmental conditions, hydrolysis occurs mainly with organic compounds. Hydrolysis is a chemical transformation process in which an organic RX reacts with water, forming a new molecule. This process normally involves the formation of a new carbon–oxygen bond and the breaking of the carbon–X bond in the original molecule:

$$RX \xrightarrow{\;H_2O\;} R\text{---}OH + X^- + H^+$$

Hydrolysis reactions are usually modelled as first-order processes, using rate constants (K_H) in units of $(\text{time})^{-1}$:

$$-d[RX]/dt = K_H[RX]$$

The rate of hydrolysis of various organic chemicals under environmental conditions can range over 14 orders of magnitude, with associated half-lives (time for one-half of the material to disappear) as low as a few seconds to as high as 10^6 years and is pH-dependent. It should be emphasized that if laboratory rate constant data are used in soil models and not corrected for environmental conditions—as is often the only choice—then model results should be evaluated with scepticism.

4.2.5. Oxidation/reduction

For some organic compounds, such as phenols, aromatic amines, electron-rich olefines and dienes, alkyl sulphides and eneamines, chemical oxidation is an important degradation process under environmental conditions. Most of these reactions depend on reactions with free radicals already in solution and are usually modelled by pseudo-first-order kinetics:

$$-d[X]/dt = K_o'[RO_2\cdot][X] = K_{ox}[X]$$

where X is the pollutant; K_o' is the second-order oxidation rate constant; $RO_2\cdot$ is a free radical; and K_{ox} is the pseudo-first-order oxidation rate constant.

4.2.6. Complexation

Complexation, or chelation, is the process by which metal ions and organic or other non-metallic molecules (called ligands) can combine to form stable metal–ligand complexes. The complex that is found will generally prevent the metal from undergoing other reactions or interactions that the free metal cation would. Complexation may be important in some situations; however, the current level of understanding of the process is not very advanced, and the available information has not been shown to be particularly useful to quantitative modelling.

4.2.7. Bioaccumulation

Bioaccumulation is the process by which terrestrial organisms such as plants and soil invertebrates accumulate and concentrate pollutants from the soil. Bioaccumulation has not been examined in soil modelling, apart from some nutrient cycle (phosphorus, nitrogen) and carbon cycle bioaccumulation attempts.

4.2.8. Biodegradation

Biodegradation refers to the process of transformation of a chemical by biological agents, usually by micro-organisms, and it actually refers to the net result of a number of different processes, such as mineralization, detoxication, co-metabolism, activation and change in spectrum. In toxic chemical modelling, biodegradation is usually treated as a first-order degradation process:

$$\mathrm{d}c/\mathrm{d}t = -K_{DE} \times c^{n}$$

where c = dissolved concentration of pollutant soil moisture ($\mu g/ml$); K_{DE} = rate of degradation (day^{-1}); and n = order of reaction ($n = 1$, first order).

4.3. Immiscible fluids

Density, viscosity, surface wetting and solubility are the important parameters governing migration of an immiscible organic liquid. According to Mackay and Roberts [57], there is extensive evidence that low-density organic liquids float on the water table. The sinking phenomenon of high-density liquids has been demonstrated in physical model experiments.

The transport of an organic liquid phase also is influenced by its viscosity and its surface-wetting properties compared with those of water. According to Mackay and Roberts [57] and the literature reviewed halogenated aliphatics tend to spread by capillary action into aquifer media and to be retained in amounts of about 0·3–5% by volume, following the passage of the organic liquid.

An organic liquid of moderately low solubility, such as PEC, will contaminate as much as 10 000 times its own volume to its solubility limit. However, organic compounds are only rarely found in groundwater at concentrations approaching their solubility limits, even when organic liquid phases are known or suspected to be present.

For the above reasons, it also is difficult to completely remove hydrocarbons from granular aquifers as the capillary attraction between smaller particles may be sufficient to prevent bubbles of, for example, gasoline passing between the grains. Figure 5 shows that a gasoline bubble will continue to move so long as $P_{gas} - P_{water}$ is greater than $P_{\sigma ap}$; otherwise, the bubble will be trapped.

Pumping or flushing will not remove all of the contaminant as water will be deflected elsewhere along less resistant paths. The trapped product will then continue to give off water-soluble components. Emphasis has, therefore, been placed upon containment of spills and the early detection of leaks.

4.4. Miscible inorganic compounds
Important processes and mechanisms of environmental interactions of inorganic materials are volatility, solubility, fast aqueous reactions, slow aqueous reactions, speciation, soil interaction and bioaccumulation. Processes not already discussed are presented below.

4.4.1. Fast reactions in aqueous media
Reactions of inorganic dissolved species in aqueous media can be classified under the following general categories:

(a) reaction/dissociation (with solvent molecules);
(b) substitution reactions (with solvent or dissolved species); and
(c) redox reactions (with dissolved gases or other ions).

These reactions can lead to formation of new complex ions (those with different ligands in the coordination sphere and those with different oxidation state of the metal centres) and/or other ions.

These fast reactions are important, since the medium the particular inorganic substance is dissolved in will determine to a great extent the speciation of the particular metal ion. Since the extent of environmental interaction (e.g. soil attenuation, uptake by plants, toxicity) will depend on the chemical form (species) of the metal, knowledge of the behaviour with regard to rapid reactions is essential.

In addition, upon mixing of solutions of these metals with the environment (e.g. leachate mixing with groundwater) chemical modification of the species will occur initially via thermodynamically favoured rapid reactions and subsequently by the possible slower reactions. Such knowledge is important in the selection and application of aquatic equilibria (or speciation) models described in a later section.

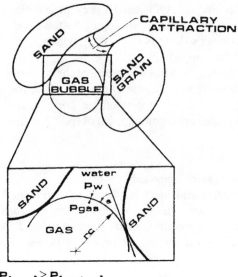

$P_{(gas)} \geq P_{(water)}$

But if $\left[P_{(gas)} - P_{(water)}\right] < P_{(cap)}$

then the gas bubble is trapped

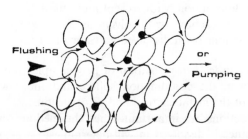

FIG. 5. Entrapment of gasoline by capillary forces (Hall and Quam [36]). (●) Trapped gas bubbles. Flushing may not remove all of the trapped products because of capillary attraction.

4.4.2. Slower reactions in aqueous media

In the category of slower reactions in aqueous media one may consider the following reactions:

(a) ligand substitution reactions of relatively kinetically inert ions;
(b) electron transfer reactions involving inner sphere mechanisms for relatively kinetically inert ions and some outer sphere reactions;

(c)   reactions with dissolved gaseous species/bacterial catalysed reduction (e.g. oxidation by dissolved oxygen);

(d)   precipitation of solids by formation of insoluble species via substitution; and/or

(e)   oxidation reactions (e.g. formation of metal hydroxides and metal sulphides).

The above slower reactions impact the speciation and concentration of the inorganic species in the aqueous phase and determine the type and extent of further interactions in the same manner as the fast reactions, although a different time frame will be required to achieve completion of some of these reactions. For the relatively kinetically inert metals (e.g. Cu^{3+}, Fe^{3+}, Cr^{3+} and other such inorganic ions), immediate modification of the speciation of the ion may not occur upon mixing with other aqueous phases (e.g. leachate with groundwater under oxidizing conditions), and the reaction may not proceed prior to the mixture reacting with soil strata.

As such, the time frame for some reactions may indicate that characterizing the speciation in the original aqueous phase (e.g. leachate) will dictate the chemical behaviour of the compound with the environment, whereas in other cases understanding the speciation under the exact conditions of environmental interaction will be required.

4.4.3. Speciation

An evaluation of the fate of trace metals in surface and sub-surface waters requires the consideration of speciation, adsorption and precipitation. These processes can affect metal solubility, toxicity, availability, physical transport and corrosion potential. As a result of a need to describe the complex interactions involved in these situations, various models have been developed to address a number of specific situations. Steps in speciation calculations are not described in this publication. Speciation computerized packages (models) are described in parts of a later section.

4.4.4. Soil and sediment interactions

Important interactions of dissolved aqueous inorganic species occurs with soils. The more important mechanisms of interaction include formation of precipitates, adsorption of components onto soil surfaces, modification of speciation by soil constituents (solid and liquid phases) and reactions induced by bacteria present in soil.

The properties of the surface of the soil particles which will dictate the adsorption behaviour can be changed by the acid/base buffer system in contact with the soil. Most of the minerals display amphoteric properties toward solutions, i.e. they behave as weak acid or bases. The buffer capacity of the soil (e.g. related to $CaCO_3$ content) is important with regard to the ability of soil water to fluctuate in pH. Micro-organisms can bring important changes in the solubility of soil minerals and other precipitates by altering solution conditions and forming films on surfaces of particles and catalysing reactions.

Conceptual models for attenuation of species onto soils have been developed which incorporate factors such as electrostatic energy and chemical bonding effects (e.g. van der Waals, dipole effects and covalent bonding). Langmuir adsorption isotherms have been generated which are consistent with this conceptual model. As previously mentioned, solution chemical equilibria will modify the adsorptive behaviour of the soil as well as the species present in solution and its tendency to be adsorbed; therefore, simplified adsorption, attenuation or speciation models have to be used with caution in inorganic fate modelling.

4.4.5. Bioaccumulation

A great deal of information exists concerning the effects of specific inorganic constituents on a large variety of organisms in specific environments. However, such information has generally been difficult to use in the development of broadly generalized techniques for estimation of effects. Estimation techniques for understanding the impact potential of inorganic compounds on organisms in aquatic or terrestrial environments have not at present been developed.

4.4.6. Processes related to terrestrial plants

If there are exceptions to the general lack of estimation techniques and models relating to inorganic chemical concentrations, biological effects and fate, they are likely to be found in agricultural theory and modelling. Application of soil amendments such as fertilizers for macronutrients (e.g. phosphates, nitrates) or micronutrients (e.g. Mn), or for improving the soil chemical environment to release nutrients (liming), is a relatively well-developed science. At the very least, there exists an understanding of the optimum ranges of concentrations of agriculturally related chemicals and growth of crop species.

5. MATHEMATICAL MODELLING—OVERVIEW

Mathematical models can greatly assist decision-makers in determining the importance of pollutant pathways in the environment, as long as they are used properly and with an understanding of their limitation. The use of models has grown dramatically over the past decade, but models are not meant to substitute for good judgement and experience.

Pollutant fate mathematical modelling in soil systems is an area of current intensive work, because of the numerous problems originating at hazardous waste sites. The variety of models has dramatically increased during the last decade, but although the variety of models appears to be large only very few 'different' modelling concepts exist and very few physical or chemical processes are modelled.

In general, soil/groundwater modelling concepts deal mainly with point source pollution and can be categorized into (1) unsaturated soil zone (or soil), (2) saturated soil zone (groundwater), (3) geochemical and (4) ranking. The first two categories follow comparable patterns of mathematics and approach, the third enters into chemistry and speciation modelling, and the fourth follows a screening approach. The first two modelling categories are designed to handle either organic or inorganic dissolved pollutants, or immiscible fluids for non-aqueous phase (NAPL) chemicals.

Soil modelling is a very complex issue and a major characteristic of a soil sub-compartment—as compared to a water or an air sub-compartment—is that the temporal physical and the chemical behaviour of this sub-compartment is governed by 'out-compartmental' forces such as precipitation, air temperature and solar radiation. This characteristic is also one of the main reasons why soil mathematical modelling can be much more complex than water or air modelling where the dynamics of the compartment are governed by 'in-compartmental' forces. Groundwater modelling can handle a limited number of chemical processes; therefore, a number of aquatic equilibrium models under development are aimed to fill this chemistry gap.

Finally, soil compartment models (watershed, soil, groundwater) are used to evaluate pollution originating from various sources, such as hazardous waste sites. Because detailed soil modelling is not always feasible or desirable, a category of 'ranking' models is known in the literature for 'screening' severity of environmental impacts originating from waste activities.

There is no scientific reason for a soil model to be an unsaturated soil model only, and not to be an unsaturated (soil) and a saturated soil (groundwater) model. Only mathematical complexity mandates the differentiation, because such a model would have to be three-dimensional (e.g. Narasimhan [70]) and very difficult to operate. Most of the soil models account for vertical flows; groundwater models for horizontal flows.

In general, models can be classified into deterministic, which describe the system as a cause/effect relationship, and stochastic, which incorporate the concept of risk, probability or other measures of uncertainty. Deterministic and stochastic models may be developed from observation, semi-empirical approaches and theoretical approaches.

In developing a model, scientists attempt to reach an optimal compromise among the above approaches, given the level of detail justified by both the data availability and the study objectives. Deterministic model formulations can be further classified into simulation models which employ a well-accepted empirical equation that is forced, via calibration coefficients, to describe a system, and analytic models in which the derived equation describes the physics/chemistry of a system.

Without a solution, formulated mathematical systems (models) are of little value. Four solution procedures are mainly followed: the analytical; the numerical (e.g. finite different, finite element); the statistical; and the iterative. Numerical techniques have been standard practice in soil quality modelling. Analytical techniques are usually employed for simplified and idealized situations. Statistical techniques have academic respect, and iterative solutions are developed for specialized cases. Both the simulation and the analytic models can employ numerical solution procedures for their equations. Although the above terminology is not standard in the literature, it has been used here as a means of outlining some of the concepts of modelling.

Generally speaking, a deterministic or stochastic soil quality model consists of two major parts or modules:

(1)   the flow module or moisture module, or hydrologic cycle module— aiming to predict flow or moisture behaviour (i.e. velocity, content) in the soil; and

(2)   the solute module—aiming to predict pollutant transport, transformation and soil quality in the soil zone.

The above two modules form the soil quality model. The flow module drives the solute module. It is important to note that the moisture module can be absent from the model, and in this case a model user has to input to the solute module information that would have been either produced by a moisture module or would have been obtained from observed data at a site.

At this stage of intensive research in soil and groundwater quality modelling it may be reasonable to group the prevailing modelling concepts of the literature into three major categories: the 'Traditional Differential Equation' (TDE) modelling, the 'compartmental' modelling, and 'other' types. This terminology is again not a standard practice, but is employed here for reasons of communication. TDE modelling applies to both the flow or moisture module and the solute module, and a modelling package may consist of one TDE module (e.g. moisture) and a compartmental solute module, or vice versa.

The following sections aim to clarify some key issues related to soil and groundwater models. The following documents provide an overview of this area of science: the publication of Abriola and Pinder [1], a state-of-the-art publication on multi-phase approach in porous media; the series of articles by Mercer and Faust (1981) describing groundwater modelling concepts which are equally applicable to unsaturated soil zone modelling; the monograph of Bachmat *et al.* [8] listing various models; the work of Bonazountas and Wagner [16] introducing the compartmental soil quality modelling concept and geochemical modelling; and the reference book of Freeze and Cherry [33].

Reference to the above sources is not meant to exclude other excellent publications presented in reputable scientific journals; rather it indicates selected basic sources employed to draft the following sections.

In the following sections, information is presented on:

(1)   dissolved (miscible) pollutant modelling and models;
(2)   immiscible (NAPL) contaminant modelling and models; and
(3)   aquatic equilibrium modelling.

6.   DISSOLVED POLLUTANT MODELLING

In this section more emphasis is placed on the unsaturated soil zone than on groundwater modelling. This emphasis can be justified by the fact that similar modelling concepts govern both environments.

6.1.   Unsaturated soil zone (soil) modelling

Soil modelling follows three different mathematical formulation patterns: (1) Traditional Differential Equation (TDE) modelling; (2) compartmental modelling; and (3) stochastic modelling. Some researchers may categorize models differently as, for example, into numerical or analytic, but this categorization applies more to the techniques employed to solve the formulated model, rather than to the formulation *per se*. A model has a flow (moisture) module and a quality module.

6.1.1. TDE modelling

The TDE moisture module (of the model) is formulated from three equations: (1) the water mass balance equation, (2) the water momentum and (3) the Darcy equation, and also other equations such as the surface tension of potential energy equation. The resulting differential equation system describes moisture movement in the soil and is written in a one-dimensional, vertical, unsteady, isotrophic formulation as

$$\partial[K(\psi)(\partial\psi/\partial z + 1)]/\partial z = C(\psi)\,\partial\psi/\partial t + S \qquad (1)$$

$$v_z = -K(z, \psi)\,\partial\phi/\partial z \qquad (2)$$

where z = elevation (cm); ψ = pressure head, often called soil moisture tension head in the unsaturated zone (cm); $K(\psi)$ = hydraulic conductivity (cm/min); $C(\psi) = \mathrm{d}\theta/\mathrm{d}\psi$ = slope of the moisture (ϕ) versus pressure head (ψ) (cm^{-1}); t = time (min); S = water source or sink term (min^{-1}); $\phi = z + \psi$; and $|v_z|$ = vertical moisture flow velocity (cm/s). The moisture module output provides the parameters v and θ as input to the solute module.

The TDE solute module is formulated with one equation describing the pollutant mass balance of the species in a representative soil volume, $\mathrm{d}V = \mathrm{d}x\,\mathrm{d}y\,\mathrm{d}z$. The solute module is frequently known as the dispersive, convective differential mass transport equation, in porous media, because of the wide employment of this equation, that may also contain an adsorptive, a decay and a source or sink term. The one-dimensional formulation of the module is

$$\partial(\theta c)/\partial t = [\partial(\theta \times K_0\,\partial c/\partial z] - [\partial(vc)/\partial z] - [\rho \times \partial s/\partial t] \pm \sum P \qquad (3)$$

where θ = soil moisture content; c = dissolved pollutant concentration in soil moisture; K_0 = apparent diffusion coefficient of compound in soil-air; v = Darcy velocity of soil moisture; ρ = soil density; s = adsorbed concentration of compound on soil particles; $\sum P$ = sum of sources or sinks of the pollutant within the soil volume; and z = depth.

Models like the above have been presented by various researchers of the US Geological Survey (USGS) and the academia. The above equation has been solved principally (a) numerically over a temporal and spatial discretized domain, via finite difference or finite element mathematical techniques (e.g. Mackay [56]); (b) analytically, by seeking exact solutions for simplified environmental conditions (e.g. Enfield *et al.* [27]); or (c) probabilistically (e.g. Schwartz and Growe [82]).

At this point it is important to note that the flow model (a hydrologic cycle model) can be absent from the overall model. In this case the user has to input to the solute module (i.e. eqn (1)) the temporal (t) and spatial (x, y, z) resolution of both the flow (i.e. soil moisture) velocity (v) and the soil moisture content (θ) of the soil matrix. This approach is employed by Enfield *et al.* [27] and other researchers. If the flow (moisture) module is not absent from the model formulation (e.g. Huff [37]), then users are concerned with input parameters that may be frequently difficult to obtain. The approach to be undertaken depends on site specificity and available monitoring data.

Some principal modelling-specific deficiencies when modelling solute transport via the TDE approach are:

(1)   Only diffusion, convection, adsorption and possibly decay can be modelled, whereas processes such as fixation or cation exchange have to be either neglected or represented with the sources and sinks term of the equation because of mathematical complexity.

(2)   The equation system is applicable mainly to pollutant transport of organics, whereas transport of metals which can be strongly affected by other processes cannot be directly modelled.

(3)   The equation system can predict volatilization only implicitly via boundary diffusion constraints; however, experimental studies have frequently demonstrated an over-estimation or under-estimation of the theoretical volatilization rate unless a 'sink' or source term is included in the equation.

(4)   No experimental or well-accepted equation for a process (e.g. volatilization) can be incorporated since the model has its own predictive mechanism.

(5)   Pollutant concentrations are estimated only in the soil-moisture and on soil-particles, whereas pollutant concentrations in the soil-air are omitted.

(6)   The discretized version of the equation in the case of numerical solutions has a pre-set temporal and spatial discretization grid that results in high operational costs (professional time, computer time) of the model, since input data have to be entered into the model for each node of the grid.

In a modelling evaluation effort, Murarka [68] reports that the currently available coupled or uncoupled models of hydrologic flow and the geochemical interactions are adversely affected by the following factors: difficulties in establishing consistency between the theoretical frameworks, laboratory experiments and field research; limited basic knowledge about non-equilibrium conditions and phase relations; inadequate existence of geochemical sub-models to couple with the hydrologic transport sub-models; uncertainties in input data, particularly for dispersion and chemical reaction rate coefficients; and numerical difficulties with model solution techniques.

6.1.2. Compartmental modelling

Compartmental soil modelling is a new concept and can apply to both modules. The solute fate module, for example, consists of the application of the law of pollutant mass conservation to a representative user specified soil element (Fig. 6). The mass conservation principle is applied over a specific

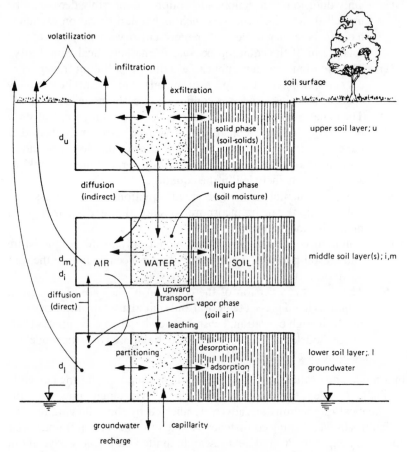

FIG. 6.   Schematic of phases in soil matrix (Bonazountas [13]).

time step, either to the entire soil matrix or to the sub-elements of the matrix such as the soil-solids, the soil-moisture and the soil-air. These phases can be assumed to be in equilibrium at all times; thus, once the concentration in one phase is known the concentration in the other phases can be calculated. Single or multiple soil compartments can be considered, whereas phases and sub-compartments can be interrelated with transport, transformation and interactive equations.

Compartmental models may bypass the deficiencies of the TDE modelling because they may handle geochemical issues in a more sophisticated way if required, but this does not imply that compartmental models are 'better' than TDE models. They are simply different. Compartmental models reflect the personal 'touch' of their developers and cannot be formulated under generalized guidelines or concepts.

The moisture module (i.e. driving element) of a compartmental solute model can be either incorporated into the overall model, or can be an independent module, as for example a TDE module of the literature. At this stage of scientific research a developed soil compartment model appears to be SESOIL: Seasonal Soil Compartment model (Bonazountas and Wagner [16]). SESOIL consists of a dynamic compartment moisture module and a dynamic compartmental solute transport module. The following sections present a demonstration of the basic mathematical equations governing compartmental soil quality modelling. This information has been abstracted from the SESOIL model.

The law of pollutant mass concentration for a representative element can be written over a small time step as

$$\Delta M = M_{in} - M_{out} - M_{trans} \tag{4}$$

The solute (dissolved) concentration of a compound can be related to its soil-air concentration via Henry's law:

$$c_{sa} = c \times H/R \times (T + 273) \tag{5}$$

where c_{sa} = pollutant concentration in soil-air; c = dissolved pollutant concentration; H = Henry's law constant; R = gas constant; and T = temperature in °C.

The pollutant concentration of the soil (i.e. solids) can be determined from the sum of the concentrations of the pollutant adsorbed, cation exchanged and/or otherwise associated with the soil particles, e.g. via adsorption isotherms. One commonly used adsorption isotherm equation is the Freundlich equation:

$$s = K \times c^{1/n} \tag{6}$$

where s = adsorbed concentration of compound; K = partitioning coefficient; c = dissolved concentration of compound; and n = Freundlich constant.

The total concentration of a chemical in a soil matrix can be calculated from the concentration of pollutant in each phase and the related volume of each phase by

$$c_o = (n - \theta)c_{sa} + (\theta) \times c + (\rho_b) \times s \qquad (7)$$

where c_o = overall (total) concentration of pollutant in soil matrix; n = soil (total) porosity; θ = soil moisture content; $(n - \theta)$ = soil-air content or soil-air filled porosity; c_{sa} = pollutant concentration in soil-air; c = pollutant concentration (dissolved) in soil-moisture; ρ_b = soil bulk density; and s = pollutant concentration on soil particles.

The above expressions are input terms to eqn (4), which is then applied for each time step, each sub-compartment and each compartment of the user specified matrix (Fig. 6). The term M_{in} may reflect input pollution from rain (upper layer), from soil-moisture from an upper layer and from a lower layer. The term M_{out} reflects pollution exports from the individual compartment, whereas the term M_{trans} reflects all transformation and chemical reactions taking place in the compartment. All terms can be normalized to the soil moisture concentration via interconnecting equations such as (5) and (6), which can describe processes such as volatilization, cation exchange, photolysis, degradation, hydrolysis, fixation, biological activity, etc. The solution of the resulting system of equations is a complicated issue and may require—for computational efficiency and other reasons—development of new numerical solution techniques or algorithms (e.g. SESOIL model).

6.1.3. Stochastic, probabilistic and other modelling concepts

Stochastic or probabilistic techniques can be applied to either the moisture module or the solution of eqn (3)—as for example the model of Schwartz and Growe [82]—or can lead to new conceptual model developments, as for example the work of Jurry [44]. Stochastic or probabilistic modelling is mainly aimed at describing 'breakthrough' times of overall concentration threshold levels, rather than individual processes or concentrations in individual soil compartments. Coefficients or response functions for these models have to be calibrated to field data, since major processes are studied via a black-box or response function approach and not individually. Other modelling concepts may be related to soil models for solid waste sites and specialized pollutant leachate issues (Schultz [81]).

6.1.4. Physical, chemical and biological processes modelled
Modellers should be fully aware of the range of applicability and processes considered by a computerized package. There exists some disagreement among soil modellers as to whether there is a need for increased model sophistication, since almost all soil modelling predictions have to be validated with monitoring data, given the physical, chemical and biological processes that affect pollutant fate in soil systems. Because of the latter consideration, many simplified models may provide excellent results, assuming accurate site-specific calibration is achieved. Nevertheless, model sophistication is reflected in the processes modelled, but model selection is mandated by the project needs and data availability.

The important physical processes of a typical soil model are:

(1) The hydrologic cycle or moisture cycle—that may encompass the process of rain infiltration in the soil, exfiltration from the soil to the air, surface run-off, evaporation, moisture behaviour, groundwater recharge and capillary rise from the groundwater. All these processes are interconnected and are frequently referred to as the hydrologic cycle components.

(2) The pollutant or solute cycle—that may encompass the processes of advection, diffusion, volatilization, adsorption and desorption, chemical degradation or decay, hydrolysis, photolysis, oxidation, cation or anion exchange, complexation, chemical equilibria, nutrient cycles and others.

(3) The biological cycle—that may encompass processes of biological transformation, plant uptake, bioaccumulation, soil organism transformation and others.

Models in the literature can handle one or more of the above processes and for various pollutants. In general, however, soil models tend to handle:

(1) From the hydrologic cycle: temporal resolution of soil moisture, surface run-off and groundwater recharge components, by inputing to the model the 'net' infiltration rate into the soil column.

(2) From the pollutant and biological cycles: the processes of advection, diffusion, volatilization (diffusion at the soil–air interface), adsorption or desorption (equilibrium) and degradation or decay, which are also the most important chemical processes in the soil zone. All other processes can be 'lumped' together under the source or sink term of eqn (3).

Fortunately—and not unfortunately—no one model exists as yet which simulates all of the physical, chemical and biological processes associated with pollutant fate in soils. We say fortunately, because such a package would be very data-intensive and difficult to use. Intensive research is required to accomplish the above objective and the value of the overall product may be questioned by users. A later section presents selected models.

6.2. Saturated soil zone (groundwater) modelling

Saturated soil zone (or groundwater) modelling is formulated almost exclusively via a TDE system consisting of two modules, the flow and the solute module. The two modules are written as (Bachmat *et al.* [8])

$$\bar{V}(\rho k/\mu)(\bar{V}p - \rho g \bar{V}Z) - q = \partial(\phi\rho)/\partial t \tag{8}$$

$$\bar{V}[\rho C(k/\mu)(\bar{V}p - \rho g \bar{V}Z)] + \bar{V}(\rho E)\bar{V}C = \partial(\rho\phi C)/\partial t \tag{9}$$

where C = concentration, mass fraction; E = dispersion coefficient; g = acceleration due to gravity; k = permeability; p = pressure; q = mass rate of production or injection of liquid per unit volume; t = time; Z = elevation above a reference plane; ϕ = porosity; ρ = density; and μ = viscosity.

Mathematical groundwater modelling has been the least problematic in its scientific formulation, but has been the most problematic model category when dealing with applications, since these models have to be calibrated and validated as described later. Actually we have only TDE and some other (e.g. stochastic) formulations. The proliferation of literature models is mainly due to different model dimensionalities (zero, one, two, three); model features (e.g. with adsorption, without absorption terms); solution procedures employed (e.g. analytic, finite difference, finite element, random walk, stochastic) for equation systems (7) and (8); sources and sinks described; and the variability of the boundary conditions imposed. Some of the principal modelling deficiencies discussed in the previous section (soil models) apply to groundwater models also. In general, (1) there exists no 'best' groundwater model and (2) for site-specific applications groundwater models have to be calibrated.

The two principal solution methods for eqns (8) and (9) that result in different model categories with substantially different impacts on the level of effort required to run a package are (1) analytical models and (2) numerical models. Employment of the first method results in formulation of expressions applicable to the nodes or the elements of a domain, the number of nodes or elements being user specified. In analytical modelling only averaged data for the entire domain have to be input to the model. The numerical modelling data have to be input for all nodes or elements of the model, a fact that frequently results in high model cost runs, in terms of both professional and computational time. Common numerical solution techniques are the finite difference, the finite element, the method of characteristics, the random walk and their variations. Interested readers are referred to the references of this chapter.

6.3. Selected models

Table 3 lists selected soil and groundwater models and their main features. Table 4 lists limitations and advantages of major model categories. Models listed in Table 3 are documented, operational and very representative of the various structures, features and capabilities. For example:

(1) PESTAN (Enfield *et al.* [27]) is a dynamic TDE soil solute (only) model, requiring the steady-state moisture behaviour components as user input. The model is based on the analytic solution of eqn (3) and is very easy to use, but has also a limited applicability, unless model coefficients (e.g. adsorption rate) can be well estimated from monitoring studies. Moisture module requirements can be obtained by any model of the literature.

(2) SCRAM (Adams and Kurisu [2]) is a TDE dynamic, numerical finite difference soil model, with a TDE flow module and a TDE solute module. It can handle moisture behaviour, surface run-off, organic pollutant advection, dispersion, adsorption, and is designed to handle (i.e. no computer code has been developed) volatilization and degradation. This model may not have received great attention by users because of the large number of input data required.

(3) SESOIL (Bonazountas and Wagner [16]) is a dynamic soil compartmental model, with a hydrologic cycle and a pollutant cycle compartmental structure, that permits users to tailor the model temporal and spatial resolution to the study objectives. The model estimates the hydrologic cycle components (including moisture behaviour) from available NOAA, USDA and USGS data, and simulates the pollutant cycle by accounting for a number of chemical processes for both inorganic (metal) and organic pollutants.

(4) PATHS (Nelson and Schur [71]) is mainly an analytical groundwater model that provides a rough evaluation of the spatial

Table 3
Selected models and features (Bonazountas [13])

Column groups — Model type: *Unsaturated zone, Groundwater, Aquatic equilibrium, Ranking* · Model formulation: *Flow module, Solute module, TDE approach, Compartmental, Statistical, other* · Mathematics: *Analytical, Numerical, Statistical* · Chemistry issues: *Organics, Inorganics, Metals, Gaseous phase, Increased chemistry* · User concerns: *Input data requirements, Calibration, Level of effort, Application study*

Model acronym	Unsaturated zone	Groundwater	Aquatic equilibrium	Ranking	Flow module	Solute module	TDE approach	Compartmental	Statistical, other	Analytical	Numerical	Statistical	Organics	Inorganics	Metals	Gaseous phase	Increased chemistry	Input data requirements	Calibration	Level of effort	Application study	Contact/information
PESTAN	×					×	×			×			×					L	L	L		Enfield; EPA (405) 332-8800
SCRAM	×	×				×	×				×		×					H	H	H	×	EPA Report PB-259933 (Adams and Kurisu [2])
SESOIL	×				×	×	×	×	×	×	×		×	×	×	×	×	M	M	M	×	Bonazountas; ADL (617) 864-5770
AT123D	×	×				×	×			×			×	×	×			M	L	M	×	Yeh; ORNL (615) 574-7285
PLUME		×				×	×			×			×	×	×			L	L	L	×	Wagner; OSU (405) 624-5280
PATHS		×				×	×		×	×			×	×				L	M	M	×	Nelson; Battelle (509) 376-8332
MMT/VVT	×	×			×	×	×				×		×					H	H	H	×	Cole; Battelle (509) 376-8451
FEMWASTE	×	×				×	×				×		×					H	H	H	×	Yeh; ORNL (615) 574-7285
R.WALK		×			×	×	×				×		×					H	H	H	×	Prickett; ILIOS (217) 344-2277
USGS Models	×	×			×	×	×				×		×			×		H	H	H	×	Appel; USGS (703) 860-6892
GEOCHEM			×					×			×		×	×	×		×	H	L	L	×	Mattigod; UCR (714) 787-1012
MITRE/JRB	×	×		×																L	×	MITRE (703) 827-6000; JRB (703) 821-4873
ADL/LeGrand	×	×		×														M	M	M	×	ADL (617) 864-5770; (919) 787-5855

This is a *partial* list of available well-documented models. Other models are PRZM (US EPA, Athens Research Laboratory, Georgia) and models of research institutions (e.g. USGS, Universities) presented in the journals *Groundwater, Water Resources Research, Environmental Science and Technology, EE/ASCE,* etc.

Table 4

Most important characteristics of major model categories (Bonazountas [13])

Model category	Advantages	Disadvantages	Comments
Soil and groundwater models			
TDE type	Clear formulation/capabilities	Rigid model structure, limited capabilities	This has been the traditional computerized modelling approach
Analytic	Easy model use; limited calibration possibilities; limited input data requirements; desk computer use	Rough averaged predictions of pollutant fate, limited application capabilities	To be used as an overall fate (screening) tool
Numerical	Wide range of applications; detailed spatial, temporal resolutions; increased chemistry capabilities	Extensive calibration requirements; input data intensive (nodes, elements, time); require computer use and related skills	Recommended for site-specific applications
Compartmental type	Can be tailored to user's requirements; increased chemistry capabilities; can better meet spatial and temporal domain requirements	Expected user interaction and problem understanding	Today's scientific tendency
Aquatic equilibrium model	Increased chemistry capability	Data intensive, parameters may not be available	Models at a developmental stage
Ranking models	Easy to use with available data	Simplistic approach, output reflects user's intuition	Employed by the EPA, US Army, Air Force and Navy

and temporal status of a pollutant fate. The model has its own structure and features, and is a deviation from the TDE, or the compartmental, or stochastic approaches.

(5)   AT-123D (Yeh and Ward [101]) is a series of soil or groundwater analytical sub-models, each sub-model addressing pollutant transport; in one, two or three dimensions; for saturated or unsaturated soils; for chemical, radioactive waste heat pollutants; and for different types of releases. The model can provide up to 450 sub-model combinations in order to accommodate various conditions analytically.

(6)   MMT (Foote [32]) is a one- or two-dimensional solute transport numerical groundwater model, to be driven off-line by a flow transport such as VTT (Variable Thickness Transport). MMT employs the random-walk numerical method and was originally developed for radionuclide transport. The model accounts for advection, sorption and decay.

The remaining models of Table 3 follow the scientific basic patterns described above with small variations. All models handle one species at a time, and two soil models (SESOIL, AT-123D) can handle gaseous pollutants also. The US Geological Survey (USGS) has been very active for a number of years in TDE soil and groundwater quality model development. Models of the USGS are well documented and available in the public domain.

6.4. Ranking modelling

Ranking models are aimed at assessing environmental impacts of waste disposal sites. The first ranking models focussed on groundwater contamination; later models had a wider scope (e.g. health considerations). These models rank or rate contaminant migration at different sites, as it is affected by hydrogeologic, soil, waste type, density and site design parameters. These models are based on questions and answers, and on weighting factors the user has to specify. They are very subjective in their use and have received wide dissemination, because they are easy to use and do not require use of computers. Well-known models are LeGrand [51], Silka and Swearingen [83], JRB [43], MITRE (1980) and Arthur D. Little, Inc. [6]. Interested readers should refer to the original publications.

6.5. Modelling issues

Model selection, application and validation are issues of major concern in mathematical soil and groundwater dissolved pollutant modelling. For the model selection, issues of importance are: the features (physics, chemistry) of the model; its temporal (steady state, dynamic) and spatial (e.g. compartmental approach resolution); the model input data requirements; the mathematical techniques employed (finite difference, analytic); monitoring data availability; and cost (professional time, computer time). For the model application, issues of importance are: the availability of realistic input data (e.g. field hydraulic conductivity, adsorption coefficient); and the existence of monitoring data to verify model predictions. Some of these issues are briefly discussed below.

Input data have to be compiled and input to the model from site-specific investigations and analyses (e.g. leaching rates of pollutants, soil permeability); national data bases (e.g. climatological data from the NOAA; and other sources (e.g. diffusion rates of pollutants from handbooks). Compilation of input data for site-specific computer runs are model-specific, geohydrology and chemistry-specific. Some data categories are pollutant source data, climatological data, geographic data, particulate transport data and biological data.

Exact knowledge of the physics of the soil system—although essential— is impossible prior to employing any modelling package. Numerical (e.g. finite difference) TDE soil models, for example, require the net infiltration rainfall rate after each storm even as an input parameter to their moisture module. The rate can be either a user input or can be generated by another model. The same models require the soil conductivity as a function of the soil moisture content as an input parameter. Its value can be obtained either from field investigations or from laboratory data, or from references, but much uncertainty exists in this area of input data gathering.

Numerical soil models (time, space) provide a general tool for quantitative and qualitative analyses of soil quality, but require time-consuming applications that may result in high study costs. In addition, input data have to be given for each one or element of the model, which model has to be run twice, the number of rainfall events. On the other hand, analytic models obtained from analytic solutions of eqn (3) are easier to use, but can simulate only averaged temporal and spatial conditions, which may not always reflect real world situations. Statistical models may provide a compromise between the above two situations.

Model output 'validation' is essential to any soil modelling effort, although this term has a broad meaning in the literature. For the purpose of this section we can define validation as 'the process which analyses the validity of final model output', namely the validity of the predicted pollutant concentrations or mass in the soil column (or in groundwater), to groundwater and to the air, as compared to available knowledge of measured pollutant concentrations from monitoring data (field sampling).

A disagreement of course in absolute levels of concentration (predicted versus measured) does not necessarily indicate that either method of obtaining data (modelling, field sampling) is incorrect or that either data set needs revision. Field sampling approaches and modelling approaches rely on two different perspectives of the same situation.

Important issues in groundwater model validation are the estimation of the aquifer physical properties, the estimation of the pollutant diffusion and decay coefficient. The aquifer properties are obtained via flow model calibration (i.e. parameter estimation) and by employing various mathematical techniques such as kriging. The other parameters are obtained by comparing model output (i.e. predicted concentrations) to field measurements; a quite difficult task, because clear contaminant plume shapes do not always exist in real life.

Three major input data categories are required for soil and groundwater modelling efforts: climatologic or hydrologic data, soil data and chemistry data. These data are used as input to models and to validate model output. Climatologic data can be obtained from site-specific investigations or from NOAA or USGS records. Soil data can be obtained from site-specific investigation (e.g. soil hydraulic conductivity) or from USDA data information documents. Chemistry data can be obtained from reference books (e.g. Lyman *et al.* [55]) or from laboratory analyses (e.g. adsorption coefficient). Data are model-specific and environment-specific.

6.6. Model applications

Numerous applications are available in the literature. Two studies are presented in this section when employing the compartmental approach outlined in Section 6.1.2 of this review.

The two studies presented are related to leachate mobilization and migration at uncontrolled waste sites: (1) leachate migration to groundwater from land treatment practices (Bonazountas *et al.* [17]), and (2) leachate migration to the atmosphere of solvents leaking from barrels buried in the soil zone (Wagner and Bonazountas [98]).

6.6.1. Leachate to groundwater—land treatment practices

Fate of organics. A site in Montana receives petroleum refinery wastes. Effluents from a number of refinery processes are collected in the wastewater treatment ponds. Sludge from the pond bottom is removed periodically and transported to the land treatment site. At the site the sludge is spread to a depth of several centimetres and is allowed to dry for 2–3 weeks. The dried sludge is mixed with the soil to a depth of 15–20 cm by a tractor-drawn rototiller.

The waste composition is reported by the site operators to contain 65% water, 25% oil, 7% solids and 3% other constituents. Considerable variation in the chemical composition of waste over time is considered likely by plant personnel. A chemical analysis of the waste was performed. The waste was found to contain significant amounts of organic compounds, primarily polycyclic nuclear aromatic hydrocarbons (PAH). The sampling program from which these data were taken did not routinely analyse soil samples for individual organic species. However, a gas chromatography mass spectrometer analysis was performed for one soil sample from the first year. The result of the analysis is presented in Table 5.

Waste had been applied during several periods starting in 1973. Some

Table 5
Analysis of soil core samples for PAHs[a]

Species[b]	Concentrations	
	Soil sample[c] (Year 1, depth 0–15 cm) ($\mu g/g$ dry soil)	Waste sample ($\mu g/ml$)
Naphthalene	3·7	76
Acenaphthalene	ND[d]	26
Acenaphthene	2·2	9·8
Fluorene	3·4	24
Phenanthrene	4·9	60
Anthracene	2·4	20
Fluoranthene	3·5	15
Pyrene	2·3	24
Chrysene	10·3	24

[a] The soil core samples were analysed GC/mass spectrophotometrically with the standard method outlined in EPA-600/7-79-191. An HP fused-silica capillary GC column coated with OV-101 was used for analyses.
[b] Higher molecular weight PAH (fluoranthenes, benzpyrenes, etc.) were detected but at too low a value ($<1\ \mu g/g$) for reliable data.
[c] Sample was composite of extractions from all soil samples.
[d] Not detected.

applications were to the whole site; others were 'partial' applications to subsections of the site.

The soil in the land treatment area selected for study is a silty clay, 157 cm in depth. Soil permeability is less than 0·15 cm/h (4·2 × 10 cm/s), which corresponds to a saturated intrinsic permeability of $4·2 × 10^{-10}$ cm^2 (Freeze and Cherry [33], p. 29). Depth to groundwater is less than 60 m. The general topography in the vicinity of the land treatment area is characterized by hills with 0–4% and occasionally 15–35% slopes. The land treatment area selected for study is nearly flat and level so little or no run-off occurs. Waste is applied during freeze-free periods in the spring and fall. No other activities take place on the land treatment area during the remainder of the year.

The 40-year average annual rainfall at the site is about 37 cm. The 5-year average (1975–80) is 41 cm and the 1-year average (July 1979 through June 1980) is 35 cm. Nearly 70% of the average annual rainfall occurs in April through September. May and June are the wettest months, accounting for about 20% of the average annual rainfall, with a secondary peak in September and October. The hottest months are usually July and August with mean daily maximum temperatures near 31°C. The coldest months are usually January and February; the long-range daily minimum temperatures are near 11°C. The freeze-free period averages 129 days a year.

Soil samples were taken from the site in October 1979 and in August 1980. Soil core samples were collected from two depths—0–15 cm below grade and 15–30 cm below grade—from both the waste application area and a control area. The control area was as nearly identical as possible to the waste application area except that no waste had been applied. In both the waste application and control areas, samples were collected from 20 equally spaced locations approximately 13·7 cm apart within square-shaped areas of 3000 m^2.

Soil core samples were air-dried prior to analysis. Trace metal analyses were performed on nitric–perchloric acid digests of representative aliquots of the respective soil samples so that reported results do not differentiate between adsorbed and dissolved analyte. Analysis results are expressed as micrograms of analyte per gram of air-dried soil.

Pollutant quantities originating from the site can be input to the model in several ways. If the pollutant is assumed to be present as a concentrated mass, a leaching rate (as a percentage of solubility) can be specified. If the pollutant depends on the rate of precipitation (i.e. acid rain), a rainfall pollutant concentration can be specified. If the pollutant is already mixed into the soil, a total pollutant quantity in each layer can be given.

The appropriate type and amount of loading were determined for each disposal site. The loading specified was a function of site history, type of waste, disposal method, etc.

Input data have been compiled from the literature (pollutant data) and from site investigations (climate, soil data) in order to simulate 2 years for each site. At the site, the waste had been applied several times over a 7-year period and had involved considerable variation of both waste composition and application area. The percentage of oil in the soil after the most recent loading (June 1979) was reported by plant personnel to be 7% by weight. Using this weight percentage and waste composition data, an organic pollutant loading rate was calculated. Loadings ranged from $560\,\mu g/cm^2$ (naphthalene) to $150\,\mu g/cm^2$ (anthracene).

For the site, the model SESOIL was run on a monthly basis from October 1978 to September 1980 for two of the most prevalent organic pollutants found at the site: naphthalene and anthracene. For ease of SESOIL application and as a test of the model's behaviour, the model was loaded at one time with the total amount of pollutant known to be present in June 1978, although at the site the waste had been applied several times to reach that total. SESOIL predicted concentrations in soil and soil moisture for a given depth were converted to a total g/g dry soil for the layer.

SESOIL predicted concentrations and average experimental measured

Table 6
Validation results—organics; petroleum waste site[a]
(all concentrations in $\mu g/ml$ (ppm))

	SESOIL prediction[b] October 1979	Experimental values[c,e] October 1979	Percentage difference	SESOIL prediction August 1980	Predicted percentage change
Naphthalene					
Upper soil layer	18·3	3·69	80	11·8	−35
Lower soil layer	0·003 12	NC[d]	—	0·012 6	+300
Anthracene					
Upper soil layer	5·55	2·36	57	5·39	−2·9
Lower soil layer	0·000 048 1	NC	—	0·002 79	+57

[a] Waste applied to 'clean site'—April 1979.
[b] Model depths: upper 0–20 cm; lower 20–5000 cm.
[c] Experimental depths: upper 0–15 cm; lower 15–30 cm.
[d] Lower depth results not directly comparable.
[e] Experimental concentrations calculated as (concentration in application area − concentration in control area).

concentrations are presented in Table 6. SESOIL predictions for 1 year later (August 1980) are also given. Again the lower layer results are presented only for completeness, since the experimental lower layer (15–30 cm) and the SESOIL lower layer (20–5000 cm) are not directly comparable. However, the upper layer for SESOIL was specified so as to be comparable with the experimental upper layer so these upper layer results are expected to agree.

The results for both organics are in modest agreement. The SESOIL results are expected to be higher than those found by the laboratory analysis since no biodegradation was modelled. Biodegradation was expected to occur at this site, but no rate data were available as input to the model.

Fate of inorganics/metals. The land treatment site considered is the property of a plastic manufacturing plant. Manufacturing process wastes are treated in a secondary wastewater treatment system at the facility. Sludge from the wastewater treatment system is centrifuged to yield a sludge whose content is 5–10% solids; the resulting sludge is disposed of by land treatment.

In July 1979, 5400 kg/ha of sludge were incorporated into the soil of a clean (i.e. not previously land cultivated) area of the site. The sludge was injected 12·7–20·3 cm below the soil surface and was subsequently mixed with the soil by ordinary farming methods.

The soil in the land treatment area is silt-loam, with a spatial intrinsic permeability of $7·05 \times 10^{-9}$ cm^2 and a surface slope of 3%. Depth of groundwater is reported to be 30–70 m. The 40-year (1940–80) average annual rainfall is about 85 cm. The 7-year rainfall (1973–80) is about 84 cm and the July 1979–August 1980 rainfall was 79 cm. The average time of rain varies between 0·18 and 0·20 day for the above period. The area receives 84–110 rainstorms per year. The rainy season is 365 days per year. The annual average temperature is 14°C. Almost no surface run-off occurs at the site, due to both the climatic and soil conditions.

Waste application occurred in the spring of 1979. In July 1979, and a year later in August 1980, soil core samples were collected from two depths, 0–15 cm below grade and 15–30 cm below grade, at both the waste application area and the control area. The control area soil was nearly identical to the soil of the waste application area, except that no waste had been applied.

Soil core samples were air-dried prior to analysis. Analyses were performed on nitric–perchloric acid digests of representative aliquots of the respective soil samples, so that reported results represent total metal concentrations and do not differentiate between adsorbed and dissolved analyte.

Chemical data and model parameters have been obtained from the literature and site-specific investigations. No calibration has been attempted, but only a minor adjustment of parameters within chemically justifiable limits was undertaken. Predicted concentrations agree reasonably well with those values measured chemically, considering the uncertainty of all parameters affecting pollutant migration in soils. Results are presented in Table 7. A sensitivity analysis has been performed to study soil contamination impacts from sludge application rates, climatologic, soil and chemistry parameters.

6.6.2. Migration to the atmosphere—buried solvents

The purpose of this research was to understand via a mathematical modelling effort the long-term potential fate of the leachate of six solvents leaking from buried barrels disposed in soil systems. A barrel was assumed to leak in 1 year. For this investigation, six halogenated organic solvents have been examined:

Perchloroethylene (tetrachloroethene)
Methylchloroform (1,1,1-trichloroethene)
Methylene chloride (dichloromethane)
Carbon tetrachloride (tetrachloromethane)
Freon 113
Trichloroethylene (1,1,2-trichloroethane)

Since it was not the intention to conduct a site-specific study, a number of hypothetical scenarios covering a wide range of US climates, soils and

Table 7
Predicted concentration of inorganic pollutants

Compound	Calculated amount of compound applied (April 1979) ($\mu g/cm^2$)	Concentrations (ppm)—July 1979			
		Measured		Predicted	
		0–15 cm	15–30 cm	0–15 cm	15–30 cm
Chromium	2·92	0·80	1·0	0·15	0·005
Copper	3·24	0·20	0·0	0·16	0·000 3
Sodium	$8·64 \times 10^3$	0·89	114·0	85·6	142·0

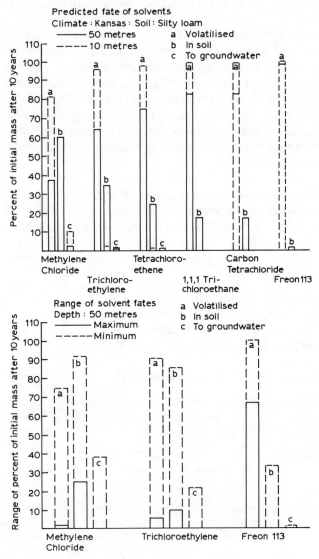

F<small>IG</small>. 7. Predicted ranges of solvents fates—summary of simulations.

solvents were considered. The methodology developed for the overall assessment is of general use and can be employed for similar and site-specific studies and classes of pollutants.

The actual quantities of pollutant mass removed by each pathway are strongly affected by the climate and soil type. Summaries of the pathways for all six chemicals for a 10-year simulation period, a moderate climate, a silty-loam soil and three depths of a soil column are presented in Fig. 7 and Table 8, and are summarized as follows:

—Of all chemicals studied, freon is most easily volatilized, whereas methylene chloride is least easily transported to the atmosphere.
—Methylene chloride contributes the most mass to groundwater; freon 113 contributes the least pollutant mass to groundwater.

Table 8
Quantitative fate of solvents (Bonazountas [13])

Chemical	Percentage volatilized	Percentage remaining in soil column	Percentage leached to groundwater
Depth to groundwater: 50 m			
Tetrachloroethene	74·9	24·7	0·4
1,1,1-Trichloroethane	82·4	17·3	0·3
Methylene chloride	37·4	60·2	2·5
Carbon tetrachloride	82·4	17·2	0·3
Freon 113	98·5	1·5	0·1
Trichloroethene	64·4	34·7	0·9
Depth to groundwater: 20 m			
Tetrachloroethene	88·0	10·6	1·3
1,1,1-Trichloroethane	94·2	4·9	0·9
Methylene chloride	57·0	33·8	9·2
Carbon tetrachloride	94·3	4·9	0·8
Freon 113	99·6	0·3	0·01
Trichloroethene	82·4	14·6	3·0
Depth to groundwater: 10 m			
Tetrachloroethene	97·8	1·1	1·1
1,1,1-Trichloroethane	99·3	0·1	0·6
Methylene chloride	81·2	10·3	8·5
Carbon tetrachloride	99·3	0·1	0·6
Freon 113	99·9	0·01	0·01
Trichloroethene	96·0	1·6	2·3

Percentage of mass after 10 years, moderate climate, silty-loam soil. Totals may not add up to 100%.

—The other solvents have fates intermediate between freon 113 and methylene chloride and are fairly similar to one another. Under moderate conditions, 99–64% of their mass volatilized and 0·01–3% of their mass reached the groundwater. The remaining mass was captured in the soil column.

—Leaching to groundwater increases for chemicals with low Henry's law constant, low diffusion coefficients and low absorption coefficients. Leaching is generally favoured by high rainfall and permeable soils.

—Volatilization is favoured for chemicals with high Henry's law constants and high diffusion rates. It is generally enhanced by dry conditions in porous soil. Decreasing soil column depth generally results in increasing volatilization rates up to a certain depth.

7.   IMMISCIBLE CONTAMINANT MODELLING

7.1. General

In an effort to quantify non-aqueous phase liquid (NAPL), or immiscible contaminant migration, from waste sites and oil spills noted with increasing frequency, various mathematical models were developed in the 1980s, and considerable research is continuously directed towards improving our understanding of these immiscible fluid processes in the soil and the groundwater zones of an area.

As reported by Mercer [65], many of the recently developed theoretical concepts and modelling approaches pertaining to this problem had originated in the petroleum industry. However, because of the different physical environment of deep petroleum reservoirs compared to the shallow aquifers, as well as different incentives and areas of concern in the petroleum industry, there is a great need to adopt and extend this work.

According to Abriola and Pinder [1], Van Dam presented the first detailed analysis of hydrocarbon pollution of groundwater as a two-phase problem. He examined the stages of contaminant infiltration and incorporated a capillary pressure term in his expression for fluid potential. Many researchers followed his work, among them Faust [28] and others reported in the reference section of this chapter.

Schwille [91] provides a review on physical, chemical and biological parameters affecting NAPL migration in porous media with an emphasis on mineral oil products and chlorohydrocarbons.

7.2. Modelling concepts

Four phases separated by distinct interfaces exist in a soil matrix: solid (soil), water, gas and contaminant. In the real physical system, each phase could possibly be formed by a number of chemical components or species, and mass transfer could occur across phase boundaries. In that respect, a contaminant may be available in four phases: adsorbed, dissolved, gaseous and NAPL. Migration of such a contaminant can be modelled as a three-phase flow process.

Basically there exist two concepts (ways) in modelling three-phase flow:

(1) by employing the three-phase flow (air, moisture, NAPL) equations in porous media jointly with a set of relationships; and

(2) by employing the governing equations of a two-phase flow in porous media (e.g. water–NAPL, or air–NAPL, or air–NAPL/ water) and by adjusting the coefficients (e.g. relative permeability of fluid) of the equations to reflect the specific problem.

The second approach eliminates the need for an equation governing the third phase (e.g. Faust [28]), but introduces the need for determining the 'relative' permeability of the 'second' phase (e.g. NAPL–water) to the 'first' phase (e.g. air). The model presented by Abriola and Pinder [1] follows the first concept. The model developed by Faust [28] follows the second concept.

Stone (1973) has proposed a model for estimating three-phase relative permeabilities based on data for two-phase relative permeabilities. In general, two-phase relative permeabilities are determined from laboratory tests on cores, then these results are often fitted to polynomial functions of saturation. Two-phase permeabilities can also be estimated from analytic functions based upon water (or dissolved chemical) characteristics, surface tension, porosity, intrinsic permeability and other parameters.

As shown from the literature sources of this review, a variety of computer codes now exist. Most of the codes consider three phases: air, water and NAPL; but the necessary input data to drive models do not exist. According to Mercer [65], for example, only one set of relative permeability curves is known in the literature for immiscible contaminants, and this is for the chemical TCE. This type of data does not exist for most solvents and chlorinated hydrocarbons found at spill sites or landfills (see model applications of Section 6.6). In addition to characterizing sites where NAPL exist, *in situ* water and NAPL saturation need to be measured or determined. Such data do not exist at these sites, or may not be available in the public domain. Therefore, the next major advance in the area of model development may not come until this type of data becomes available (Mercer [65]).

The following soil and fluid parameters are necessary to predict flow rate and pattern of immiscible fluids in soil matrices:

—chemical and physical properties of the fluids, as for example polarity, molecular weight, density, contaminants in fluid (e.g. metal species), viscosity and solubility in compressed water (water with density of approximately $1 \cdot 2 \, \text{g/cm}^3$); and

—chemical and physical properties of soil, as for example soil structure, soil texture, soil horizontation and depths, clay mineralogy, surface and area content, water-holding capacity at $\frac{1}{3}$, 1 and 15 bar, effective porosity and pore size distribution.

7.3. Selected models/codes

The US Geological Survey in Reston, Virginia, has developed a series of computer codes for NAPL fluids. These codes are available in the public domain and can be easily obtained. Many researchers have developed proprietary codes, information on which is obtained from the publication of their research, as reported below.

Abriola and Pinder [1] recently presented a research on multi-phase modelling of organic contaminants in porous media. The researcher developed a model to describe simultaneous transport of chemical contaminant in three physical forms: as a non-aqueous phase, as a soluble component of an aqueous phase, and as a mobile fraction of a gas phase. The contaminant may be composed of at most two distinct components, one of which may be volatile and slightly water-soluble, and another of which is both non-volatile and insoluble in water.

Equations which describe the system of Abriola and Pinder are derived from basic conservation of mass principles by the application of volume averaging techniques and the incorporation of various constitutive relations and approximations. Effects of matrix and fluid compressibilities, gravity, phase composition, interphase mass exchange, capillarity, diffusion and dispersion are considered. The resulting mathematical model consists of a system of three non-linear partial differential equations subject to two equilibrium constraints. The three equations describe mass conservation of the water phase, mass conservation of the inert organic species and mass conservation of the volatile organic species. The equations have five unknowns: two capillary pressures and three mass fractions. To handle the solution of the resultant system, a Newton–Raphson iteration scheme is employed.

In order to apply the finite difference one-dimensional model of Abriola and Pinder to a specific problem, a number of parameters must be evaluated. They include three-phase relative permeabilities, saturation pressure relations, partition coefficients, mixture densities and viscosities.

Faust [28] has developed a numerical model that describes the simultaneous flow of water and a second immiscible fluid under unsaturated and saturated conditions in porous media. Example applications of the model are performed to demonstrate both proper model function related to solving governing equations and model sensitivity to fluid properties. No field model application or validation is presented, since data such as relative permeabilities and capillary pressures are not available for the type of NAPL and sites considered.

7.4. Modelling applications

Faust [28] demonstrates the impact of an undetected leak of a NAPL on a surficial unit considered to be an aquitard. Two base cases were simulated:

— a NAPL more dense than water, where we expect gravity effects to be dominant and create a downward migration; and
— a NAPL less dense than water, where we expect the contaminant to pool near the water table.

The above two cases and expected conditions are observed in the results of the simulation cited below.

Figures 8 and 9 show contour plots of non-aqueous phase saturations superimposed on a cross-section. In one year the dense fluid has long (about 3 months) reached the underlying aquifer that drains the aquitard. A relatively uniform saturation profile has been established below the source. For the non-aqueous fluid that is less dense than water, contaminant saturations are higher (0·44 versus 0·19) and do not extend nearly as far into the saturated zone. The lighter contaminants have also migrated farther laterally. For this particular example, it is interesting that a large amount of the contaminant remains in the soil zone. The results of Mercer also indicate that the lighter contaminants do not necessarily form a distinct lens above the water table. This example shows that the lens concept may not strictly apply for all possible types of contaminants and hydrogeologic settings.

Figures 10 and 11 illustrate additional migration characteristics for the two cases. Each figure shows a sequence of three-phase saturations versus depth below the source grid block. For the dense NAPL (Fig. 10), a stable saturation profile is established after about three months of leakage. For the higher contaminant, a stable profile is not achieved even after one year.

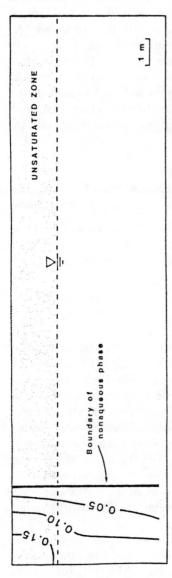

FIG. 8. Volumetric saturation of non-aqueous phase with density greater than water, 1200 kg/m³. Time = 1·16 years (Faust [28]).

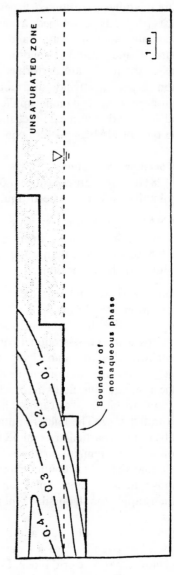

FIG. 9. Volumetric saturation of non-aqueous phase with density less than water, 950 kg/m³. Time = 1·07 years (Faust [28]).

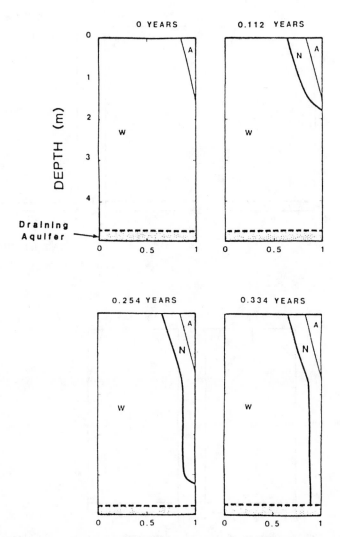

FIG. 10. Time sequence of three-phase saturation plots for grid blocks below source. Example application with NAPL density greater than water (Faust [28]).

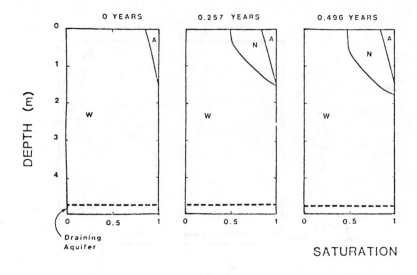

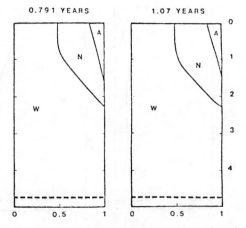

FIG. 11. Time sequence of three-phase saturation plots for grid blocks below source. Example application with NAPL of density less than water, $950 \, kg/m^3$ (Faust [28]).

Rather, the two-phase zone continues to expand downward (laterally also, although not shown) (Faust [28]).

In addition to the two cases above, the sensitivity to density and viscosity was investigated with other simulations by Faust [28]. These results are illustrated in Fig. 12. The same data used in the previous cases were used, except density and viscosity of the NAPL were varied. This figure shows

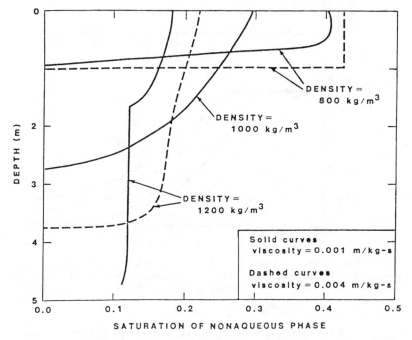

FIG. 12. Saturation profile of NAPL for grid below the source (Faust [28]).

depth saturation profiles of the NAPL for different density and viscosity conditions after 0·317 years of leakage. Both increased viscosity and decreased density of the NAPL have a similar effect, i.e. increasing the NAPL saturation near the source. (Note that the results show the profile under the source, Faust [28].)

Abriola and Pinder [1] similarly demonstrate their model capabilities with the simulation of the fate of one- and two-component organic contaminants in soils. Propagation of the organic liquid front with time is illustrated in their publication (Fig. 13), which the reader is referred to.

(a)

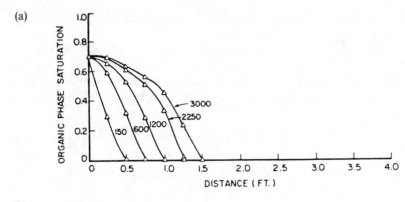

(b)

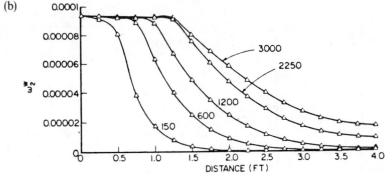

FIG. 13. Propagation of (a) organic phase and (b) species 2 in the water phase as a function of time elapsed—time for each curve is indicated in seconds (Abriola and Pinder [1]).

8.   AQUATIC EQUILIBRIUM MODELLING

8.1. General

An evaluation of the fate of inorganic compounds in soil and groundwater requires more detailed consideration of the chemical and biological processes involved, such as complexation, adsorption, coagulation, oxidation–reduction, chemical speciation and biological activity, as schematically shown in Fig. 14, in the control of free trace metal concentration in soil solutions. The referenced processes can affect, for example, species solubility, availability, physical transport and corrosion potential.

As a result of a need to describe the complex interactions involved in

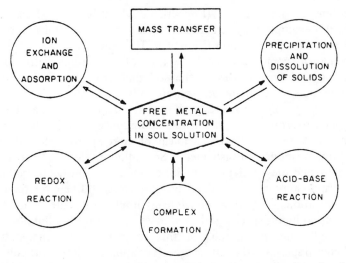

FIG. 14. Principal controls on free trace metal concentration in soil solutions (Mattigod *et al.* [61]).

these situations, various models have been developed to address specific needs. We have, for example, adsorption models, surface complexation models, constant capacitance models, cation-exchange models and overall fate modelling packages which account for one or more geochemical processes.

As Cederberg *et al.* [21] report, until recently only a single specific reaction such as ion-exchange or sorption for a small number of reacting solutes has been incorporated into mass transport models. It has also been assumed that the solutes being modelled act independently of the bulk solution composition. Because most contamination sources are actually multi-component solutions, the need is apparent for models capable of simulating chemical interaction processes.

One of these model categories is aiming at providing distribution of inorganic species in the soil water. These models are called 'equilibrium' or 'speciation' models. Chemically-based computer models of soluble trace metal speciation are being employed increasingly in decision making, in studies related to sewage and effluent when applied, for example, to agricultural land (Sposito *et al.* [89]).

The transport of 'species' in terrestrial systems is of interest to a variety of scientists, since measurement or reporting of 'total' concentration of a particular inorganic compound in the soil may be very misleading in many environmental managements (Emmerich *et al.* [26]). Toxic effects of trace metals, for example, may be determined more by their chemical form than by their total concentration (Florence [31]; Allen *et al.* [3]); therefore, mathematical computer models capable of simulating distribution of inorganic pollutant species in the soil and groundwater systems are valuable tools for contaminant soil/groundwater pathway analyses.

At this stage of intensive research in terrestrial chemical modelling it is reasonable to group the prevailing modelling concepts of the literature into two major categories: the 'geochemical' or more appropriately 'dissolved pollutant fate models', investigated in Soil Solutions (Mattigod *et al.* [61]), Section 6; and the 'speciation equilibria models' for inorganic pollutants. This terminology is not standard, but it is employed here for reasons of communication. Speciation models can simulate the fate of individual dissolved pollutant species and total mass of dissolved pollutants.

The following paragraphs of this chapter present the status of geochemical speciation equilibria models as applied to inorganic pollutants in soils. It has to be noted that speciation computer codes do also account for additional processes, such as redox reactions (Lindberg and Runnells [52]), adsorption (Langmuir [50]), complexation (Kirkner *et al.* [48]) and others (James and Rubin [38]), since they are related to speciation equilibria (Lyman *et al.* [55]). Speciation equilibria models are based on chemical thermodynamic reactions and principles, a research area receiving great progress in recent years. A number of computer codes are available for speciation calculations in soil and groundwater aqueous systems, the most well known of which are presented below.

Excellent state-of-knowledge reviews on chemical equilibria codes (models) of inorganic pollutants in soils are presented by Sposito [85], Cederberg *et al.* [21], Kincaid *et al.* [46], Miller and Benson [66], Jennings *et al.* [42], Theis *et al.* [93], Mattigod and Sposito [60], Jenne *et al.* [39], Nordstrom *et al.* [73] and many others. The following paragraphs have benefited from the reviews of the above researches, and particularly from the work of Sposito [85]. Readers interested in details should refer to the original publications.

8.2. Modelling concepts

Although geochemical speciation modelling was developed relatively recently, more than a dozen comprehensive computer codes are available, as depicted in Fig. 15. Kincaid *et al.* [46] provide an overview of the geochemical code history in which they group models into four major families according to their evolution stage.

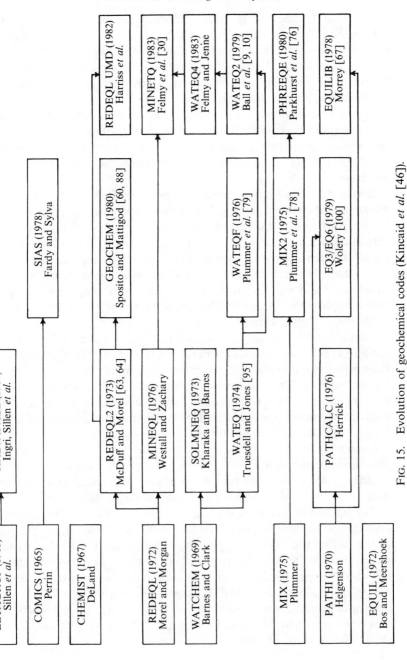

FIG. 15. Evolution of geochemical codes (Kincaid *et al.* [46]).

The four families of speciation models can be grouped into two categories: (1) 'speciation' codes which account for speciation equilibria of inorganic pollutants for a terrestrial water compartment, and (2) 'coupled speciation' codes which can simulate both speciation equilibria and solute transport of individual species in the terrestrial (soil, groundwater) environment, both in time and space.

Several investigators have recently addressed the problem of modelling speciation and transport of a multi-component solution in equilibrium with a soil or groundwater system. Cederberg *et al.* [21], Miller and Benson [66] and Jennings *et al.* [42] review the subject and provide detailed literature.

In summary, equilibrium chemistry models are formulated around two mathematical methods. (1) The first method interfaces the computer code for equilibrium distributions of species with the code of transport and performs calculations in two steps; the first step is species estimation and the second step is species transport. This procedure is repeated in time. (2) The second method consists of solving simultaneously a system of equations describing chemical reactions, advective dispersive transport with interphase mass transfer. The solution employed does not concern a code user to a great extent.

Coupled speciation codes such as TRANQL (Cederberg *et al.* [21]), FIESTA (Theis *et al.* [93]; Kirkner *et al.* [48]) and CHEMTRN (Miller and Benson [66]) are the subject of current research and are the extension of the speciation codes. Although coupled codes appear to be very powerful tools, they are also very large and require extensive input data. As a result, their use is complex and frequently inefficient (input data requirements, effort) for practical solutions to problems.

Coupled codes are not further analysed in this chapter, rather emphasis is placed on speciation equilibria (geochemical simple compartment) codes because these models have been more extensively applied and validated, with laboratory and field activity coefficients (input data).

The significant advances made in aquatic chemistry during the past decade (Stumm and Morgan [90]) attest to the availability of an approach to the soil solution that runs parallel with that taken for other natural waters (Sposito [85]). In that respect there exist many similarities between models developed for soil or groundwater and water environments. For example, the model GEOCHEM (Sposito and Mattigod [88]) has been developed from the water bodies speciation model, REDEQL2, a program created originally by Morel and Morgan and their co-workers at Caltech (McDuff and Morel [62, 63]).

Basic geochemical and aquatic equilibria processes modelled by existing computer codes are adsorption/desorption, precipitation/dissolution, reduction/oxidation, hydration and ion-interaction, and aqueous speciation in soil solution. Various mathematical expressions are used to model the above processes.

For the aqueous speciation in soil solution, Sposito [85] reports that 'chemical modelling' can be accomplished in two distinct ways: via a 'specific interaction' approach or via an 'ion-association' approach.

In the first case, the composition of an electrolyte solution is described with the total number of molalities of the stoichiometric components of neutral solutes, whereas the thermodynamic properties of the solution are expressed with 'mean ionic' activity coefficients for neutral solutes introduced into standard chemical thermodynamic equations.

In the second case, the composition of an electrolyte solution is described with the molalities of molecular 'species' presumed to exist in solution, whereas the thermodynamic properties of the solution are expressed with 'single-species' activity coefficients for the assumed molecular constituents. The ion-association approach appears to be adopted universally by soil chemists, despite its more tenuous relationship with rigorous thermodynamics. For additional details, the reader is referred to the original work of Sposito [85].

Geochemical data bases associated with each model and the above processes are the most important features of a computer code. Data bases can provide typical values of kinetic and adsorption rates at typical temperatures (e.g. 25°C). Reasonably good data bases are available for GEOCHEM, MINTEQ and other codes.

Model selection is a major issue in mathematical geochemical modelling. Factors of importance are study objectives, the features of the model, model input requirements, monitoring data available to validate predictions, model documentation and cost for an application. It is very difficult to recommend a universal computer code that can handle all of the above factors, therefore decision-makers have to first review a number of codes that may meet project requirements and consequently select the code which best suits their study objectives.

8.3. Selected models/codes

Table 9 lists features of known computer codes suitable for geochemical modelling. Of the twelve codes summarized, six are very capable in the areas of aqueous speciation, adsorption/desorption and precipitation/dissolution (Kincaid *et al.* 1983). In addition, these codes are documented,

Table 9

Summary of geochemical code capabilities, adaptability and availability
(Kincaid et al. [46])

Criterion	GEOCHEM	REDEQL UMD	MINTEQ	PHREEQDE	EQUILIB	EQ3/EQ6	SAIS	SOLMNEQ	CHEMIST
Code classification	III	III	III	IIIR	II	IIR	I	II	I
Number of elements	36(94)	3	32	19(26)	26	~20	10	25	1
Aqueous speciation									
Number of ligands/species	69	65	16/373	8/120·	186	~20	10 (200)	162	(169)
Number of redox species	48(60)	20	22	6	20			10	Variable
Activity coefficient correction	DA,HN	DA,SHM	TI,DA	EDH,TJ	HN	HN	None	HN	None
Calculation of pH	Yes	Yes	Yes	Yes	Yes	Yes	No	No	No
Method of iteration	NR	NR	NF	NR,CF	Pred.	NR	BS	BS	No
Adsorption									
Model	JH	Swiss	Three mdls	Two mdls	None	None	None	None	None
Number of species	10	20	No limit	Variable	None	None	None	None	None
Number of surfaces	6	5	3	NA	None	None	None	None	None
Number of species	No limit	None		None	None	None	None	None	None
Precipitation/dissolution									
Number of minerals	~500	<500	238	24	200	250	None	158	None
Quantitative mass transfer	Yes	Yes	Yes	Yes	Yes	Yes	None	No	No
Automatic selection of minerals	Yes	Yes	Yes	No	Yes	Yes	None	No	No
Capability for solid solutions	No	No	No	No	None	Yes	None	No	No

Gas generation									
Ammonia	No	No	No	No	Yes	No	No	No	No
Oxygen	Yes	No	Yes	Yes	Yes	Yes	No	No	No
Hydrogen	No	No	No	Yes	Yes	Yes	No	No	No
Carbon dioxide	Yes	Yes	Yes	Yes	Yes	Yes	No	Yes	No
Code structure									
Size (32-bit words, K)	434	62	68·1	>64	20·5	363	34		25
Modularity	Yes	Yes	Yes	Yes	Yes	Yes	No	No	No
Language	FORTRAN IVG	FORTRAN IV	FORTRAN IV	FORTRAN IVH	FORTRAN IV	FORTRAN 4·6	FORTRAN H	PL/1	FORTRAN
System	IBM 4314 VAX	CDC CYBER 171	UNIVAC 1144 PDP 11/70 VAX	Amdahl DEC VAX	CDC 7600 PDP VAX	CDC 7600 CDC 6600 VAX	IBM/360/65	IBM/360	IBM/360
Other criteria									
Latest documentation date	1980	1982	1982	1980	1978	1979	1978	1973	1968
Data base									
Temperature range	No	No	Yes	Yes	Yes	Yes	No	Yes	No
Easily modified	Yes	Yes	Yes	Yes	Yes	Yes	US	No	US

( ) = maximum, US = user supplied, JH = James–Healy, NR = Newton–Raphson, BS = back substitution, Pred. = predictor/back substitution, DA = Davies, HN = Helgeson–Nigrini, SHM = Sun–Harriss–Mattigod, EDH = extended Debye–Hückel, I = chemical speciation, II = I + mass transfer by precipitation or dissolution, III = II + adsorption and ion exchange, IIR, IIIR = pseudo kinetics.

Table 10
Reasons for selection or rejection of specific geochemical codes
(Kincaid *et al.* [46])

Code	Selected	Reason
GEOCHEM	Yes	Available in public domain. Documentation marginal. Recently updated. Models important processes. Includes adsorption. Data base probably the largest available. Modularity not yet evaluated
REDEQL UMD	No	Should be available in the public domain soon. Well documented, although not in final form. Recently updated. Models important processes. Data base and modularity not yet evaluated. Not selected because not available soon enough
MINTEQ	Yes	Available in the public domain. Well documented. Models important processes. Includes adsorption. Modular construction. Data base is the best documented; one of largest available
PHREEQE	Yes	Available in the public domain. Documented Models precipitation/dissolution. Includes adsorption. Data base supplied by user. Modular construction
EQUILIB	Yes	EPRI proprietary code. Well documented. Models precipitation, not adsorption. Data base quite extensive, reasonably well documented. Can use other data bases. Construction is modular. Method of solution involves unique elements
EQ3/EQ6	Yes	Publicly available. Documentation available for EQ3 only. Updated version of PATHI construction. Contains precipitation, no adsorption but structured for inclusion. Contains capability to model paths of chemical changes
SOLMNEQ	No	Publicly available. Documentation old. Does not contain unique characteristics. Data base not easily modifiable
CHEMIST	No	Publicly available. Old and sketchy documentation. No precipitation/adsorption. No unique modelling characteristics
CHEMCSMP	No	Publicly available. Old and very inadequate documentation. May have unique kinetic characteristics but tied to IBM system
SIAS	No	Update of an old program. Documentation sketchy. SIAS capabilities covered in better documented, more inclusive codes

available in the public domain in the United States, recently updated and frequently accompanied with a sizeable data base. Table 10 provides user-related information for the codes presented in Table 9. Many other codes exist in the literature (e.g. BALANCE, Parkhurst *et al.* [75]), which are not reported here.

Two of the best known geochemical and speciation codes are GEOCHEM and MINTEQ. Coupled geochemical/speciation and transport (codes receiving increased attention) are TRANQL and CHEMTRN. These codes are briefly presented below.

GEOCHEM is based on the computer program REDEQL2 for calculating equilibrium speciation of chemical elements in a soil solution. The component species are identified as uncomplexed metal cations, the free proton, uncomplexed ligands and the free electron. Single-species activity coefficients are calculated in the program. The model contains critical thermodynamic data for soils, a method for calculating cation exchange and a correcting method for non-zero ionic strength up to 3 mole. Currently the model stores thermodynamic data for 36 metals and 69 ligands, which form more than 200 soluble complexes and solids. Adsorbed metal species are described in the model whereas adsorbed ligand species will be considered in the future. A new version of GEOCHEM is under way (Sposito [85]), therefore interested scientists should contact model developers.

MINTEQ is similar in structure to MINEQL which is similar in overall structure to GEOCHEM, since both models originate from REDEQL. MINTEQ is formed from MINEQL and the data base of WATEQ (Ball *et al.* [10]). The model includes ion speciation, redox equilibria, calculation of activity coefficients, solubility, adsorption and mass transfer. The model and its large data base are well documented, the latter involving more than 35 metals and 60 ligands.

As reported in a comparative analysis by Sposito [85], a principal difference between MINEQL (and consequently MINTEQ) and GEOCHEM is that MINEQL can accept the concentration of any free ionic species, soluble complex or dissolved gas as input data to be held fixed during a calculation, whereas GEOCHEM can do this only for the activities of H^+, e^-, $CO_2(g)$ and $N_2(g)$. Therefore, the concentration of $Cu^{2+}(aq)$ can be specified as a fixed input datum, in MINEQL if desired, a feature providing the code user with a great flexibility in speciation calculations.

CHEMTRN is a one-dimensional geochemical transport/speciation (coupled) model for solutes in a saturated porous medium. The model includes dispersion/diffusion, advection, ion exchange, formation of complexes and speciation in the aqueous phase, and the dissociation of water. The mass action, transport and site constraint equations are expressed in differential/algebraic form and are solved simultaneously. This coupled model is at a developmental stage, therefore no data base accompanies the computer code.

TRANQL is a groundwater mass transport and equilibrium chemistry model for multi-component systems. The equilibrium interaction chemistry is posed independently of the mass transport equations which leads to a set of algebraic equations for the chemistry coupled to a set of differential equations for the mass transport. Significant equilibrium chemical reactions such as complexation, ion exchange, competitive adsorption and dissociation of water may be included in the model. In a recent application, a finite solution is presented first for cadmium, chloride and bromide transport in a one-dimensional column where complexation and sorption are considered. Second, binary and ternary ion exchange are modelled and compared to the results of other investigations. According to model developers, results show the model to be a versatile multi-component transport model, with potential for extension to a wide range of equilibrium reactions.

8.4. Model applications.
Reports in the literature on the chemical form of heavy metals in soil solutions are limited (Emmerich et al. [26]; Mahier et al. [58]), which may be due, first, to the large number of chemical forms in which metals exist and the analytical problems associated with their determination (Norvell [74]) and, second, to only the recent concerns on modelling metal species in soil solutions.

The use of chemical equilibrium models in decision making is reported by Emmerich et al. [26], Behel et al. [12], Matthews [59], Westall and Hohl [99], Miller and Benson [66], Jennings et al. [41], Mattigod and Sposito [60], Theis et al. [93], Dowdy and Volk [23], Kirkner et al. [47], Jennings and Kirkner [40] and many others. The above researchers have employed and validated models (e.g. GEOCHEM) with data from field (e.g. Behel et al. [12]) and laboratory (Emmerich et al. [26]) analyses.

A verbatim presentation of the work of Behel et al. [12] dealing with the validation of GEOCHEM in an application to assess heavy metal equilibria in sewage sludge in land-treated soil is as follows.

An experiment was initiated in the fall of 1971 at the Tennessee Valley Authority (TVA) Agricultural Research Farm in Muscle Shoals, Alabama, on a Sango silt-loam soil (Glossic Fragiudult) with a pH of 4·9 and cation-exchange capacity (CEC) of 7 cmol (NH_4^+) kg^{-1}.

Anaerobically digested, air-dried sewage sludge from Tuscumbia, Alabama, was applied at rates of 50, 100 and 200 t/ha, and was incorporated into the soil by disking to a depth of 15 cm. Soil pH after sludge application ranged from 5·3 to 5·6. A randomized complete block design with four replications was used in the study. Original plot dimensions were 3·6 × 15·0 m.

In the fall of 1972, all plots were sub-divided into areas 3·6 × 7·5 m. One-half of each plot received additional sludge applications in 1972, 1973 and 1974 at the same rate used for the initial application. The other half of each plot received no additional sludge, so that residual effects from the original applications could be monitored. The plots have been cropped continuously with sweet corn (*Zea mays* L., cv. Silver Queen) since 1972.

A collection and analysis program was initiated at the site, during which soil samples were collected in the fall of 1977, air-dried and crushed to less than 2 mm. Water was added to 150-g samples to bring the soil water potential to −0·33 bars, and the samples were then incubated for 7 days at 25°C. Soil solutions were recovered by centrifugation (Gillman [35]). The solutions were analysed for total Cd by flameless atomic absorption spectrophotometry, using background correction.

The soil solutions were also analysed for total Zn, Mn, Fe, Al, Cu, Ni, Pb, Ca, Mg, Na and K by flame atomic absorption or flame emission spectrophotometry; inorganic P by a molybdenum-blue method (Murphy and Riley [69]); Cl by a Mohr titration method (Bower and Wilcox [19]); SO_4-S by a turbidimetric method (Tabatabai [92]); organic C by the method of Mebius [64]; and conductivity and pH by standard conductivity and pH meters, respectively. All values reported are the average of at least duplicate determinations on soil solutions obtained from three or four replicate field plots.

To predict the equilibria of the metal in solution, the GEOCHEM model was used, and three approaches were employed to estimate the importance of metal interactions with soluble organic C (e.g. fulvates):

(i) no organic ligands were included in the calculations;
(ii) a mixture model consisting of nine organic acids with known stability constants for each metal (Mattigod and Sposito [60]); the concentration of organic acids in the mixture model is equivalent to 226 mg C/litre, and the molar concentration of each organic acid

was scaled to measured values of organic C in the soil solutions; and

(iii) a fulvate model based on experimentally determined values for the conditional stability constants of sludge-derived fulvate with Ca, Cd, Cu and Pb (Sposito *et al.* [87, 88]). Conditional constants were calculated by correlation techniques for Mg, Mn, Fe, Zn and Ni; it should be realized that such conditional constants are strictly valid only at pH 5 and 0·1 M ionic strength. Fortunately, the soil solutions used in this study were at pH = 5·0.

The chemical composition of soil solutions from untreated and sludge-treated soils is shown in Table 11. Concentrations of Zn, Cd, Mn, Ca, Mg, Na, K, PO_4–P, SO_4–S and Cl in the soil solution tended to increase with sludge application, particularly in plots receiving multiple sludge applications. Soil solution concentrations of Cu, Ni, Pb and Fe were generally low (less than 10^{-6} M) at all sludge application rates; however, soluble Cu and Ni concentrations in soil treated with sludge at 800 t/ha were markedly higher than those in soil receiving lesser amounts of sludge.

The predominant species of Zn in the soil solutions was Zn^{2+} (Table 12). Even though sludge applications increased total soluble Zn (Zn_t) from 4×10^{-6} M to more than 1×10^{-4} M, from 87–97% of the Zn_t was present as Zn^{2+}. For most sludge treatments, SO_4 complexes accounted for 5–8% of Zn_t. Zinc complexes containing PO_4, CO_3, Cl and OH were all less than 0·1% of Zn_t. Fulvate complexes (mixture model) with Zn tended to increase with sludge applied, but in all cases less than 2% of Zn_t was associated with soluble organic C.

The distribution of Zn species was essentially the same after application of 200 t/ha in a single treatment or in four annual 50 t/ha increments. Soluble Zn_t increased to a greater extent with the single 200 t/ha sludge application ($1·7 \times 10^{-4}$ M) when compared with four increments of 50 t/ha ($1·2 \times 10^{-4}$ M).

The activity of Zn^{2+} in the soil solutions was similar to that supported by several Zn solid phases. The activity of Zn^{2+} ranged from approximately 10^{-5} to 10^{-3} M solutions of pH 5 to 6 (Table 12). Based on stability diagrams (Lindsay [53]), the Zn^{2+} in soil solution could be controlled by one or more of the following solids: (i) $ZnFe_2O_4$ in equilibrium with soil Fe; (ii) soil Zn; and (iii) Zn_2SiO_4 in equilibrium with amorphous Si. At similar pH values, Zn^{2+} activities obtained from this study were greater than those previously reported for soils incubated with sewage sludge (Emmerich *et al.* [26]). The pH range of the sludge-treated soils was too narrow to evaluate

Table 11

Chemical composition of soil solutions from untreated and sludge-treated soils
(Behel *et al.* [12])

Parameter	Units	Sludge applied (t/ha^a)							
		A				B			
		0	50	100	200	0	200	400	800
Zn	$\times 10^{-4}$ M	0.35	7.94	10.72	16.98	0.40	11.75	16.60	36.31
Cd	$\times 10^{-5}$ M	0.98	2.24	3.09	79.4	1.26	3.02	58.9	144.5
Cu	$\times 10^{-6}$ M	<1.0	<1.0	<1.0	3.16	<1.0	<1.0	<1.0	6.31
Ni	$\times 10^{-6}$ M	<1.0	<1.0	<1.0	<1.0	<1.0	<1.0	<1.0	6.76
Pb	$\times 10^{-6}$ M	<1.0	<1.0	<1.0	<1.0	<1.0	<1.0	<1.0	<1.0
Mn	$\times 10^{-4}$ M	3.72	3.09	3.89	3.55	3.63	5.37	6.46	7.41
Fe	$\times 10^{-6}$ M	<1.0	<1.0	<1.0	<1.0	<1.0	<1.0	<1.0	<1.0
Al	$\times 10^{-4}$ M	1.86	1.86	1.86	<1.0	1.86	1.86	<1.0	<1.0
Ca	$\times 10^{-1}$ M	1.95	3.63	3.72	4.07	2.19	3.89	4.68	9.12
Mg	$\times 10^{-4}$ M	1.20	5.89	4.07	4.37	0.95	4.37	4.68	6.31
K	$\times 10^{-4}$ M	7.59	19.95	10.23	9.55	8.13	10.0	10.72	11.75
Na	$\times 10^{-1}$ M	3.31	3.89	3.89	3.98	2.51	8.32	8.51	9.12
PO$_4$–P	$\times 10^{-6}$ M	2.6	2.9	4.2	15.49	2.57	9.33	26.92	60.26
SO$_4$–S	$\times 10^{-1}$ M	0.26	0.85	0.98	1.15	0.44	0.91	1.66	3.39
Cl	$\times 10^{-1}$ M	0.60	1.55	1.29	1.23	0.51	0.72	1.07	1.35
Organic C (g m^{-1})		160	130	160	190	170	150	180	250
Conductivity (dS m^{-1})		1.54	2.25	2.14	2.08	1.56	2.03	2.38	3.88
pH		4.7	4.9	5.1	5.6	4.6	5.0	5.2	5.5

[a] A denotes a single application in 1971; B denotes repeat applications of initial rate in 1972, 1973 and 1974.

the pH dependence of Zn^{2+}, as predicted by the solubility of known Zn solid phases.

To facilitate interpretation of the data obtained, GEOCHEM calculations were performed using three sets of conditions: (i) metal–organic complexes were ignored; (ii) metal–organic complexes were approximated by a mixture model (Mattigod and Sposito [60]); and (iii) metal–organic complexes were modelled using stability constants obtained experimentally (Sposito *et al.* [87]) for a sludge-derived fulvic acid (fulvate model). The mixture model predicted that less than 16% of the Cd_t, Zn_t and Mn_t in

Table 12

Concentration of Zn and Cd in soil solution as affected by sludge rate and frequency of application
(Behel *et al.* [12])

Sludge applied	Number of applica-tions[a]	Total Zn	Free Zn^{2+}	Complexed Zn^b		Total Cd	Free Cd^{2+}	Complexed Cd^b		
				SO_4	MM^c			SO_4	Cl	MM^c
t/ha/y		$\times 10^{-5}$ M	—— % of total ——			$\times 10^{-8}$ M	—— % of total ——			
0	0	0·35	97	2·2	0·5	0·98	92	2·7	3·6	1·7
	0	0·40	96	3·5	0·4	1·3	91	4·2	2·9	1·5
50	1	7·9	94	5·3	0·4	2·2	85	6·0	7·5	1·5
	4	11·8	94	5·0	0·7	3·0	89	5·9	3·3	1·9
100	1	10·7	93	6·0	0·9	3·1	84	6·9	6·2	2·4
	4	16·6	91	8·1	1·0	59	83	9·4	4·4	2·8
200	1	17·0	91	6·7	1·7	79	82	7·7	5·6	4·6
	4	36·3	87	11·3	1·3	144	79	13·0	4·6	3·6

[a] A single application in 1971; repeat applications of initial rate in 1972, 1973 and 1974.
[b] Complexes not given represented <0·1% of total soluble metal.
[c] MM denotes mixture model for fulvate (Mattigod and Sposito [60]).

solution was complexed with organic and inorganic ligands; therefore, it was not surprising that exclusion of the mixture model from GEOCHEM had a minor effect on the speciation of Cd, Zn and Mn (Table 13).

In summary, GEOCHEM calculations were close to monitored values, indicating that Zn^{2+} and Cd^{2+} are the predominant species present in the soil solution of acid soils treated with sludges at metal additions approaching the maximum recommendations for growth of agronomic crops.

Computer modelling of trace metal equilibria, therefore, can provide a useful framework to understand the nature of the multitude of reactions that take place in soil systems.

Table 13
Effects of excluding organic ligands from consideration by GEOCHEM on percentage of free metal in solution
(Behel *et al.* [12])

Sludge applied[a]		Organic ligands included	Cd		Zn		Cu		Ni		Mn	
A	B		A	B	A	B	A	B	A	B	A	B
t/ha			calculated % of total present as free metal ion									
0	0	Fulvate model	67	67	79	80	—[b]	—	—	—	69	73
		Mixture model	92	91	97	96	—	—	—	—	98	97
		None	94	93	98	96	—	—	—	—	98	97
50	200	Fulvate model	69	72	84	84	—	—	—	—	80	81
		Mixture model	85	89	94	94	—	—	—	—	95	95
		None	86	90	95	95	—	—	—	—	96	96
100	400	Fulvate model	66	67	80	81	—	—	—	—	74	78
		Mixture model	84	83	93	91	—	—	—	—	95	84
		None	86	86	94	92	—	—	—	—	95	93
200	800	Fulvate model	64	63	79	78	69	69	—	72	74	76
		Mixture model	82	79	91	87	19	19	—	87	77	80
		None	86	82	93	88	93	88	—	90	94	90

[a] A denotes one application in 1971; B denotes repeat applications of initial rate in 1972, 1973 and 1974. Fulvate model—calculations based on equilibrium constants for fulvic acid extracted from sewage sludge (Sposito *et al.* [87]). Mixture model— mixture model for soluble fulvic acid (Mattigod and Sposito [60]).
[b] Total concentrations below detection limits of analytical methods used.

Model output 'validation' is essential to any soil modelling effort, although this term has a broad meaning in the literature. For the purpose of this section we can define validation as 'the process which analyses the validity of final model output', namely the validity of the predicted pollutant concentrations or mass in the soil column (or in groundwater), as compared to the available knowledge of measured pollutant concentrations from monitoring data (field sampling).

A disagreement of course in absolute levels of concentration (predicted versus measured) does not necessarily indicate that either method of obtaining data (modelling, field sampling) is incorrect or that either data set needs revision. Field sampling approaches and modelling approaches rely on two different perspectives of the same situation (Bonazountas [13]).

9. PARAMETERS AFFECTING MIGRATION

A quantitative evaluation of critical parameters affecting dissolved organic contaminant migration through soils when conducting mathematical modelling is presented by Tucker *et al.* [96].

Critical parameters affecting inorganic species and NAPL migration are discussed in the corresponding chapters.

ACKNOWLEDGEMENTS

Information contained in this chapter has been compiled under the direct or indirect support provided by Melanie Byrne, Alan Eschenroeder, Warren Lyman, Annett Nold, Joanne Perwak, Hans Sachdeva, Kate Scow, Janet Wagner and others. Their contributions are appreciated. In addition, this material will serve as a background for a Handbook on Soil and Groundwater Pollutant Fate Mathematical Modeling, in preparation for the American Society of Civil Engineers, Waste Management Committee.

REFERENCES

1. ABRIOLA, L. M. and PINDER, G. F. (1985). A multiphase approach to modeling of porous media contamination by organic compounds: 1. equation development, 2. numerical solution. *Water Resources Research*, **21**, No. 1, 11–32.
2. ADAMS, R. T. and KURISU, F. M. (1976). Simulation of pesticide movement on small agricultural watersheds. Final Report, Environmental Research Laboratory, US Environmental Protection Agency, Athens, GA.
3. ALLEN, H. E., HALL, R. H. and BRISHIN, T. D. (1980). Metal speciation: effects on aquatic toxicity. *Environ. Sci. Technol.*, **14**, 441–3.
4. ANDERSON, M. P. (1984). Movement of contaminants in groundwater: chemical processes. In: *Groundwater*, National Academy Press, Washington, DC.
5. APPEL, C. A. and BREDEHOEFT, J. D. (1978). Status of Groundwater Modeling in the US Geological Survey, US Department of the Interior, Washington, DC.
6. ARTHUR D. LITTLE, INC. (1981). *Prepared Revisions to MITRE Model*, Arthur D. Little, Inc., Cambridge, MA.
7. ARTHUR D. LITTLE, INC. (1982). Capillary Pressure and its Effects on Non-Aqueous Phase Liquid Migration and Contamination. Report to Commercial Client, prepared for USEPA, State of New York. Arthur D. Little, Inc., Cambridge, MA.

8. BACHMAT, Y., BREDEHOEFT, J., ANDREWS, B., HOLZ, D. and SEBASTIAN, S. (1980). *Groundwater Management: The Use of Numerical Models*, Water Resources Monograph, American Geophysical Union, Washington, DC.
9. BALL, J. W. and JENNE, E. A. (1979). WATEQ2—a computerized chemical model for trace and major element speciation and mineral equilibria in natural waters. In: *Chemical Modeling in Aqueous Systems* (Ed. E. A. Jenne), American Chemical Society Symposium, Series No. 93, American Chemical Society, Washington, DC.
10. BALL, J. W., JENNE, E. A. and NORDSTROM, D. K. (1979). WATEQ2—a computerized chemical model for trace and major element speciation and mineral equilibria of natural waters. In: *Chemical Modeling in Aqueous Systems, Speciation, Sorption, Solubility, and Kinetics* (Ed. E. A. Jenne), American Chemical Society Symposium, Series No. 9, American Chemical Society, Washington, DC.
11. BEAR, J. (1979). *Hydraulics of Groundwater*, McGraw-Hill, New York.
12. BEHEL, D. Jr, NELSON, D. W. and SOMMERS, L. E. (1983). Assessment of heavy metal equilibria in sewage sludge-treated soil. *J. Environ. Qual.*, **12**, No. 2, 181.
13. BONAZOUNTAS, M. (1983). Soil and groundwater fate modeling: a review. In: *Fate of Chemicals in the Environment, Compartmental and Multimedia Models for Predictions* (Eds R. Swank and A. Eschenroeder), American Chemical Society Symposium, Series No. 225, American Chemical Society, Washington, DC.
14. BONAZOUNTAS, M. (1985). *Mathematical Modeling of Inorganic Contaminants in the Environment: Air, Soil, Water*. US Army Contract Research Effort in Progress. Arthur D. Little, Inc., Cambridge, MA.
15. BONAZOUNTAS, M. and WAGNER, J. (1984). Modeling mobilization and fate of leachates below uncontrolled hazardous waste sites. *Proceedings, The 5th National Conference on Management of Uncontrolled Hazardous Waste Sites*, November 7–9, HMCRI, VA.
16. BONAZOUNTAS, M. and WAGNER, J. (1984). SESOIL: A Seasonal Soil Compartment Model, UESPA/OTS Contract No. 68-01-6271. Report by Arthur D. Little, Inc., Cambridge, MA.
17. BONAZOUNTAS, M., WAGNER, J. and GOODWIN, B. (1984). Evaluation of Seasonal Soil/Groundwater Pathways via SESOIL, Final Report, USEPA/MDSD Contract No. 68-01-5949(9). Report by Arthur D. Little, Inc., Cambridge, MA.
18. BONAZOUNTAS, M. and FIKSEL, J. (1982). ENVIRO: Environmental Mathematical Pollutant Fate Modeling Handbook and Catalogue, EPA Contract No. 68-01-5146. Draft Report, Arthur D. Little, Inc., Cambridge, MA.
19. BOWER, C. A. and WILCOX, L. V. (1965). Soluble salts. In: *Methods of Soil Analysis* (Ed. C. A. Black), American Society of Agronomy, Madison, WI.
20. CANSFIELD, P. E. and RACZ, G. J. (1978). Degradation of hydrocarbons in the soil. *Can. J. Soil Sci.*, **58**, 339–45.
21. CEDERBERG, G. A., STREET, R. L. and LECKIE, J. O. (1985). A groundwater mass transport and equilibrium chemistry model for multicomponent systems. *Water Resources Research*, **21**, No. 8, 1095–104.
22. CONWAY, R. A. (Ed.) (1982). *Environmental Risk Analysis for Chemicals*, Van Nostrand Reinhold Co., New York.

23. DOWDY, R. H. and VOLK, V. V. (1983). Movement of heavy metals in soils. In: SSSA Special Publication No. 11 (Eds Nelson *et al.*), Soil Science Society of America, American Society of Agronomy, Madison, WI.

24. EHRENFELD, J. and BASS, J. (1983). *Handbook for Evaluating Remedial Action Technology Plans*, USEPA Report No. EPA-600/2-83-076, Aug. 1983, USEPA, Cincinnati, OH.

25. EMMERICH, W. E. (1980). Chemical forms of heavy metals in sewage sludge-amended soils as they relate to movement through soils. PhD dissertation, University of California, Riverside.

26. EMMERICH, W. E., LUND, L. J., PAGE, A. L. and CHANG, A. C. (1982). Predicted solution pharc. forms of heavy metals in sewage sludge-treated soils. *J. Environ. Qual.*, **11**, No. 2, 182.

27. ENFIELD, C. G., CARSEL, R. F., COHEN, S. Z., PHAN, T. and WALTERS, D. M. (1980). Approximating pollutant transport to groundwater, USEPA, RSKERL, Ada, OK (unpublished paper).

28. FAUST, C. R. (1984). Transport of Immiscible Fluids Within and Below the Unsaturated Zone—A Numerical Model, Geotrans Report No. 84-01, Geotrans, Herdon, VA.

29. FEDDES, R. A. *et al.* (1978). *Simulation of Field Water Use*, John Wiley, New York.

30. FELMY, A. R., GIRVIN, D. and JENNE, E. A. (1983). *MINTEQ: A Computer Program for Calculating Aqueous Geochemical Equilibria*, US Environmental Protection Agency, Washington, DC.

31. FLORENCE, T. M. (1977). Trace metal species in fresh waters. *Water Resources Research*, **11**, 681–7.

32. FOOTE, H. P. (1982). For Information: Battelle Pacific Northwest Laboratories, PO Box 999, Richland, VA.

33. FREEZE, R. A. and CHERRY, J. A. (1979). *Groundwater*, Prentice-Hall, Englewood Cliffs, NJ.

34. FRIED, J. J., MUNTZER, P. and ZILLIOX, L. (1979). Ground-water pollution by transfer of oil hydrocarbons. *Groundwater*, **17**, No. 6, 586.

35. GILLMAN, G. P. (1976). A Centrifuge Method for Obtaining Soil Solution, Division of Soils Report No. 16. Commonwealth Scientific and Industrial Research Organization, Melbourne, Australia.

36. HALL, P. L. and QUAM, H. (1976). Countermeasures to control oil spills in Western Canada. *Groundwater*, **14**, No. 3, 163.

37. HUFF, D. D. (1977). *TEHM: A Terrestrial Ecosystem Hydrology Model*, Oak Ridge National Laboratory, Oak Ridge, TN.

38. JAMES, R. V. and RUBIN, J. (1979). Applicabilities of local equilibrium assumption to transport through soils of solutes affected by ion exchange. In: *Chemical Modeling in Aqueous Systems* (Ed. E. A. Jenne), American Chemical Society Symposium, Series No. 93, American Chemical Society, Washington, DC.

39. JENNE, E. A. (1979). Chemical modeling—goals, problems, approaches and priorities. In: *Chemical Modeling in Aqueous Systems* (Ed. E. A. Jenne), American Chemical Society Symposium, Series No. 93, American Chemical Society, Washington, DC.

40. JENNINGS, A. A. and KIRKNER, D. J. (1984). Instantaneous equilibrium

approximation analysis. *Journal of Hydraulic Engineering*, American Society of Civil Engineers, New York.
41. JENNINGS, A. A., KIRKNER, D. J. and THEIS, T. L. (1982). Multicomponent equilibrium chemistry in groundwater quality models. *Water Resources Research*, **18**, No. 4, 1089–96.
42. JENNINGS, A. A., KIRKNER, D. J. and THEIS, T. L. (1982). Multicomponent equilibrium chemistry in groundwater quality models. *Water Resources Research*, **18**, No. 4, 1089–96.
43. JRB ASSOCIATES (1980). *Methodology for Rating the Hazard Potential of Waste Disposal Sites*, JRB Associates, McLean, VA.
44. JURRY, W. A. (1982). Simulation of solute transport using a transfer function model. *Water Resources Research*, **18**(2), 363–8.
45. KAZMANN, R. (1984). Guest Editorial on groundwater modeling. *Groundwater*, Summer 1984, 4.
46. KINCAID, C. T., MORREY, J. R. and ROGERS, J. E. (1984). *Geochemical Models for Solute Migration*, Vol. 1: *Process Description and Computer Code Selection*, Report EA-3417, Electric Power Research Institute, Palo Alto, CA.
47. KIRKNER, D. J., REEVES, H. W. and JENNINGS, A. (1984). Finite Element Analysis of Multicomponent Contaminant Transport Including Precipitation–Dissolution Reactions. *Proceedings of the 5th International Conference, Burlington, Vermont* (Eds Laible *et al.*), Springer-Verlag, New York.
48. KIRKNER, D. J., THEIS, T. L. and JENNINGS, A. A. (1984). Multicomponent solute transport with sorption and soluble complexation. *Adv. Water Resources*, **7**, 120.
49. KNISEL, W. G. (Ed.) (1984). *CREAMS: A Field Scale Model for Chemicals, Runoff, and Erosion from Agricultural Management Systems*, US Department of Agriculture, Washington, DC.
50. LANGMUIR, D. (1979). Techniques in estimation of thermodynamic properties for some aqueous complexes of geochemical interest. In: *Chemical Modeling in Aqueous Systems* (Ed. E. A. Jenne), American Chemical Society Symposium, Series No. 93, American Chemical Society, Washington, DC.
51. LEGRAND, H. E. (1980). *A Standard System for Evaluating Waste Disposal Sites*, National Water Well Association, Washington, DC.
52. LINDBERG, R. D. and RUNNELLS, D. D. (1984). Groundwater redox reactions: an analysis of equilibrium state applied to E_h measurements and geochemical modeling. *Science*, **225**, 925.
53. LINDSAY, W. L. (1979). *Chemical Equilibria in Soils*, John Wiley, New York.
54. LIU, D. (1980). Fate of petroleum hydrocarbons in sewage sludge after land disposal. *Bull. Environ. Contam. Toxicology*, **25**, 616–22.
55. LYMAN, W. *et al.* (1982). *Chemical Property Estimation Methods*, McGraw-Hill, New York.
56. MACKAY, D. M. (1979). Finding fugacity feasible. *Environ. Sci. Technol.*, **13**, 1218–23.
57. MACKAY, D. M. and ROBERTS, P. V. (1985). Transport of organic contaminants in groundwater. *Environ. Sci. Technol.*, **19**, No. 5, 384.
58. MAHIER, R. J., BINGHAM, F. T., SPOSITO, G. and PAGE, A. L. (1980). Cadmium-enriched sewage sludge application to acid and calcareous soils: relation

between treatment, cadmium in saturation extracts, and cadmium uptake. *J. Environ. Qual.*, **9**, 359–64.

59. MATTHEWS, P. J. (1984). Control of metal application rates from sewage sludge utilization in agriculture. *CRC Critical Reviews in Environmental Control*, **14**, No. 3.

60. MATTIGOD, S. V. and SPOSITO, G. (1979). Chemical modeling of trace metal equilibria in contaminated soil solutions using the computer program GEOCHEM. In: *Chemical Modeling in Aqueous Systems* (Ed. E. A. Jenne), American Chemical Society Symposium, Series No. 93, American Chemical Society, Washington, DC, pp. 837–56.

61. MATTIGOD, S. V., SPOSITO, G. and PAGE, A. L. (1981). Factors affecting the solubilities of trace metals in soils. In: *Chemistry in the Environment*, American Society of Agronomy, Soil Science Society of America, ASA Special Publication No. 40.

62. McDUFF, R. E. and MOREL, F. M. M. (1974). Description and Use of the Chemical Equilibrium Program REDEQL-2, Tech. Rep. EQ-73-02, W. M. Keck Lab., California Institute of Technology, Pasadena, CA.

63. McDUFF, R. E. and MOREL, F. M. M. (1973). Description and Use of the Chemical Equilibrium Program REDEQL-2, EQ-73-02, California Institute of Technology, Pasadena, CA.

64. MEBIUS, L. J. (1960). A rapid method for the determination of organic carbon in soil. *Anal. Chim. Acta*, **22**, 120–4.

65. MERCER, J. W. (1984). Miscible and Immiscible Transport in Groundwater, EOS, USGS, p. 691.

66. MILLER, C. W. and BENSON, L. V. (1983). Simulation of solute transport in a chemically reactive heterogeneous system: model development and application. *Water Resources Research*, **19**, No. 2, 381–91.

67. MORREY, J. R. (1978). *Operator's Manual for EQUILIB: A Computer Code for Predicting Mineral Formation in Geothermal Brines*, Electric Power Research Institute, Palo Alto, CA.

68. MURARKA, L. (1982). Planning Workshop on Solute Migration from Utility Solid Waste, Publication EA-2415, Electric Power Research Institute, Palo Alto, CA.

69. MURPHY, J. and RILEY, J. P. (1962). A modified single solution method for determination of phosphate in natural water. *Anal. Chim. Acta*, **27**, 31–6.

70. NARASIMHAN, T. N. (1975). A unified numerical model for saturated and unsaturated groundwater flow. PhD thesis, University of California, Berkeley, CA.

71. NELSON, R. W. and SCHUR, J. A. (1980). *Assessment of Effectiveness of Geologic Oscillation Systems: PATHS Groundwater Hydrologic Model*, Battelle, Pacific Northwest Laboratory, Richland, WA.

72. NELSON, D. W., ELRICK, D. E. and TANJI, K. K. (Eds) (1983). Chemical Mobility and Reactivity in Soil Systems. SSSA Special Publication No. 11, Soil Science Society of America, American Society of Agronomy, Madison, WI.

73. NORDSTROM, D. K., PLUMMER, L. N., WIGLEY, T. M. L., WOLERY, T. T., BALL, J. W., JENNE, E. A., BASSETT, R. L., CRERAR, D. A., FLORENCE, T. M., FRITZ, B., HOFFMAN, M., HOLDREN, G. R., LAFON, G. M., MATTIGOD, S. V., McDUFF, R. E., MOREL, F., REDDY, M. M., SPOSITO, G. and THROIL KILL, J. (1979). A

comparison of computerized chemical models for equilibrium calculations in aqueous systems. In: *Chemical Modeling in Aqueous Systems* (Ed. E. A. Jenne), American Chemical Society Symposium, Series No. 93, American Chemical Society, Washington, DC.

74. NORVELL, W. A. (1972). Equilibria of metal chelates in soil solution. In: *Micronutrients in Agriculture* (Eds J. J. Mortvedt, P. M. Giordano and W. L. Lindsay), Soil Science Society of America, Madison, WI.

75. PARKHURST, D. L., PLUMMER, L. N. and THORSTENSON, D. C. (1982). BALANCE—A Computer Program for Calculating Mass Transfer for Geochemical Reactions in Ground Water, Report USGS/WRI-82-14, US Geological Survey, Reston, VA.

76. PARKHURST, D. L., THORSTENSON, D. C. and PLUMMER, L. N. (1980). PHREEQE: A Computer Program for Geochemical Calculations, US Geological Survey, Water Resources Investigations, pp. 80–96.

77. PEREZ, A. I. (1974). A Water Quality Model for Conjunctive Surface and Groundwater Systems, Office of Water Research and Technology, US Environmental Protection Agency, Washington, DC.

78. PLUMMER, L. N., PARKHURST, L. and KOUIR, D. R. (1975). MIX2: A Computer Program for Modeling Chemical Reactions in Natural Waters, US Geological Survey, Water Resources Investigations, pp. 61–75.

79. PLUMMER, L. N., JONES, B. F. and TRUESDELL, A. H. (1976). WATEQF—A FORTRAN IV Version of WATEQ, A Computer Program for Calculating Equilibrium of Natural Water, US Geological Survey, Water Resources Investigations, Vol. 76.

80. PRICKETT, T. A., NAYMIK, T. G. and LONNQUIST, C. G. (1981). A Random Walk Solute Transport Model for Selected Groundwater Quality Evaluations, Illinois State Water Survey, Bulletin 65.

81. SCHULTZ, D. (1982). Land disposal of hazardous waste, *Proceedings of the 8th Annual Research Symposium*, F. J. Hutchell, Kentucky, March 8–10, US EPA, Cincinnati, OH.

82. SCHWARTZ, F. W. and GROWE, A. (1980). A Deterministic Probabilistic Model for Contaminant Transport, US NRC, NUREG/CR-1609, Washington, DC.

83. SILKA, L. R. and SWEARINGEN, T. L. (1978). A Manual for Evaluating Surface Impoundment Contamination, EPA 570/9-78-003.

84. SPOSITO, G. (1981). *The Thermodynamics of Soil Solutions*, Clarendon Press, Oxford.

85. SPOSITO, G. (1985). Chemical models of inorganic pollutants in soils. *CRC Critical Reviews in Environmental Control*, **15**, No. 1.

86. SPOSITO, G., BINGHAM, F. T., YADAV, S. S. and INOUYE, C. A. (1982). Trace metal complexation by fulvic acid extracted from sewage sludge: II. Development of chemical models. *Soil Sci. Soc. Amer. J.*, **46**, 51–6.

87. SPOSITO, G., HOLTZCLAW, K. M. and LeVESQUE-MADORE, C. S. (1981). Trace metal complexation by fulvic acid extracted from sewage sludge: I. Determination of stability constants and linear correlation analysis. *Soil Sci. Soc. Amer. J.*, **45**, 465–8.

88. SPOSITO, G. and MATTIGOD, S. V. (1980). GEOCHEM: A Computer Program for the Calculation of Chemical Equilibria in Soil Solutions and Other

Natural Water Systems, Dept of Soil and Environmental Science, University of California, Riverside.

89. SPOSITO, G., BINGHAM, F. T., YADAV, S. S. and INOUYE, C. A. (1982). Trace metal complexation by fulvic acid extracted from sewage sludge: II. Development of chemical models. *Soil Sci. Soc. Amer. J.*, **46**, 51.

90. STUMM, J. J. MORGAN (1981). *Aquatic Chemistry: An Introduction Emphasizing Chemical Equilibria in Natural Waters*, Ind. Ed. John Wiley, New York.

91. SCHWILLE, F. (1981). Groundwater pollution in porous media by fluids immiscible with water. *The Science of the Total Environment*, **21**, 173–85.

92. TABATABAI, M. A. (1974). A rapid method for determination of sulfate in water samples. *Environ. Letters*, **7**, 237–43.

93. THEIS, T. L., KIRKNER, D. J. and JENNINGS, A. A. (1982). Multi-solute Subsurface Transport Modeling for Energy Solid Wastes, Technical Progress Report, Department of Civil Engineering, University of Notre Dame, Notre Dame, Indiana.

94. THEIS, D. L., KIRKNER, D. J. and JENNINGS, A. A. (1982). Multi-solute Subsurface Transport Modeling for Energy and Solid Wastes, Department of Civil Engineering, University of Notre Dame, Notre Dame, Indiana.

95. TRUESDELL, A. H. and JONES, B. F. (1974). WATEQ—a computer program for calculating chemical equilibria of natural water. *J. Res. US Geol. Surv.*, **2**.

96. TUCKER, W. A., DOSE, E. V., GENSHEIMER, G. J., HALL, R. E., KOLTUNIAK, D. N., POLLMAN, C. D. and POWELL, D. H. (1985). Evaluation of critical parameters affecting contaminant migration through soils. *National Conference on Environmental Engineering, July 1–3, 1985*, Northeastern University, Boston, Massachusetts, USA. Unpublished: available from Dr W. A. Tucker, Environmental Science and Engineering, Inc., Gainesville, Florida, USA.

97. VAN GENUCHTEN, M. T., PINDER, G. F. and SAUKIN, W. P. (1977). Modeling of Leachate and Soil Interactions in an Aquifer, EPA-600/9-77-026, pp. 95–103.

98. WAGNER, J. and BONAZOUNTAS, M. (1983). Potential Fate of Halogenated Solvents via SESOIL, USEPA/OTS Contract No. 68-01-6271. Draft Report by Arthur D. Little, Inc., Cambridge, MA.

99. WESTALL, J. and HOHL, H. (1980). A comparison of electrostatic models for the oxide/solution interface. *Advances in Colloid and Interface Science*, **12**, 265–94.

100. WOLERY, T. J. (1979). Calculation of Chemical Equilibrium Between Aqueous Solution and Mineral, the EQ3/6 Software Package, UCRL-52658, Lawrence Livermore Laboratory, Livermore, CA.

101. YEH, G. T. and WARD, D. S. (1981). FEMWASTE: A Finite-Element Model of Waste Transport Through Saturated–Unsaturated Porous Media, Oak Ridge National Laboratory, Environmental Sciences Division, Publication No. 1462, ORNL-5601, 137 pp.

SESSION IV

Conclusions, Recommendations and Research Needs

Research Needs in Soil Protection: The Federal Ministry of Research and Technology's Soil Research Programme

G. KLOSTER

Ecology Division of Biology, Ecology and Energy,
Nuclear Research Centre Jülich, D-5170 Jülich,
Federal Republic of Germany

SUMMARY

The aim of the Soil Research Programme run by the Federal Ministry of Research and Technology is to bring together researchers from different disciplines to lay an application-oriented scientific foundation for optimizing environmental protection measures, ensuring that they remain effective in the long-term and making more sparing use of the resource soil. The research needs were taken from the Federal Government's Soil Protection Plan and have been announced to all groups concerned.

The main points covered by this research programme are:

A. Ecosystems function of characteristic soils.
B. Quantification of soil pollution.
C. Soil damage by erosion and land use.
D. Burden placed on soil by land use requirements.
E. Evaluation concepts; recommendations and instructions for action.

The idea is for research teams from different disciplines to join forces, either for work at a specific site or on a specific topic, and thus lay the foundations for an ecosystems research programme to be defined in greater detail in due course.

1. INTRODUCTION

Soil protection is one of the top priorities of the Federal Government's environment policy. The Federal Government set out its objectives in this area in its Soil Protection Plan dated 7 March 1985 [1]. This plan gives a comprehensive survey of current soil protection measures and knowhow in the sciences with a bearing on soil protection and specifies which research the Federal Government feels is still needed.

To satisfy the needs identified in the Soil Protection Plan and to give a firmer shape to the 'Soil Pollution and Water Resources' section of the Federal Ministry of Research and Technology's 1984–87 'Environmental Research and Technology Programme' [2], on 17 May 1985 the Ministry issued an announcement to all groups concerned of a new priority entitled 'Soil Pollution and Water Resources (Soil Research)' [3].

2.   THE NEED FOR A SOIL RESEARCH PROGRAMME

Two central forms of action are addressed in the Soil Protection Plan (1):

(1)   minimization of emissions from industry, commerce, transport, agriculture or the domestic sector that cause problems either because of their quality or because of the quantities involved;
(2)   changes in the pattern of land use.

These demands were the result of an analysis of the available data on the load on soil resources. The Federal Republic of Germany is exceptionally densely populated, with 247 people per km^2. Since pollutants are carried long distances in the air and water and the land is intensively used, there are hardly any areas left in the Federal Republic to compensate for this [1].

According to the 1982 figures, every year a wide range of sources emit 0·7 million tonnes of dust, 3·0 million tonnes of sulphur dioxide, 3·1 million tonnes of nitrogen oxides, 8·2 million tonnes of carbon monoxide and 1·6 million tonnes of organic compounds. Every year 44 000 million m^3 of effluent and cooling water is discharged and 47 million tonnes of sewage sludge is produced at sewage treatment works. Households in turn generate around 32 million tonnes of waste a year, to which industry adds a further 50 million tonnes, including 3 or 4 million tonnes of toxic wastes [1].

Year after year some 30 000 tonnes of pesticides (60% of them herbicides) and 3·5 million tonnes of nutrients (including 1·5 million tonnes of nitrogen) are spread, predominantly on intensively-farmed agricultural land [1].

In the recent past, on average around 113 ha of land a day was used for housing. By 1981 there was still no sign of any slowdown [1].

In 1983 the Federal Republic lay fourth in the world with 1·96 km of roads per km² Transport infrastructure covered 4·7% of the total area of the country. In all, no more than 100 or 120 or so 10 km by 10 km plots are left without a main road or railway track cutting through it [1].

These many different loads and demands on the land are all a drain on the same scarce, unrenewable resource: the soil. In view of the many different types of soil and of pollution factors, there is a genuine need to assess the ecological impact of each of these loads on each type of soil. After all, all too often the diversity, geographical spread and time scale of changes in the ecosystem, combined with the irreversibility of so many of the processes, make it impossible, with the knowedge currently available, to predict when, how and through which synergistic effects an accumulation of loads will seriously endanger the functions of the soil. Consequently, in addition to the ongoing research projects on specific measures by the individual Federal and Land authorities, a broad-based scientific research programme has been drafted by the Project Management for Biology, Ecology and Energy at the Nuclear Research Centre Jülich, which will be coordinating the project, in conjunction with the Federal Ministry of Research and Technology, outside experts and the Federal authorities responsible for implementing the lines of action set out in the Soil Protection Plan. The central objective of this research is to lay a scientific foundation for more sparing use of the non-renewable, indispensable resource soil and for preventive soil protection measures.

The idea is to achieve this objective by:

(i)   determining the function of characteristic soils within the ecosystem and the biological communities which depend on these soils;

(ii)   monitoring the effect on the environment from the levels of soil damage caused by immissions, erosion and land use;

(iii) developing methods and models for evaluating the results recorded, however incomplete, as an aid to decision-making on soil protection measures.

In the light of these objectives, the Soil Research Programme must be seen as a programme of preventive research to protect the soil. Even though still too little is known about ecological inter-relationships, this must not mean postponing environmental protection measures until all the scientific details have been clarified. Preventive measures must be taken and reliable environmental protection standards set even if the complex ecological inter-relationships have yet to be unravelled once and for all. But further

advances in environmental research can and should make the environmental protection measures as effective as possible, even in the long term.

3. THE FEDERAL MINISTRY OF RESEARCH AND
TECHNOLOGY'S SOIL RESEARCH PROGRAMME

The earlier ecological research funded by the Ministry already covered
certain aspects of soil research. For instance, in the past the unit which has
now become the Project Management for Biology, Ecology and Energy at
the Nuclear Research Centre Jülich was asked by the Ministry to
coordinate and supervise the programme on:

(i)   methods for the ecotoxicological evaluation of chemicals; and
(ii)   forest damage caused by air pollutants.

Both the Ministry of Research and Technology and the Project
Management took due account of the experience gained in the course of
these two programmes when they came to drafting the Soil Protection
Programme. Particular importance was attached to ensuring that wherever
possible projects to be funded bring together as many different disciplines
as possible to work on a specific topic or at a specific site cover the topics of
interest.

To match the objectives outlined in Section 2, the Federal Ministry of
Research and Technology's Soil Research Programme, coordinated and
supervised by the Project Management for Biology, Ecology and Energy at
the Nuclear Research Centre Jülich, has been subdivided into five areas:

A: Ecosystems function of characteristic soils.
B: Quantification of soil pollution.
C: Soil damage by erosion and land use.
D: Burden placed on soil by land use requirements.
E: Evaluation concepts; recommendations and instructions for
     action.

On 17 May 1985 the Federal Ministry of Research and Technology
published this detailed announcement of the funding scheme [3]: 'The
Federal Minister for Research and Technology *Announcement of 17 May
1985 on the funding of research and development projects on soil pollution and
water resources (soil research).*'

The Federal Government notified the Bundestag:

(i) in April 1984, of the Federal Minister of Research and Technology's 1984–87 'Environmental Research and Technology Programme' (Official Journal of the Bundestag 10/1280);

(ii) in March 1985, of the Federal Government's Soil Protection Plan (Official Journal of the Bundestag 10/2977).

Under the programmes set out in these papers, ecological research into soil pollution and water resources (soil research) is to be one of the priority areas for support in the future. This research is to lay a scientific foundation for more sparing use of an indispensable and scarce resource currently being put at risk by pollution and land use: the soil.

Initially, support will be given to projects in the following fields.

3.1. Ecosystems function of characteristic soils

3.1.1. *Determination of the current situation*

Breakdown of the function of soil within the ecosystem, by area, time and function.

Establishment of typical indicators which, together or individually, could allow a meaningful assessment of the structural integrity of all or part of the ecosystem (e.g. inter-relationships, interspecific stabilization processes and elasticity).

3.1.2. *Impact of soil burden on the ecosystem*

1. Impact of changes in specific soil qualities, and in particular in soil air and soil moisture content or texture (compression), and of changes in chemical, physical and biological processes brought about by immissions and various types of land use.

2. Evaluation, with supporting evidence, of the minimum number of individuals is a population for key communities of fauna and flora and of the minimum land area needed to keep ecosystem compartments intact (assessment of measures which further sub-divide plots of land; environmental impact of land conversion or clearance schemes; and ecological assessment).

3. Impact of soil pollutants on ecosystems.

3.1.3. *Measures to safeguard and restore the ecosystems function of the soil*

1. Means of compensating for any detrimental change in the functions of the soil within the ecosystem.

2. Criteria for differential assessments of the load on the edaphon.

3.2. Quantification of soil pollution

3.2.1. *Pollutant categories*

(i)   Potentially ecotoxic elements.
(ii) Organic pollutants.
(iii) Noxious gases such as SO_2, NO_x, photochemical oxidants, etc.
(iv) Nutrients (such as fertilizers) or pesticides.

3.2.2. *Quantification*

Harmonization of methods of sampling and analysis for the above-mentioned pollutant categories to achieve greater comparability.

Calculation of data on rates of deposition (input), accumulation and leaching (output) of the above-mentioned pollutant categories and on their mobility in typical soils (demonstration sites for specific types of use and degrees of pollution).

Impact of the above-mentioned pollutant categories on physical, chemical and biological soil parameters and assessment of the consequences.

Immobilization, mobilization and bioavailability of the above-mentioned pollutant categories; assessment of the relevance of pollutants bound in the soil.

Load placed on groundwater resources should the above-mentioned pollutant categories exceed the buffer and filter capacity of the soil.

Development of models to predict pollution loads at the demonstration sites, including feedback to the measurements for calibration and adjustment purposes.

Criteria for deriving limit values.

3.3. Soil damage by erosion and land use

1. Extent (calculation of erosion factors; modelling of individual slopes and catchment areas) and mechanism of erosion by water and wind. Drafting of a hazard map for the entire Federal Republic, from remote sensing and surface data.

2. Development of protection measures/strategies.

3. Impact of the type of land use and crop rotation on susceptibility to erosion and on the humus balance.

4. Damage to soil (changes of texture, etc.) caused by agricultural cultivation or sealing of land.

3.4. Burden placed on soil by land use requirements

1. Models to analyse the gross load on the soil to check land use

objectives and limit values in the event of accumulation of pollutants in given regions (linking soil protection, land use and pollution data).

2. Comparison of different land use plans and alternative planning strategies including ecological (see Section 3.1) criteria for planning and safeguarding non-built up areas. Integrated recording of the impact of all land uses at both local and regional levels and indication of the ecologically sustainable limit values.

3. Criteria to assess the prospects of recycling land and rolling back land use (ecological significance of resealing and reclaiming soil).

4. Criteria for detecting and assessing derelict toxic waste sites (detection, mapping and assessment).

3.5. Evaluation concepts; recommendations and instructions for action

1. Assessment of available know-how, including German and non-German literature.

2. Survey of the current situation and need to substantiate evidence for a regional and local soil register.

3. Formulation of evaluation strategies summing up all the individual points mentioned and formulation of coexistence concepts.

4. Formulation of soil use plans, based on the properties, suitability and priority of the soil, for different land use requirements.

5. Measures and procedures to protect undamaged soil or lowly loaded soils.

6. Measures and procedures to reclaim soil.

For further details of the tasks and procedure, and for application forms and advice on how to apply, contact:
Jülich Nuclear Research Centre,
Biology, Ecology and Energy Project Coordinating Unit,
Post Box 1913,
5170 Jülich,
Tel. 02461/61-3298 or -3729.

Applications and project summaries suitable for consideration will be accepted at any time.
Bonn, 17 May 1985
325-7276-33-19/85
For the Federal Minister of Research and Technology Schmitz.

The Project Management appraises the applications received, on behalf of the Federal Ministry of Research and Technology, by peer review and gives the Ministry recommendations on which should be funded. The

Federal Ministry of Research and Technology then takes the final decision on allocation of the research projects.

4. RESPONSE IN EACH OF THE AREAS SELECTED TO DATE

Since publication of the announcement in May 1985, the Project Coordinating Unit has received well over 100 proposals. With the aid of the committee of outside experts (peer reviews) and of the Federal authorities primarily responsible for implementing the lines of action set out in the Soil Protection Plan, the Coordinating Unit has so far singled out more than 40 of the projects as qualifying for funding and recommended these to the Federal Ministry of Research and Technology. Many others are now being appraised to see whether they too qualify.

To gain fresh insight, some of the projects on the *ecosystems function of characteristic soils* are investigating short-term and long-term changes in terrestrial ecosystems on agricultural land. To this end, new mobile instruments are being developed for measuring biotic and abiotic factors in terrestrial ecosystems. Another research project is exploring the significance of mycorrhiza as an indicator of site characteristics of agricultural ecosystems and for sustaining soil fertility. A series of projects on non-agricultural ecosystems are dealing with the consequences of soil pollution for dry and semi-dry grassland, with synecological surveys of wet grassland habitats or with surveys on the impact of disused land in towns and cities on soil ecology. One other important area of research into measures to protect habitats and provide a basis for assessing the impact of human intervention in nature and landscape is covered by a project to determine the minimum land area required by specific communities of fauna.

On *soil pollution* a series of projects are designed to investigate nutrient (particularly nitrogen), pesticide, and heavy metal inputs, under various soil conditions and for different types of land use in agriculture or forestry. For instance, studies are being conducted to assess and construct models of the ecological impact of nitrate loads on intensively farmed land with high liquid manure input or on vineyards. Alongside a comprehensive joint project to develop suitable assessment criteria for registering the ecological impact of long-term pesticide use, strategies are also being formulated to minimize herbicide use on special agricultural crops. Furthermore, a survey is being conducted to ascertain the amounts and types of pesticides spread on farm and forest land. Another comprehensive joint project is studying the impact of long-term exposure to heavy metals from urban waste and sewage sludge on soil, soil organisms and plants including speciation of the metals bound. One project is trying to devise a generally applicable method of analysis for organic pollutants in soils. Work is also under way on the degradation of chlorinated phenols in groundwater. Lastly, suitable models are being developed to determine and assess the links between water balance, dynamic behaviour of the substances and pollutant tolerance, on the one hand, and climate, soil properties and type of application on the other.

The research into *soil damage by erosion and land use* is concentrating primarily on agricultural land. For instance, some projects are looking into the impact of farming methods on the susceptibility of various soils to water and wind erosion and into the impact of mulch layers as a means of preventing erosion. Erosion hazard maps are also being drawn.

Several extensive joint projects are assessing the long-term impact of various ways of soil tillage on the soil ecosystem, the load, strain and compression of the soil by farm machinery and the impact of the various ways of soil tillage on the chemical properties of organic matter in the soil.

To learn about the *burden placed on soil by land use* requirements, surveys are being conducted to assess the impact of infrastructure schemes and ways of controlling land planning. One other topical, application-oriented field of research in parallel with the development work to clean up derelict toxic sites is the work to construct models to assess such hazardous sites adequately.

Many of the above-mentioned research topics partly overlap with the work on *evaluation concepts, recommendations and instructions for action.* As a result, there are very few projects in this field alone. Amongst other things, they are looking into the basic requirements for local soil protection plans, soil pollution in conurbations and the ways open to the public authorities to set aside and safeguard ecologically important land.

The projects approved so far will be presented in greater detail at the status seminar planned in early December 1986.

5. OUTLOOK

One typical feature of the environmental problems which industrialized countries have to face in particular is that often they no longer take the form of damage triggered by a cause-and-effect relationship which can be traced back to a single source, but, increasingly, are due to the effects of chronic, low doses which endanger entire ecosystems and public health [4].

In the past, environment policy and environmental research have scored notable successes by means of measures or scientific studies confined to specific sectors or media. But all too often the scientists involved have not looked at integral parameters of ecosystems, namely the interdependence and feed-back control of partial processes.

Findings to date from the research into forest damage caused by air pollutants [5, 6] have demonstrated how little is known about the normal state of major ecosystems and their natural fluctuation and stability limits. Long-term research is needed, in the form of a joint multidisciplinary or interdisciplinary project bringing together as many different lines of research as possible in order to make statements about the function of forest ecosystems that stand up to closer scrutiny.

The Soil Research Programme described in this chapter pursues much the same objectives: to conduct joint multidisciplinary or interdisciplinary research in order to make statements about terrestrial ecosystems which stand up to closer scrutiny. Since soil performs so many functions and has to meet so many demands, it is an ideal field of research for pioneering the transition from environmental research to ecosystems research.

In the future, ecosystems research, i.e. the establishment of element and energy fluxes as well as description of the structure, dynamic behaviour, control mechanisms and stability criteria in ecosystems, will play two key roles in any preventive environment policy designed to solve the current environmental problems.

1.   The clearer understanding which it will give of the function of ecosystems can help to direct environmental protection measures towards strategic areas and to adapt the load limits to ecological needs.
2.   Greater insight into the principles of function and regulation in ecosystems can give important clues as to ways of making future technologies more environmentally acceptable [4].

At the moment the Federal Ministry of Research and Technology is drafting a plan for ecosystems research along these lines, covering several different media.

REFERENCES

1. Bodenschutzkonzeption der Bundesregierung, Bundestags-Drucksache 10/2977, 07.03.1985; W. Kohlhammer Verlag, Stuttgart, 1985.
2. Umweltforschung und Umwelttechnologie, Programm 1984–1987, Bundesminister für Forschung und Technologie, Bonn, 1984.

3. Förderung von Forschungs- und Entwicklungsvorhaben zum Schwerpunkt Bodenbelastung und Wasserhaushalt (Bodenforschung), Bundesminister für Forschung und Technologie, Bonn, 1985.

4. VON OSTEN W., RAMI, B., Ökosystemforschung, eine notwendige Weiterentwicklung der Umweltforschung. *Allg. Forstzeitschrift*, **41**, 535 (1986).

5. Umweltforschung zu Waldschäden, 2. Bericht (Eds.: F. Führ, G. Kloster, H. Papke, B. Scheele, E. Stüttgen), Bundesminister für Forschung und Technologie, Bonn, 1985.

6. Forschungsbeirat Waldschäden/Luftverunreinigungen der Bundesregierung und der Länder: 2. Bericht (May 1986), Kernforschungszentrum Karlsruhe, 1986.

Report on Session I: Soil Protection—A Need for a European Programme?

P. BULLOCK

Soil Survey of England and Wales, Rothamsted Experimental Station, Harpenden, Herts AL5 2JQ, UK

SUMMARY

The first day of the Symposium was concerned mainly with an introduction to the soils of the CEC and an overview of the threat to their quality. Several speakers drew attention to the sensitive nature of some of the soils and examples were given of their degradation. The need for more monitoring to determine the extent of the reduction in soil quality was identified. More extensive databases for many of the problems endangering soil quality are necessary. Research is needed in some areas of soil erosion and soil pollution and there is a need to develop more adequate predictive models. Attempts should be made to foster more collaboration between institutes in the CEC and to standardise methods as much as possible.

1. INTRODUCTION

Soil and water are probably the world's most important natural resources. It is generally accepted that such natural resources should be safeguarded for present and future generations. Yet there are increasing signs that the quality of some soils in the European Community is becoming degraded. It is therefore timely that the Commission of the European Communities (CEC) should organise this Symposium so as to take stock of the current state of soil degradation.

There have been several periods in history when changes in land use have led to deterioration in soil quality but pressure on the soil since World War

II has been unprecedented. Intensification of crop production aimed at preventing future food shortages in Europe which, having the desired effect of increasing food supplies, has had detrimental effects on some soils. Increasing use of heavy machinery, loss of organic matter and decreasing biological activity has led to compaction and soil erosion.

The soil in this same period has received large quantities of fertiliser, particularly nitrogen, and has acted not only as a medium for uptake of nutrients by the plant but also as a pathway by which unused fertiliser components find their way to the groundwater. There is increasing concern, for example, about the levels of nitrates in the groundwater in many areas.

The soil has also been used more extensively than hitherto as a receptacle for waste from cities and industry. Sewage sludge and fly ash, for example, have been added to the soil, often without sufficient knowledge of the suitability of soils for this disposal. The consequence is that in some areas the soils have accumulated large amounts of heavy metals, sometimes in toxic quantities.

In addition to the above problems concerning soil quality, urbanisation continues to increase and remove land, often of good quality, from future availability for agricultural or forestry production.

The Symposium formed an important forum at which scientists and policy makers could assess the extent of the problems, and the measures that might be necessary to control and ameliorate them.

2. THE SOILS OF THE EUROPEAN COMMUNITY

An introduction to the soils of the Community was given by Professor Tavernier, one of the principal driving forces behind the publication of the 1:1 Million Soil Map of the European Communities. A range of physiography, geology, climate and vegetation ensures an intricate pattern of soils in the Community and the map contains some 312 units. Lee (see pp. 29–63) reported on the suitability of the main soil units for specific purposes, particularly arable cropping and grassland. Only 20% of the arable land is Class 1. The corresponding value for grassland is 28%.

The 1:1 Million Soil Map provides a basis for making very broad assessments of the suitability of the soils of the Community for particular uses. It is a broad planning tool and is unsuitable for use when more detail is required at a national level.

Most of the member countries of the Community have carried out some detailed soil surveys. A summary of progress on these is given in the book that accompanies the 1:1 M Soil Map. The distribution of detailed soil information is uneven, as is the scale and level of detail. Yet it is so important that each member country has available comprehensive information on the soil types, their distribution, properties and potential uses. The CEC is in a good position to encourage and stimulate the completion of a full cover of detailed soil surveys in each member country. Only when this is achieved will there be an adequate framework for the transfer of information and experience in one area to other areas of similar soils within the Community. It also provides a basis for establishing monitoring programmes, for identifying the sensitive soils of the Community and for predicting the implications of a given legislation or a change in land use on the soil.

3. IDENTIFICATION OF THE SENSITIVE SOILS IN THE COMMUNITY

It would be wrong to give the impression that all practices associated with intensive agriculture are harmful to the soil. Many practices, in fact, give rise to beneficial conditions (Tavernier) compared to those in the natural state, e.g. liming to combat leaching, drainage to overcome wetness.

Some practices are, however, detrimental especially when carried out on a regular basis on sensitive soils. Lee reported that intensive agriculture is being carried out on sensitive soils in several areas. Cambisols and Luvisols, which together occupy some 60% of the EEC-12 area, are subject to erosion hazard under certain conditions, e.g. in Spain and France. Soils with gleyic and vertic characteristics are also intensively cultivated and subject to deterioration, e.g. in England and Italy. Coarse textured podzols with a high sensitivity to nutrient leaching, groundwater pollution and windblow are among the most intensively farmed soils.

Both Fedoroff (see pp. 67–85) and Yassoglou (see pp. 87–122) in their respective papers recognised the need to identify the sensitive soils of the Community. Fedoroff expressed the view that each soil system reponded uniquely to intensive cultivation and that soils with high productive potential were more prone to degradation than other less productive soils. Some highly productive soils under intensive arable cultivation have undergone degradation of soil structure, loss of organic matter, reduction in biological activity and as a result have become fragile, sensitive, and liable to erosion. Yassoglou assessed the sensitivity of the main soil taxonomic units of Southern Europe to degradation. He identified the following main causes of serious soil degradation: loss of soil volume; degradation of soil structure; loss of organic matter and biological activity; chemical degradation; and soil fertility degradation. Soil erosion is considered to be the main degrading process, becoming intense where heavy rainfall and vegetative cover are out of phase, especially on steep slopes. Yassoglou examined in detail the sensitivity of five pedons.

Both the papers by Fedoroff and Yassoglou and references within other papers during the first day of the Symposium made a strong case for the need to identify the sensitive soils within the Community. This can be done very broadly using the 1:1 M Soil Map but can be achieved much more accurately using the more detailed soil maps available in each member country. Given a strong soil information base, such as that derived during national detailed soil mapping programmes, it is possible to predict how a particular event, e.g. changing from arable cultivation to a particular type of forestry, will affect the soil.

4. SOIL EROSION

The aspect of soil degradation most extensively dealt with on Day 1 of the Symposium was soil erosion. Morgan (see pp. 147–54) gave a good introduction to the problem in a European Community context. He called for a more effective method of survey combining techniques of erosion assessment with erosion rate prediction. There is clearly a lack of adequate information with which to develop realistic models for predicting various aspects of erosion. The USLE model, widely used for several years, is now considered to be unreliable for predicting rates of erosion. The Stocking and Pain model used by Morgan suffers from dependence on soil loss tolerance assessments. There has been insufficient research to establish a sound base for soil loss tolerance prediction.

The concept of 'ultimate soil degradation' is an interesting one but it needs to be defined and researched better before it becomes useful practically.

A number of questions about erosion remain unanswered. Firstly, it is necessary to determine the scale of the problem. There have been several reports of serious erosion within the European Community but so far there is no informed overview of the extent of the problem. This leaves policy makers without an adequate basis from which to judge the severity of the problem.

It is important to establish the rates of erosion. As Morgan pointed out, there are very few data based on observations over a number of years.

There is much research experience from the United States on which to draw, although there seem to be no clear answers yet to the following questions: Is there a productivity loss associated with erosion? If so, at what stage in the erosion process does a decline in productivity become apparent? To what extent is the increase in soil erosion related to current farming practice? At what stage does the situation become irretrievable? Answers to these questions would go some way towards making Morgan's concept of 'ultimate soil degradation' a more useful one.

Too little attention has been given to ground research on which to base sensible, useful models and better models based on sound scientific information are essential. In particular, attention needs to be directed toward improving estimates of soil loss tolerance.

Finally on the question of soil erosion, there is a need to assess conservation measures applicable to the European Community both north and south of the Alps. There is much experience of soil conservation in countries outside the European Community. This should be assessed before embarking on an extensive research programme. Much of the erosion occurring in the Community could be reduced by the adoption of a few simple conservation measures.

It is well known that erosion south of the Alps has been extensive and severe. The extent to which this relates to historical events or to current environmental conditions is not well established. North of the Alps erosion is less severe but there is growing concern at the amount that has been initiated in the last 40 years. In both areas, erosion needs to be put into perspective and this can only be achieved by a sound monitoring and research programme.

5. SOIL POLLUTION

The theme was not considered in detail on the first day of the Symposium, apart from the interesting paper by Fränzle (pp. 123–45) on 'Sensitivity of European soils related to pollution'. In it the author outlined strategies for determining sensitivity-related soil variables. These strategies can be accomplished by combining the use of representative soils as test media and representative reference chemicals as components of test systems. The author used comparative leaching experiments in laboratory lysimeters and on experimental plots to define the relative importance of sensitivity related soil variables. An analysis was made of the 2, 4-D impact on a gleyic Luvisol and a ferric-humic Podzol.

As with soil erosion, there is a need to monitor the levels of various pollutants in the soils of the European Community. A database needs to be established which could be updated on a regular basis.

Much research and development needs to be carried out on pollutants and their interaction with the soil. Research into the fate and behaviour of many pollutants in the soil and their effects on the physical, chemical and biological properties of soils should be encouraged. Soil quality needs to be better defined in terms of such soil properties.

6. THE NEED FOR COHERENCE, QUANTIFICATION AND STANDARDISATION

On the first day of the Symposium and throughout subsequent days, three pleas were commonly expressed—for coherence, quantification and standardisation.

Most speakers identified the need for better collaboration between institutes so as to avoid wasteful duplication and overlap of research effort and to foster discussion between scientists with similar interests. The CEC has recognised this need and has made some funds available for exchange of visits between scientists. Further such stimulus is needed.

The second plea concerned quantification. Too often, qualitative statements are made and there is little or no support from a well researched sound database. Although there are numerous reports suggesting that soil quality in the European Community is at risk, there are often insufficient data with which to ascertain the scale of the problem. Policy makers have a right to and a need for a sound database on which to act. The CEC is encouraged to support the collection of quantitative information on the main problems facing soil quality and to develop databases which can be used to provide on-line assessments of the scale of the problem with regular updatings.

Finally, standardisation of methodology is a prerequisite to good international collaboration. So often different laboratories use different techniques making comparison of results difficult, if not impossible. It would be valuable if the CEC could take some initiative in stimulating the standardisation of procedures.

7. SUMMARY OF RECOMMENDATIONS

1. CEC member countries should be supported in their endeavours to provide complete detailed soil maps as a basis for determining the overall

soil resources, the distribution of sensitive soils, the state of land degradation and pollution and for predicting the range of land uses and management systems for a given area of land.

2. Soil erosion is a serious problem both north and south of the Alps. It is important to establish how much of the erosion is recent and the factors responsible for it. There is a need for more realistic models based on erosion research and development. Yassoglou has identified five areas of research pertinent to Southern Europe. The question of soil loss tolerance and the relationship between productivity and erosion need to be addressed. Conservation measures tried and tested elsewhere should be examined for their relevance to the Community before embarking on extensive new research.

3. A monitoring programme needs to be established to determine the extent of soil pollution and a flexible database set up. Research is required on the influence on and the fate of pollutants in the soil.

4. Attention has been directed towards chemical and physical properties of soils but biological properties have been neglected. There is a need for more information on biological activity in the main soils and the effects of change in land use or the addition of pollutants on it.

5. Collaboration between institutes in the CEC should be further supported and attempts should be made to standardise methods. There is a need to provide a stronger quantitative base for the studies in the CEC.

6. It is important that better bridges are established between the scientist and the policy maker so that the latter can make better use of current and future scientific information.

Report on Session II: Assessment of Impacts on the Soil Environment

G. VIGNA-GUIDI

CNR, Institute for Soil Chemistry,
Via Corridoni 78, 56100 Pisa, Italy

and

M. SHERWOOD

An Foras Taluntais, Johnstown Castle Research Centre,
Wexford, Ireland

The number of different topics covered by papers presented in this session shows the multiplicity of the possible interactions which can occur between man and the soil.

Unfortunately the various human activities are seldom without any effect on the soil and their environmental impact must be evaluated. Since the soil is a natural resource, largely not renewable, it must be protected against any risk of irreversible deterioration.

It was highly appropriate that the review papers asked for from the speakers by the Commission were divided into three broad areas, each of which represents one of the possible ways in which man interacts with the soil, i.e. (i) through the production of food, fibre and wood (agriculture and forestry), (ii) through the extraction and transformation of natural resources (mining and industry), and (iii) through man's mere existence and lifestyle (towns, infrastructures, recreational areas).

1. AGRICULTURE AND FORESTRY

The first paper presented by Balloni and Favilli reviewed the effects of some agricultural practices on biological soil fertility. All the parameters related to the soil structure are modified by agricultural practices especially by tillage and crop rotation. The few data available seem to indicate an increase of microbial population which is accompanied by a more rapid release of nutrients. On the contrary monoculture depresses soil biological fertility which is maintained at a good level by a microbial biomass of 10 t/ha. It is concluded that agricultural practices should try to keep in the soil a good biological fertility for its beneficial effect both on soil structure and on plant health and development.

The second aspect taken into consideration in this session was the impact on the soil environment by the use of sewage sludge and agricultural wastes. Sauerbeck affirmed that these materials must be considered not only for the toxic elements, inorganic and organic, they carry into the soil but also for the nutrients, an excess of which may cause pollution of soils and eutrophication of waters. Concerning the addition and accumulation in the soil of organic and inorganic pollutants, in Sauerbeck's view it is not yet possible to separate clearly safe and unsafe concentrations of such compounds in soil. While researches to clarify the unknown aspects of soil pollution are needed the concentration of hazardous compounds in soil should remain under the present limits whenever possible. Compounds of ecological concern should therefore be eliminated at the origin and should not reach waste waters and sewage sludge.

The wrong or excessive use of mineral fertilizers and pesticides may cause chemical degradation of soil. Regarding mineral fertilizers, de Haan stressed that imbalance between the crop's requirements and the supply of nutrients is one of the main reasons for their accumulation in the soil. Pesticides can spoil quality through their degradability, toxicity and mobility. In de Haan's opinion the chemical degradation caused by the use of mineral fertilizers and pesticides is less important than the damage caused by industrial and agricultural waste disposal, and pollution from the air. In order to better define the quality of the soil and its protection, there is a need of further researches on soil buffering capacity, compound speciation, soil heterogeneity and bio-availability of pollutants.

Miller considered the effect of forestry practices on the physical, chemical and biological properties of soil. The agricultural practices on forest soils are lighter than in agricultural soils and neither fertilizers nor pesticides are generally used. N, P, Mg, Ca requirements were discussed for some different plant species. Miller's conclusion was that attention should be given not to cause erosion through clear cutting, fire and road construction.

The use of agricultural practices such as irrigation and drainage for the improvement of wet and saline soil was tackled by Pereira, Souza and

Pereira. After a general review on the management of wet and saline soils Pereira spoke about methods for solving and minimizing problems in the Portuguese conditions. The author's general feeling is that too little is known about crop growth in such different environments.

After the first group of papers the discussion focused on two main points.

The first point was the need of further research on ecological and environmental aspects of forestry especially in the Mediterranean area where problems related to forest fires should also be studied in greater detail.

The second point was that the knowledge of the total amount of heavy metals in soil seemed not to give a true appreciation of the level of soil pollution and the use of chemical methods more related to metal bio-availability was requested.

2. MINING AND INDUSTRY

As first aspect of the impacts exerted on the soil by industrial activities, Franzius spoke about the impact of hazardous waste disposal on the soil environment. Because of the potential risk of environmental pollution of soil and water surrounding abandoned hazardous sites, this problem is considered of first priority in the Federal Republic of Germany. Franzius reported that in Germany the estimated cost of action to solve the problem of the 5000 abandoned sites would cost about 20 milliard DM. While the monitoring of abandoned sites is relatively simple the assessment of potential risks is much more complicated.

Finnecy regarded in detail causes and nature of incidental and accidental soil pollution as a result of industrial activity. For the reclamation of incidentally and accidentally contaminated land, financial and practical considerations are however accompanied by a lack of adequate scientific knowledge in what he identified as the most important problem areas: (i) the difficulty of identifying and quantifying the nature and the extent of soil pollution, (ii) the difficulty to assign priority for action when a large number of polluted sites are present in a given area, and (iii) the generalized lack of 'trigger' levels especially for organic contaminants and the difficulty to relate these levels to human health.

The environmental impact of mining industries was seen by Laville-Timsit from the point of view of the estimation of soil contamination caused by the spoil which remains after the extraction of minerals. The main conclusion of her work, based on French data, was that the level of a given element could be high in the soil but it was generally not increased by the mining activity and depended primarily on the regional geochemical background.

The effects of air pollution on soil is one of the most important causes of concern in north and central Europe. Ulrich reported that the emission of soil pollutants of anthropogenic origin is about tenfold the natural one in central Europe. Deposition on forest soils is often twice that on agricultural soils due to the particular microclimatic conditions caused by the forests. Heavy metals accumulate mainly on the top layer because they interact with soil organic matter. The increased acidity on forest soils has caused the leaching of nutrients and the formation of acid zones near plant roots. To restore the fertility in forest soil a large scale project of liming and fertilizing is needed accompanied by a reduction of air pollutants in the industrial emissions. Crops and agricultural soils are however less sensitive to the effects of air pollutants.

The discussion that followed this paper pointed out the lack of information on the depth reached by the acidification and on how much the emission of NO_x and SO_2 should be lessened to decrease the danger of the acid deposition on forest soil.

3. MAN

The need of recreational areas is becoming an important aspect in Europe as the standard of living has increased. But often these areas are created without taking into account in a proper way the environmental damage they cause in the soil. Cassios reviewed soil characteristics such as erodibility, trafficability, texture, slope, drainage, etc., which must be considered in planning recreational and touristic activities in order to minimize risks of erosion, compaction, loss of organic layers, soil microfauna destruction, etc. Managerial, technical and engineering actions should be taken to prevent future severe damages by the users. Cassios urged the production of soil suitability maps for the best use of these areas.

Soil erosion was again the main topic of the paper presented by Gabriels on the effect of human activity on soil. Gabriels stressed the influence of road construction, residence development, dams, etc., on soil erosion and pointed out that up to now researches on this theme have been carried out mainly on agricultural lands. Drainage and a vegetative cover were proposed as practices to reduce erosion and surface runoff. To help the growth of the vegetation in the soils subjected to wind erosion, Gabriels proposed the use of hydroseeding methods.

The final paper in this session presented by Roquero was a comprehensive review about the complete cycle of irreversible soil erosion caused by the different human activities. All different steps leading to the irreversible soil erosion were discussed with a particular emphasis on the parameters which cause erosion and desertification in the Mediterranean area. Roquero submitted that erosion in the Mediterranean area must be studied better and soil conservation in this area must receive more attention by the Commission.

The discussion after the second group of papers was mainly focused on the erosion and on the different aspects that such a phenomenon can assume in north and south Europe. The use of heavy machinery, the destruction of terraces and the agricultural utilization of steep slopes were reported to be the main causes for the increase of the erosion processes in the Mediterranean area.

4. OVERALL CONCLUSIONS

From the papers. The papers dealt with a wide range of topics but were mainly concentrated on *soil erosion* and *soil pollution*.

Those papers which dealt with *soil erosion* showed that:

(1) Deforestation and overstocking;
(2) Continuous arable cropping;
(3) Fires;
(4) Construction of roads and recreational areas

are mainly responsible for soil erosion in EC countries. The areas affected by each of these activities were not quantified during the symposium.

The papers which dealt with *soil pollution* highlighted the problems which arise from:

(1) Inorganic contaminants (nitrate, phosphate, heavy metals);
(2) Organic contaminants (mainly pesticides);
(3) Acidification.

From the discussion. Attention was drawn to the need for standardization of definitions (e.g. what is soil quality) and methodologies between member states and the influence of soil type on the retention and movement of pollutants.

Overall Conclusions

D. J. KUENEN

President, Council for Environment and Nature Research (RMNO),
Huis te Landelaar 492, 2283 VJ Rijswijk, The Netherlands

Ladies and Gentlemen,

You have just heard the summaries of the Rapporteurs of the sessions of this symposium. They have summed up a number of facts, ideas and suggestions to which each of you, no doubt, would like to add your personal interpretation of all we have been hearing.

It has been a good symposium and most of you will think that it has been quite long enough. I have been asked to present a final paper and I will do so, in spite of your fatigue, using, of course, a great deal of the material which has been presented. I will frequently say things that refer to the papers and discussions of this meeting without referring to who said so. Those of you who have listened carefully—and still can listen at this late hour—will know when I am quoting.

Man can live on earth because of the possibilities the biosphere provides. This biosphere has developed in the course of more than 3000 million years. Man is a latecomer in this period and only a few million years ago organisms appeared which were clearly the ancestors of present man. Modern man, *Homo sapiens*, has been around for only about 100 000 years. In the beginning man lived in an environment to which he was reasonably well adapted. He functioned like any other fairly large omnivorous mammal, eating, moving, producing organic waste products and dying without leaving much of a trace of his existence behind.

It is only since about 15 000 years ago that he has been actively changing his environment, primarily through agriculture. That was the beginning of the development which led to such problems as we have been hearing about these last three days.

Agriculture means removing the natural vegetation and replacing it by cultivated crops. The aim was to ensure a more stable food supply. One of the consequences was excessive erosion. Not all erosion is bad. In fact all our soils are the result of that fundamental process.

Many of the early civilisations have developed in river deltas which are the result of erosion processes higher up in the mountains. With a modest population, the situation is satisfactory. Continuous cropping is possible because of the permanent supply of silt, produced by erosion. But when the population begins to increase and the surrounding hills are cleared of forests, that is when the problems begin. The clearing of forests increases the rate of run-off of water, reduces the infiltration of water into the soil, thus causing the groundwater supply to be reduced. At the same time excessive siltation enlarges the swamps around the habitable areas in the delta, which increases the chances of malaria epidemics and other water-borne parasitic diseases. The discussions about erosion yesterday afternoon do not invalidate this picture.

Around the Mediterranean we can see the catastrophic results of unwise use of soils, and much of the poverty there is due to overexploitation in ancient times. The landscape has changed fundamentally. The Greek culture could never have developed in a landscape as we now find in Argos, in Attica, and neighbouring areas.

The disastrous flood in Florence, some time ago, which did so much damage to the unique cultural heritage of that beautiful town, can be said to have been due to the Punic wars. The Romans had to build a large fleet to subdue the Carthaginians and to obtain the necessary timber they had to cut down large tracts of the Apennine forests. Subsequent loss of soil and the consequent incapacity to regulate water flow, much later, resulted in this catastrophic flood.

We can now see what the results are of the misguided efforts of those times. What we are doing now will be judged as mischievous by those who come after us. Because now we know what we are doing. We are now deliberately degrading what should remain to be the basis for life for those coming after us and not only through erosion, but also by dumping wastes, compaction, covering with buildings and asphalt and many other ways of maltreating soils.

By the way, it is curious that so little has been said about underwater soils. It was mentioned once or twice. The problem of accumulation of toxic materials in lowland rivers and harbours may be particularly serious in the Netherlands, but surely it is as bad in other places where upriver factories produce toxic waste which mingles with what gets into the water in harbours to form a reservoir of unmanageable toxic waste. All this is due to the development of industrial technology in the last 200 years with the result that all over the world soils are now being degraded at a really alarming rate.

The summaries given these three days however have not exposed all there can be said about the state of our soils, the rate of deterioration and the methods there are for improving the conditions of that source of life. But what has been said is already enough and nobody today who can read and listen can ignore the warnings. Compaction, drying, erosion, toxification, acidification, hypertrophication, covering for housing, industry and transport: we are all involved directly or indirectly in these processes. The countermeasures, though well thought out and seriously applied, are tragically insufficient to stop the process and certainly to reverse it on a large enough scale.

We know a great deal about our soils. But soil is one of the most complex of biosystems on earth. Its history is older than life on earth. Clay minerals were certainly involved in the beginnings of life as we now know it. It is in and on the soil that all the essential processes of the biosphere take place, specifically the pathway back from organic to inorganic state of the chemical elements. It is there that we find the greatest diversity in both prokaryotic and eukaryotic life and an unnumbered mass of species pass some part of their life cycle in the soil.

Classifying soils may seem a somewhat academic enterprise. But if we want to know in what state a dynamic system is, we must have some kind of reference scale. To know what the effect is of some activity in a specific area we must be able to measure it. We must also know what is the potential use of a soil before we decide whether a planned activity is compliant with the local situation. Damage to soil can only be quantified if we can compare the quality of that soil as we find it, with the potential use as we wish it.

Soil quality can be assessed in different ways: structure, composition of minerals and biological elements such as carbon dioxide production, the composition of the vegetation it supports or the numbers of certain species or groups of species we can find. Further study is essential to see which methods or combination of methods yields the best results.

Three days of papers presented on knowledge of the soils, a number of reviews of what is being done and not done in a number of countries, three summaries of the three sessions into which this symposium was divided, provide the basis for a concerted action within the European Community. The message is there. We must do something drastic if we wish our world and in particular our relatively small European world to remain functional.

The papers will, in due course be published and the organisers of this symposium will, no doubt, produce a summary of the contents of the different presentations.

The number of suggestions for action is too great for all to be undertaken. A selection will have to be made because of lack of resources.

In selecting items a few criteria can be applied. The first should be that it should approach the soil as a dynamic system. Only then can we learn what we are dealing with. The second is that it must be multidisciplinary, or interdisciplinary if you wish. We must see soil as a complex with many potential functions. Analysis is a first priority in all scientific research. Without synthesis, however, it will remain fairly sterile. In the third place international cooperation must be a criterion. Not because that sounds good to the politician. It should be that way because then the research worker broadens his horizon and sees his specific knowledge more as a part of the whole.

I suggest that five of the scientists who have been here should be called to Brussels at short notice to work out a research plan. Even if the projects do not come in until December they should prepare their recommendation well ahead of the process of decision taking.

May I ask those who are going to do the job to remember that you must know how men or women are going to act. Only if the sociological aspects of the problem, reaction to changes, acceptance of restriction, necessity of collaboration of citizens, are carefully worked into the plans, only then will anything positive come out of the efforts.

All this implies more strategic thinking which means long-term concerted effort of all kinds of research workers and policy makers.

Some urgent problems have been named by Rapporteurs. The most urgent problems seem to me:

—classification of soils;
—the relations between agricultural practice, geohydrology and water resources;
—the function of landscapes and functioning of soil biosystems;
—the treatment of excess manure;
—the handling of sewage sludge and underwater toxic silt accumulation;
—the recycling of organic material where this is functional;
—the changes in the structural factors of European agriculture in relation to the future of soils in Europe as an entity;
—further studies of toxic chemical products and ways to stop them coming out of the factory;
—standardisation of methods and modelling-coordination.

I want to add that some of the problems raised and research suggested could come better under other programmes, such as acid deposition, etc. The facts presented concern the knowledge of the functions of the soils, the degradation which is taking place, and what we should do to prevent this further continued degradation and how to improve the function of our soils upon which so much of our health and future life depend. What is known is sufficient to make it obvious to our policy makers that a number of things must be done. Further research will increase our knowledge and thus make it possible to improve methods of dealing with the problems as they present themselves.

It has been asked whether we do not know enough to go into action, and spend the available money on applying what we know. I think the answer to that is that we cannot do without continuing research. It takes 10 or more years for a good research project to be thought out carefully, for it to be funded, executed, accepted by the scientific world and by the policy makers, and used in application.

We must now begin to initiate the research which we will need in 10 years' time. If we were to stop all, that would mean that soon we would not have the reserve of knowledge to solve the problems of the future.

We have heard a number of expositions from policy makers on what is being done and we know that enormous amounts of energy and money have gone into what has already been undertaken. We also know that all this is tragically insufficient if we look at the extent, the urgency and the scale of the problem as it has been presented to us here. We must realise that most of the processes which caused the damage in the past are continuing, even if in a number of cases activities have been stopped.

It is worth while to see if we can become clear in our minds why there is a gap between what scientific research provides in the way of data and what is performed in the field through the activity of the policy makers. One of the reasons is the fundamental difference in the way of thinking between the scientist and policy maker who has to work with the results of scientific research.

For science it is essential that there be a diversity of opinion. It is because of the doubt about an answer to a question that science and scientific research proceed. The moment that doubt vanishes, that part of science stops dead. The history of natural sciences shows us how uncertainties which seemed to be resolved, later, on closer scrutiny, were shown to be uncertainties all the same. When doubt appears, science can proceed. The policy maker on the contrary expects the scientists to remove all doubts. The conscientious research worker is always dubious when asked to supply simple quantitative data to the layman. He knows that his results are imprecise and that further research will improve the quality of his results. He is also aware of the context within which his data are valid. It is his responsibility for the quality of his data which makes him reluctant to come up with exact figures. What he has to learn is to help the policy maker to interpret the data in the right way. This implies that he should formulate in an understandable way how far he could be wrong and why. He cannot and need not be absolutely certain.

It is an accepted fact in some sciences that results cannot be absolutely certain. Economic research is an example. Policy makers have learned how to deal with the economists' data. They should also talk with natural scientists and learn how to act on the data they present.

To estimate the accuracy of a result is an accepted part of scientific procedure. It will have to be applied (to the results) for practical decisions in such a way that the integrity of the research is maintained while the policy maker can use the data for his part in the process of achieving improvement in our way of life.

In the field of soil science the complexity of problems is only too evident. There is an enormous diversity of soils in Europe, there is a great variety in agricultural practices and the ways governments function are wide apart, even when the principle of democracy is the basic system. But if we concentrate on the fundamentals there are certainly overall rules which should be applied. Even a cursory look through the presentations will show us what these general principles are.

However, we are aware that what should be done probably will not be done. There is something in particular which makes it difficult for us to conceive constructive plans. That has to do with the fact that so many people believe their responsibility to be restricted to only a small field of personal concern.

Primitive man was a generalist. He and she each knew all there was to be known. As culture began to develop specialisation stepped in. Some individuals were better at certain jobs, metallurgy, weapon making, stalking prey, some specialised in religious matters, healing or artistic functions. This was of course essential for further development of all parts of culture. But with that specialisation developed a specific responsibility for that part of the total of crafts which one could perform best. With the concentration of specific knowledge in certain individuals and groups the feeling for the community as a whole receded. It is a widespread characteristic of our present society.

We all have a restricted feeling of responsibility. The farmer and his associates feel responsible for food production and do not feel responsible for the quality of drinking water or wildlife. The industrialist feels responsible for his shareholders, consumers and personnel, but not for the soil which is polluted by the unavoidable waste of the chemical processes which are necessary to manufacture his products. The conservationist feels his responsibility for wildlife and nature conservation and pays too little attention to the economic consequences of his ideas. The administrator feels responsibility for his unit which competes with other units of equal importance for the same restricted amount of money, space and influence on the organisation of which the unit is part. This results in competition within our democratic structure becoming a frustrating factor for all processes of government, be they local, national or international.

It is the environment in general and certainly the future of our soils in particular, which will in the long run be the element to suffer most from this situation, because for too many the environment still is an alien factor of secondary importance and not the main factor for the problem of our survival.

The environment requires a new way of thinking and of doing. It requires concerted action of those concerned who must be prepared to use their special knowledge and their special responsibility for the common interest. There is not room for competition and compromises which satisfy nobody; what is required is concerted action and an integrated approach which can satisfy all: that is what we must try to achieve.

These three days have given us a summary of the problems of soil management. A selected number of scientists have given us the data which should form the basis for a policy of the European Community. If we really want to be a community, it is up to the Commission now to approach the problem of our soils in a new way. A group should be formed consisting of a few scientists, administrators and economists, not more than ten or twelve in all. Each of these must be prepared to listen to others and try to follow their way of reasoning, to assimilate their argumentation, and to present their own views in such a way that others can understand their way of thinking. They should make plans on a collaborative basis and not on a competitive basis. The method has been named: 'Integrated policy studies'.

Some experience has been gathered on a small scale with this system in our country and it has been shown that such a procedure can achieve results. The problem in this case was a conflict of interests of, mainly, recreation, nature-protection, drinking water production, water management, and involved also physical planning. It concerned the use or non-use of the dunes for safeguarding the future supply of drinking water. Discussions had come to a deadlock. A group was then formed in which specialists on each of these aspects discussed all possible solutions. Through careful planning of the discussions and true communication of ideas and by weighing the arguments of all those taking part by all who took part, a solution could be found which was satisfactory for all. The responsible provincial authorities promptly decided to follow the constructive suggestion. It was a small scale attempt and it succeeded. It is worthwhile to learn from this example and to try out the method for a large scale problem.

As regards soil, a framework for soil research has recently been set up in the Bundesrepublik Deutschland and in other countries is at least being prepared. In The Netherlands, it is supported by four Ministries. Not only research but also practical application of that research is envisaged. It will be useful for the Commission of the Community to have a closer look at these plans.

In using these suggestions surely something constructive can result. Do not begin by saying that this is unrealistic. The large scale of the Community level will of course present specific problems. But we must do something. A meeting like the one we are here for will, hopefully, increase the speed of progress, but even then we will only be slowing down the destruction and not be in time to get on with reconstruction.

Some of you will tell us it will not work. There is a general answer to that kind of attitude: try it and we will know. Others will say they are too busy with the job they already have. My answer is that investing energy in a new approach will make their present work a great deal more effective. We must attempt this new approach because otherwise we will lose the battle against ourselves. We will continue to talk and do research, we will meet at symposia and write reports, we will assist in or undergo demonstrations and in the mean time our soils will be degraded and the basis for our future will continue to deteriorate.

There is a quotation from Roman history: 'Senatu deliberante Saguntum periit'. 'While the Senate debated, Saguntum was lost'. We must heed that warning. We can deliberate endlessly and in the mean time our soils will perish.

The knowledge is there, the wish is there, the ambition to achieve something worthwhile is there. By taking an essential new initiative the Commission of the European Community can do something fundamental to help to save the future.

List of Participants

Allnoch, G.
Dornier System GmbH
Postfach 1360
D-7990 Friedrichshafen

Altschuh
Gesellschaft für Strahlen- und
Umweltforschung mbH München
Ingolstädter Landstraße, 1
D-8042 Neuherberg

Arendt, F.
Kernforschungszentrum
Karlsruhe
Abt. TMU
Postfach 3640
D-7500 Karlsruhe

Arneth
Institut für Wasser-, Boden und
Lufthygiene des
Bundesgesundheitsamtes
Corrensplatz, 1
D-1000 Berlin 33

Bachmann
Umweltbundesamt
Bismarckplatz, 1
D-1000 Berlin 33

Balloni, W.
Istituto di Microbiologia
Ple delle Cascine, 18
I-50144 Florence

Barraqueta, P.
Servicio de Investigacion y
Mejora
Agraria (SIMA)
Carretera Arteaga
E-Derio (Vizcaya)

Barth, H.
Directorate-General Science,
Research and Development
Commission of the European
Communities
200, rue de la Loi
B-1049 Bruxelles

603

Bau
Umweltbundesamt
Bismarckplatz, 1
D-1000 Berlin 33

Bertram, H.-J.
Deutscher Bauernverband e.V.
Godesberger Allee, 142–148
D-5300 Bonn 2

Bode, J.
Bundesministerium für Forschung
und Technologie
Postfach 20 06 07
D-5300 Bonn 2

Boezeman, A. B. M.
Provinciale Waterstaat
Provincie Noord-Brabant
Brabantlaan, 1
NL-5216 TV S-Hertogenbosch

von Boguslawski, E.
Institut für Pflanzenbau und
Pflanzenzüchtung der Justus
Liebig-Universität
Versuchsstation
Rauischholzhausen
D-3557 Ebsdorfergrund 4

Bonazountas, M.
National Technical University
5, Iroon Polytechniou Str.
GR-15753 Zografou Athens

de Borst, A. T.
TAUW Infra Consult BV
Handelskade 11
NL-7400 Al Deventer

Bouma, J.
Bodenkunde en Geologie
Landbouw Hogeschool
PO Box 37
NL-6700 AA Wageningen

Bourdeau, Ph.
Directorate-General Science,
Research and Development
Commission of the European
Communities
200, rue de la Loi
B-1049 Bruxelles

de Brabander, K.
Instituut voor Hygiëne en
Epidemiologie
J. Wytsmanstraat, 14
B-1050 Bruxelles

Brühl, H.
Freie Universität Berlin
Fb 24, Geowissenschaften
Wichernstrasse, 16
D-1000 Berlin 33

Buch, Th.
Ministerium für Umwelt,
Raumordnung und
Landwirtschaft des Landes NW
Schwannstrasse, 3
D-4000 Düsseldorf 30

Bullock, P.
Soil Survey of England and Wales
Rothamsted Experimental Station
UK-Harpenden, Herts AL5 2JQ

Buschardt
Umweltbundesamt
Bismarckplatz, 1
D-1000 Berlin 33

Cassios, C.
Dept. of Surveying & Rural
Planning,
National Technical University
GR-15710 Athens

Cegarra, J.
Consejo Superior Investigaciones
Cientificas
Centro de Edafologia y Biologia
Aplicada del Seguro
Apartado 195
E-30003 Murcia

Chabason
Ministère de l'Environnement
Service de la Recherche
14, bvd du Général Leclerc
F-92524 Neuilly sur Seine Cedex

Chisci, G.
Istituto Sperimentale per lo
Studio e la Difesa del Suolo
Piazza M. D'Azeglio, 30
I-50121 Firenze

Christensen, H.-H.
Head of Agricultural Division
National Agency of
Environmental Protection
Centre for Terrestrial Ecology
Gyden 2
DK-2860 Soeborg

Chu, Lee-man
University of Liverpool
Dept of Botany
PO Box 147
UK-Liverpool L69 3BX

Culliton, E.
Directorate-General Agriculture
Commission of the European
Communities
200, rue de la Loi
B-1049 Bruxelles

de Haan, F. A. M.
Agriculture University
Dept of Soil and Plant Nutrition
De Dreijen 3
NL-6703 BC Wageningen

de Ploey, J.
Instituut voor
Aardwetenschappen
Redingenstraat 16 bis
B-3000 Leuven

Dekker, A. L. I.
Adviesbureau Bongaerts
Kuyper en Huiswaard
Postbus 93224
NL-2509 EA s'Gravenhage

Dekker, P.
Philips Export BV
Concern Environmental and
Energy
PO Box 218
NL-5600 MD Eindhoven

Delmhorst, B.
Bundesministerium für Umwelt,
Naturschutz und
Reaktorsicherheit
Graurheindorfer Strasse, 198
D-5300 Bonn 1

Depret, P.
Inter-Environment-Wallonie
rue d'Arlon, 25
B-1040 Bruxelles

Dilling, J.
Senator für Stadtentwicklung und
Umweltschutz Berlin -
Wasserbehörde-
Lindenstrasse, 20-25
D-1000 Berlin 61

Dingenis, K.
Nederlandse Stikstof Mij BV
c/o Norsk Hydro Belgium
Louizalaan, 149
B-1050 Bruxelles

Drechsler, W.
Umweltbundesamt
Bismarckplatz, 1
D-1000 Berlin 33

Dreissigacker
Senator für Stadtentwicklung und
Umweltschutz
Lindenstrasse, 20–25
D-1000 Berlin 61

Dyhr–Nielsen, M.
Ministry of the Environment
National Agency of
Environmental Protection
29, Strandgade
DK-1401 Copenhagen K

Eckelmann, W.
Niedersächsisches Landesamt für
Bodenforschung
Klagesmarkt, 27
D-3000 Hannover

Eiland, F.
State Laboratory for Soil and
Crop Research
Lottenborgvej, 24
DK-2800 Lyngby

Evans, D.
EG Presse- und Informationsbüro
Kurfürstendamm 102
D-1000 Berlin 31

Favilli, F.
Istituto Microbiologia Agraria e
Tecnica
Piazzale Cascine, 27
I-50144 Florenz

Fedoroff, N.
Institut National Agronomique
PG
Dépt des Sols
F-78850 Thiverval Grignon

Feigin, A.
Institute of Soils and Water
ARO, The Volcani Center
Israel-Bet Dagan

Finnecy, G.
Environmental Safety Centre
United Kingdom Atomic Energy
Authority,, Harwell
UK-Didcot, Oxfordshire OX11
0RA

Fischer, H.
Umweltbundesamt
Bismarckplatz, 1
D-1000 Berlin 33

Foppe De Walle, Ir.
SCMO/TNO
Schoemakerstraat, 79
Postbus 186
NL-2600 AD Delft

Franzius, V.
Umweltbundesamt (UBA)
Bismarckplatz, 1
D-1000 Berlin 33

Fränzle, O.
Christian-Albrechts-Universität zu
Kiel
Geographisches Institut
Olshausenstrasse, 40
D-2300 Kiel 1

Frey, W.
Senator für Stadtentwicklung und
Umweltschutz Berlin
Lindenstrasse, 20–25
D-1000 Berlin 61

Gabriels, D.
Laboratory of Soil Physics
University of Ghent
Coupure Links 653
B-9000 Ghent

Geiss, F.
Directorate-General Science,
Research and Development
Joint Research Centre
I-21020 ISPRA

Giovannini, G.
C.N.R. Institute for Soil
Chemistry
Via Corridoni, 78
I-56100 Pisa

Gomez, A.
INRA Station d'Agronomie
Centre de Recherches de
Bordeaux
Domaine de la Grande Ferrande
F-33140 Pont de la Maye

Götz, L.
Joint Research Centre
I-21020 ISPRA

Goubier, R.
ANRED
2, Square la Fayette
BP 406
F-49004 Angers Cedex

Grieve, I.
University of Stirling
UK-Stirling FK9 4LA

Gulinck, H.
Katholieke Universiteit Leuven
Kardinaal Mercierlaan, 92
B-3030 Leuven

Haberland
Umweltbundesamt
Bismarckplatz, 1
D-1000 Berlin 33

Häberli, R.
National Research Programme
'Soil'
Schweizerischer Nationalfond zur
Förderung der wissenschaftlichen
Forschung
Bundesrain, 20
CH-3003 Bern

Hafkamp, W.
Institute for Environmental Studies
Free University
PO Box 7161
NL-1007 MC Amsterdam

Hall, J.
Water Research Centre
Medmenham Lab.
Henley Road
Medmenham, PO Box 16, Marlow
UK-Buckinghamshire SL7 2HD

Hammer, R.
Gesellschaft für Landeskultur
GmbH
Friedrich-Missler-Strasse, 42
D-2800 Bremen

Hammer, S.
Thyssen Stahl AG
Abt. Thyssen Forschung
Nebenprodukte
Postfach 11 05 61
D-4100 Duisburg

Hansen, S.
A/S Soil Cleaning Technology
Helgeshoj allé 63
DK-2630 Tåstrup

Hedding, H.
'Oranjewoud' BV
PO Box 24
NL-8440 AA Heerenveen

van Helvert, P.
Wageningen Agricultural
University
PO Box 9101
NL-6700 HB Wageningen

Heyn, B.
Institut für Geologie der Freien
Universität Berlin
Altensteinstrasse, 34a
D-1000 Berlin 33

Hirn, G.
Agrarbüro die Grünen im
Bundestag
Bundeshaus Bonn
D-5300 Bonn

Hommes, R. W.
Advisory Council for Research on
Nature and Environment
Huis te Landelaan, 492
NL-2283 VJ Rijswijk

Hornung, M.
Institute of Terrestrial Ecology
Penrhos Road
Bangor
UK-Gwynedd LL57 2LQ

Hubler, K.-H.
TU Berlin
Sekr. Fer 2–7
Institut für Landschaftsökonomie
Franklinstrasse 28/29
D-1000 Berlin 10

Hutton, M.
King's College London
459A Fulham Road
UK-London SW10 0QX

Ignazzi, J.-Cl.
CdF Chimie A Z F
Tour Aurore Cedex 5
F-92080 Paris La Defense 2

Jaedtke, E.
EG Presse und Informationsbüro
Kurfürstendamm, 102
D-1000 Berlin 31

Jodice, R.
IPLA—Istituto per le Piante de
Legno e l'Ambiente
Corso Casale, 476
I-10132 Turin

Jones, K. C.
University of Lancaster
Environmental Sciences Dept
UK-Lancaster

Jørgensen, F.
Ministry of the Environment
National Agency of
Environmental Protection
29, Strandgade
DK-1401 Copenhagen K

Jürgens, S.
Landwirtschaftliche
Versuchsstation der BASF AG
D-6703 Limburgerhof

Kantzow, W.
Technische Universität Berlin Fb.
Franklinstrasse, 28/29
D-1000 Berlin 10

Kasperowski, E.
Umweltbundesamt Österreich
Biberstrasse, 11
A-1010 Wien

Kassebohm
Senator für Stadtentwicklung und
Umweltschutz Berlin
Lindenstrasse, 20–25
D-1000 Berlin 61

Kaule, G.
Universität Stuttgart
Institut für Landschaftsplanung
Keppler Strasse, 11
D-7000 Stuttgart 1

Kern
Freie Journalistin
Ahornstrasse, 22
D-1000 Berlin 37

Kloster, G.
Kernforschungsanlage Jülich
Schwerpunktprogramm
Bodenforschung
Postfach 1913
D-5170 Jülich

Koehler, H.
Universität Bremen
Biologie, Chemie (Fb 2)
Achterstrasse
D-2800 Bremen 1

Kooper, W. F.
DHV Consulting Engineers
PO Box 85
NL-3800 AB Amersfoort

Kördel, W.
Fraunhofer Institut für
Umweltchemie und
Oekotoxikologie
Grafschaft
D-5948 Schmallenberg-Grafschaft

Kratz, W.
Institut für Angewandte Zoologie
der Freien Universität Berlin
Haderslebener Strasse, 9
D-1000 Berlin 41

Krebs, H.-J.
Kernforschungsanlage Jülich
GmbH
Postfach 19 13
D-5170 Jülich

de Kruijf, H. A. M.
National Institute of Public Health
and Environmental Hygiene
PO Box 1
NL-3720 BA Bilthoven

Kuenen, D. J.
President, Council for
Environment and Nature
Research (RMNO)
Huis te Landelaan, 492
NL-2283 VJ Rijswijk

Kumpfmüller, M.
Österreichisches Bundesinstitut für
Gesundheitswesen
Stubenring, 6
A-1010 Wien

Larsen, B.
Risø National Laboratories
PO Box 49
DK-4000 Roskilde

Laville-Timsit, L.
BRGM
BP 6009
F-45060 Orleans Cedex

Lee, J.
The Agricultural Institute
Johnstown Castle Research
Centre
IRL-Wexford

von Lersner, H.
Präsident des Umweltbundesamtes
Bismarckplatz, 1
D-1000 Berlin 33

Leschber, R.
Institut für Wasser-, Boden- und
Lufthygiene des
Bundesgesundheitsamtes
Corrensplatz, 1
D-1000 Berlin 33

Lewis, W.
World Health Organization
Copenhagen
European Office
Scherfigsvej, 8
DK-2100 Kopenhagen

L'Hermite, P.
Directorate-General Science,
Research and Development
Commission of the European
Communities
200, rue de la Loi
B-1049 Bruxelles

Lieth, H.
Universität Osnabrück
FB Biologie, AG Systemforschung
Albrechtstrasse, 16
D-4500 Osnabrück

Litz, N.
Umweltbundesamt
Bismarckplatz, 1
D-1000 Berlin 33

Lucks
Umweltbundesamt
Bismarckplatz, 1
D-1000 Berlin 33

Lühr, H.-P.
Technische Universität Berlin
Institut für wassergefährdende
Stoffe
Joachimsthaler Strasse, 31
D-1000 Berlin 15

Marek, K.
INTECUS—
Ingenieurgemeinschaft für
technischen Umweltschutz
Krefelder Strasse, 20
D-1000 Berlin 21

Mathy, P.
Directorate-General Science,
Research and Development
Commission of the European
Communities
200, rue de la Loi
B-1049 Bruxelles

Matthies, M.
Gesellschaft für Strahlen-und
Umweltforschung mbH München
Ingolstädter Landstrasse, 1
D-8042 Neuherberg

Milde, G.
Institut für Wasser-, Boden- und
Lufthygiene des BGA
Corrensplatz, 1
D-1000 Berlin 33

Miller, H. G.
University of Aberdeen
Dept of Forestry
St Machar Drive
UK-Aberdeen AB9 2UU

Møller, A.
Agri Contact
Torupvejen, 97
DK-3390 Hundested

Moen, J. E. T.
Ministry of Housing, Physical
Planning and Environment
Soil division—A/318
PO Box 450
NL-2260 MB-Leidschendam

Montoro, J. A.
Centro de Edafologia y Biologia
Aplicada del Segura
Apartado 195
E-Murcia

Morgan, R. C. P.
Cranfield Institute of Technology
Silsoe College
Cranfield
UK-Bedford MK43 0RL

Moseholm, L.
Cowiconsult—Institute of Plant
Ecology
Teknikerbyen, 45
DK—2830 Virum

Moss, G.
G. Moss Associates
12, Eton Street
Richmond
UK-Surrey

Muheim, R.
Schweizerischer Nationalfond
Abt. Forschungsprogramme
Wildhainweg, 20
CH-3001 Bern

van Muilekom, R. G. H.
TAUW Infra Consult BV
Handelskade, 11
NL-7400 AL Deventer

Neuland, H.
Dornier System GmbH
Postfach 1360
D-7990 Friedrichshafen 1

Neumeier, G.
Umweltbundesamt
Bismarckplatz, 1
D-1000 Berlin 33

Niclauß, M.
Gesellschaft für Systemtechnik
mbH
Am Westbahnhof, 2
D-4300 Essen 1

Nortcliff, St.
University of Reading
Dept of Soil Science
UK-Reading RG1 5AQ

Nuesink, J. A.
Bosbureau Wageningen B. V.
PO Box 635
NL-6700 AP Wageningen

Osterkamp, G.
Institut für Geologie der Freien
Universität Berlin
Malteserstrasse, 74–100
D-1000 Berlin 46

Ott, H.
Directorate-General Science,
Research and Development
Commission of the European
Communities
200, rue de la Loi
B-1049 Bruxelles

Pereira, L.
Istituto Superior da Agronomia
Technical University of Lisbon
P-1399 Lisbon Codex

Pfirrmann, Th.
Gesellschaft für Strahlen- und
Umweltforschung mbH München
Ingolstädter Landstrasse, 1
D-8042 Neuherberg

Pflugmacher, J.
Biologische Bundesanstalt
Königin-Luise-Strasse, 19
D-1000 Berlin 33

Pimpl, M.
Kernforschungszentrum Karlsruhe
GmbH
Hauptabteilung Sicherheit
Postfach 3640
D-7500 Karlsruhe

Pluquet
Niedersächsisches Landesamt für
Bodenforschung—
Bodentechnolog. Institut
Friedrich-Mißler-Strasse, 46–50
D-2800 Bremen

Preisler-Holl, MdA
Alternative Liste, Fraktion im
Abgeordnetenhaus
John-F.-Kennedy-Platz
D-1000 Berlin 62

Puchstein, R.
Boothstrasse, 20c
D-1000 Berlin 45

Raymaekers, P. J. M.
DHV Consulting Engineers
PO Box 85
NL-3800 AB Amersfoort

Remy, J.-Cl.
Institut National de la Recherche
Agronomique
145, rue de l'Université
F-75341 Paris Cedex

Rickson, R. J.
Silsoe College Cranfield
Silsoe
UK-Bedford MK45 4DT

Robson, Ch.
Dept of the Environment
Romney House, Room A316
43, Marsham Street
UK-London SW1P 3PY

Rogaar, H.
Universität Wageningen
Fachgr, für Bodenkunde und
Geologie
PO Box 37
NL-6700 AA Wageningen

Ronzheimer
Associated Press
Kurfürstendamm, 269
D-1000 Berlin 15

Russo, G.
AGRIMONT
Piazza Repubblica, 14–16
I-Milano

Roquero, C.
Dept of Pedology
Universita Politecnica de Madrid
Escuela Tecnica Superior de
Ingenieros Agronomos
E-Madrid

Salandin, R.
IPLA—Istituto per le Piante de
Legno e l'Ambiente
Corso Casale, 476
I-10132 Turin

Rosenkranz
Umweltbundesamt
Bismarckplatz, 1
D-1000 Berlin 33

Sauerbeck, D.
Institut für Pflanzenernährung
und Bodenkunde
Bundesallee, 50
D-3300 Braunschweig

Roth-Kleyer, St.
Institut für Landschaftsbau
TU Berlin
Albrecht-Thoer-Weg, 4
D-1000 Berlin 33

Saunders, J. P. W.
Dept of the Environment
Romney House
43, Marsham Street
UK-London SW1P 3PY

Runge, M.
Senator für Stadtentwicklung und
Umweltschutz
Landesbeauftragte für
Naturschutz
Lindenstrasse, 25
D-1000 Berlin 61

Schafmeister-Spierling, M.-Th.
Institut für Geologie der Freien
Universität Berlin
Malteserstrasse, 74–100
D-1000 Berlin 46

Rupp, E.
Ministerium für Ernährung,
Landwirtschaft und Forsten
Baden-Württemberg
Marienstrasse, 41
D-7000 Stuttgart 1

Schäfer, J.
Adam & Schäfer, Ing.-büro für
Verfahrenstechnik und
Umweltschutz
Ackerstrasse, 71–76
D-1000 Berlin 65

Scheunert, I.
Gesellschaft für Strahlen- und
Umweltforschung mbH München
Institut für Oekologische Chemie
Ingolstädter Landstrasse, 1
D-8042 Neuherberg

Schierenbeck, B.
Gesellschaft für Landeskultur
GmbH
Friedrich-Mißler-Strasse, 42
D-2800 Bremen

Schilling, M.
Meininger Strasse, 4
D-1000 Berlin 62

Schlömp, F.
Senator für Stadtentwicklung und
Umweltschutz Berlin,
Wasserbehörde, Lindenstrasse,
20–25
B-1000 Berlin 61

Schmidt, G.
APROTEC GmbH
Bettinastrasse, 14–16
D-6000 Frankfurt/Main 1

Schmidt, H.
Bayrisches Staatsministerium für
Landesentwicklung und
Umweltfragen
Rosenkavalierplatz, 2
D-8000 München 81

Schneider, G.
Institut für Stadtplanung der
Technischen Universität Berlin
DO 801
Dovestrasse, 1
D-1000 Berlin 10

Schott, P.
Technische Universität Berlin
Fb 14
Franklinstrasse, 28/29
D-1000 Berlin 10

Schreiber, H.
Wissenschaftszentrum Berlin
Potsdamer Strasse, 58
D-1000 Berlin 30

Schulz
Umweltbundesamt
Bismarckplatz, 1
D-1000 Berlin 33

Schuster, G.
Senator für Justiz und
Bundesangelegenheiten
Joachimstrasse, 7
D-5300 Bonn

Schwartengräber, R.
Institut für Bodenkunde
TU Berlin
Salzufer 11–12
D-1000 Berlin 10

Sherwood, M.
An Foras Taluntais
Johnstown Castle Research
Centre
IRL-Wexford

Skola, W.
Institut für Geologie der Freien
Universität Berlin
Malteserstrasse, 74–100
D-1000 Berlin 46

Smettan, U.
Institut für Bodenkunde
Technische Universität
Salzufer, 11–12
D-1000 Berlin 10

Solbach, H. J.
Institut f. Landschaftsbau TU
Berlin
Albrecht-Thoer-Weg, 4
D-1000 Berlin 33

Sommer, S.
Centre for Terrestrial Ecology
Gyden 2
DK-2860 Soborg

Stasch, D.
Institut für Bodenkunde
Technische Universität
Salzufer, 11–12
D-1000 Berlin 10

Steen, B.
Redaktör
Äkerbyvägen 206, I
S-183 35 Täby

Steenvoorden, J. H. A. M.
Dutch Institute for Land and
Water Management Research
PO Box 35
NL-6700 AA Wageningen

Stief
Umweltbundesamt
Bismarckplatz, 1
D-1000 Berlin 33

Stolpe, H.
AG Hydrogeologie und
Umweltschutz
Bachstrasse, 62–64
D-5100 Aachen

Straßer, H.
ARSU—Arbeitsgruppe für
regionale Struktur- und
Umweltforschung
Artillerieweg, 38
D-2900 Oldenburg

Süß, A.
Bayrische Landesanstalt für
Bodenkultur und Pflanzenbau
Abt. Landwirtschaftliches
Untersuchungsw.
Lange Point, 6
D-8050 Freising

Suttner, Th.
Universität Bayreuth
Lehrstuhl für Bodenkunde
Birkengut
D-8580 Bayreuth

Swain, R.
Ministry of Agriculture, Fisheries
and Food, Woodthorne
UK-Wolverhampton WV6 8TQ

Tavernier, R.
Geological Institute
University of Ghent
Krijgslaan, 281
B-9000 Ghent

Thierbach, J.
Senator für Stadtentwicklung und
Umweltschutz Berlin
Lindenstrasse, 20–25
D-1000 Berlin 61

Thormann, A.
Hessischer Minister für Umwelt
und Energie
Dostojewskistrasse, 8
D-6200 Wiesbaden

Thornton, J.
AGRG, Dept of Geology
Imperial College London
Prince Consort Road
UK-London SW7 2BP

Tjell, J.-Ch.
Technical University of Denmark
Dept of Environmental
Engineering
Building 115
DK-2800 Lyngby

Traulsen, B.-D.
Biologische Bundesanstalt für
Land- und Forstwirtschaft
Königin-Luise-Strasse, 19
D-1000 Berlin 33

Ulrich, B.
Universität Göttingen
Institut für Bodenkunde und
Waldernährung
Büsgenweg, 2
D-3400 Göttingen

Uppenbrink, M.
Umweltbundesamt
Bismarckplatz, 1
D-1000 Berlin 33

van Dam, J. E.
Ministry of Education and
Sciences
PO Box 25000
NL-2700 LZ Zoetemeer

van de Werf, H.
Rijksuniversiteit Gent
Lab. voor Microbiele Ecologie
Coupure Links, 653
B-9000 Ghent

Verheye, W.
Geological Institute
University of Ghent
Krijgslaan, 281
B-9000 Ghent

Versino, B.
Joint Research Centre
I-21020 ISPRA

Vigna-Guidi, G.
Istituto per la Chimica del
Terreno C.N.R.
Via Corridoni, 78
I-56100 Pisa

Voget, M.
Gesellschaft für Strahlen- und
Umweltforschung mbH München
Institut für Oekologische Chemie
Schulstrasse, 10
D-8050 Freising

Vogt, H.
Université Louis Pasteur
Institut de Géographie
3, rue de l'Argonne
F-67083 Strasbourg Cedex

von Reis, J.
Projektleiter Biol., Oekologie,
Energie
Kernforschungslanlage Jülich
Postfach 1913
D-5170 Jülich

Webster, R.
Rothamsted Experimental Station
UK-Harpenden, Herts AL5 2JQ

Weidenbach, G.
Directorate-General Science,
Research and Development
Commission of the European
Communities
200, rue de la Loi
B-1049 Bruxelles

Weigmann, G.
Freie Universität Berlin
Institut für Angewandte Zoologie
Haderslebener Strasse, 9
D-1000 Berlin 41

Wessolek
Technische Universität Berlin
Institut für Oekologie,
Bodenkunde
Am Salzufer, 11-12
D-1000 Berlin 10

Westphal
Senator für Stadtentwicklung und
Umweltschutz Berlin
Lindenstrasse, 20-25
D-1000 Berlin 61

Wiesemann, M.
Teufelsseestrasse, 5
D-1000 Berlin 19

Wilke, B.-M.
Universität Bayreuth
Lehrstuhl für Bodenkunde und -
geographie
Birkengut
D-8580 Bayreuth

Williams, A. J.
ICI
Billingham Office
UK-Cleveland TS23 1LB

Winkel, P.
TNO—Division of Technology
for Society
PO Box 217
NL-2600 AE Delft

Winkler
Institut für Stadtplanung der
Technischen Universität Berlin
DO 801
Dovestrasse, 1
D-1000 Berlin 10

Wirth, St.
Institut für Pflanzenpathologie
und Pflanzenschutz
Griesebachstrasse, 6
D-3400 Göttingen

Wolf, J.
Institut für Geologie der
Technischen Universität
Braunschweig
Pockelsstrasse, 4
D-3300 Braunschweig

Worthington, P.
Directorate-General Environment,
Consumer Protection and Nuclear
Safety
Commission of the European
Communities
200, rue de la Loi
B-1049 Bruxelles

Wurzel, A.
Deutscher Rat für Landespflege
Konstantinstrasse, 110
D-5300 Bonn 2

Yassoglou, N.
Athens Faculty of Agriculture
Botanicos Iera
GR-Odos Athens

Zander, C.-L.
Institut für Geologie der
Technischen Universität
Braunschweig
Pockelsstrasse, 4
D-3300 Braunschweig

Zehfuß, G.
Vereinigte Wirtschaftsdienste
GmbH
Savignyplatz, 6
D-1000 Berlin 12

Zehnder, A.
Agricultural University
Dept of Microbiology
H.-v. Suchtelenweg, 4
NL-6703 CT Wageningen

Zierl
Nationalparkverwaltung
Berchtesgaden
Doktorberg, 6
D-8240 Berchtesgaden

Zieschank, R.
Wissenschaftszentrum Berlin
Internationales Inst. für Umwelt
und Gesellschaft
Potsdamer Strasse, 58
D-1000 Berlin 30

Zoetelief, J.
Ministry of Agriculture and
Fisheries
Bezuidenhoutseweg, 73
Postbus 20401
NL-2500 Ek Den Haag

Zubr, J.
Agriculture University
Institute of Agricultural
Engineering
Rolejheds, 23
DK-1958 Copenhagen

Index